Intermetallic Matrix Composites II

MATERIALS RESEARCH SOCIETY SYMPOSIUM PROCEEDINGS VOLUME 273

Intermetallic Matrix Composites II

Symposium held April 27-30, 1992, San Francisco, California, U.S.A.

EDITORS:

D.B. Miracle

Air Force Wright Laboratory
Dayton, Ohio, U.S.A.

D.L. Anton

United Technologies Research Center
East Hartford, Connecticut, U.S.A.

J.A. Graves

Rockwell International Science Center
Thousand Oaks, California, U.S.A.

MATERIALS RESEARCH SOCIETY
Pittsburgh, Pennsylvania

This work relates to Department of Navy Grant N00014-92-J-1145 issued by the Office of Naval Research. The United States Government has a royalty-free license throughout the world in all copyrightable material contained herein.

Single article reprints from this publication are available through University Microfilms Inc., 300 North Zeeb Road, Ann Arbor, Michigan 48106

CODEN: MRSPDH

Published by:

Materials Research Society
9800 McKnight Road
Pittsburgh, Pennsylvania 15237
Telephone (412) 367-3003
Fax (412) 367-4373

Library of Congress Cataloging in Publication Data

Intermetallic matrix composites II: symposium held April 27-30, 1992, San Francisco, California, U.S.A./editors, D.B. Miracle, D.L. Anton, J.A. Graves.
p. cm. -- (Materials Research Society symposium proceedings, ISSN 0272-9172; v. 273)
Includes bibliographical references and indexes.
ISBN 1-55899-168-9
1. Metallic composites--Congresses. 2. Intermetallic compounds--Congresses. I. Miracle, D.B. II. Anton, D.L. III. Graves, J.A. IV. Title: Intermetallic matrix composites 2. V. Series: Materials Research Society symposium proceedings; v. 194.
TA481.K64 1992
620.1'6--dc20

92-31550
CIP

Manufactured in the United States of America

Contents

*Invited Paper

PART II: NICKEL ALUMINIDE COMPOSITES

*Invited Paper

PART III: $MoSi_2$ COMPOSITES

PART IV: INTERFACE PROPERTIES, PROCESSING, AND ADVANCED INTERMETALLIC COMPOSITES

*Invited Paper

*Invited Paper

Preface

These are the proceedings of the second MRS symposium dedicated to current coverage of research and development of intermetallic composites and multiphase materials with intermetallic constituents. The vigorous discussions and technical interchange both during and after the symposium sessions reflected the high level of interest in this interdisciplinary field. This symposium focused on the basic science and engineering development of the materials and approaches currently being pursued to achieve the requirements for advanced high temperature structural materials.

While it is a relatively new field of endeavor, the increasing breadth of research reflects the vitality in the area of intermetallic composites. A strong emphasis on processing, chemical compatibility and first tier mechanical properties in the preceding MRS symposium on intermetallic composites (Intermetallic Matrix Composites, V. 194) has been supplemented in this current symposium by a more broad treatment of first and second tier properties, conventional and novel processing approaches, and testing techniques to quantify both interface and composite properties. A dynamic interchange fostered by a more complete integration of the Ti aluminide metal matrix composite community and the advanced intermetallic community was evident during the sessions. The poor off-axis properties of continuously-reinforced Ti-aluminide composites emphasize the need for improved interfaces in this important class of intermetallic composites. The limitations of the fiber pushout test were discussed in detail at the meeting, and the need for simple, quantitative test techniques to measure interface properties in metallic composites was evident. The variety of processing approaches discussed (deformation processing, directional solidification of intermetallic eutectics, physical vapor deposition, melt infiltration, and solid state displacement reaction) point to this as a fruitful area for future advancements.

This symposium was sponsored by the Office of Naval Research, Rockwell International Science Center, Los Alamos National Laboratory, and NASA Lewis Research Center. The support of these institutions is graciously acknowledged. We would also like to thank Linda Butler for her help in arranging the front and back matter for the proceedings. Finally, we would like to express our appreciation for the contribution of the session chairs and the individuals who served as reviewers for the manuscripts. Their efforts were vital to the successful conduct of the symposium and the rapid publication of these proceedings.

D. Miracle
D. Anton
J. Graves

August 1992

Volume 239—Thin Films: Stresses and Mechanical Properties III, W.D. Nix, J.C. Bravman, E. Arzt, L.B. Freund, 1992, ISBN: 1-55899-133-6

Volume 240—Advanced III-V Compound Semiconductor Growth, Processing and Devices, S.J. Pearton, D.K. Sadana, J.M. Zavada, 1992, ISBN: 1-55899-134-4

Volume 241—Low Temperature (LT) GaAs and Related Materials, G.L. Witt, R. Calawa, U. Mishra, E. Weber, 1992, ISBN: 1-55899-135-2

Volume 242—Wide Band Gap Semiconductors, T.D. Moustakas, J.I. Pankove, Y. Hamakawa, 1992, ISBN: 1-55899-136-0

Volume 243—Ferroelectric Thin Films II, A.I. Kingon, E.R. Myers, B. Tuttle, 1992, ISBN: 1-55899-137-9

Volume 244—Optical Waveguide Materials, M.M. Broer, G.H. Sigel, Jr., R.Th. Kersten, H. Kawazoe, 1992, ISBN: 1-55899-138-7

Volume 245—Advanced Cementitious Systems: Mechanisms and Properties, F.P. Glasser, G.J. McCarthy, J.F. Young, T.O. Mason, P.L. Pratt, 1992, ISBN: 1-55899-139-5

Volume 246—Shape-Memory Materials and Phenomena—Fundamental Aspects and Applications, C.T. Liu, H. Kunsmann, K. Otsuka, M. Wuttig, 1992, ISBN: 1-55899-140-9

Volume 247—Electrical, Optical, and Magnetic Properties of Organic Solid State Materials, L.Y. Chiang, A.F. Garito, D.J. Sandman, 1992, ISBN: 1-55899-141-7

Volume 248—Complex Fluids, E.B. Sirota, D. Weitz, T. Witten, J. Israelachvili, 1992, ISBN: 1-55899-142-5

Volume 249—Synthesis and Processing of Ceramics: Scientific Issues, W.E. Rhine, T.M. Shaw, R.J. Gottschall, Y. Chen, 1992, ISBN: 1-55899-143-3

Volume 250—Chemical Vapor Deposition of Refractory Metals and Ceramics II, T.M. Besmann, B.M. Gallois, J.W. Warren, 1992, ISBN: 1-55899-144-1

Volume 251—Pressure Effects on Materials Processing and Design, K. Ishizaki, E. Hodge, M. Concannon, 1992, ISBN: 1-55899-145-X

Volume 252—Tissue-Inducing Biomaterials, L.G. Cima, E.S. Ron, 1992, ISBN: 1-55899-146-8

Volume 253—Applications of Multiple Scattering Theory to Materials Science, W.H. Butler, P.H. Dederichs, A. Gonis, R.L. Weaver, 1992, ISBN: 1-55899-147-6

Volume 254—Specimen Preparation for Transmission Electron Microscopy of Materials-III, R. Anderson, B. Tracy, J. Bravman, 1992, ISBN: 1-55899-148-4

Volume 255—Hierarchically Structured Materials, I.A. Aksay, E. Baer, M. Sarikaya, D.A. Tirrell, 1992, ISBN: 1-55899-149-2

Volume 256—Light Emission from Silicon, S.S. Iyer, R.T. Collins, L.T. Canham, 1992, ISBN: 1-55899-150-6

Volume 257—Scientific Basis for Nuclear Waste Management XV, C.G. Sombret, 1992, ISBN: 1-55899-151-4

MATERIALS RESEARCH SOCIETY SYMPOSIUM PROCEEDINGS

Volume 258—Amorphous Silicon Technology—1992, M.J. Thompson, Y. Hamakawa, P.G. LeComber, A. Madan, E. Schiff, 1992, ISBN: 1-55899-153-0

Volume 259—Chemical Surface Preparation, Passivation and Cleaning for Semiconductor Growth and Processing, R.J. Nemanich, C.R. Helms, M. Hirose, G.W. Rubloff, 1992, ISBN: 1-55899-154-9

Volume 260—Advanced Metallization and Processing for Semiconductor Devices and Circuits II, A. Katz, Y.I. Nissim, S.P. Murarka, J.M.E. Harper, 1992, ISBN: 1-55899-155-7

Volume 261—Photo-Induced Space Charge Effects in Semiconductors: Electro-optics, Photoconductivity, and the Photorefractive Effect, D.D. Nolte, N.M. Haegel, K.W. Goossen, 1992, ISBN: 1-55899-156-5

Volume 262—Defect Engineering in Semiconductor Growth, Processing and Device Technology, S. Ashok, J. Chevallier, K. Sumino, E. Weber, 1992, ISBN: 1-55899-157-3

Volume 263—Mechanisms of Heteroepitaxial Growth, M.F. Chisholm, B.J. Garrison, R. Hull, L.J. Schowalter, 1992, ISBN: 1-55899-158-1

Volume 264—Electronic Packaging Materials Science VI, P.S. Ho, K.A. Jackson, C-Y. Li, G.F. Lipscomb, 1992, ISBN: 1-55899-159-X

Volume 265—Materials Reliability in Microelectronics II, C.V. Thompson, J.R. Lloyd, 1992, ISBN: 1-55899-160-3

Volume 266—Materials Interactions Relevant to Recycling of Wood-Based Materials, R.M. Rowell, T.L. Laufenberg, J.K. Rowell, 1992, ISBN: 1-55899-161-1

Volume 267—Materials Issues in Art and Archaeology III, J.R. Druzik, P.B. Vandiver, G.S. Wheeler, I. Freestone, 1992, ISBN: 1-55899-162-X

Volume 268—Materials Modification by Energetic Atoms and Ions, K.S. Grabowski, S.A. Barnett, S.M. Rossnagel, K. Wasa, 1992, ISBN: 1-55899-163-8

Volume 269—Microwave Processing of Materials III, R.L. Beatty, W.H. Sutton, M.F. Iskander, 1992, ISBN: 1-55899-164-6

Volume 270—Novel Forms of Carbon, C.L. Renschler, J. Pouch, D. Cox, 1992, ISBN: 1-55899-165-4

Volume 271—Better Ceramics Through Chemistry V, M.J. Hampden-Smith, W.G. Klemperer, C.J. Brinker, 1992, ISBN: 1-55899-166-2

Volume 272—Chemical Processes in Inorganic Materials: Metal and Semiconductor Clusters and Colloids, P.D. Persans, J.S. Bradley, R.R. Chianelli, G. Schmid, 1992, ISBN: 1-55899-167-0

Volume 273—Intermetallic Matrix Composites II, D. Miracle, J. Graves, D. Anton, 1992, ISBN: 1-55899-168-9

Volume 274—Submicron Multiphase Materials, R. Baney, L. Gilliom, S.-I. Hirano, H. Schmidt, 1992, ISBN: 1-55899-169-7

Volume 275—Layered Superconductors: Fabrication, Properties and Applications, D.T. Shaw, C.C. Tsuei, T.R. Schneider, Y. Shiohara, 1992, ISBN: 1-55899-170-0

Volume 276—Materials for Smart Devices and Micro-Electro-Mechanical Systems, A.P. Jardine, G.C. Johnson, A. Crowson, M. Allen, 1992, ISBN: 1-55899-171-9

Volume 277—Macromolecular Host-Guest Complexes: Optical, Optoelectronic, and Photorefractive Properties and Applications, S.A. Jenekhe, 1992, ISBN: 1-55899-172-7

Volume 278—Computational Methods in Materials Science, J.E. Mark, M.E. Glicksman, S.P. Marsh, 1992, ISBN: 1-55899-173-5

Prior Materials Research Society Symposium Proceedings available by contacting Materials Research Society

PART I

Titanium Aluminide Composites

AN OVERVIEW OF POTENTIAL TITANIUM ALUMINIDE COMPOSITES IN AEROSPACE APPLICATIONS

JAMES M. LARSEN, WILLIAM C. REVELOS, AND MARY L. GAMBONE
Materials Directorate, Wright Laboratory, WL/MLL
Wright-Patterson AFB, OH 45433

ABSTRACT

High-temperature, light-weight materials represent enabling technology in the continued evolution of high-performance aerospace vehicles and propulsion systems being pursued by the U.S. Air Force. In this regard, titanium aluminide matrix composites appear to offer unique advantages in terms of a variety of weight-specific properties at high temperatures. However, a key requirement for eventual structural use of these materials is a balance of mechanical properties that can be suitably exploited by aircraft and engine designers without compromising reliability. An overview of the current capability of titanium aluminide composites is presented, with an effort to assess the balance of properties offered by this class of materials. Emphasis is given to life-limiting cyclic and monotonic properties and the roles of high-temperature, time-dependent deformation and environmental effects. An attempt is made to assess the limitations of currently available titanium aluminide composites with respect to application needs and to suggest avenues for improvements in key properties.

INTRODUCTION

Titanium aluminide composites [1,2], continuously reinforced with silicon carbide (SiC) fibers, have recently received considerable attention due to their potential to replace titanium and nickel-base alloys in aerospace systems, such as advanced turbine engines [3] and hypersonic vehicles [4] where specific strength and stiffness at high temperatures are critical.

Figure 1 is a schematic of a possible application of titanium aluminide composite technology in an advanced turbine engine. Rotating components in compressor sections that utilize advanced composites can have significantly reduced weight, allowing important improvements in design that are not possible using conventional nickel-base superalloys [5]. This weight savings translates into higher thrust-to-weight ratios and lower specific fuel consumption in the engine. Other rotating, as well as static, applications of titanium aluminide composites have lately been the subject of increased attention. These include shafts, blades, vanes, stators, actuators, struts, and nozzles. The majority of these applications will utilize unidirectionally reinforced composite architectures. If these composites are to see application in advanced engines, however, progress must be made in several areas, including the improvement of off-axis tensile and creep properties, low cycle fatigue resistance, environmental and burn resistance for those exposed to gas flow paths, and life-management and damage tolerant design methods.

Hypersonic vehicle applications require composites with good specific strength and stiffness at elevated temperature, combined with improved transverse properties. As shown in Figure 2, these composites will be used primarily as hat-channel stiffened panels in hypersonic airframes, where balanced in-plane properties in the thin-sheet form are important. This requirement has fueled interest in cross-ply and quasi-isotropic reinforced laminates, which present new challenges in the areas of materials development, application, and life prediction.

Ultimately, the use of titanium aluminides in structural applications will depend on a number of factors which include life-cycle cost, producibility, a range of mechanical properties, and reliability and maintainability in service. To qualify a

Mat. Res. Soc. Symp. Proc. Vol. 273.

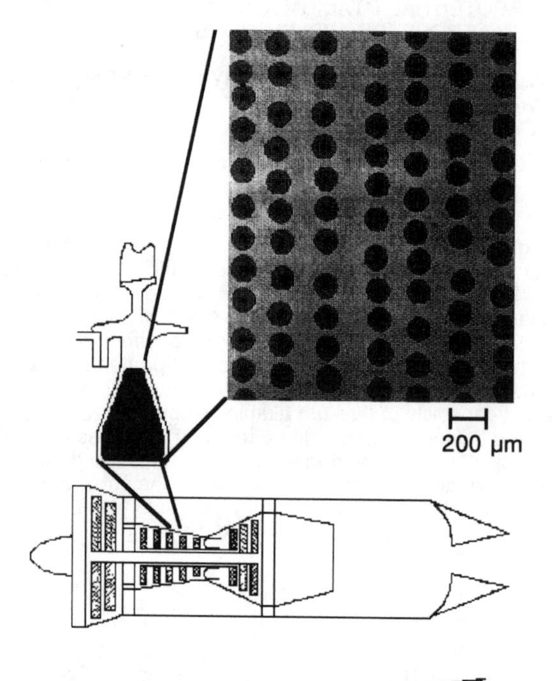

Figure 1. Potential application of hoop-wound titanium aluminide composite in a turbine engine compressor.

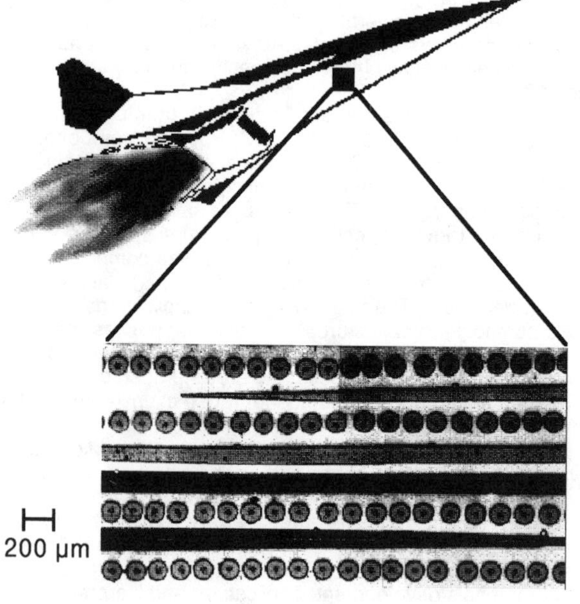

Figure 2. Potential application of titanium aluminide composite in a generic hypersonic vehicle.

new material for use in a critical structural component, a comprehensive testing and analysis program must be conducted, and this can only be justified if the payoff for using the material is significant. Titanium aluminide composites, for example, must demonstrate mechanical properties that exceed, at least in key areas, the capability of less-costly monolithic materials. Coupled with this requirement is the need for high reliability in extended use and the ability to predict component life accurately.

To examine some of these issues, a series of mechanical property comparisons between the first-generation titanium aluminide composite SCS-6/Ti-24Al-11Nb, produced by the foil-fiber-foil method, and other advanced materials is presented below. This material is representative of the general class of titanium aluminide composites, and it is the only material from this class that has been evaluated in various laminate configurations to characterize a broad range of mechanical properties. The SCS-6/Ti-24Al-11Nb data to be presented are from a U.S. Air Force contract with Allison Gas Turbine Division of General Motors, reported by Gambone [6,7], as well as from efforts by the authors and our colleagues on identical material produced under the contract. This composite material contained a nominal fiber volume fraction of 0.33, and the equiaxed microstructure of the matrix alloy contained approximately 90 percent α_2 phase (ordered $D0_{19}$ structure) surrounding small islands of disordered β phase [8]. The capability of the composite will be compared with a variety of monolithic materials: the nickel-base superalloy IN100 [9,10], which is used in turbine disks, the high temperature titanium alloy Ti-1100 [11,12,13], and a relatively new titanium aluminide alloy [14]. All of the property comparisons are for tests conduced in air. The data are presented on a density-corrected basis because of the anticipated use of the material in turbine engine and aircraft applications. The densities of the various materials are listed in Table 1.

Table 1 – Fiber, alloy, and composite densities used in calculating weight-specific properties.

Material	**Density (Mg/m^3)**
SCS-6 fiber	3.045
Ti-24Al-11Nb	4.67
SCS-6/Ti-24Al-11Nb (V_f = 0.33)	4.13
IN100	7.87
Ti-1100	4.5
Ti-24Al-17Nb-1Mo	4.93
Ti-6Al-2Sn-4Zr-2Mo	4.54

COMPOSITE MECHANICAL PROPERTIES

Figure 3 presents data of density-corrected (specific) strength of four composite laminates and the fiberless matrix alloy, compared with the yield strengths of the monolithic materials. The yield strengths of the composite were generally undefined, as these specimens failed before achieving 0.2 percent plastic strain. A similar plot of specific ultimate strength of all the materials is presented in Fig. 4. As shown in both figures, the unidirectional material, $[0]_8$, exhibits outstanding strength, but as will be discussed later, the competitiveness of the other laminates is less clear, particularly on an ultimate-strength basis.

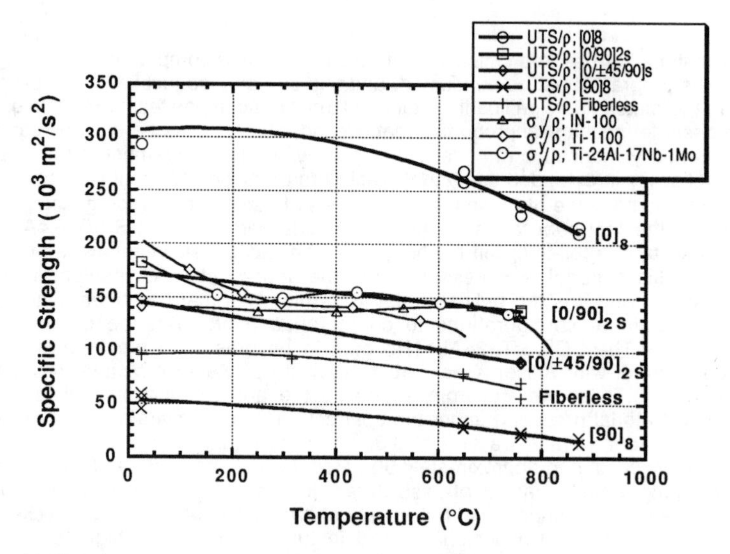

Figure 3. Density-corrected ultimate strengths of SCS-6/Ti-24Al-11Nb laminates [7] and fiberless Ti-24Al-11Nb [6] compared with the yield strengths of alternative monolithic materials.

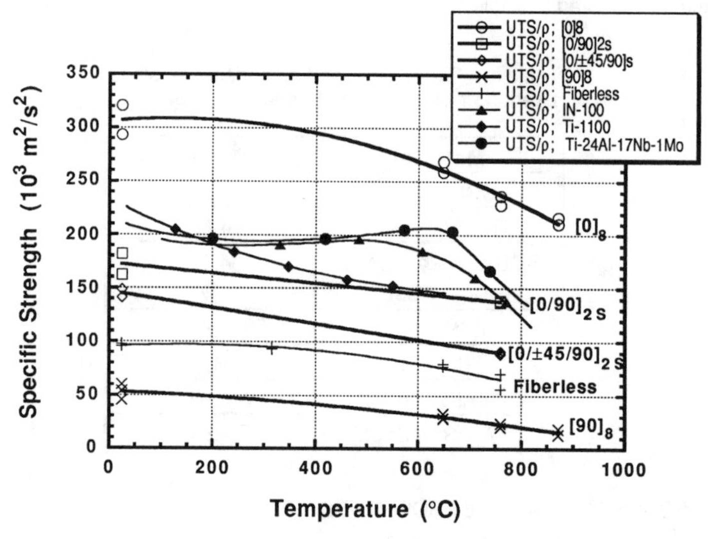

Figure 4. Density-corrected ultimate strengths of SCS-6/Ti-24Al-11Nb laminates [7] compared with fiberless Ti-24Al-11Nb [6] and alternative monolithic materials.

As shown in Fig. 4, the ultimate strength of SCS-6/Ti-24Al-11Nb is strongly dependent on laminate architecture. This effect is more easily visualized by plotting the strength against a geometric parameter, such as the percent of [0] plies in the laminate, as illustrated in Fig. 5 for tests at room temperature and 760°C. With the

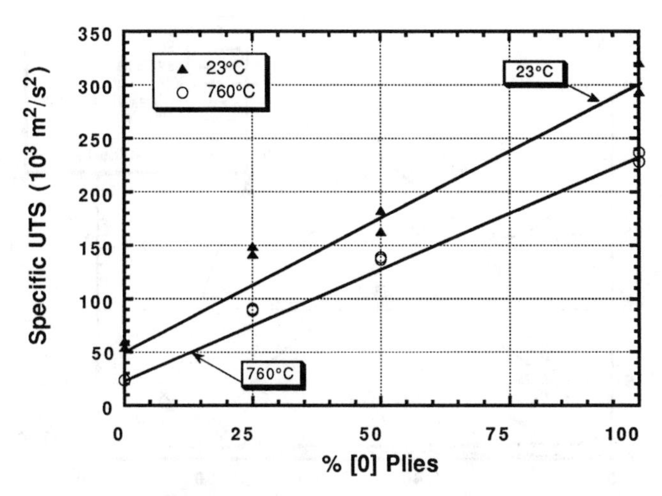

Figure 5. SCS-6/Ti-24Al-11Nb laminate strength plotted versus the percentage of plies oriented along the axis of loading.

exception of the quasi-isotropic laminate, $[0/\pm45/90]_S$, the ultimate strength is directly proportional to the percent of [0] plies. The low strength of the $[90]_8$ material reflects the very weak fiber/matrix bond that exists in this material. As shown, the [±45] plies in the quasi-isotropic laminate contribute slightly to the strength of this material at room temperature and to a lesser extent at 760°C.

Figure 6 illustrates measurements of specific Young's modulus for the various materials, and only the $[0]_8$ material exhibits outstanding behavior. The specific stiffness of the cross-ply and $[90]_8$ laminates is slightly higher than that of conventional materials. The tensile elongations at failure for these tests are shown in Fig. 7, indicating the extremely limited ductility of the composite.

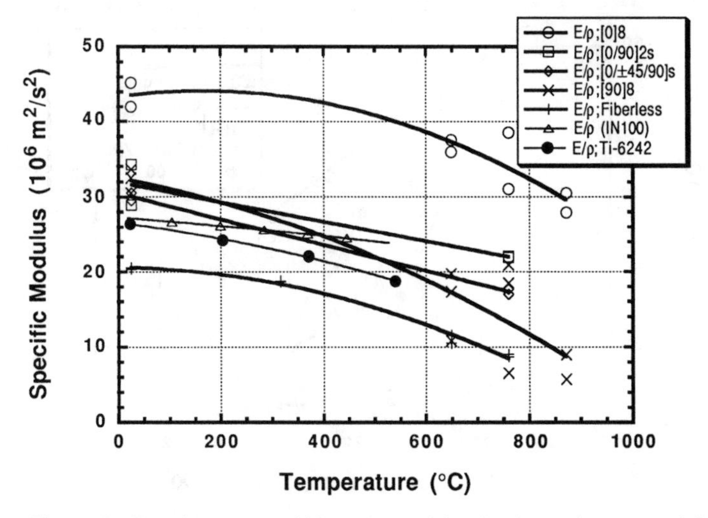

Figure 6. Density-corrected Young's modulus for the various materials.

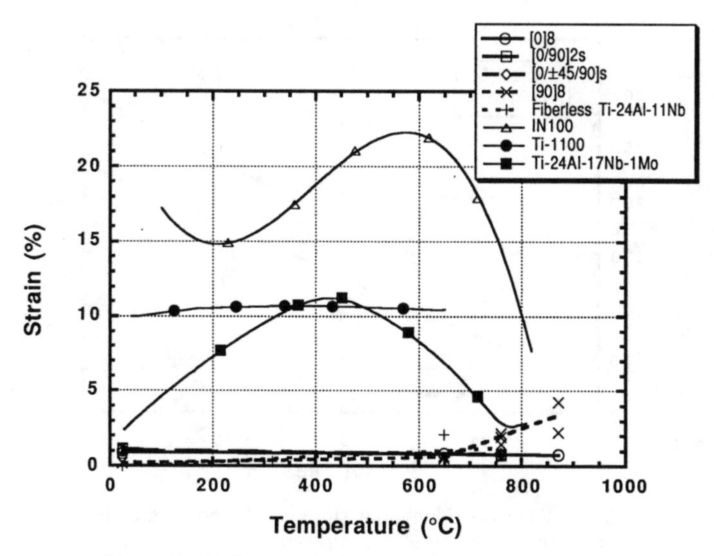

Figure 7. Failure strain for the various materials.

Moreover, it should be noted that the data plotted for the composites in Fig. 7 represent total strain to failure, which is mostly elastic, while plastic elongation is plotted for the other materials.

To provide a measure of long-term, high-temperature capability of the various materials, the Larson-Miller parameter was used to correlate the stress-rupture lives for a range of test conditions. The density-corrected Larson-Miller plot, shown in Fig. 8, illustrates a trend seen previously in the tensile results. The $[0]_8$ and $[90]_8$ composite exhibit the best and the worst capability, respectively, and cross-ply laminates fall in the intermediate range.

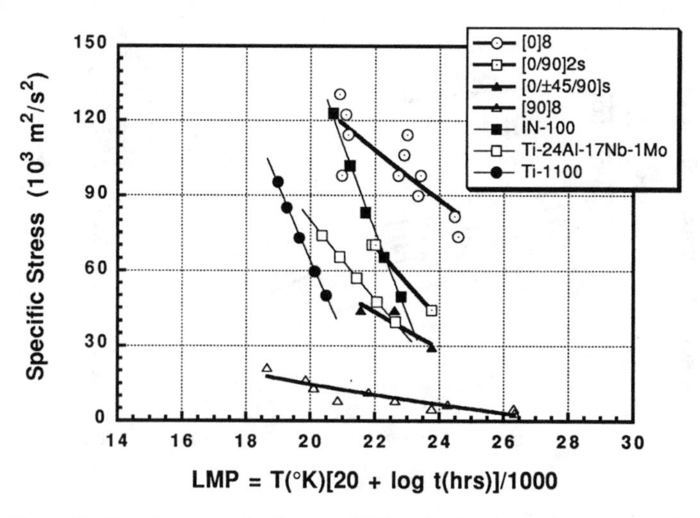

Figure 8. Density-corrected Larson-Miller plot for the various materials.

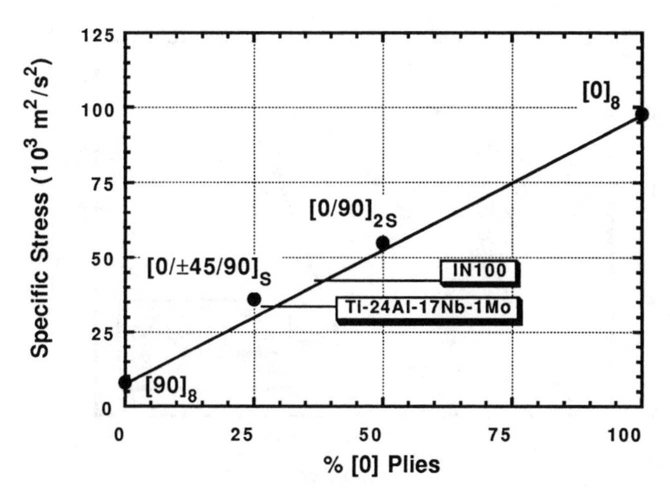

Figure 9. Specific rupture stress corresponding to a Larson-Miller parameter of 23 in Fig. 8.

The IN100, Ti-24Al-17Nb-1Mo, and Ti-1100 are obviously competitive with the cross-ply laminates, and as shown in Fig. 9, the creep capability of the various laminates is proportional to the percent of [0] plies. The [0] plies are primarily responsible for the measured creep resistance, while the [90] and [±45] plies contribute very little to the creep resistance.

Assessing the fatigue capability of titanium aluminide composites is a complex process, considering the wide range of potential usage conditions which include both thermal and mechanical cycling. Pertinent references on fatigue of SCS-6/Ti-24Al-11Nb include [7,15,16,17]. For purposes of the current comparison, Fig. 10 shows density-corrected data for tests at 650°C, indicating that the $[0]_8$ composite is outstanding, while the $[90]_8$ material falls well below all of the monolithic materials.

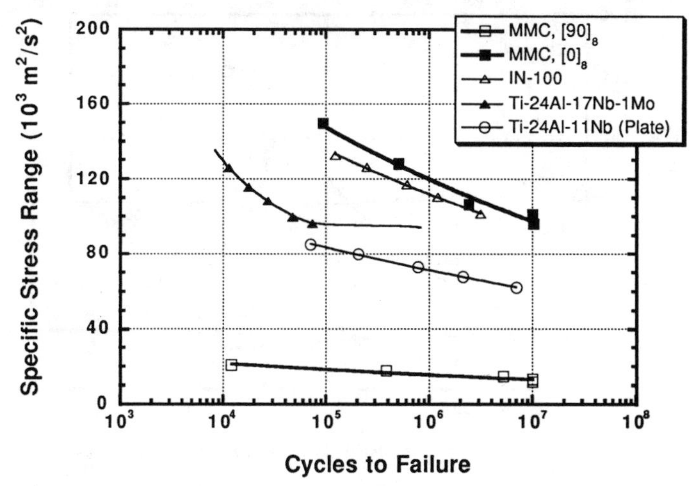

Figure 10. Density-corrected fatigue stress range versus cycles to failure; 650°C, frequency ≈ 30 Hz, R ≈ 0.

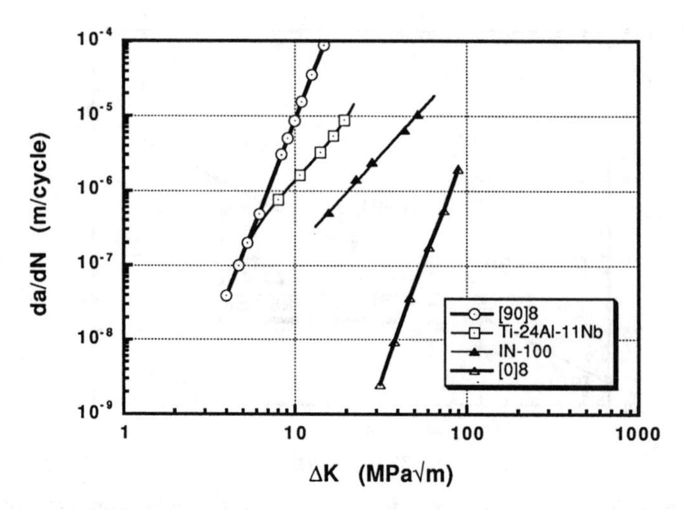

Figure 11. Comparison of fatigue crack growth behavior, 650°C, frequency ≈ 0.2, R= 0.1

Based on the trends shown earlier, it might be anticipated that the performance of the $[0/90]_{2S}$ composite would be approximately an average of the $[0]_8$ and the $[90]_8$. Thus the fatigue capability of the $[0/90]_{2S}$ composite would be expected to be approximately equal to the monolithic Ti-24Al-11Nb plate material [18] and would fall substantially below monolithic IN100 and Ti-24Al-17Nb-1Mo.

Fatigue crack propagation data for the various materials tested at 650°C, R = 0.1, and an approximate frequency of 0.2 Hz are presented in Fig. 11. The frequency of the tests of the composite was actually 3.33 Hz, giving somewhat lower crack growth rates than would be expected for 0.2 Hz testing. Recognizing

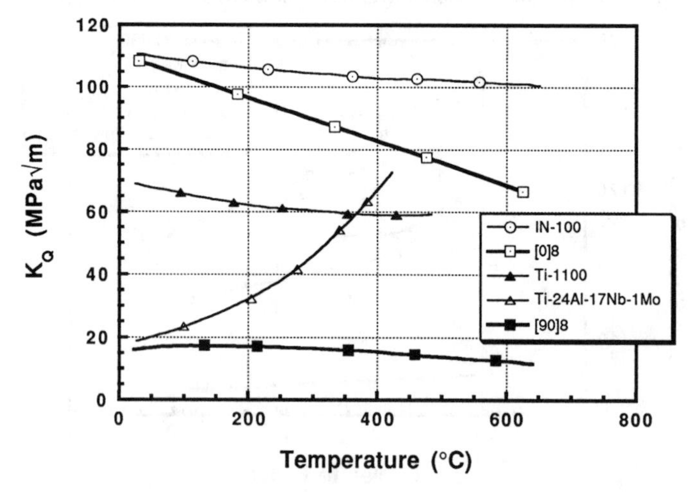

Figure 12. Comparison of apparent fracture toughness.

this discrepancy, it is clear that the transverse orientation of the composite exhibited the highest crack growth rates of the materials tested, followed by Ti-24Al-11Nb plate. Substantially greater crack growth resistance is offered by IN100 under the noted test conditions, while the $[0]_8$ composite material was by far the most resistant to crack propagation when the data are plotted against nominal ΔK.

The trends in apparent fracture toughness of the various materials are presented in Fig. 12, but only the data from Ti-1100 and Ti-24Al-17Nb-1Mo represent valid plane-strain fracture toughness, K_{IC}, tests. The highest observed stress intensity factor at the termination of fatigue crack growth testing was used as an operationally defined fracture toughness, K_Q, for IN100 and both orientations of the composite. As shown in the figure, IN100 and the $[0]_8$ composite have the highest apparent fracture toughness, followed by Ti-1100; Ti-24Al-17Nb-1Mo exhibits a relatively low toughness at room temperature. The $[90]_8$ composite has by far the lowest apparent fracture toughness.

The high K_Q for the $[0]_8$ composite is apparently the result of extensive fiber bridging in the wake of the crack. This phenomenon also appears to be responsible for the excellent da/dN-ΔK behavior of this material. Figure 13 shows examples from Jira [19] of cracking from a notch in the same SCS-6/Ti-24Al-11Nb composite showing significant crack bridging in room temperature tests. Similar behavior has been documented at elevated temperature.

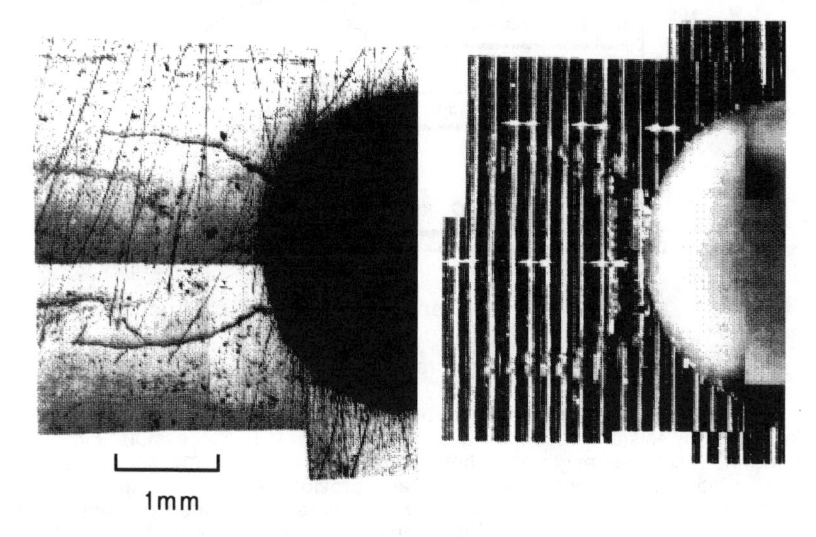

Figure 13. Fatigue cracking from a center hole in SCS-6/Ti-24Al-11Nb showing the specimen as-tested and after chemical removal of the outer matrix material.

Thermal Fatigue/Environmental Effects

The cyclic temperature excursions anticipated for titanium aluminide composites in service pose a unique problem due to the high sensitivity of the α_2 matrix to interstitial embrittlement and oxidation, as well as the two-to-one coefficient of thermal expansion (CTE) mismatch between matrix and fiber [20,21]. This mismatch creates significant residual stresses during cool down from processing temperature, and damaging cyclic strain ranges can occur when these composites are thermally cycled. A number of recent investigations have identified the overwhelming contribution of the environment to damage under thermal fatigue [15,22,23,24].

Figure 14 presents data of the post-cycling room-temperature residual strength of the composite as a function of the number of thermal cycles. Several features of this plot are noteworthy. The residual strength of the $[0]_4$ composite cycled in air drops precipitously in as little as 500 cycles, while the same material cycled in vacuum and inert environment retains almost all of its uncycled strength after 500 cycles. This environmental effect is also reflected in the difference in residual strength between $[0]_4$ and $[0]_8$ specimens cycled in air. Because of the dominance of surface-initiated matrix cracking that tends to affect the outer plies of the composite, specimens with thicker cross-sections retain a higher percentage of their original strength.

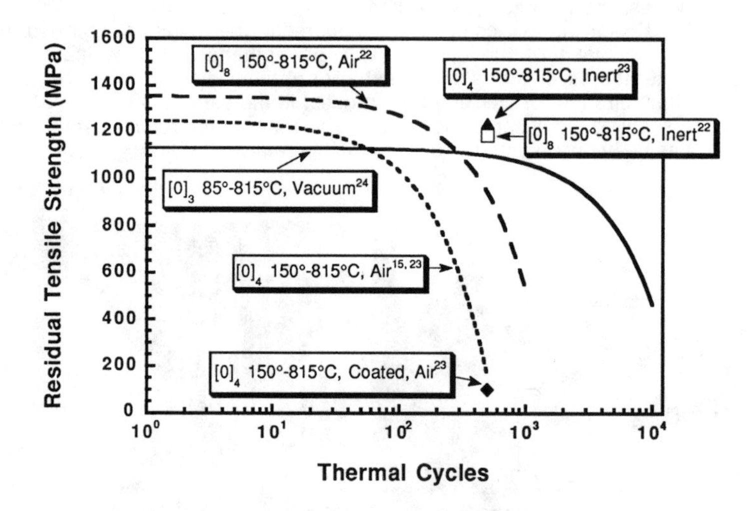

Figure 14. Thermal fatigue response of SCS-6/Ti-24Al-11Nb.

Attempts to provide environmental protection to titanium aluminide composites via oxidation-resistant coatings have thus far proved unsuccessful. Figure 14 shows the residual strength after thermal fatigue in air of a composite coated with an alumina-forming coating that normally provides excellent static oxidation resistance to the monolithic alloy. However, when the coated composite was subjected to thermal fatigue in air, this brittle coating cracked quickly and exposed the alloy substrate to concentrated environmental attack. The residual strength of this material was even lower than the uncoated composite cycled in air [23].

Crossweave Effects

Much of the SCS-6/Ti-24Al-11Nb composite fabricated to date has incorporated molybdenum crossweave wire to maintain fiber spacing during fabrication. Superior room temperature tensile properties of composites consolidated with this crossweave was the basis for its choice; however, in-depth characterization of mechanical properties in a number of studies has revealed the detrimental nature of the Mo crossweave [25,26,27]. Not only does Mo form brittle intermetallic reaction products with Ti-24Al-11Nb [28], it also forms MoO_3 which sublimes readily above 700°C [29]. When present on exposed composite surfaces in air at high temperatures, these crossweave wires oxidize and provide a direct path for environmental attack of the fibers. This attack is manifested by the presence of molybdenum on many fracture surfaces of specimens tested at elevated temperature in air under tension, creep, and thermal and thermomechanical fatigue. Figure 15 presents an example of composite damage due to oxidation of the Mo crossweave.

Ideally, if a crossweave is to be utilized, it should be manufactured from the same alloy as the matrix. However, progress in the processing of ribbon and thin gage wire must be made before identical alloys for both matrix and crossweave can be incorporated into intermetallic matrix composites. The use of alternative crossweave materials, in spite of their improved performance over molybdenum, are generally considered interim solutions and a transition to composite processing techniques that avoid the use of crossweave materials altogether is an option currently being considered [30].

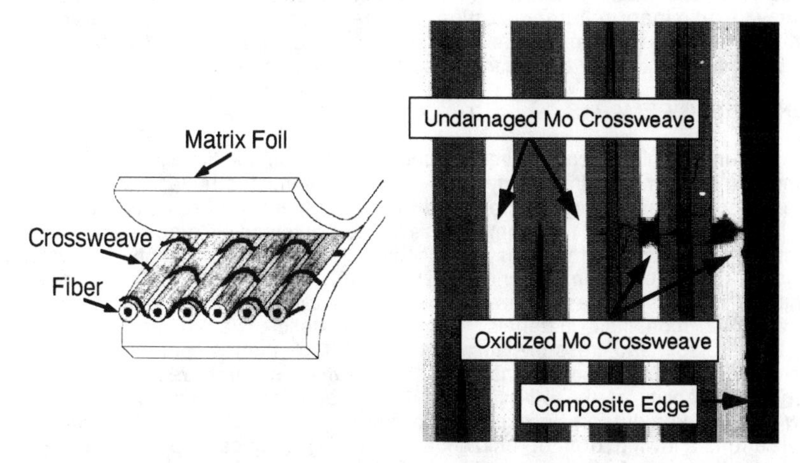

Figure 15. Crossweave used to hold fiber mats together during MMC fabrication [22]. (a) Schematic of crossweave incorporation into MMC, (b) cross-sectional view of first-fiber layer of an SCS-6/Ti-24Al-11Nb composite isothermally exposed for two hours at 815°C in air showing effect of molybdenum crossweave exposed to environment at composite edge. The crossweave wire in this micrograph has oxidized a distance of two fiber rows into the composite, damaging the matrix as well as the fibers and their interfaces. Note the undamaged portion of the crossweave wire further into the interior of the specimen [22].

FUTURE TRENDS

The future of titanium-based composites for use in the 650°-760°C range requires development of composite constituents (fiber, matrix, and interface) as a *system* where each constituent's role in the composite is fully understood and carefully controlled. The practice of combining an alloy such as Ti-24Al-11Nb, developed as a monolithic alloy, with the only suitable reinforcement has proved to be an inefficient method for advancing metal matrix composite technology.

Recent advances in innovative chemical vapor deposition reactor designs that allow a variety of interface coatings to be applied to fibers during production, coupled with the identification of several new fiber/matrix interface coating systems, show promise for higher quality fibers and improved composite properties [31]. For example, work by Marshall et al [32] has shown that the use of duplex, ductile interfacial layers in titanium aluminide composites can double the transverse strength of the interface, when compared to composites without such interfacial layers, while maintaining adequate toughness in the longitudinal direction.

Presently, significant gains in the development of titanium aluminide composites using an integrated "systems" approach are targeting improvement in matrix alloys to circumvent the environmental and ductility limitations of first-generation alloys such as Ti-24Al-11Nb. Titanium alloys that contain the orthorhombic Ti_2AlNb phase exhibit significantly higher room-temperature ductility, toughness, and specific strength than Ti-24Al-11Nb [33,34]. Smith et al [35] chose this class of alloys (specifically, Ti-22Al-23Nb) as an improved matrix material based upon knowledge of the limitations of α_2 titanium aluminides. Differences reported between SCS-6/Ti-22Al-23Nb and SCS-6/Ti-24Al-11Nb include reduced reaction product formation at the interface and the lack of a β-denuded zone around the SiC fiber. Preliminary characterization of these composites show dramatic improvements in thermal fatigue resistance when compared to SCS-6/Ti-24Al-11Nb and appear to offer significant possibilities for future development.

CONCLUDING REMARKS

Titanium aluminide composites overcome a number of the shortcomings of monolithic titanium aluminide alloys and offer the potential to replace conventional materials of higher density. As revealed by the comparison of selected materials, the $[0]_8$ SCS-6/Ti-24Al-11Nb composite exhibited outstanding properties, but the same material subjected to transverse loading ($[90]_8$) was consistently the worst-rated of the comparison materials. The $[0/90]_{2S}$ and $[0/\pm45/90]_S$ laminates exhibited many properties that were not significant improvements over the more conventional monolithic materials. Many of the deficiencies in the mechanical behavior of $[90]_8$ and cross-ply laminates appear to be the result of the very weak fiber/matrix interface under transverse stress, which severely restricts the load carrying capability of the off-axis plies. By proper construction of laminates, however, anisotropy can be exploited to suit the needs of a specific structural application, allowing greater utilization of the [0] composite properties. For example, if engine designs can be developed to accommodate the low transverse properties of the composite, then opportunity exists for development of unidirectional hoop-wound turbine engine rings that greatly exceed current capabilities. In addition, it should be possible to tailor cross-plied laminates to suit the needs of specific airframe applications, facilitating maximum utilization of the [0] properties.

It should be remembered that SCS-6/Ti-24Al-11Nb represents a first-generation titanium aluminide composite, and future materials offer important potential improvements in terms of environmental resistance, as well as matrix strength at high temperatures and ductility at low temperatures. Moreover, efforts are underway to improve transverse creep and tensile properties by developing fiber coatings to improve fiber/matrix bonding without sacrificing the fracture toughening benefits that exist under [0] loading. More research in this area is, however, required. Another potential avenue for improvement in properties, particularly for cross-ply laminates, is to increase the fiber volume fraction (V_f). However, careful selection and processing of the matrix material, combined with innovative consolidation methods, will probably be required to achieve a meaningful increase in V_f.

From the point of view of composite usage and reliability in critical structural applications, it is clear that design philosophies must be developed for using materials that have very limited ductility. In addition, methods for nondestructive evaluation of initial and service-induced damage must be developed and demonstrated, and much work is needed to develop mature approaches for life prediction of these materials in the complex thermomechanical environment in which they are required to operate.

Ultimately, the widespread use of titanium aluminide composites will depend on a combination of factors which include life-cycle cost, fabricability, consistent material properties that significantly exceed capabilities of current materials, and reliability in long-term service. Although a number of important advancements are in progress, much work remains to be done, and unprecedented cooperation among the materials-development and mechanical-design communities will be required to attain the full potential of these novel materials.

ACKNOWLEDGMENTS

This overview was complied by the authors, who are members of the Materials Behavior and Processing Branch, Materials Directorate, Wright Laboratory, Wright-Patterson Air Force Base, OH and was funded by the Air Force Office of Scientific Research under project 2302P101. We would like to acknowledge the helpful efforts of A. L. Hix and E. J. Dolley who aided in collection, analysis, and presentation of the data.

REFERENCES

1. J.M. Larsen, K.A. Williams, S.J. Balsone, and M.A. Stucke in High Temperature Aluminides and Intermetallics, S.H. Whang, C.T. Liu, D.P. Pope, J.O. Stiegler, Eds. (The Minerals, Metals & Mater. Soc., Warrendale, PA, 1990) pp. 521-556.
2. R.A. MacKay, P.K. Brindley, and F.H. Froes, JOM, pp. 23-29 (May 1991).
3. Wright Laboratory (WL/POT), "Integrated High Performance Turbine Engine Technology Initiative," Wright-Patterson AFB, OH, 1988.
4. T. M. F. Ronald, Advanced Materials & Processes, 29-37, (1989).
5. S.W. Kandebo, Aviation Week & Space Technology, February 24, 1991.
6. M. L. Gambone, "Fatigue and Fracture of Titanium Aluminides, Vol. I," U.S. Air Force report WRDC-TR-89-4145.I, Wright-Patterson Air Force Base, OH, 1989.
7. M. L. Gambone, "Fatigue and Fracture of Titanium Aluminides, Vol. II," U.S. Air Force report WRDC-TR-89-4145.II, Wright-Patterson Air Force Base, OH, 1989.
8. Smith, P. R., Rhodes, C. G., and Revelos, W.C., in Interfaces in Metal-Ceramic Composites, R.Y. Lin, et. al., eds.(TMS Warrendale, PA, 1990) pp. 35-58.
9. Larsen, J.M., Schwartz, B.J., and Annis, C.G., Jr., "Cumulative Damage Fracture Mechanics Under Engine Spectra," Air Force Materials Laboratory Report AFML-TR-79-4159, Wright-Patterson Air Force Base, OH, 1979.
10. B.A. Cowles, private communication, Pratt & Whitney Aircraft, West Palm Beach, FL, 1989.
11. P.J. Bania, JOM, 20-22 (1988)
12. P.J. Bania in Space Age Metals Technology, SAMPE, 1988, 286-297.
13. P.J. Bania in Proc. Sixth World Conference on Titanium, Cannes, France, 1988, P. Lacombe, R. Tricot, and G. Beranger, Eds., (Societe Francaise de Metallurgie, Les Editions de Physique, Les Ulis, France, 1989)
14. M. J. Blackburn and M. P. Smith, "Development of Improved Toughness Alloys Based on Titanium Aluminides," U.S. Air Force report WRDC-TR-89-4045, Wright-Patterson Air Force Base, OH, 1989.
15. S.M. Russ, Metall. Trans. A, 21, 1595 (1990).
16. T. Nicholas, and S. M. Russ, "Elevated Temperature Fatigue Behavior of SCS-6/Ti-24Al-11Nb," in press J. Mat. Sci. Eng.
17. S.M. Russ T. Nicholas, M. Bates, and S. Mall, "Thermomechanical Fatigue of SCS-6/Ti-24Al-11Nb Metal Matrix Composite," in press, ASME proceedings.
18. D. P. DeLuca and B. A. Cowles, "Fatigue and Fracture of Titanium Aluminides" U.S. Air Force report WRDC-TR-89-4136, Wright-Patterson AFB, OH, 1989.

19. J.R. Jira and J.M. Larsen, presented at the meeting of ASTM Committee E24.07, San Diego, CA, 1991 (unpublished).
20. S.J. Balsone and W.C. Revelos, presented at the 1991 TMS Annual Meeting, New Orleans, LA, 1991 (unpublished).
21. B.N. Cox, M.R. James, D.B. Marshall, R.C. Addison, Metall. Trans. A, 21, 2701 (1990).
22. W.C. Revelos and I. Roman, "Laminate Thickness and Orientation Effects on an SCS-6/Ti-24Al-11Nb Composites Under Thermal Fatigue," in this proceedings.
23. W.C. Revelos and P.R. Smith, Metall. Trans. A, 23, 587 (1992).
24. P.K. Brindley, R.A. MacKay, and P.A. Bartolotta in Titanium Aluminide Composites, P.R. Smith, S.J. Balsone and T. Nicholas, Eds., WL-TR-91-4020, Wright-Patterson AFB, OH, 1991, pp. 484-496.
25. M. Khobaib, in Proceeding of the American Society for Composites, edited by S.S. Sternstein (Technomic Publishing, Lancaster, PA 1991) pp. 638-647.
26. S.M. Russ and T. Nicholas, ibid. Ref. 23, pp. 450-467.
27. B.R. Kortyna and N.E. Ashbaugh, ibid. Ref. 23, pp. 467-483.
28. G. Das, presented at the 1991 TMS Annual Meeting, New Orleans, LA, 1991 (unpublished).
29. M.G. Fontana and N.D. Green, Corrosion Engineering, McGraw Hill, New York, 1978, p. 179.
30. M. Mrazek, S. M. Russ, and J.M. Larsen, presented at the 1990 American Institute for Aeronautics and Astronautics Mini-Symposium, Dayton, OH, June 1990 (unpublished).
31. K. Marnoch, V.R. Fry, W.E. Beyermann and D. Varnon, "Advanced CVD Fiber for Metal Matrix Composites" (Interim Report No. 1, Wright-Patterson Air Force Base, OH, 1991).
32. D.B. Marshall, M.C. Shaw, and W.L. Morris, "Measurement of Interface Properties in Intermetallic Matrix Composites," in this proceedings.
33. S. Ashley, Mechanical Engineering, 49-52 (Dec. 1991).
34. R.G. Rowe, Advanced Materials & Processes,. 33-35 (Mar 1992).
35. P.R. Smith, J. Graves, and C.G. Rhodes, "Evaluation of a SCS-6/Ti-22Al-23Nb Orthorhombic Composite," in this proceedings.

CHARACTERIZATION OF FIBER/MATRIX INTERFACES BY TRANSMISSION ELECTRON MICROSCOPY IN TITANIUM ALUMINIDE/SIC COMPOSITES

CECIL G. RHODES
Rockwell International Science Center, 1049 Camino Dos Rios, Thousand Oaks, CA 91360

ABSTRACT

This paper presents examples of the use of transmission electron microscopy to characterize matrix/reinforcement interaction in titanium aluminide matrix composites reinforced with continuous SCS-6 type SiC. As a result of the high temperature required for consolidating this type composite, reaction products form in the interface. Using diffraction and x-ray energy dispersive spectroscopy techniques, reaction products in Ti_3Al and Ti_2AlNb alloy matrix composites have been identified. TiC_{1-x} and Ti_5Si_3 compounds are common in these composites, with $AlTi_3C$ also present depending on consolidation temperature and matrix composition. Residual stress calculations indicate that these reaction products may be subject to cracking during cooling from consolidation temperatures.

INTRODUCTION

There are materials from which certain significant information can only be gained by the use of transmission electron microscopy (TEM). Among those special materials are composites, in which interphase interfaces contribute strongly to composite properties.

In the case of composites in which the matrix and reinforcement react chemically during processing, TEM is a useful tool for characterizing reaction products. In those composites in which matrix and reinforcement do not react, the structure of the interface can be analyzed by TEM. For instance, interfacial dislocations and ledges, which can contribute to the strength of interfaces and thus to the 'composite', can only be characterized by TEM.

This paper presents some examples of the use of TEM to characterize matrix/reinforcement interaction in titanium aluminide matrix composites reinforced with continuous SiC fiber. The significance of these results in the tailoring of titanium aluminide matrix composites is also discussed.

INTERFACES IN TITANIUM ALUMINIDE MATRIX COMPOSITES

The state-of-the-art reinforcement for titanium aluminides, whether Ti_3Al, Ti_2NbAl, or TiAl based, is SCS-6 SiC. SiC was selected for its high strength and modulus and the carbon-rich coating (SCS) was developed for improved post-consolidation strength of the fiber[1,2].

The titanium aluminides, like all intermetallics, require very high temperature for solid state composite consolidation. As a result of the consolidation cycle, two conditions develop. First, there is diffusional reaction between the titanium aluminide and SiC at the consolidation temperature. Reaction products, on the order of 0.5 to 2.0 μm in breadth, form in the matrix/fiber interfaces[3-9]. Second, differences in coefficients of thermal expansion result in thermal mismatch stresses during cooling from the consolidation temperature. Both of these conditions lead to a complex fiber/matrix interfacial region.

In order to control the interface condition, and thus the properties of the composite, it is critical to know the extent and compositions of the reaction products. Transmission electron microscopy is a valuable tool for determining the composition and dimension of each of the species in the interface. The contribution of the reaction zone phases to residual stresses and ultimately to mechanical properties of the composite can be calculated. Steps can then be taken to adjust the extent of interface reaction, by altering processing parameters or by including diffusion barrier materials in the interface.

EXAMPLES OF TEM IN TITANIUM ALUMINIDE MATRIX COMPOSITES

Types of Titanium Aluminide Matrices

There are currently three types of titanium aluminide intermetallics of interest for use as a matrix in a SiC reinforced composite. The first group comprises those based on Ti_3Al, the alpha-two phase, which generally have significant amounts of Nb added for improved ductility. This addition of beta-stabilizing Nb results in a microstructure that consists of alpha-two plus beta phases. Two of those to be described here are listed in Table I.

The second group is made up of those alloys based on Ti_2AlNb, the orthorhombic phase [10]. The alloys of interest in this class have non-stoichiometric levels of Al and Nb, leading to two or three phase microstructures (orthorhombic, beta, and alpha-two phases). Table I includes the one composition to be discussed here.

The third group consists of those based on TiAl, gamma phase. Because of its limited ductility and CTE mismatch with SiC, gamma matrix composites have not yet been successfully consolidated and will

Table I. Titanium Aluminide Matrices in Composites Studied in this Paper

Alloy Type	Composition, at. pct.
$\alpha_2 + \beta$	Ti-24Al-11Nb
	Ti-25Al-10Nb-3V-1Mo
orthorhombic	Ti-22Al-23Nb

not be discussed here.

Alpha-two + Beta Titanium Aluminides/SCS-6 SiC

Typical examples of $\alpha_2 + \beta$/SCS-6 SiC composites are shown in Figure 1. The major difference in the two aluminides is the increased amount of beta phase in Ti-25Al-10Nb-3V-1Mo compared to Ti-24Al-11Nb. Scanning electron microscopy (SEM) reveals the structure at the fiber/matrix interface, Figure 2. Reaction zone widths are about the same in both composites, on the order of 1-2 μm. In addition, there appear to be two layers in the reaction zone. The matrix immediately

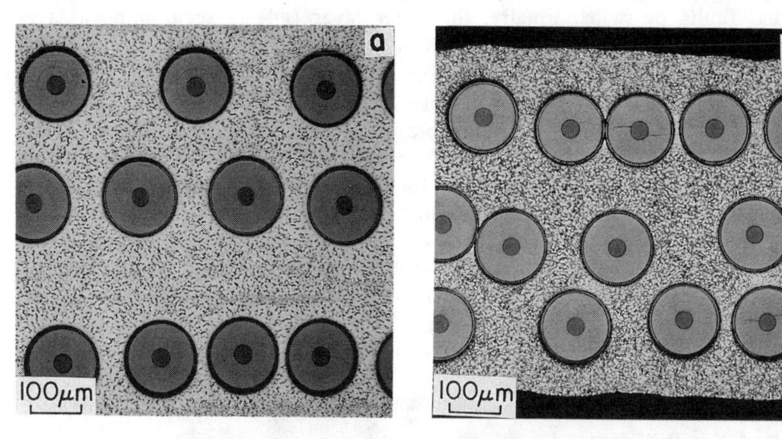

Figure 1. Metallographic cross sections of (a) Ti-24Al-11Nb/SCS-6 SiC, and (b) Ti-25Al-10Nb-3V-1Mo/SCS-6 SiC.

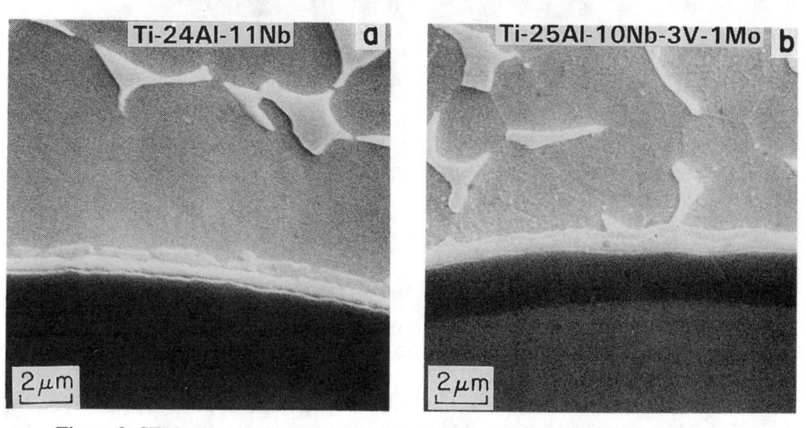

Figure 2. SEM cross sections showing fiber/matrix interface in (a) Ti-24Al-Nb/SCS-6 SiC, and (b) Ti-25Al-10Nb-3V-1Mo/SCS-6 SiC.

adjacent to the fiber shows a depeletion of beta phase in the Ti-24Al-11Nb, but not in the Ti-25Al-10Nb-3V-1Mo.

Thin-foil TEM provides a closer view of the interface reaction zones, Figure 3. Four distinct layers can now be seen, compared to what looked like 2 layers by SEM (Fig. 2). Adjacent to the fiber is a layer (#1) that is present in all titanium alloy matrix/SCS-6 SiC composites. This inner layer is about 0.1μm thick and consists of very fine equiaxed grains (≈20-50 nm). The next layer (#2) out from the fiber is about 0.5μm wide consisting of equiaxed grains about the size of the layer. Further towards the matrix, the third layer (#3) ranges in width from 0.5 to 0.7μm and is made up of large equiaxed grains characterized by internal faults or twins. Finally, the outer layer (#4) is seen to be a layer of elongated grains about 0.3 to 0.5 μm wide.

A typical selected area electron diffraction (SAD) pattern from a grain in the outer (#4) layer of the Ti-25Al-10Nb-3V-1Mo is shown in Figure 4. This pattern is indexable as [1123] hexagonal close-packed zone with lattice parameters a=0.75nm and c=0.51nm. The parameters, Table II, are close to those reported for Ti_5Si_3. EDS analyses, as listed in Table II, reveals that the major matrix alloying elements are included in the silicide. The phase in each of the two composites is near the compostion Ti_5Si_3 if partitioning of the elements is as follows: Nb,V, and Mo occupy Ti lattice sites, while Al is evenly distributed between Ti and Si lattice sites.

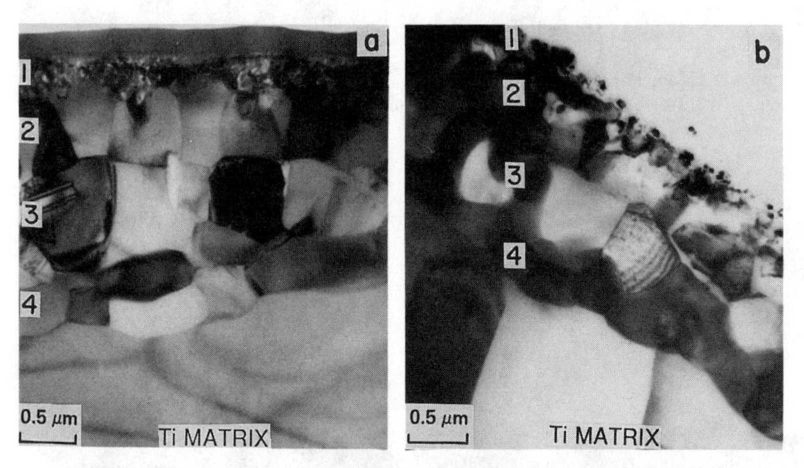

Figure 3. TEM cross sections showing fiber/matrix interfaces in (a) Ti-24Al-11Nb/SCS-6 SiC, and (b) Ti-25Al-10Nb-3V-1Mo/SCS-6 SiC.

A diffraction pattern from the third layer (#3, Figure 3) is given in Figure 5. This pattern is indexable as [110] fcc zone. Based on this and several other patterns, the structure was deduced to be ordered FCC with a lattice parameter of a ≈ 0.413nm, Table III. This is consistent with the carbide, $AlTi_3C$. EDS analysis does not give reliable quantitative analysis of carbon, so the values reported in Table III are modified by assuming a value of 20 at.pct. carbon. The carbide in both composites is only slightly rich in Al, if Nb, V, Mo, and Si occupy Ti lattice sites.

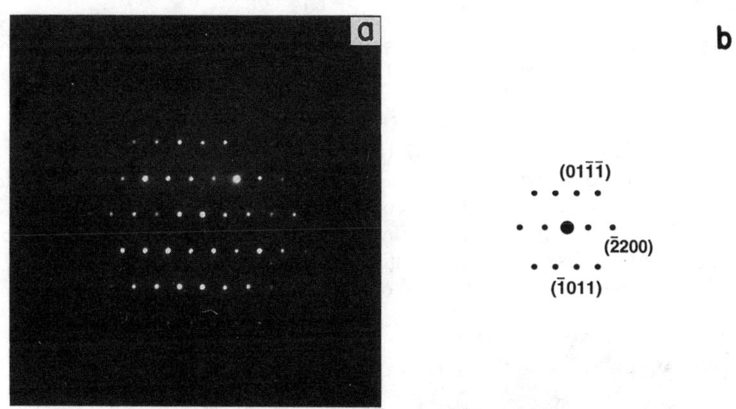

Figure 4. Selected area electron diffraction pattern from outer reaction zone layer in Ti-25Al-10Nb-3V-1Mo/SCS6 SiC. (a) pattern, (b) schematic, indexed as [1123] hcp.

Table II. TEM Analyses of Matrix/ SCS6 SiC Fiber Outer Reaction Zone Layer

	Latttice Parameters, SAD		Composition, at. pct., EDS					
Matrix	a, nm	c, nm	Ti	Al	Nb	V	Mo	Si
standard Ti_5Si_3	0.744	0.513	62.5	--	--	--	--	37.5
Ti-24Al-11Nb	0.78	0.50	40	12	18	--	--	30
Ti-25Al-10Nb-3V-1Mo	0.75	0.51	38.8	12	12.5	2.6	0.5	33.5

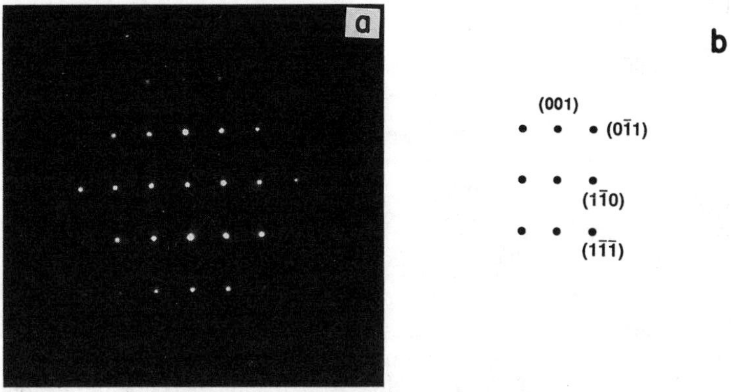

Figure 5. Electron diffraction pattern from reaction zone layer #3 in Ti-25Al-10Nb-3V-1Mo/SCS6 SiC. (a) pattern, (b) schematic analysis indexed as [110] fcc.

Table III. TEM Analyses of Matrix/ SCS6 SiC Fiber Reaction Zone Layer #3

	Latttice Parameter, SAD	Composition, at. pct., EDS						
Matrix	a, nm	Ti	Al	Nb	V	Mo	Si	C*
standard $AlTi_3C$	0.415	60	20	--	--	--	--	20
Ti-24Al-11Nb	0.40	54	21	4	--	--	1	20
Ti-25Al-10Nb-3V-1Mo	0.41	54	22	2	1	--	1	20

* - carbon assumed to be 20 at. pct.

A selected area electron diffraction pattern from a grain in reaction zone layer #2 in Ti-25Al-10Nb-3V-1Mo/SCS6 SiC (Fig. 3b) is shown in Figure 6. The pattern is indexable as [011] fcc, with lattice parameter, $a_o \approx 0.427$, which is consistent with TiC_{1-X}, Table IV. In addition, there are superlattice reflections at (1/2, 1/2, 1/2) reciprocal lattice positions. These reflections suggest ordering of the C sublattice in fcc TiC_{1-X}, as has been reported by others[11].

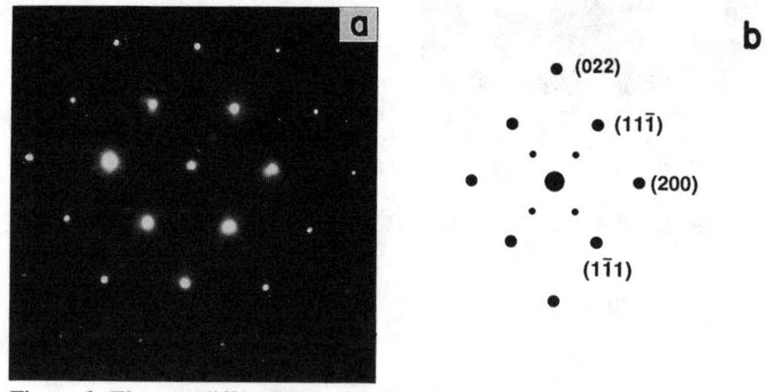

Figure 6. Electron diffraction pattern from reaction zone layer #2 in Ti-25Al-10Nb-3V-1Mo/SCS6 SiC. (a) pattern, (b) schematic indexed as [011] fcc.

Table IV. TEM Analyses of Matrix/ SCS6 SiC Fiber Reaction Zone Layer #2

	Latttice Parameter, SAD	Composition, at. pct., EDS						
Matrix	a, nm	Ti	Al	Nb	V	Mo	Si	C*
standard TiC_{1-X}	0.426-.432	55	--	--	--	--	--	45
Ti-24Al-11Nb	0.42	48	1	6	--	--	1	45
Ti-25Al-10Nb-3V-1Mo	0.43	50	1	3	1	--	1	45

* - carbon arbitrarily assumed to be 45 at. pct.

The inner reaction zone layer, #1 in Fig. 3, is deduced to be a fine-grained mixture of two reaction zone products, TiC_{1-X} and $(Ti,Nb,Al)_5(Si,Al)_3$. This conclusion is based on dark field imaging of either a TiC_{1-X} or a $(Ti,Nb,Al)_5(Si,Al)_3$ grain in which many of the layer #1 grains are imaged. Figure 7, which is taken from Ref. 12, shows the effect in an alpha Ti matrix composite reinforced with SCS6 SiC. Although this is not a titanium aluminide matrix, the phenomenon is the same.

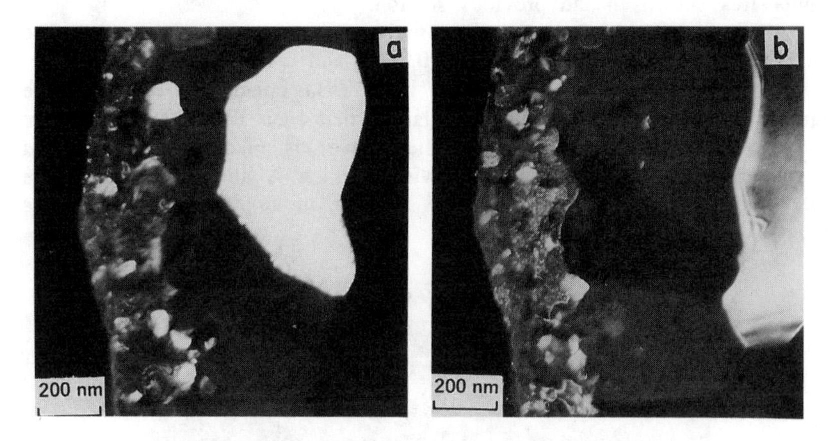

Figure 7. Dark field image of reaction zone layer #1 . (a) imaged with TiC_{1-x} reflection, (b) imaged with $(Ti,Al)_5(Si,Al)_3$ reflection.

Orthorhombic Titanium Aluminides/SCS-6 SiC

A typical example of an orthorhombic (O) + beta titanium aluminide matrix composite is shown in Figure 8. The Ti-22Al-23Nb matrix can be processed to produce a non-equilibrium micro-structure of O + β + α_2 phases or the equilibrium mixture of O + β prior to consolidation. The former non-equilibrium structure is present in the composite in Figure 8. An SEM image of the fiber/matrix interface,

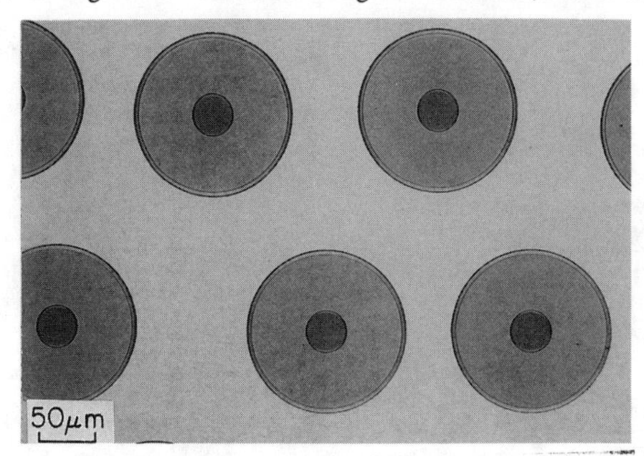

Figure 8. Metallographic cross section of Ti-22Al-23Nb/SCS-6 SiC composite.

Figure 9, reveals only a slight reaction zone in this composite which had a consolidation temperature lower than that for the $\alpha_2+\beta$ matrix composites shown in the previous section.

Thin-foil TEM of the interface region is presented in Figure 10. The total reaction zone width is about 0.3μm and comprises three distinct layers. Adjacent to the fiber is a layer (#1) consisting of very fine equiaxed grains (≈ 20-50 nm) similar to that seen in the $\alpha_2+\beta$ matrix composites (Fig. 3). The intermediate layer is made up of equiaxed grains about the size of the layer width, which is about 150 nm. The outer layer, adjacent to the matrix, is a string of slightly elongated grains about 100 nm wide.

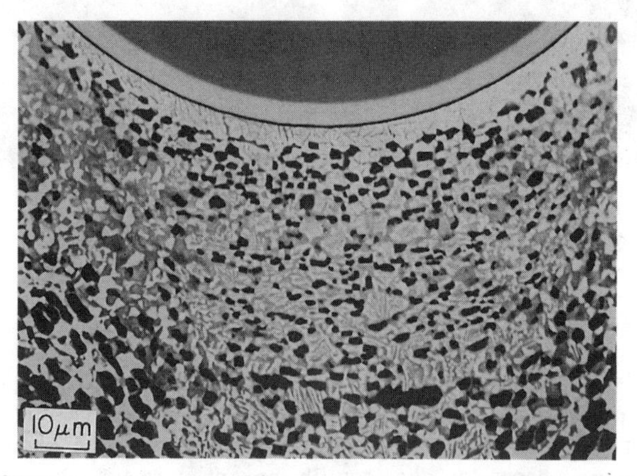

Figure 9. SEM cross section showing fiber/matrix interface in Ti-22Al-23Nb/SCS-6 SiC.

Figure 10. TEM cross section showing fiber/matrix interface in Ti-22Al-23Nb/SCS-6 SiC, consolidated at low temperature.

Electron diffraction and EDS techniques have been used to identify the reaction zone products, Table V. The reaction products are the same as those identified in the $\alpha_2 + \beta$ matrix composites, except that $AlTi_3C$ is not present in this O-phase matrix composite. This condition is not so much a function of the matrix composition as it is the consolidation temperature. When this Ti-22Al-23Nb matrix is consolidated at a higher temperature, the $AlTi_3C$ type carbide appears in the interface, Figure 11. In this case, the total reaction zone width is about 0.7μm. So, although

Table V. TEM Analyses of Reaction Zone Layers in Ti-22Al-23Nb/SCS-6 SiC

Layer	Product Species	Lattice Parameter a, nm	c, nm	Composition, at. pct., EDS Ti	Al	Nb	Si	C
Outer	$(Ti,Nb,Al)_5(Si,Al)_3$	0.72	0.50	36	6	20	38	--
Inter-mediate	$(Ti,Nb,Si)C_{1-X}$	0.42		43	0	5	7	45*
Inner	$TiC_{1-X} + Ti_5Si_3$**							

* Assumed to be 45 at. pct.
** Deduced based on dark-field imaging

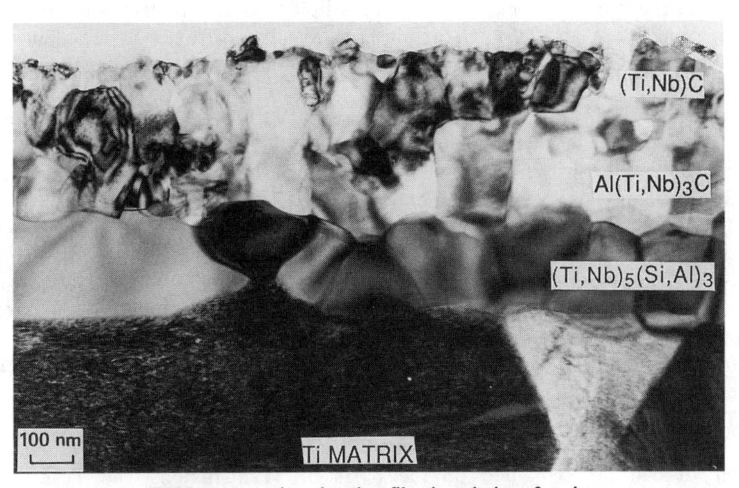

Figure 11. TEM cross section showing fiber/matrix interface in Ti-22Al-23Nb/SCS-6 SiC consolidated at high temperature.

the consolidation temperature was about 100°C higher, this latter O-phase matrix composite had less fiber/matrix reaction than the $\alpha_2 + \beta$ matrix composites.

DISCUSSION

Reaction Zone Formation

Titanium silicide found in the fiber/matrix interfaces of the titanium aluminide matrix composites is typical of all SiC-reinforced titanium alloys[3-9,12-14]. The silicide is generally present as the outermost, i.e., nearest the matrix, reaction zone product. The TiC-type carbide phase is always found nearer the fiber and, when present, the $AlTi_3C$ type carbide is generally in an intermediate position in the reaction zone. Because this arrangement is present in all SiC-reinforced titanium alloys, it can be assumed that relative diffusion rates of Si, C, and Ti control growth of the reaction zone phases.

With the titanium aluminide matrix in contact with the C-rich SCS coating on the fiber, there is interdiffusion of Ti into the fiber surface layer and C and Si into the titanium aluminide matrix. Because there is more C than Si in the SCS coating, the titanium aluminide becomes saturated with C more quickly than with Si and TiC precipitates first. The growth of the layers then appears to be controlled primarily by the diffusion of Ti, C, and Si through the TiC.

C diffuses faster than Ti through TiC, with Nb diffusing even faster than C[15,16]. The presence of titanium silicide at the matrix side of the reaction zone leads to the conclusion that Si also diffuses faster than Ti through the TiC. The Si atoms diffusing outward through the TiC eventually reach the matrix and react to form Ti_5Si_3. Most titanium atoms diffusing into the TiC eventually react with the C atoms moving outward resulting in growth of the TiC layer. This scenario suggests that the reaction zone grows somewhat more rapidly into the matrix than into the fiber, as has been documented [13,17].

The rapid diffusivity of Nb in TiC accounts for the presence of Nb in all the reaction zone products. Al appears to be less rapidly diffusing in the reaction zone, as virtually none is detected in the inner TiC layer. However, as the titanium aluminide matrix initially reacts to form TiC, the matrix at the reaction zone is enriched in Al. This Al enrichment leads to the formation of $AlTi_3C$, which is the stable carbide in the Ti-Al-C system at Ti_3Al [18].

The presence of the narrow layer of mixed carbide and silicide, which is found in all SiC reinforced titanium alloys, seems unusual. However, it can be explained on the basis of Ti diffusion. The Ti atoms that diffuse through the TiC without reacting reach the fiber surface and are free to react with both C and Si atoms. Mostly TiC is formed (because of the excess C in the coating), but some small amount of Ti_5Si_3 is also formed, resulting in the narrow layer. With time, the TiC is consumed by the growing TiC layer. The small Ti_5Si_3 grains are trapped within the TiC, as are observed in composites that have been exposed to longer thermal treatments.

TiC_{1-x} Ordering

Ordering of TiC_{1-x} is characterized by the presence of faint reflections at (1/2, 1/2, 1/2) reciprocal lattice positions in electron diffraction patterns. They have been observed in several titanium aluminde matrix composites[5,12]. Baumann et al.[5] attributed the super-reflections to ordering of either Nb on the Ti sublattice or vacancies on the C sublattice. The latter is most likely because ordering in TiC_{1-x} containing no Nb has been reported [12].

Vacancy ordering in TiC_{1-x} has been observed when the C concentration ranges from 32 to 36 at. pct.[11]. In the present work no precise measurements of composition (either by EDS, EELS, or CBED) of the TiC_{1-x} have been made. Furthermore, simulations of electron diffraction patterns using Diffract computer code have not produced the super-reflections observed in the patterns. So there is no direct evidence of vacancy ordering. Baumann et al. [5] associated the ordering with Nb because they detected a Nb gradient that corresponded to an ordering gradient in the TiC_{1-x}. However, the Nb gradient could also parallel a C gradient, i.e., less carbon (and more Nb) in the region of ordering. Such a reduction in C concentration could place the TiC_{1-x} in the range of 32-36 at. pct., thereby promoting vacancy ordering.

APPLICATION OF RESULTS

Properties of Reaction Zone Products

The formation and growth of reaction products in the fiber/matrix interface can degrade properties of the composite. The reaction products can crack under the stresses that develop during cooling from consolidation. These cracks then act as nuclei for further cracking into the matrix under external loading [8]. By identifying the reaction zone products, calculations can lead to correlations of reaction zone width with residual stresses.

Using the co-axial model developed at NRL [19], radial, tangential, and axial stresses have been calculated for an SCS-6 SiC reinforced Ti-24Al-11Nb composite. The approximate stresses in the reaction zone layer are given in Table VI. The tangential and axial stresses are quite

Table VI. Calculated Residual Stresses in Reaction Zone Products in SCS-6 SiC Reinforced Ti-24Al-11Nb

Reaction Product	Radial Stress	Tangential Stress	Axial Stress
TiC_{1-x}	- 39 ksi	230 ksi	165 ksi
$AlTi_3C$	- 38	290	230
Ti_5Si_3	- 38	310	260

high and could easily exceed the tensile strengths of the compounds. Standard tensile strengths of the carbides or silicide are not available for direct comparisons. An estimate of Ti_5Si_3 is about 300 ksi [20], so the tangential residual stress is of a magnitude that could initiate cracking. The carbides may also be approaching their tensile strength. Radial stresses are, of course, compressive and do not pose a threat of crack initiation.

The presence of reaction zone products can be detrimental to properties of the composite. One approach to reducing or eliminating the formation of these undesirable components is the introduction of barrier layers in the fiber/matrix interface. The added layer should prevent chemical interaction between Ti matrix and SiC fiber, while providing a reasonable interface bond. This interface tailoring is being explored by several investigators[9,21], and will be aided in development by the use of transmission electron microscopy in its characterization.

ACKNOWLEGEMENTS

I am pleased to acknowledge the technical assistance of Mr. R. Spurling of the Rockwell Science Center. The orthorhombic matrix composites were supplied by Dr. J. Graves of the Rockwell Science Center and Mr. P. Smith of Materials Directorate, Wright Laboratory. The alpha-2 matrix composites were supplied by Mr. C. Rosen of Rockwell International. This work was supported by Rockwell International IR&D funding.

REFERENCES

1. J.A. McElman, in Engineering Materials Handbook: Composites, vol. 1 (ASM International, Metals Park, OH, 1987) pp. 858-866.
2. X.J. Ning and P. Piroux, J. Mater. Res. 6, 2234 (1991).
3. P.K. Brindley, in High Temperature Ordered Intermetallic Alloys II, edited by N.S. Stoloff et al. (Mater. Res. Soc. Proc. 81, Pittsburgh, PA 1987) pp. 419-424.
4. D.M. Bowder, S.M.L. Sastry, and P.R. Smith, Scripta Metal. 23, 407 (1989).
5. S.F. Baumann, P.K. Brindley, and S.D. Smith, Metall. Trans. 21A, 1559 (1990).
6. C.G. Rhodes, C.C. Bampton, and J.A. Graves in Intermetallic Matrix Composites, edited by D.L. Anton et al. (Mater. Res. Soc. Proc. 194, Pittsburgh, PA 1990) pp. 349-354.
7. D.R. Baker, P.J. Doorbar, and M.H. Loretto in Interfaces in Composites, edited by C.G. Pantano and E.J.H. Chen (Mater. Res. Soc. Proc. 170,

Pittsburgh, PA 1990) pp. 85-90.
8. P.R. Smith, C.G. Rhodes, and W.C. Revelos in Interfaces in Metal-Ceramics Composites, edited by R.J. Lin et al. (TMS, Warrendale, PA 1989) pp. 35-58.
9. A.M. Ritter, E.L. Hall, and N. Lewis in Intermetallic Matrix Composites, edited by D.L. Anton et al. (Mater. Res. Soc. Proc. 194, Pittsburgh, PA 1990) pp. 413-421.
10. R.G. Rowe in Microstructure/Property Relationships in Titanium Aluminides and Alloys, edited by Y-W Kim and R.R. Boyer (TMS, Warrencale, PA 1990) pp. 387-398.
11. H. Goretzki, Phys. Stat. Sol. 20, K141 (1967).
12. C.G. Rhodes and R.A. Spurling in Developments in Ceramic and Metal-Matrix Composites, edited by K. Upadhya (TMS, Warrendale, PA, 1992), pp. 99-113.
13. C.G. Rhodes and R.A. Spurling in Recent Advances in Composites in the United States and Japan, edited by J.R. Vinson and M. Taya (ASTM STP 864, Philadelphia, PA, 1985) pp. 585-599.
14. C.G. Rhodes, A.K. Ghosh, and R.A. Spurling, Metall. Trans. 18A, 2151 (1987).
15. F.J.J. van Loo and G.F. Bastin, Metall. Trans. 20A, 403 (1989).
16. S. Sarian, J. Appl. Phys. 39, 3305 (1968).
17. J-M Yang and S.M. Jeng, Scripta Metall. 23, 1559 (1989).
18. J.C. Schuster, H. Nowotny, and C. Vacarro, J. Sol. Stat. Chem. 32, 213 (1980)
19. T. Hahn, Naval Research Laboratory, Washington D.C.(private communication).
20. G. Reynolds, MSNW, Inc., San Marcos, CA (private communication).
21. S.M. Arnold, V.K. Arya, and M.E. Melis, NASA TM 103204, 1990.

EVALUATION OF A Ti-22Al-23Nb "ORTHORHOMBIC" ALLOY FOR USE AS THE MATRIX IN A HIGH TEMPERATURE Ti-BASED COMPOSITE

J.A. Graves*, P.R. Smith+, and C.G. Rhodes*

* Rockwell International Science Center, Thousand Oaks, CA 91360
+ Materials Directorate, Wright-Laboratories, Wright-Patterson AFB, OH 45433

ABSTRACT

The production of titanium aluminide intermetallic compound foil represents a significant manufacturing challenge. Cold rolling, which imparts excellent thickness uniformity and surface finish characteristics that are of benefit in composite fabrication, is especially difficult with these alloys. However, recent modifications in Ti aluminide alloy compositions and advances in thermomechanical processing have made it possible to produce foil of thickness less than 100 μm, having the microstructure and mechanical property characteristics required for composite fabrication and improved performance. This paper describes the properties of a new Ti aluminide alloy, of nominal composition Ti-22Al-23Nb (at.%), comprising a three phase microstructure of α_2 (Ti_3Al), an ordered orthorhombic phase (Ti_2AlNb) and an ordered beta phase. The discussion emphasizes the processing of this alloy through cold rolling to foil, and the associated microstructures and mechanical property characteristics that are relevant to the use of this foil to form a composite matrix.

INTRODUCTION

Fabrication of Ti-matrix composites using foil-fiber laminate consolidation techniques typically requires foil of thickness 75-150 μm. The use of this processing approach with α_2 (Ti_3Al) base alloys has been limited largely by difficulties in producing foil of the quality required for the composite matrix. These difficulties have arisen primarily from the poor ductility of the precursor sheet and problems associated with the cold rolling of the sheet to thin foil gauges [1]. While multi-component two-phase, $\alpha_2 + \beta$, alloys having improved ductility and toughness have been developed for monolithic application, even the most advanced of these materials has proved difficult to cold roll to foil. In addition, they suffer from severe environmental embrittlement when exposed to high temperature air environments. In spite of these difficulties, much progress has recently been made in developing the cold rolling and heat treatment techniques required to produce foil with high surface finish qualities and excellent thickness uniformity in Ti intermetallic alloy systems [2].

The objective of the present work was to determine the cold rolling characteristics and associated foil microstructures and mechanical property behavior for a new class of Ti alloy designed for composite matrix applications in the 540°-760°C range. The alloy of specific interest in this work is Ti-22Al-23Nb (at.%), whose microstructure comprises an ordered orthorhombic (O) phase (Ti_2AlNb), an ordered beta (β_o) phase and an ordered α_2 phase [3]. For simplicity, this alloy will be termed an "orthorhombic alloy", due to the large volume fraction of this phase and its potential impact on the mechanical properties of the material. This paper is designed to provide a basic description of the processing and properties of the orthorhombic alloy foil, as processed for composite matrix applications. The processing and characteristics of multi-ply SiC reinforced composite panels produced from this foil are presented in a companion paper [4].

EXPERIMENTAL PROCEDURE AND RESULTS

Processing of Orthorhombic Sheet and Foil

The material used in this study was of target composition Ti-22Al-23Nb (at.%). The primary ingot and sheet production was carried out by Timet Corp. (Henderson, NV), with subsequent processing to foil conducted by Texas Instruments, Inc. (Attleboro, MA) . The processing history for the sheet material is summarized in Table I. Following the final conditioning treatment, this sheet represented the input "precursor material" for subsequent cold-rolling to foil gauge thickness.

Table I
Processing History of Ti-22Al-23Nb (at.%) Sheet

- Double-melt VAR used to produce 10.2 cm diameter ingot weighing 36.2 kg
- 36.2 kg ingot forged to 7.6 cm thick x 12.7 cm wide slab at 1260°C (2300°F)
- 7.6 cm slab forged to 5.0 cm thick x 12.7 cm wide slab at 1150°C (2100°F)
- 5.0 cm thick slab hot-rolled to 1.0 cm thick x 12.7 cm wide sheet at 1038°C (1900°F)
- 1.0 cm sheet "beta annealed" at 1175°C (2150°F) for 0.5 hours and air cooled
- 1.0 cm sheet hot-rolled to 0.76 cm thick x 12.7 cm wide sheet at 982°C (1800°F)
- 0.76 cm sheet cross-rolled to 0.20 cm thick sheet at 982°C (1800°F)
- 0.20 cm sheet annealed at 1010°C (1850°F) for 0.5 hours and air cooled
- 0.20 cm sheet conditioned for further thermomechanical processing to foil

Sheet thickness variations significantly influence the cold rolling characteristics of foil, particularly for intermetallic alloys having low ductility at ambient temperature. Precursor sheet gauge variations in this study ranged from 70 μm to 120 μm (4-8%) at different locations on individual sheets. This implies the introduction of various percentage reductions along the length of each sheet and additional stressing of the material as it is processed to foil. In spite of these thickness variations in the precursor sheet material, foil of final thickness 127 μm and 89 μm was produced through cold-rolling and intermittent vacuum annealing operations. These two thicknesses were required for the subsequent fabrication of composite panels [4]. Chemical compositions of the starting ingot, the hot-rolled sheet and the cold-rolled foil are summarized in Table II. A similar summary of interstitial content is provided in Table III. Note that the oxygen level is relatively high compared to those typically targeted for Ti aluminide alloys. This interstitial content can significantly influence the phase stability, transformation kinetics and properties of the alloy.

Table II
Chemical Composition of Orthorhombic Material (at.%)

Alloy Lot Number	Ti	Al	Nb	Fe
Ingot (Lot No. V 7382)	(bal.)	21.2	21.9	0.060
Sheet (0.15 cm)	(bal.)	20.0	22.6	0.120
Foil (127 μm)	(bal.)	21.1	22.3	0.100
Foil (89 μm)	(bal.)	21.1	22.8	0.020

Table III
Interstitial Content of Orthorhombic Material (wt.%)

Alloy Lot Number	O	N
Ingot (Lot No. V 7382)	0.123	0.007
Sheet (0.15 cm)	0.136	0.013
Foil (127 μm)	0.167	0.007
Foil (89 μm)	0.179	0.011

The introduction of significant plastic deformation through the rolling process typically leads to the formation of crystallographic textures and corresponding anisotropy in the properties of the sheet and foil. The influence of crystallographic orientation on the yield properties for the orthorhombic sheet is summarized in Figure 1, with a comparison made to an $\alpha_2+\beta$ alloy (Ti-24Al-11Nb, at.%). The yield loci, measured using the Knoop hardness method, were quite different [5,6]. This is most readily apparent when compared to the von Mises isotropic loci. The O-alloy exhibited nearly isotropic behavior, while the $\alpha_2+\beta$ alloy showed a yield locus noticeably elongated relative to the von Mises behavior. This latter behavior is representative of a material which has a strong normal anisotropy (i.e., a yield strength considerably higher in the through-thickness direction as compared to the in-plane directions). Thus, the effect of crystallographic texturing of the O-alloy, although significant, may be less pronounced than in the conventional $\alpha_2+\beta$ alloy.

Microstructure of Orthorhombic Sheet and Foil

In understanding the behavior of the matrix alloy it is important to begin with the microstructure of the sheet and foil material. Figure 2 presents optical (OM) and

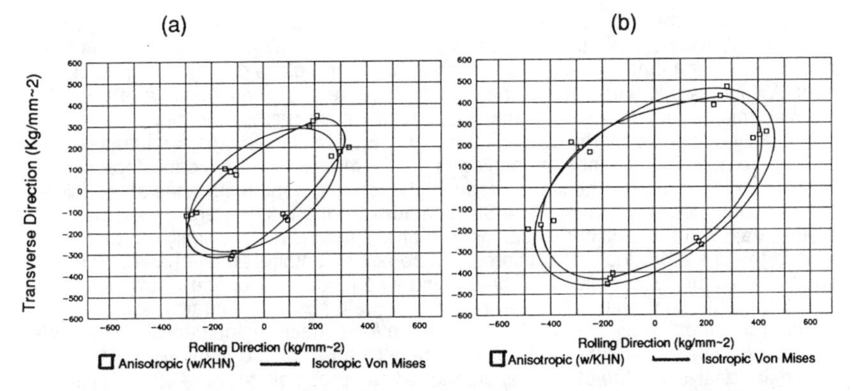

Figure 1. Knoop hardness yield loci for (a) $\alpha_2+\beta$; and (b) orthorhombic sheet material. The measured yield loci are compared to the yield locus for an isotropic (von Mises) material.

scanning electron (SEM) micrographs of the as-rolled orthorhombic sheet. The microstructure appears to be relatively homogeneous in grain size and shape. There are, however, banded areas wherein the composition of the sheet appears to be inhomogeneous. It should be noted that there was not a suitable master alloy with high enough Nb content to directly produce an ingot of the desired alloy chemistry. Thus, pure Nb had to be added in order to approach the target level of 23 at.% Nb. Although double VAR was utilized, there is the possibility that some areas of high Nb concentration remain after solidification of the ingot. In spite of the large amount of deformation at elevated temperature in processing the ingot to sheet, segregation may be maintained due to the sluggish diffusion kinetics in these ordered phases. The SEM photomicrograph (right) provides evidence of three phases in the hot-rolled sheet material: an approximately 40/60 ratio of an equiaxed dark gray phase and a mixture of a light gray and white phases. The dark gray phase has a low aspect ratio and is on the order of 3- 5 μm in size. The two-phase mixture consists of a continuous white phase with a lenticular light gray phase precipitate. As described in detail below, the dark gray phase was later determined to be α_2, while the two-phase mixture comprised a continuous β_o (white) with O-phase precipitates (light gray).

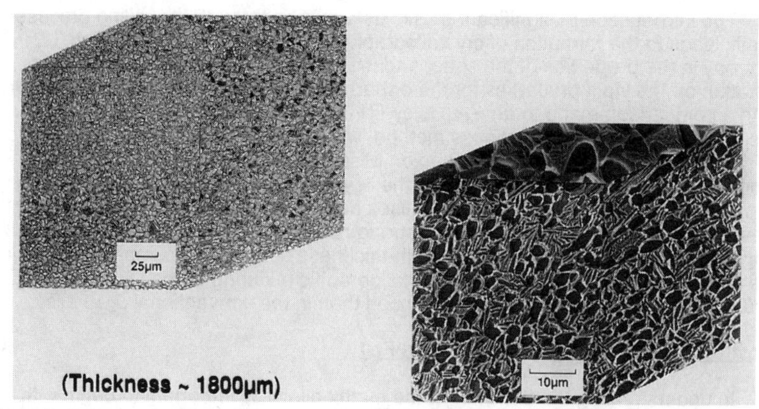

Figure 2. Optical and scanning electron micrographs of hot-rolled orthorhombic sheet.

Extending the microstructural analysis to the cold-rolled/annealed foil, Figure 3 presents OM and SEM photomicrographs of the as-rolled 127 μm foil product. In the OM it can be seen that significant deformation has taken place during rolling of the sheet. While the microstructure in the short transverse direction exhibits a relatively equiaxed microstructure, in the longitudinal directions the grains have been dramatically elongated by the deformation. Two additional microstructural features can also be noticed: evidence of banding in the foil microstructure is considerably more pronounced than in the sheet product; the top and bottom surfaces of the foil exhibit a continuous dark layer. The SEM photomicrograph (right) again indicates that the majority of the foil comprises a three phase microstructure similar to that found in the sheet product. This microstructure again contains an α_2 phase and a two-phase mixture of a lenticular O-phase and a continuous β_o phase. However, in the foil, the banded areas tend to have a smaller grain size and contain a lower volume fraction of the α_2 phase. In addition, the foil surface region can be discerned to consist of only the two-phase O+ β_o mixture, with no evidence of the α_2 phase. In a similar analysis of the 89 μm foil product, all features noted for the 127 μm foil are also present, however there was less banding present in the microstructure.

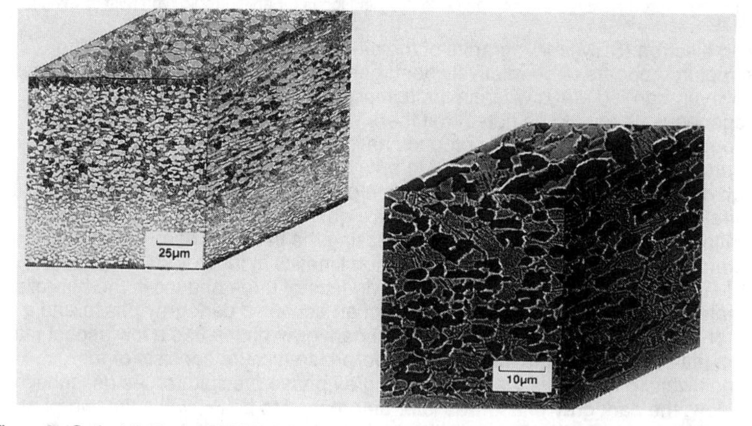

Figure 3. Optical (left) and SEM (right) micrographs of cold-rolled 127 μm orthorhombic foil.

In order to quantitatively assess the chemistry differences between the microstructural constituents, Wavelength Dispersive Spectroscopy (WDS) analysis was conducted on the foil cross-sections. A representative SEM micrograph of the 127 µm thick foil, along with corresponding WDS data, is provided in Figure 4. It can be seen that the regions comprising the two-phase, O+β_o, mixture in the internal areas and foil surfaces are very similar in both appearance and in composition. These two-phase regions contain a much higher percentage of Nb (26 at.%) than the dark gray phase (13 at.%). Furthermore, the two-phase mixture contains lower levels of Ti (56 at.%) and Al (18 at.%) than the dark gray phase (63 at.% and 24 at.%, respectively). It should be noted that, due to the fine scale of the O-phase precipitate, the compositional analysis for these regions represents an average of the two phases. Alternatively, the α_2 phase is of large enough size that compositional measurements of this individual constituent may be made.

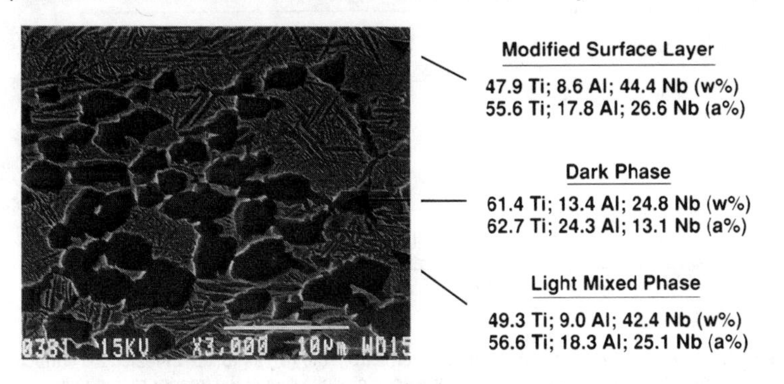

Figure 4. Scanning electron micrographs and corresponding WDS composition data for cold-rolled 127 µm orthorhombic foil.

In order to fully characterize the individual constituent phases in the Ti-22Al-23Nb foil, both transmission electron microscopy (TEM) and x-ray diffraction (XRD) techniques were employed. Beginning with the as-rolled/annealed foil, XRD analysis indicated that the near-surface region (up to ~3 µm in depth) comprised a two-phase mixture of orthorhombic and beta phase. After chemically etching approximately 10 µm from both surfaces of the foil, three phases were detected by XRD: orthorhombic, beta and alpha-two. Representative XRD patterns for both the foil surface and interior are presented in Figure 5. Based upon this data alone, it was difficult to draw conclusions concerning the degree of order in the these phases at this point in the study. As such, a more detailed characterization of the constituent phases was conducted through TEM analysis.

TEM analysis of the foil surface region indicated that this two-phase mixture comprises primarily an ordered beta phase (β_o) with O-phase lath precipitates dispersed throughout. The crystal structure, and presence of ordering, for the phases were determined by electron diffraction techniques. These same two phases were also found in the foil interior, with each having similar morphologies. However, consistent with the XRD analysis, α_2 was also present in the interior regions. A representative micrograph of the three phase microstructure is presented in Figure 6. The α_2 phase was heavily faulted and textured due to the deformation experienced during cold rolling. While the lath morphology of the O phase implied that it precipitated directly from beta (or β_o), in some instances the O phase also appeared to be a transformation product from a parent α_2 phase. The relative stability of each of these phases in the Ti-22Al-23Nb base alloy remains an area of active study.

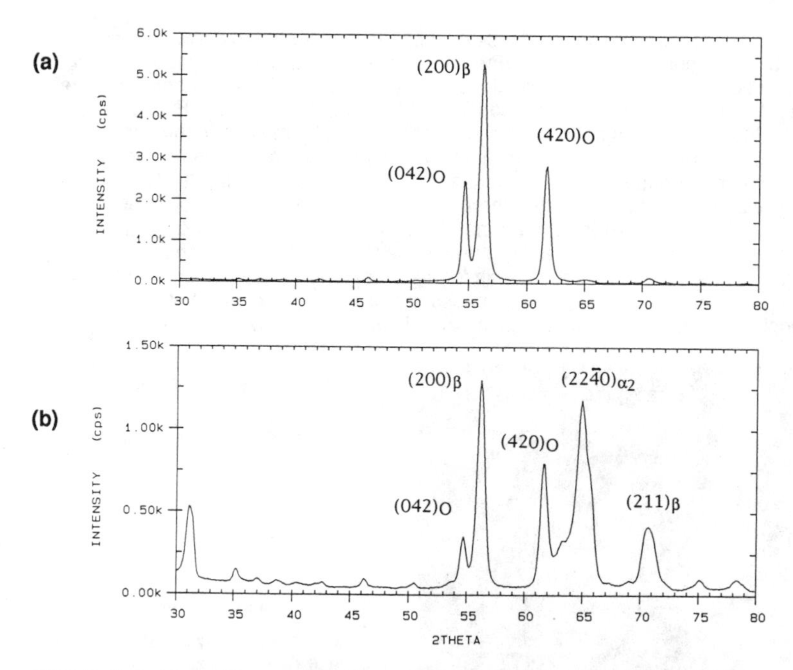

Figure 5. Comparison of the x-ray diffraction patterns from the (a) as-rolled/annealed foil surface and (b) foil interior region. Note the presence of alpha-two in the interior region only.

Figure 6. TEM micrograph showing a three phase mixture of faulted alpha-two grains with an ordered beta phase and orthorhombic lath precipitates in the as-rolled/annealed Ti-22Al-23Nb foil (127 μm thickness).

Energy Dispersive Spectroscopy (EDS) analysis provides insight into phase identification by comparing Nb levels and Al/Nb ratios for each of the constituents. Using electron diffraction to first determine the crystal structure of the constituent phase, EDS was then used to measure the chemistry of that phase. Table IV summarizes the composition of each constituent. While there is some degree of statistical variation, the composition of the individual O and β_o phases is similar for both the surface region and the foil interior. The only significant change was the presence of α_2 in the foil interior. These EDS and XRD analyses imply that the surface region has been depleted in Al. One possible explanation for the presence of this Al-lean region is that the rolling/annealing processing sequence used to fabricate the foil resulted in Al vaporization from the surface. This results in a shift in local phase equilibria to produce a surface region comprising only the two-phase O+β_o mixture, with the foil interior comprising the three phase, O+β_o+α_2, mixture.

Table IV
EDS Analysis of Constituents In 127 μm Thick Orthorhombic Foil

Phase	Titanium (at.%)	Aluminum (at.%)	Niobium (at.%)
Beta (β_o)	43	10	47
Orthorhombic	47	26	27
Alpha-Two	54	28	18

In characterizing the microstructure of a new alloy and evaluating processing pathways, a key element is the temperature at which phase transformations occur (i.e., the transus temperature). In the case of the orthorhombic alloy in this study, the transus temperature of interest is the temperature above which the α_2 and O phase are no longer stable. To experimentally determine the transus temperature, 1 cm x 1 cm samples were cut from the foil, wrapped in a Ta foil, and heat treated in 1×10^{-6} torr vacuum for 30 min. at four different temperatures: 1177°C, 1149°C, 1121°C, and 1093°C. All of the samples were furnace-cooled at a controlled rate of 100°C/min. following the thermal treatments. Similar heat treatments were also conducted for 1 hr at 1204°C, 1177°C, 1149°C, 1121°C, 1093°C, 1066°C, and 1038°C. The heat treated samples were then characterized by optical microscopy and SEM to determine the transus temperature.

The foil samples which were heat treated for 30 min. showed microstructural variation across their thickness. This was attributed to the microstructural inhomogeneity of the starting foil. Such variation was not observed in the samples which were heat treated for 1 hr, and the results of these longer heat treatments were used for estimating the transus temperature of the alloy. Analysis of the samples heat treated at 1093°C and higher temperatures showed coarse prior beta grains with some coarse grain boundary α_2 allotriomorphs and idiomorphs, in addition to fine transformed beta structure. In contrast, heat treatment at 1066°C and 1038°C produced a duplex equiaxed structure with no evidence of prior beta grain boundaries. Based upon these observations, the transus temperature of the orthorhombic alloy in this study is between 1066°C and 1093°C.

Nanoindentation Hardness of Constituent Phases

As the mechanical properties of this class of alloys are a sensitive function of the constituent phase characteristics, it is reasonable to assume that the surface region (O+β_o) and the interior region (O+β_o+α_2) of the foil may behave quite differently. In order to begin to understand the characteristics of the individual constituent phases and how they influence the behavior of the foil, nanohardness measurements were made on each of the distinct microstructural regions in the foil.

The microhardness of the α_2 phase relative to that of the mixed β_o and O phases in the as-received foil was determined using an instrumented, ultra low-load microhardness indenter. This measurement is of interest in that the yield stress of most metals is related to their hardness [7], so that by determining the microhardness of the mixed O+β_o phases, their relative yield stresses may be compared. During each indentation experiment the nanoindenter simultaneously measured the applied force and resultant penetration depth with a force and displacement resolution of 0.3 μN and 0.16 nm, respectively [8]. As a result of this sensitivity it was possible to investigate the microhardness of the surface regions ~ 5 - 10 μm in dimension. Unlike conventional microhardness machines where the final area of the indentation is measured subsequent to penetration, the nanoindenter system determines the contact area of the indenter continuously during an indentation experiment. This is accomplished by previously calibrating the area of the indenter tip as a function of penetration depth. The microhardness is then, $H = F / A$, where H is the nanoindentation hardness (N/m^2), F is the applied force (N) and A is the area (m^2) of the indenter tip actually in contact with the specimen surface [7]. The nanoindenter system thus provides a continuous measure of the microhardness as a function of distance from the surface. In the present analysis the microhardness in microstructurally different areas was evaluated at the same nominal plastic depth (160 nm), thus providing a direct comparison of the microhardnesses of the phases.

Of approximately 200 indentations performed in this study, only those which were located well within the boundaries of a given phase were selected for analysis. As summarized in Figure 7, areas identified as α_2 exhibited a higher hardness than areas identified as predominantly β_o, which in turn exhibited a higher hardness than areas identified as mixed β_o and O phases. A small range of plastic depths was observed within the hardness results of each phase partly because of minor fluctuations in microhardness, and partly because of the digital data acquisition system which does not collect data at precisely the same load in each experiment (Here, ± ~ 1 mN). However, the variation in microhardness within each phase over the range of plastic depths (158 - 162 nm) was small compared to the difference in average hardness between the different phases. Thus, the two-phase mixture of β_o and O phases which dominates the foil surface region is likely to provide a relatively ductile/low strength zone at the external surface and, if retained through fabrication, in the areas adjacent to the fiber surface in the resulting composite .

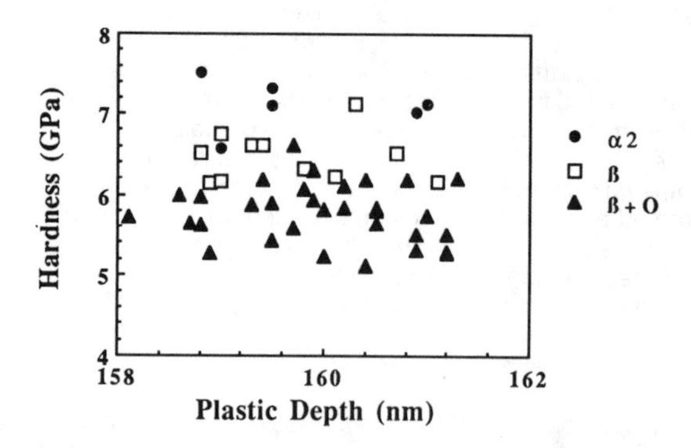

Figure 7. Microhardness of the alpha-2, beta and mixed beta and orthorhombic phases in the as-received foil, as determined by nanoindentation. All microhardnesses were evaluated at a nearly constant plastic depth (160 nm).

Mechanical Properties of Orthorhombic Sheet and Foil

Following characterization of the microstructures, an initial assessment of the mechanical property characteristics of the orthorhombic foil was made. For composite application it is important to consider the properties of the material following elevated temperature exposure. This condition is most representative of the material as it will ultimately exist as a composite matrix. The appropriate thermal condition to consider therefore is a simulated composite consolidation cycle. This cycle may be followed by various heat treatments to simulate service environments, as appropriate to the specific application of interest.

In beginning the mechanical property assessment for this study, an evaluation of the tensile properties of the as-rolled sheet and as-rolled/annealed foil specimens was completed. Following assessment of the as-rolled properties, specimens of the foil were exposed to a simulated composite consolidation cycle of 940°C/4 hr/vacuum (1×10^{-6} torr) prior to testing. Both as-rolled and heat treated materials were machined into tensile specimens with reduced gage sections measuring 15 mm x 5 mm, using wire EDM. All specimens were tensile tested at ambient temperature in laboratory air using a nominal strain rate of 1.6×10^{-4} s^{-1}. A summary of the ambient temperature tensile property data is provided in Table VI, along with data for a conventional $\alpha_2+\beta$ alloy (Ti-24Al-11Nb) for comparison.

Table VI
Ambient Temperature Tensile Properties for Orthorhombic Sheet & Foil

Material Type	Product/ Condition	0.2% YS (MPa)	UTS (MPa)	Modulus (GPa)	Plastic El. (%)
$\alpha_2+\beta$	Sheet/ As-Rolled	413	579	83	7.0
Orthorhombic	Sheet/ As-Rolled	793	979	117	7.0
$\alpha_2+\beta$	127 µm Foil/ Annealed	352	490	62	9.2
Orthorhombic	127 µm Foil/ Annealed	984	1159	--	4.1
Orthorhombic	127 µm Foil/ 940°C/4hr	863	1077	--	5.6
Orthorhombic	89 µm Foil/ Annealed	1034	1172	90	4.8

To build upon this initial assessment of the tensile behavior and stability of the orthorhombic foil, various heat treatments were conducted to simulate elevated temperature service exposures. The alloy foil (127 µm) was heat treated at 940°C/4 hr/FC in 1×10^{-6} torr vacuum to simulate the MMC consolidation cycle. Following this exposure, foil samples were wrapped in Ta foil, encapsulated in evacuated vycor tubing, and aged at six different temperatures in the range 315°-815°C for 100 hr., and furnace cooled. As the microstructural analysis had shown a degree of texture existed in the foil, a preliminary assessment of the influence of this texture on tensile properties was also begun. While most of the tensile specimens for this study were machined with the tensile axis parallel to the rolling direction (L-specimens), some specimens were also prepared transverse to the rolling direction (T-Specimens).

In assessing the influence of heat treatment on tensile properties, only L-specimens were considered. Most of the tensile specimens which were subjected to 100 hr aging heat treatments in the temperature range 538°-815°C failed outside the gage sections or in the grip regions, and showed elongations in the range 0.75-1.5% prior to fracture. One specimen aged at 649°C for 100 hr had an elongation of 2.5%. The elongation values increased following 100 hr aging at 427°C and 315°C, to 3.6% and 9.3%, respectively.

In comparing the L- and T-specimen behavior, the foil exhibited anisotropic tensile properties with the L-specimens showing higher elongations than the T-specimens. In the as-rolled condition, the L-specimens showed elongation values >4%, while T-specimens showed values of approximately 1%. Following the heat treatment simulating the consolidation cycle, the alloy retained high elongation (>5.5%) parallel to the rolling direction.

DISCUSSION

Based upon the results presented above, the orthorhombic alloy, comprising primarily an $O+\beta_o$ microstructure, is an excellent candidate for evaluation as a composite matrix. Regarding the processing of the alloy to foil, several items warrant discussion and further consideration. While the properties of the material appear to be attractive, it is clear that virtually all of the material evaluated in this study suffered from some degree of chemical segregation (banding). While this inhomogeneity was reduced through processing to the finest foil gauge thickness, this approach has limited potential. Clearly, melting practices must be improved and a master-alloy should be developed to reduce this tendency toward segregation. A reliance upon solution treatments to reduce segregation following primary processing may be difficult due to the presence of the ordered beta (β_o) phase at elevated temperature. A positive aspect of this phase being present, however, is the likely maintenance of strength at elevated temperature. Through improved melting practice, controlled ingot breakdown and hot-rolling parameters, this tendency toward segregation should not be a major problem to overcome. Once accomplished, the rolling of foil to the dimensions required for composite fabrication should not represent an extraordinary challenge. Indeed, the success demonstrated by TI in rolling foil from the banded sheet material used in this program bodes well for further improvements in the processing of foil. Control of texture development and residual stresses in the foil throughout the processing cycle remains a key area to be addressed.

The relative stability of the phases (O, β_o , and α_2) for all alloys in this orthorhombic class must be further studied to provide for better control of the phase distribution and morphology in the material. Control of these constituent characteristics will allow for improved processibility of foil, and the resulting mechanical behavior of the material. While it is clear that little control was exerted over the constituent characteristics in this study, the significant improvements in tensile properties for the orthorhombic alloy foil over the more conventional $\alpha_2+\beta$ alloy illustrate the potential for an improved composite matrix. Texture effects in this alloy are clearly pronounced, even with this relatively limited data base. Control of these rolling textures requires a much larger degree of understanding of phase stability (including interstitial element effects) than now exists.

In the case of the orthorhombic alloys, the loss of Al at the surface of the foil which accompanies vacuum annealing under certain circumstances, may be used to effectively stabilize a higher volume fraction of beta at the surface. An important characteristic of beta-rich alloys is an increased tolerance to interstitial contamination and improved flow characteristics compared to other Ti-based alloys. The presence of this layer at the surface of the composite and in the interface region around the fiber reinforcement may improve the environmental resistance and the fatigue response of the composite system. It should be pointed out, however, that this layer is not in chemical equilibrium with the bulk of the foil and thus the layer will be reduced in extent through prolonged interdiffusion with the underlying material. The kinetics of this interdiffusion, however, are likely to be quite slow at the service temperatures targeted for application of these materials, due to the presence of atomically ordered constituents. This to remains an active area for further research.

CONCLUSIONS

Based upon the results presented and discussed in this study, the following conclusions may be drawn:

- the orthorhombic-alloy, Ti-22Al-23Nb (at.%), can be effectively cold-rolled to produce foil of thickness 127 μm and 89 μm, even with oxygen levels as high as 1700 ppm (by weight);

- the rolling anisotropy was lower for this orthorhombic alloy than for Ti-24Al-11Nb (at.%);

- chemical inhomogeneity (banding) was detected in this material and is a concern for this high-Nb alloy class due to a lack of master alloy and processing experience;

- the microstructure of this Ti-22Al-23Nb (at.%) alloy comprises a three phase mixture of $O+\beta_o+\alpha_2$ phases and exhibits a high degree of texture following unidirectional cold-rolling;

- the near-surface region of the foil comprises a two phase mixture of $O+\beta_o$ which may be created by Al depletion during vacuum annealing operations;

- the two phase mixture of $O+\beta_o$ at the surface of the foil has a lower hardness than the other regions of the microstructure;

- the strength of the Ti-22Al-23Nb (at.%) sheet and foil was consistently higher than that of the Ti-24Al-11Nb (at.%) material, however the ductility was somewhat lower.

ACKNOWLEDGMENTS

The support of M. Shaw (Rockwell Science Center), S. Krishnamurthy (Metcut Materials Research Group), I. Sukonnik and A. Pandey (Texas Instruments) for support in conducting and analyzing experiments, and in the fabrication of foil is gratefully acknowledged.

REFERENCES

1. J.M. Larsen, et. al., "Titanium Aluminides for Aerospace Applications", Proceedings of the Symposium High Temperature Aluminides and Intermetallics, S.H. Whang, C.T. Liu, D.P. Pope and J.O. Stiegler (eds.), TMS-AIME, Warrendale, PA (1990) 521.

2. S.C. Jha, et. al., Advanced Materials and Processes, 139, 87 (1991).

3. G. Rowe, "Recent Developments in Ti-Al-Nb Titanium Aluminides", Proceedings of the Symposium High Temperature Aluminides and Intermetallics, S.H. Whang, C.T. Liu, D.P. Pope and J.O. Stiegler (eds.), TMS-AIME, Warrendale, PA (1990) 375.

4. P. R. Smith et. al., "Evaluation of an SCS-6/Ti-22Al-23Nb "Orthorhombic" Composite", Proceedings of This Symposium.

5. D. Lee, et. al., Trans. AIME, 239, 1467 (1967).

6. I.M. Sukonnik, S.L. Semiatin and M. Haynes, Scripta Met., 26 (1992) 993-998.

7. D. Tabor,"Indentation Hardness and Its Measurement: Some Cautionary Comments" pp. 129 - 159 in ASTM STP No. 889, Microindentation Techniques in Materials Science and Engineering, P. J. Blau and B. R. Lawn (eds.), American Society for Testing and Materials, Philadelphia, PA, 1985.

8. W. C. Oliver, R. Hutchings and J. B. Pethica, "Measurement of Hardness at Indentation Depths as Low as 20 Nanometers," pp. 90 - 108 in ASTM STP No. 889, Microindentation Techniques in Materials Science and Engineering, P. J. Blau and B. R. Lawn (eds.), American Society for Testing and Materials, Philadelphia, PA, 1985.

EVALUATION OF AN SCS-6/TI-22AL-23NB "ORTHORHOMBIC" COMPOSITE

P.R. Smith*, J.A. Graves** and C.G. Rhodes**
* Materials Directorate, Wright Laboratory, Wright-Patterson AFB, OH 45433-6533
** Rockwell International Corp. Science Center, Thousand Oaks, CA. 91360-0085

ABSTRACT

A study was undertaken to examine the attributes of utilizing a high niobium-containing titanium aluminide (orthorhombic) composition, specifically Ti-22Al-23Nb (a%), for use as a matrix for a continuously reinforced metal matrix composite. Both unreinforced "neat" panels and 35v% 4-ply unidirectional ([0]4) SiC (SCS-6) panels were fabricated by HIP'ing using a foil / fiber / foil approach. The microstructure of these panels were examined via optical, scanning electron and transmission electron microscopy. Reaction zone kinetics including both the primary reaction product and any beta-depleted zone growth were determined at 982°C. Analytical electron microscopy was employed to identify fiber / matrix interfacial compounds as well as local phases and their associated chemistries. Preliminary mechanical properties were obtained which included: longitudinal and transverse tensile, matrix thermal stability, thermal fatigue, thermal mechanical fatigue and transverse composite creep. The results were compared with panels fabricated from the baseline matrix composition, Ti-24Al-11Nb.

INTRODUCTION

Future aerospace systems embodying advanced gas turbine engines as well as aircraft structure, are chiefly reliant upon the development of high temperature lightweight materials. Consequently, titanium aluminides (principally those based upon the intermetallic compound, Ti_3Al) and continuous reinforced composites thereof, have been the subject of a significant number of research activities over the past several years [1]. These materials in their metal matrix composite form have been shown to exhibit high specific tensile strength and stiffness that are superior to nickel-base superalloys in the temperature range from room temperature to 760°C [2]. Unfortunately, matrix compositions within this class suffer from severe sensitivity to interstitial embrittlement when subjected to high temperatures in an air environment [3] In fact, the alloy composition within this class to receive the most thorough examination, Ti-24Al-11Nb (a%), has been shown to suffer extreme environmental embrittlement during thermal cycling, such that when cycled in composite form 500 times from 315°C to 815°C in air, the tensile strength degraded from the as-consolidated level of 1238 MPa to a cycled value of 176 MPa [4,5]. In addition, this alloy also suffers from a relatively significant chemical reactivity with commercial SiC fibers such as Textron's SCS-6 filament [6,7,8]. These investigations have demonstrated that the result of this chemical reaction was the formation of a brittle fiber/matrix reaction zone as well as a brittle zone extending into the matrix which is void of the ductilizing/toughening beta (beta-depleted zone) phase. Furthermore, this alloy and others within this class, exhibit relatively poor ductility and fracture resistance at low temperature [9].

The objective of this study was to examine a new class of titanium alloys which contain significantly higher levels of niobium and which exhibit a third ordered "orthorhombic" phase based upon the composition Ti_2AlNb [10]. This alloy class has been the subject of recent research activities in its monolithic form [11]. These studies have indicated that a good balance of properties can be obtained with such compositions including high tensile strength at temperatures up to 760°C and improved fracture toughness at low temperature. The subject study sought to utilize the attributes offered by this alloy class by examining the continuous reinforced metal matrix composite system SCS-6 / Ti-22Al-23Nb.

EXPERIMENTAL/RESULTS

Ingot Processing

An 8.2 kg ingot was triple vacuum arc remelted of the nominal composition of Ti-22Al-23Nb. This ingot was then forged, hot packed rolled and ground as input material for cold rolling to 125μm (5 mil) foil. A more detailed description of this processing has been provided by Sukonnik et al. [12]. Bulk chemical analysis was performed on each product form up to and inclusive of the foil (Table I).

Table I. Chemical Analysis of Intermediate Product Forms

	Ti	Al	Nb	Fe	O	N
Ingot	Bal	10.8(21.3)	37.9(21.7)	0.06	0.123	0.007
Sheet	Bal	10.0(20.0)	38.9(22.6)	0.12	0.136	0.013
Foil(5 mil)	Bal	10.6(21.1)	39.1(22.6)	0.10	0.167	0.007
Foil(3.5mil)	Bal	10.6(21.1)	39.4(22.8)	0.02	0.179	0.011

Reaction Zone Studies

Small monolayer composite samples were made of both the SCS-6/Ti-24Al-11Nb and SCS-6/Ti-22Al-23Nb systems by vacuum hot pressing foils of the respective matrices around woven SCS-6 fiber mats. Conservative consolidation parameters were chosen to ensure complete metal flow and diffusion bonding of metal-metal interfaces. Coupons of each system were wrapped in tantalum foil and vacuum (10^{-6}torr) encapsulated in quartz tubing. The samples were then isothermally exposed at 982°C for times up to 100 hours in duration. Secondary electron imaging via scanning electron microscopy (SEM) was utilized to measure the growth of both the primary reaction zone as well as the beta-depleted zone (if present). Figure 1 shows an SEM micrograph of each composite system in their as-consolidated condition. It can be noticed that the Ti-24Al-11Nb composite exhibits a relatively large primary reaction zone (>1μm) as well as a large zone (>5μm) which has been depleted of the beta (light) phase. Conversely, the Ti-22Al-23Nb composite has a much smaller primary reaction zone (<0.5μm) and no beta-depleted zone.

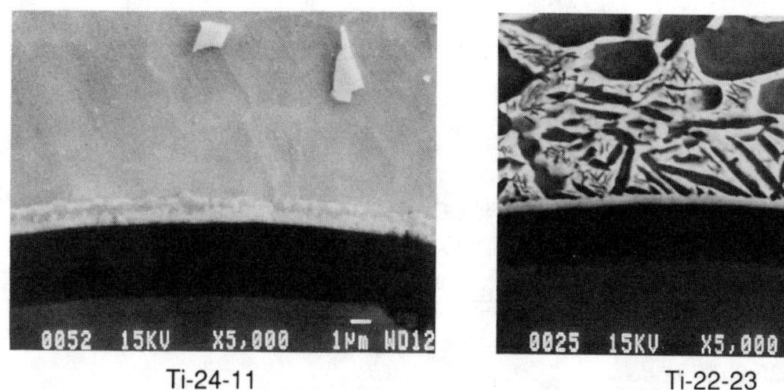

Figure 1. SEM micrographs of the Ti-24Al-11Nb and Ti-22Al-23Nb composites in the as-consolidated condition.

Figure 2 contains plots of both the primary reaction zone and beta-depleted zone growth for the two composite systems as well as two other intermediate level niobium-containing compositions (Ti-25Al-17Nb and Ti-25Al-17Nb-1Mo). It was noticed that as the niobium content increased, the reaction rate constant decreased. In the case of the beta-depleted zone, the growth rate is extremely high for the Ti-24Al-11Nb composite, but no such growth was found for the Ti-22Al-23Nb system.

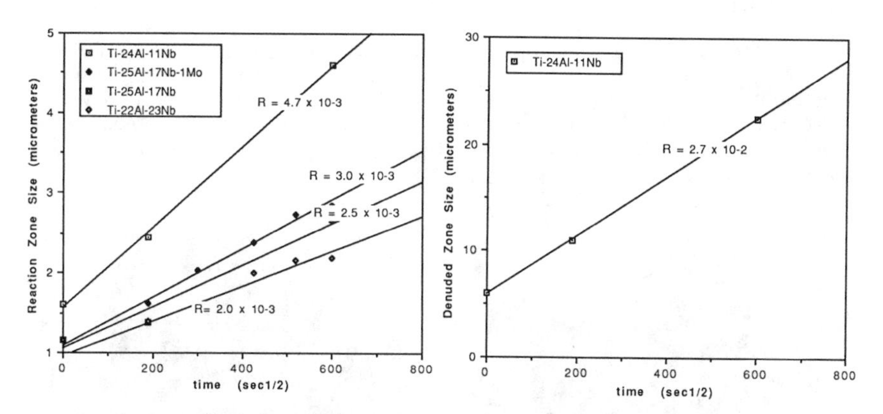

Figure 2. Primary reaction zone and beta-depleted zone growth kinetics of SCS-6/Ti-24Al-11Nb vs. SCS-6/Ti-22Al-23Nb at 982°C.

Selected area (SAD) and convergent beam (CBED) electron diffraction techniques, were used in conjunction with x-ray energy dispersive spectroscopy (EDS), to identify the individual reaction zone products. The as-consolidated SCS-6/Ti-22Al-23Nb composite reaction product actually consisted of four individual reaction compounds whose total thickness was 0.6 μm (Table II).

Table II. Reaction Zone Products in SCS-6 Reinforced Ti-22Al-23Nb &Ti-24Al-11Nb

Layer Position	Assigned Structure	Ti-22Al-23Nb Size (μm)	Ti-24Al-11Nb Size (μm)
Adjacent to Fiber	$TiC + Ti_5Si_3$*	0.05	0.2
Inner Layer	FCC; (Ti,Nb)C	0.15	0.4
Intermediate	FCC; L12, $Al(Ti,Nb)_3C$	0.15	1.0
Adjacent to Matrix	HCP; $(Ti,Nb)_5(Si,Al)_3$	0.25	0.2
		Total = 0.6	Total = 1.6

Matrix Microstructure

Figure 3 contains SEM micrographs for the SCS-6/Ti-24Al-11Nb and the SCS-6/Ti-22Al-23Nb systems. Previous studies [7] had indicated the former system consisted of approximately 90v% alpha-2 (DO19) and 10v% disordered beta (BCC). X-ray diffraction analysis of the input Ti-22Al-23Nb foil [12] indicated that the interior (bulk) consisted of three ordered phases: alpha-2(α_2), ordered beta(B2) and ordered orthorhombic(O). However, a 5μm modified layer consisting only of B2 + O which was observed on the input foil surface, was seen to persist after the consolidation cycle. In fact, this layer was found not only on the composite surface, but also on the foil-foil bond line, and at most of the fiber/matrix interfaces. TEM analysis reveals that this layer is primarily B2 with "O" phase plates dispersed throughout .

By comparison, the interior of the matrix exhibited a microstructure distinctly different than that of the surface layer. In this particular composite, there was a gradient of microstructures through the thickness of the input foil [12] whose microstructure was directly transferrable to the composite interior. Most of the matrix consisted of a mixture of deformed (or faulted) α_2 phase grains, B2 phase grains, and both equiaxed and plate-like "O" phase.

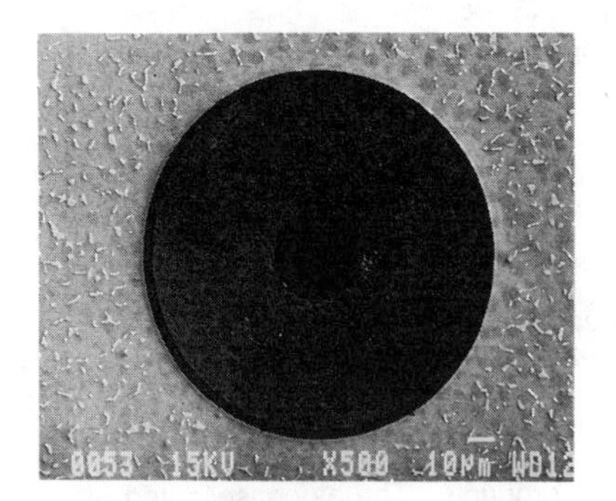

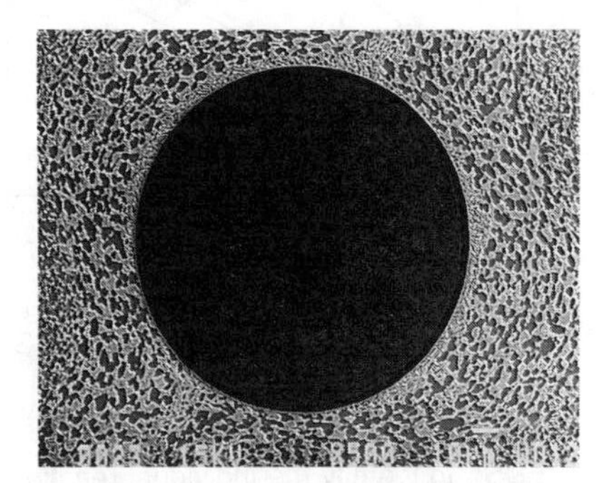

Figure 3. SEM micrograph of the Ti-24Al-11Nb and Ti-22Al-23Nb composite systems illustrating the differences in matrix microstructural features for the as-consolidated condition.

<u>Mechanical Behavior</u>

Four [0]4 35v% SCS-6/Ti-22Al-23Nb composite panels and two unreinforced "neat" Ti-22Al-23Nb panels were fabricated by foil-fiber-foil processing for evaluation of mechanical properties. A series of mechanical property screening tests were conducted that included: tensile (longitudinal and transverse; RT-815°C), thermal stability (neat panel), transverse creep (650°C, 760°C), thermal fatigue, thermomechanical fatigue and residual tensile (after 815°C hold). Due to restrictions on manuscript length, only the thermal stability, transverse composite tensile, and thermal fatigue data will be included in any detail in the subject paper. However, the remaining information including specimen preparation, test procedures and other properties may be found in Smith et al. [13]. It should be noted that the longitudinal composite strength in the as-fabricated condition demonstrated rule-of-mixtures levels [13] over the temperature range of room temperature to 815°C.

The as-consolidated matrix material (from the 0.9mm [3.5 mil] foil) was examined for its thermal stability by isothermally exposing tensile specimens in vacuum at temperatures between 315°C-815°C for 100 hours. The specimens were wrapped in tantalum foil (to getter oxygen) and vacuum encapsulated in quartz tubes before exposing in an air furnace After exposure, the specimens were tensile tested at room temperature and selected optical/scanning electron microscopy and associated fractography performed. Figure 4 contains a plot of the longitudinal tensile strength and ductility as function of thermal exposure temperature. It could be seen that the ultimate tensile strength does not appear to be sensitive (900-975MPa) to the thermal exposures examined. However, the tensile ductility did decrease from an initial value of approximately 8.0% to a reasonably constant value of about 4.5% upon all thermal exposure conditions examined. This ductility decrease was the subject of additional studies [14] which were not complete at the time of this writing.

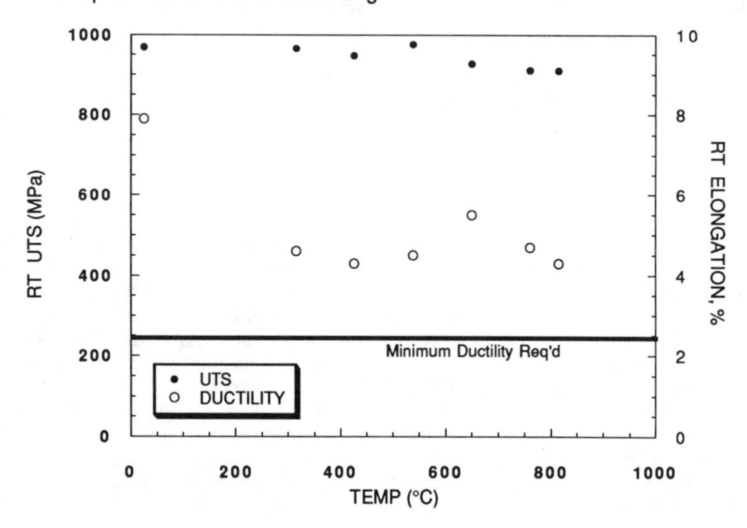

Figure 4. Plot indicating the thermal stability of the longitudinal UTS and % elongation for the as-consolidated matrix as a function of isothermal exposure in vacuum for 100 hr.

The transverse composite strength is much more dependent upon the matrix strength than in the longitudinal case in that the contribution from the fiber is negligible in these weakly bonded systems [15]. This was borne out in Figure 5, which depicts the transverse tensile strength of the composite as a function of temperature. Here, it could be noticed that the higher strength of the Ti-22Al-23Nb matrix was resulting in higher composite strengths, when compared to the Ti-24Al-11Nb system [16], particularly at the higher temperatures. The room temperature results were somewhat suspect and additional data will need to be collected.

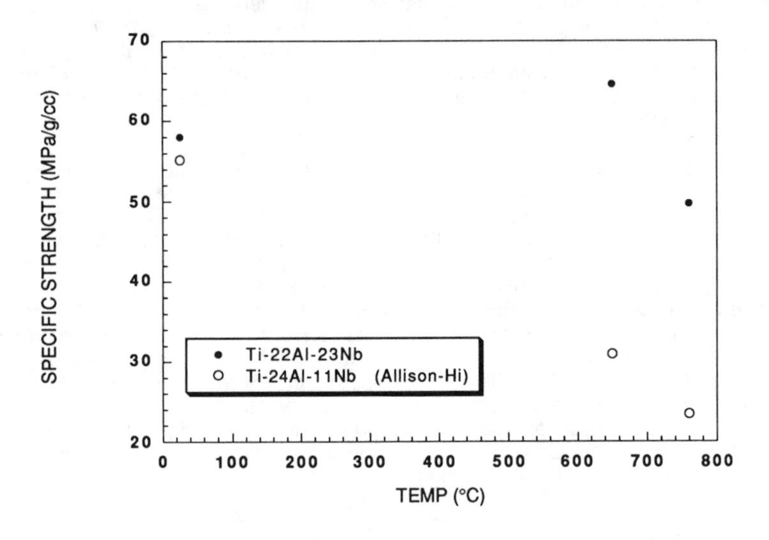

Figure 5. Transverse tensile strength for the SCS-6/Ti-22Al-23Nb composite as a function of temperature.

As previously noted earlier, work by Russ [4] had shown that the longitudinal tensile strength of the SCS-6/Ti-24Al-11Nb system was significantly degraded when subjected to thermal cycling in an air environment. Further work wherein completely analogous cycling in vacuum was performed [17], conclusively demonstrated that the environment played the primary role in the thermal fatigue debit. In this work an SCS-6/Ti-24Al-11Nb composite was cycled 500 times from 150°C-815°C with no measurable reduction in the composite tensile properties. Based on this information, it was decided to examine the subject matrix in composite form for its resistance to thermal fatigue degradation. The thermal cycling data for the SCS-6/Ti-22Al-23Nb system is plotted against the SCS-6/Ti-24Al-11Nb system obtained under the same conditions in Figure 6. It can be seen that even after 500 cycles from 150°C-815°C in an <u>air</u> environment, that this composite maintained essentially all of its as-fabricated tensile strength.

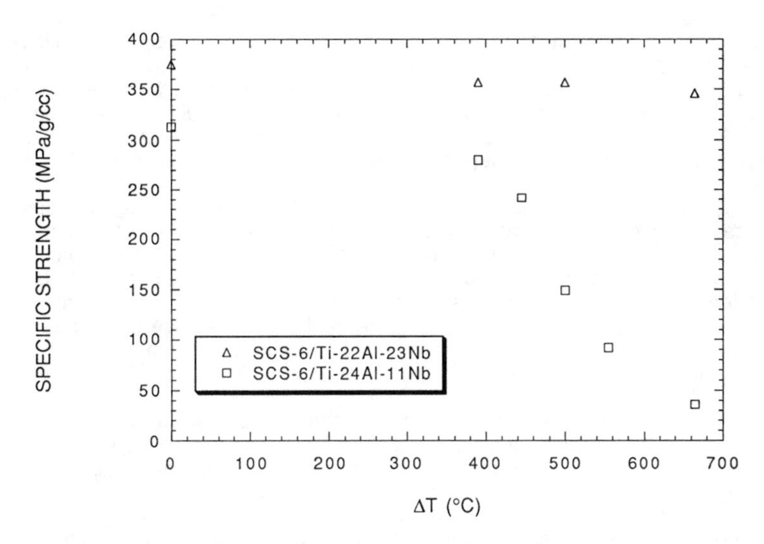

Figure 6. Comparison of thermal fatigue results for SCS-6/Ti-22Al-23Nb vs. SCS-6/Ti-24Al-11Nb in an air environment.

DISCUSSION

The formation of completely analogous fiber/matrix reaction compounds between SCS-6 reinforced Ti-22Al-23Nb and Ti-24Al-11Nb is not very surprising in that the reacting species are the same in each case (ie., Ti, Al, Nb, Si, and C). The reduction in fiber/matrix reaction zone kinetics for the SCS-6/Ti-22Al-23Nb system is consistent with previous results of reduced kinetics with increasing levels of niobium (18). In addition, there was no formation of a beta-depleted zone (found for the Ti-24Al-11Nb composite system) adjacent to the primary reaction zone for the Ti-22Al-23Nb composite system. Possible explanations for these observations include: 1) increased levels of a slow diffusing element (ie. Nb), 2) decreased activity of titanium and/or 3) slower kinetics associated with ordered phases. The potential significance of less fiber/matrix interaction may be in an increase in those mechanical properties which are matrix dominated such as fatigue crack initiation.

The microstructure that was present in the input foils [12] was for the most part directly transferred to the composite. The 5-10 μm surface layer (O + B2) was found at the composite surface, down the foil-foil bond line and at least partially surrounding the fiber. Although no cyclic mechanical properties were evaluated, it is speculated that the strategic incorporation of this softer/fracture resistant layer at critical locations in the composite may have a positive effect on fatigue crack initiation, foil-to-foil bonding and resistance to interstitial embrittlement. The remainder of the matrix consisted primarily of 40v% discontinuous α_2 and a 60v% mixture of O + B2. The decreased level of the low fracture resistant alpha-2 and the increased level (continuous) of the softer (potentially more fracture resistant) O + B2 mixture may provide increased resistance to crack propagation in the matrix.

The thermal stability of the Ti-22Al-23Nb matrix over the temperature range of 315°C-815°C appeared to be very reasonable especially with respect to ultimate tensile strength. Although the ductility did decrease from an average value of around 8% in the as-consolidated condition, to approximately 4.5% after exposures for 100 hours at 315°C, this latter value is thought to be above the minimum required (2.5%) to enable the matrix to: 1) overcome residual stresses due to CTE mismatch during composite manufacture and 2) effectively load the fiber [19].

The transverse tensile strength of the SCS-/Ti-22Al-23Nb composite significantly exceeded that for the SCS-6/Ti-24Al-11Nb system, particularly at 650°C and 760°C. This is thought to have been directly related to the increase strength of the Ti-22Al-23Nb matrix [13]. The transverse composite strength is not dependent on fiber strength to any significant extent, since the fiber/matrix interface has been shown to be extremely weak. The negligible difference found between the two composite systems at 25°C is thought to be due to experimental error, and additional tests are planned to verify this postulation.

The result of no measurable reduction in the residual tensile strength of the SCS-6/Ti-22Al-23Nb system after thermal cycling 500 times in an air environment between 150°C-815°C is thought to be very significant for those applications requiring both thermal and/or mechanical loading in combination with large temperature excursions. The SCS-6/Ti-24Al-11Nb system was shown under analogous cycling to degrade by environmental embrittlement such that almost no tensile strength remained. The main difference between the two matrices appears to be the matrix microstructure developed. The Ti-24Al-11Nb matrix contains 90v% alpha-2 and only 10v% beta, while as noted above, the Ti-22Al-23Nb matrix consists of only 40v% alpha-2 and a 60v% mixture of orthorhombic + ordered beta. The reduction in the low fracture resistant and interstitially sensitive alpha-2 phase (Ti-22Al-23Nb) such that it is not present at the composite surface and it is no longer the continuous phase for uninterrupted crack propagation is thought to be the main reason for the improvement in thermal fatigue response.

CONCLUSIONS

1. The fiber/matrix reaction zone products found for the SCS-6/Ti-22Al-23Nb composite were completely analogous in terms of composition to those found previously for the SCS-6/Ti-24Al-11Nb composite, those being: TiC + Ti_5Si_3,(Ti,Nb)C, $Al(Ti,Nb)_3C$ and $(Ti,Nb)_5(Si,Al)_3$.
2. The reaction zone growth kinetics at 982°C were much slower for the Ti-22Al-23Nb composite system than for the Ti-24Al-11Nb composite. The Ti-24Al-11Nb composite also exhibited a beta-depleted zone whose growth kinetics were relatively large, while the Ti-22Al-23Nb composite did <u>not</u> exhibit such a zone.
3. The bulk of the matrix microstructure for the Ti-22Al-23Nb composite consisted of approximately 40v% alpha-2 (α_2) and a 60v% mixture of orthorhombic (O) and ordered beta (B2).
4. The 5-10 µm O + B2 surface layer present on the foil used to fabricate the composites, was translated to the composite and was present at the composite surface, at the foil-foil bond line and at the fiber/matrix interface.
5. The tensile strength of the Ti-22Al-23Nb consolidated matrix was consistently higher than that for the Ti-24Al-11Nb matrix over the temperature range of 25°C - 815°C, even on a density corrected basis.
6. The Ti-22Al-23Nb consolidated matrix indicated no signs of thermal instability with respect to tensile strength after exposures of 100 hr. in vacuum at temperatures between 315°C - 815°C. The ductility of the matrix dropped from

8% to an average value of 4.5% upon all thermal exposures in this range. However, this value is still above that required to overcome residual stresses and to properly load the fiber (ie. 2.5%).
7. Rule-of-mixtures longitudinal tensile properties were demonstrated for $[0]_4$; 35v% SCS-6/Ti-22Al-23Nb composites in the temperature range of 25°C - 815°C.
8. The transverse tensile properties of the Ti-22Al-23Nb composite were improved when compared to the Ti-24Al-11Nb composite.
9. The Ti-22Al-23Nb composite system maintained its as-consolidated tensile properties when thermally cycled in air, while the Ti-24Al-11Nb composite exhibited a tremendous reduction in such properties, when cycled under completely analogous conditions. Thus, it appears the Ti-22Al-23Nb matrix is much more resistant to environmental degradation than is the Ti-24Al-11Nb matrix composition.

ACKNOWLEDGEMENTS

The authors wish to thank Dr. M. James, Dr. J. Porter, Dr. G. Rowe, Dr. I. Sukonnik and Mr. T. Broderick for their many helpful technical discussions, Mr. R. Spurling, Mr. E. Fletcher and Mr. C. Begg for microscopy assistance, and Mr. E. Fletcher for sample preparation, and Mr. K. Goecke and Mr. T. Johnson for mechanical testing and Mr. A. Kraus for photographic assistance. Special thanks Dr. A. Kumnick for the fabrication of composite panels.

REFERENCES

1. R.A. MacKay, P.K. Brindley and F.H. Froes, "Continuous Fiber-Reinforced Titanium Aluminide Composites," JOM 43 (5) (May 1991), pp. 23-29.
2. J.M. Larsen et al., "Titanium Aluminides for Aerospace Applications," High Temperature Aluminides and Intermetallics, ed. S.H. Whang et al. (Warrendale, PA: TMS, 1990) pp. 521-556.
3. S.J. Balsone, "The Effect of Elevated Temperature Exposure on the Tensile and Creep Properties of Ti-24Al-11Nb," Oxidation of High Temperature Intermetallics, ed. T. Grobstein and J. Doychak (Warrendale PA: TMS, 1989) pp. 219-234.
4. S.M. Russ: Metall. Trans. A, 1990, vol 21A, pp. 1592-1602.
5. W.C. Revelos and P.R. Smith: Metall. Trans. A, 1992, vol 23A, pp. 587-595.
6. S.F. Baumann, P.K. Brindley and S.D. Smith, "Reaction Zone Microstructure in a Ti_3Al + Nb/SiC Composite," Metall. Trans. A, 1990, vol 21A, pp. 1559-1569.
7. P.R. Smith, C.G. Rhodes and W.C. Revelos, "Interfacial Evaluation in a Ti-25Al-17Nb/SCS-6 Composite," Interfaces in Metal-Ceramics Composites, ed. R.Y. Lin et al. (Warrendale PA: TMS, 1989), pp. 35-58.
8. A.M. Ritter, E.L. Hall and N. Lewis,"Reaction Zone Growth in Ti-Base/SiC Composites," Intermetallic Matrix Composites, ed. D.L Anton et al. (Pittsburgh PA: MRS, 1990), pp. 56-59.
9. H.A. Lipsitt, "Titanium Aluminides-An Overview," High Temperature Ordered Intermetallic Alloys, ed. C.C. Koch, C.T. Liu and N. S. Stoloff (Pittsburgh PA: MRS, 1985) pp. 351-364.
10. R.G. Rowe, "The Mechanical Properties of Titanium Aluminides Near Ti-25Al-25Nb," Microstructure/Property Relationships in Titanium Aluminides and Alloys, ed. Y.-W. Kim and R.R. Boyer (Warrendale PA: TMS, 1991), pp. 387-398.
11. R.G. Rowe, "Ti_2AlNb-based Alloys Outperform Conventional Titanium Aluminides," Adv. Mat. and Proc., 1992, vol 141 (3), pp. 33-35.

12. I. Sukonnik et al., "Advances in Thermomechanical Processing of Ti-Aluminide "Orthorhomibic" Foils," Proceedings of this Conference.
13. P.R. Smith et al. "Evaluation of an SCS-6/Ti-22Al-23Nb "Orthorhombic" Composite," to be published in Proceedings of Titanium Matrix Composite Workshop, Orlando FL., 6-8 November 1991.
14. S. Krishnamurthy, P.R. Smith and K.E. Goecke, "Thermal Stability of a Ti-22Al-23Nb Matrix," Presented at the AIAA mini-symposium, Dayton OH, March 1992.
15. D.B. Marshall et al., "Mechanical Properties of Ceramic and Intermetallic Matrix Composites," Proceedings 2nd Int'l Ceramic Science and Technology Congress, Amer. Cer. Soc., Orlando FL, November 1990.
16. M.L. Gambone, "Fatigue and Fracture of Titanium Aluminides," Final Report-Vol 1, WRDC-TR-4145; Wright-Patterson AFB, OH (February 1990).
17. P.R. Smith and W.C. Revelos, "Environmental Aspects of Thermal Fatigue in an SCS-6/Ti-24Al-11Nb Composite," Fatigue 90, ed. H. Kitagawa and T. Tanaka (Birmingham, UK: MCE Publications Ltd, 1990) pp. 1711-1716.
18. D.M. Bowden, S.M.L. Sastry and P.R. Smith, "Reactions Between Ti3Al and Mixed Second Phases," Scripta Metal., 1989. vol 23, pp. 407-410.
19. R.A. Amato and D.R. Pank, "Advanced Titanium Matrix Alloy Development," to be published in Proceedings of Titanium Matrix Composite Workshop, Orlando FL., 6-8 November 1991.

LAMINATE ORIENTATION AND THICKNESS EFFECTS ON AN SCS-6/Ti-24Al-11Nb COMPOSITE UNDER THERMAL FATIGUE

WILLIAM C. REVELOS* AND ITZHAK ROMAN**
*Materials Directorate, Wright Laboratory, WL/MLLN
Wright-Patterson AFB, OH 45433
**The Hebrew University of Jerusalem, Jerusalem, Israel 91904

ABSTRACT

A SiC/Ti-24Al-11Nb (at. %) composite (30-35 vol. %) was thermally cycled in air and an inert environment between 150°C and 815°C for various cycle counts. Various hold times at maximum temperature were employed to determine time-dependent effects on composite integrity. Laminate orientations investigated included: $[0]_4$, $[0]_8$, $[90]_4$, $[90]_8$ and $[0/90]_{2S}$. Acoustic emission produced during thermal fatigue of selected specimens was employed to monitor damage progression. Post-cycling room temperature tension tests as well as optical and scanning electron microscopy were used to document damage, which was particularly acute when hold times at temperature were employed on tests performed in air. The roles of the environment, composite thickness, and off-axis fibers during thermal fatigue on the composite strength and integrity are discussed.

INTRODUCTION

Continuously reinforced titanium aluminide metal matrix composites have potential for application in advanced engineering systems where specific strength and stiffness, environmental stability, and durability at elevated temperature are paramount. Specific applications under consideration include advanced gas turbine engines and hypersonic vehicles.

Recent investigations of the thermal fatigue behavior of the α_2-based class of titanium aluminide composites, specifically, SCS-6/Ti-24Al-11Nb (at. %) have demonstrated their susceptibility to acute damage when tested in air.[1, 2] This damage, measured by severe reductions in post-cycling room temperature tensile strength, is absent when these composites are thermally cycled in vacuum or an inert environment.[2, 3] It has been proposed that the sensitivity of the α_2 matrix to interstitial embrittlement and oxidation is the main vehicle for damage accumulation when these composites are cycled in air.[1, 2] Residual stresses caused by the coefficient of thermal expansion (CTE) mismatch between fiber and matrix is thought to contribute as well; however, the environmental component must be present for significant losses in residual strength to be observed. For composites cycled in air, cracks tend to initiate at the composite surface and propagate in a direction perpendicular to the fiber direction, reach the fiber matrix interface, and allow environmental attack of the interface and subsequent damage to the silicon carbide (SiC) reinforcement. Degradation of the SiC fiber strength appears to be primarily responsible for the loss in residual composite strength after thermal cycling in air.[2]

To develop a more complete picture of these phenomena, the present work investigated effects of laminate thickness, orientation, and hold times at maximum temperature during thermal cycling.

MATERIAL AND EXPERIMENTAL PROCEDURE

The composite used in this study consists of Textron's SCS-6 SiC fiber as the continuous reinforcement and Ti-24Al-11Nb (at.%) as the matrix. The composite was fabricated by hot isostatic pressing of alternating layers of matrix foil and fiber mats. The four-ply material, provided by General Electric, utilized Ti-6Al-4V (wt.%) crossweave to hold the fiber mats together while the eight-ply composite, provided by Allison Gas Turbine Division, incorporated molybdenum crossweave.

Tensile specimens having reduced gage sections were machined using abrasive water jet cutting. Dimensions measured 101.6 mm x 12.7 mm with a 5.08 mm wide gage section. Fiber orientations investigated included $[0]_4$[1,2], $[0]_8$, $[90]_4$, $[90]_8$, and $[0/90]_{2S}$. The four- and eight-ply specimen thickness averaged 0.986 mm and 1.978 mm, respectively. Details of the thermal cycling apparatus are described elsewhere.[2] Specimens were cycled in both laboratory air, and for selected tests, a low pressure inert environment (10^{-3} mm 99.9999% Ar). A typical thermal cycle lasted four or eight minutes (the longer time for the inert environment), and incorporated a 0.2 minute soak at minimum and maximum temperatures and a 1.5 minute heat/cool time. The temperature throughout the gage section at 815°C was uniform to ±6°C.

A temperature range of 150°-815°C was selected for cycle counts of 100, 250, 500 and 1000. Hold times of 100 and 1000 seconds were employed at 815°C for the case of 500 cycles. Specimens were examined in the as-fabricated and after thermally-cycled conditions. Damage assessment included microstructural examination and room temperature tension testing.

Acoustic emission (AE) activity was monitored during a 500 cycle, 150°-815°C test on a $[0]_8$ specimen in air using quartz lamp heating. A resonant transducer with a nominal center frequency of 250 KHz (MICRO 30, Physical Acoustics Corporation (PAC)) was coupled via high vacuum grease to the grip. Transducer outputs were amplified first by 40 dB using a preamplifier (Model 1220A, PAC) with a bandpass filter of 100-400 KHz and then by 20 dB at the main amplifier (Locan AT, PAC). The rms voltage of the amplified outputs were measured using an rms voltmeter (Model 3400A, HP).

RESULTS AND DISCUSSION

Figure 1 is a plot comparing the normalized room temperature residual tensile strength of $[0]_4$ and $[0]_8$ SCS-6/Ti-24Al-11Nb composites as a function of thermal cycles in air and an inert environment. The normalized strength is the residual strength of the cycled composite divided by its strength in the as-fabricated condition. Displaying the data in this fashion allows for a more direct comparison of different specimens.

Immediately obvious is the large difference in residual strength between the $[0]_4$ and $[0]_8$ composites cycled in air. For example, at a cycle count of 500, the $[0]_8$ composite retained close to 70% of its original strength while the $[0]_4$ retained less than 20%. Microstructural examination of specimens cycled in air revealed the presence of surface-initiated cracks that reach the outer ply fiber/matrix interfaces. In the $[0]_4$ composite, these cracks affect half of the four plies present (since they initiate on both surfaces of the composite) leaving the two inner plies undamaged.[1,2] In the same manner, cracks affect the outer two plies (one from each side) of the $[0]_8$ composite, but in this case, six plies remain intact. It is thought that this difference is primarily responsible for the differences in residual strength between the composites of two thicknesses.

As reported by Revelos and Smith, the $[0]_4$ composite retained all of its original strength after cycling in an inert atmosphere despite a large degree of scatter.[2] No cracks were found upon microstructural examination. The $[0]_8$ composite retained only 82% of its original strength, yet still performed better than specimens cycled in air. This difference could be attributable to scatter, as microstructural examination revealed no evidence of damage that would lead to a reduction in strength of this magnitude. Measurable reductions in post-cycling residual strength in similar composites cycled 10,000 times in vacuum, however, have been reported.[3]

Molybdenum crossweave present in these composites may also be playing a role in damage accumulation. Molybdenum was noted on many of the eight-ply composite fracture surfaces, and evidence of molybdenum oxidation and associated matrix cracking was found in polished cross sections of all eight-ply specimens. The four-ply material, which contained Ti-6Al-4V crossweave, did not display evidence of detrimental crossweave effects. If molybdenum is contributing to the damage accumulation process in the eight-ply material, then one would expect the residual strength of the eight-ply composite to be greater than observed. Molybdenum has been found to play a detrimental role in other investigations of similar composites.[4]

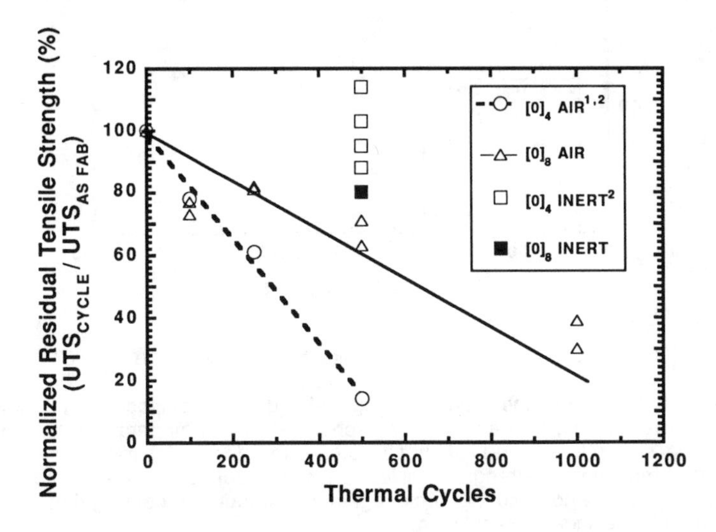

Figure 1. $[0]_4$ vs. $[0]_8$ thermal fatigue results, SCS-6/Ti-24Al-11Nb, 150°-815°C.

Figure 2 shows residual strength data for composites in the [90] orientation. In this case, the reduction in strength up to 1000 cycles is much less significant for both thicknesses. A large degree of scatter in the data exists as well. The relative insensitivity of residual tensile strength as a measure of damage in composites of this fiber orientation is thought to be a direct result of the fact that the major cracks that develop propagate in a direction perpendicular to the fiber direction and, when tested in residual tension, are parallel to the loading axis and have little effect on the strength of the composite. Therefore, differences in thickness between these two composites play a minimal role in damage accumulation as measured by residual tensile strength in contrast to their [0] counterparts.

Perhaps more significant is the susceptibility of the fiber/matrix interface to environmental attack as the test progresses. Microstructural examination of polished cross-sections revealed evidence of a large number of damaged interfaces. It is thought that this damage occurs by environmental attack of interfaces from both surface matrix cracks that reach the outer ply interfaces and the edge of the specimen gage section where the [90] fibers are exposed.

Since the residual tension test measures the strength of a weak fiber/matrix interface in this orientation, environmentally assisted damage to these interfaces will most likely contribute to subtle weakening of the composite and could explain the slight reductions in residual strength observed. The $[90]_8$ composites cycled 500 times in an inert atmosphere saw no measurable reduction in tensile strength. Microstructural examination of these specimens support these findings.

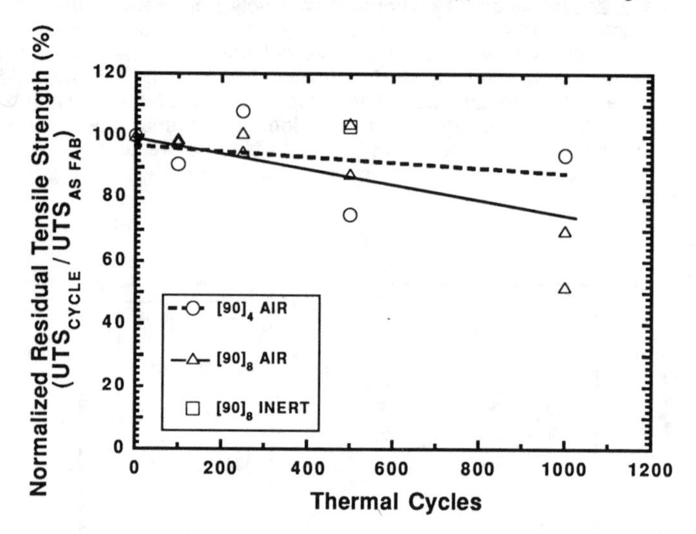

Figure 2. $[90]_4$ vs. $[90]_8$ thermal fatigue results, SCS-6/Ti-24Al-11Nb, 150°-815°C.

Figure 3 compares the residual strength of a $[0/90]_{2S}$ composite as a function of thermal cycles in air and an inert atmosphere. While a moderate reduction of post-cycling strength is found for these composites up to 500 cycles in air, no measurable reduction in strength was found for identical specimens cycled in an inert atmosphere, underscoring once again the important contribution of the environment to damage accumulation.

Interesting evidence of the mode of damage accumulation for specimens cycled in air was found via microstructural examination. As with all of these composites, surface initiated cracks were oriented perpendicular to the outer fiber direction. These cracks reached the first fiber/matrix interface and allowed for its degradation by the environment as seen before. Many of the interfaces of the off-axis plies were damaged by environmental attack in the same manner seen in the [90] orientation with one major difference: oxidized cracks emanating radially from these interfaces were found to be prevalent. The direction of propagation of all of these cracks was perpendicular to the adjacent fibers in the [0] orientation. It is thought that these environmentally assisted cracks form in this way due to the interaction of the axial and hoop components of residual stress in the matrix from the [0] and [90] fibers. These cracks were in the same orientation as the surface-initiated cracks (perpendicular to the tensile loading axis) and therefore most likely helped to contribute to the loss in residual strength.

Figure 4 shows room temperature residual strength vs. total time at temperature for cyclic vs. isothermal exposures on the $[0]_8$ composite. The cyclic data displayed are for 500 thermal cycles in air with no hold, 100-second and 1000-second hold times per cycle at 815°C. These data are compared to residual strength measurement of specimens subjected to isothermal exposure at 815°C in air prior to tension testing.

Focusing on the cyclic data, the dramatic reduction in residual strength for longer times at temperature illustrates the importance of the time dependent contribution to thermal cycling damage. For example, a 1000-second hold time at 815°C during each thermal cycle reduced the composite residual strength to near zero. When one compares the residual strength of the composite specimen cycled with no hold time to the strength of the specimen that experienced isothermal exposure at 815°C with no thermal cycles, the reduction in strength is nearly the same. As longer times at temperature are experienced, however, the cyclic contribution to damage is more readily apparent.

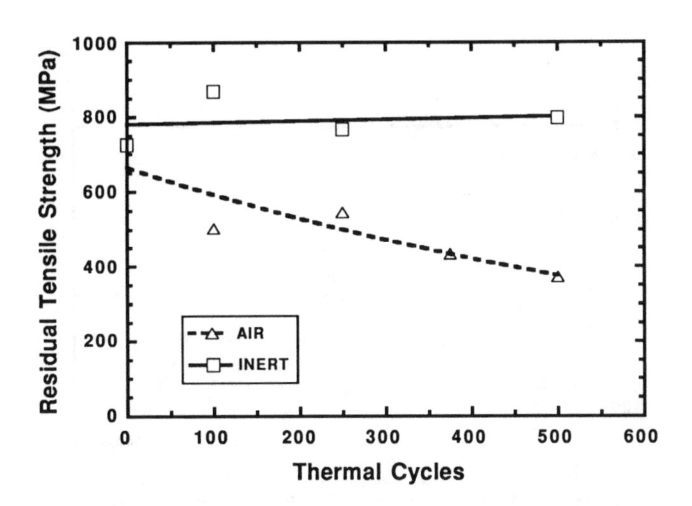

Figure 3. $[0/90]_{2S}$ thermal fatigue results, SCS-6/Ti-24Al-11Nb, 150°-815°C.

The molybdenum crossweave present in these composites, as previously stated, could have a significant effect on the data presented. Although interstitial embrittlement of the matrix material is probably a contributing factor, oxidation of the molybdenum crossweave material, which is exposed on the composite surfaces and wraps around the SCS-6 fiber, may be playing a dominant role.

An attempt was made to employ acoustic emission (AE) monitoring techniques to determine its effectiveness in determining when damage initiates during the course of a thermal fatigue test and at what portion of the thermal cycle damage occurred.

In the latter stage of this 150°-815°C 500 cycle test, AE activity significantly increased on the cooldown portion of the thermal cycle and was concentrated in the temperature regime from about 650°C to 450°C. This activity was characterized by several large rms voltage peaks and was very repeatable for the duration of the 500 cycle test. Attempts to duplicate this test failed to yield useful information. Further tests need to be performed to determine the effectiveness of AE monitoring techniques in understanding the thermal fatigue behavior of these materials.

CONCLUSIONS

(1) Large differences in room temperature residual strength of $[0]_4$ and $[0]_8$ specimens of SCS-6/Ti-24Al-11Nb thermally cycled in air underscore the importance of the environmental aspects of thermal fatigue in the titanium

aluminide composites. Residual strength reduction of $[90]_4$ and $[90]_8$ specimens under the same conditions was minimal up to 1000 cycles. $[0/90]_{2S}$ composites showed degradation similar to that of the $[0]_8$ composites when cycled in air and did not exhibit any measurable loss of residual strength when cycled in an inert atmosphere.

(2) The importance of considering total time at temperature under cyclic, and to a lesser degree, isothermal exposure conditions is demonstrated by a reduction in residual strength of the $[0]_8$ composite with increasing hold time at maximum temperature.

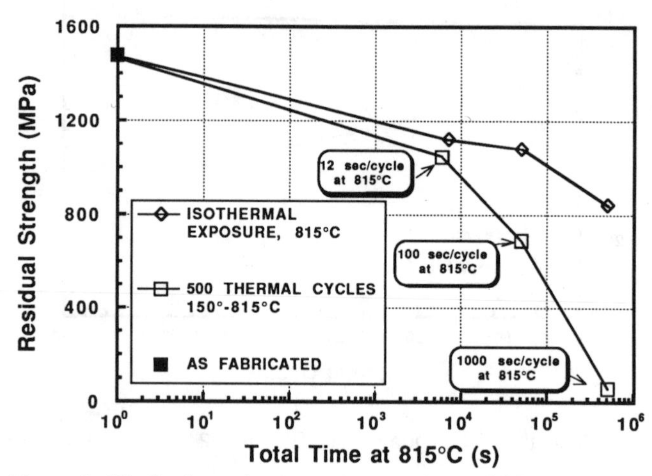

Figure 4. $[0]_8$ Cyclic vs. isothermal comparison, SCS-6/Ti-24Al-11Nb.

ACKNOWLEDGEMENTS

This work was funded by the Air Force Office of Scientific Research under project 2302P101. The authors would like to thank Mr. Tim Johnson and Mr. Evan Dolley for assistance in the thermal fatigue experiments and preparation and examination of test specimens. Special thanks to Dr. James Larsen and Dr. Theodore Nicholas for their technical expertise and input.

REFERENCES

1. S.M. Russ, Metall. Trans. A, 21, 1595 (1990).
2. W.C. Revelos and P.R. Smith, Metall. Trans. A, 23, 587 (1992).
3. P.K. Brindley, R.A. MacKay, and P.A. Bartolotta in Titanium Aluminide Composites, edited by P.R. Smith, S.J. Balsone and T. Nicholas, WL-TR-91-4020, Wright-Patterson AFB, OH, February 1991, pp. 484-496.
4. M. Khobaib in Proceeding of the American Society for Composites, edited by S.S. Sternstein (Technomic Publishing, Lancaster, PA 1991) pp. 638-647.

INTERDIFFUSION AND PHASE RELATIONS IN TI-ALUMINIDE/NB/TI-ALUMINIDE COMPOSITE STRUCTURES

Z. Ma[a], M.A. Dayananda[b], and L.H. Allen[a]

a) Department of Materials Science and Engineering, University of Illinois, Urbana, IL 61801.

b) School of Materials Engineering, Purdue University, West Lafayette, IN 47907.

ABSTRACT

We have studied interdiffusion and phase relations in two titanium aluminide/Nb/titanium aluminide composite structures at 1100 °C using solid state diffusion approach. The composite structures were prepared with two practical intermetallic alloys, γ (Ti-48Al-2Nb-0.3Ta, in at.%) alloy and super-α_2 (Ti-25Al-10Nb-3V-1Mo, in at.%) alloy, and pure Nb sheet by diffusion bonding technique. The interfacial microstructures evolved during thermal diffusion annealing were characterized using optical microscopy, SEM, TEM and X-ray diffraction. The effect of Nb on the phase stabilities of two intermetallic alloys and the phase relations among different phases formed in the diffusion zone due to interdiffusion were analyzed in terms of rules of diffusion paths on the basis of the observed diffusion structures and experimental diffusion paths.

INTRODUCTION

Titanium aluminides with addition of a small amount of Nb have been of great technological interest for years due to their high temperature strength, low density, better creep and oxidation resistance relative to conventional titanium alloys[1]. The development of new titanium aluminide based intermetallic composites with optimum properties requires a full understanding of their interface structures and phase stabilities. On the other hand, synthesis of new titanium aluminide based intermetallic alloys needs a complete knowledge of Ti-Al-Nb ternary phase diagrams at temperatures where these materials undergo heat treatment. Although a number of systematic investigations have been carried out on the phase transitions[2] and phase relations[3] in this system, phase equilibria at some temperatures of interest are still not quite clear. Also, there is a lack of kinetics data in the literature concerning the interdiffusion of Ti, Al, and Nb in this system.

For these reasons, we have investigated interdiffusion in titanium aluminide/Nb/titanium aluminide composite structures using solid state diffusion approach. The objectives of this work are two-fold: (1) to elucidate the evolution of interfacial structures in these composite materials during diffusion annealing at elevated temperatures and to establish their phase relationships based upon the observed diffusion structures and experimental diffusion paths; (2) to extract useful interdiffusion data for Ti, Al, and Nb in some important intermetallics. In this paper, we only present the first part of this work. The theoretical background for diffusion analysis in multicomponent system and resultant interdiffusion data will appear elsewhere.

EXPERIMENTAL PROCEDURE

The materials used in this study were two commercially available intermetallic alloys, γ-base alloy (nominal Ti-48Al-2Nb-0.3Ta, designated A here) and super-α_2 alloy (Ti-25Al-10Nb-3V-1Mo, designated B), and pure Nb sheet. The composite structures, A/Nb/A and B/Nb/B, were prepared at 1000 °C and 1000 psi for 2 to 4 hours using diffusion bonding (DB) technique. These structures naturally constitute solid-solid diffusion couples needed for diffusion studies. Before diffusion annealing, the sandwiched couples were examined by optical microscopy, SEM and X-ray diffraction for grain size and phase identifications of the terminal alloys. The couples were placed in quartz capsules, evacuated and sealed under high vacuum. The capsules were then annealed in a furnace at 1100 °C for various times and ice-water quenched after diffusion annealing.

The annealed diffusion couples were characterized by optical microscopy and SEM-EDS for interfacial structures and concentration profiles within the diffusion zone. Selected couples were also analyzed by X-ray diffraction and TEM for phase identifications.

RESULTS

Microstructures of Couples A/Nb/A and B/Nb/B prior to Diffusion Annealing

The microstructures of the couples A/Nb/A and B/Nb/B prior to diffusion annealing are shown in Figs.1(a) and (b), respectively. The composite structures exhibit good interfacial bonding with negligible evolution of diffusion structures. Alloy A consists of primary γ (TiAl) phase and an eutectic structure of ($\gamma + \alpha_2$). Alloy B is also a two-phase mixture with α_2 (Ti_3Al) phase surrounded by disordered β or ordered B2 phase. The growth fronts are indicated by arrows.

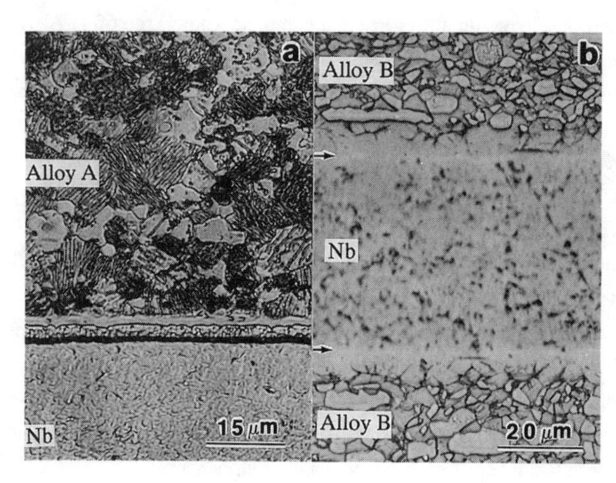

Fig.1.Photomicrographs for the couples prior to annealing (a)A/Nb/A; (b)B/Nb/B.

Diffusion Structures for Couple A/Nb/A after Diffusion Annealing

Fig.2(a) shows the diffusion structures developed in the diffusion zone after annealing at 1100 °C for 12 days. Several single-phase layers such as δ (Nb_3Al), σ (Nb_2Al), and α_2 (Ti_3Al) were formed as a result of mutual interdiffusion of terminal materials. A two-phase layer ($\sigma+\alpha_2$) was observed between σ and α_2. This is more clearly seen in the backscattering electron (BSE) image presented in Fig.2(b). The σ single phase layer shows nice columnar structure grown along the direction of diffusion. The contrast difference shown in the two-phase region is due to atomic number difference for Ti and Nb. The β/δ and δ/σ interfaces are planar, while the others are nonplanar.

The concentration profiles for this couple is shown in Fig.2(c). No attempt has been made to measure the concentration of Ta. The profiles exhibit steps in concentrations at the interfaces between different intermetallic phases formed in the diffusion zone. The Matano plane (x_o) is calculated from the mass balance of the diffusing species.

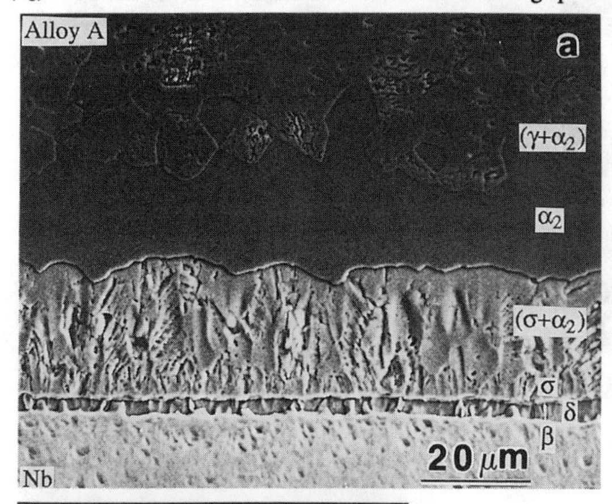

Fig.2.(a)SE image;(b)BSE image for the diffusion zone;(c)concentration profiles for the couple A/Nb/A annealed at 1100°C for 12 days.

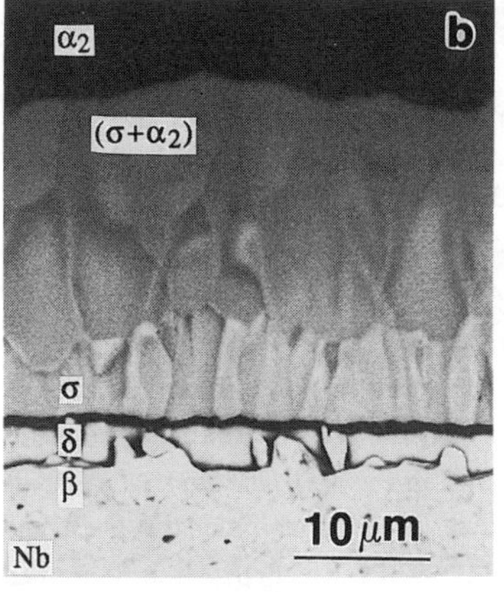

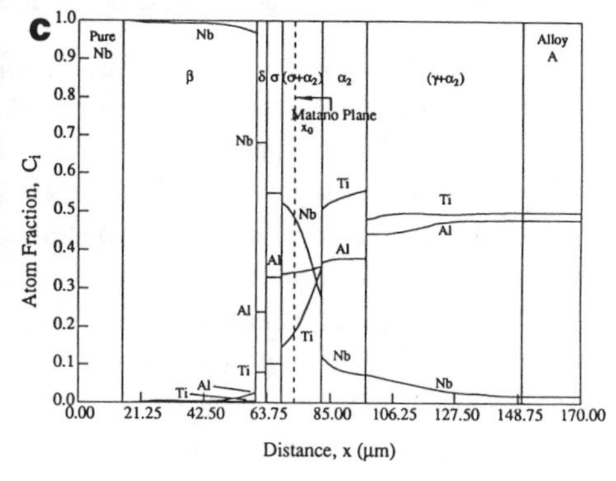

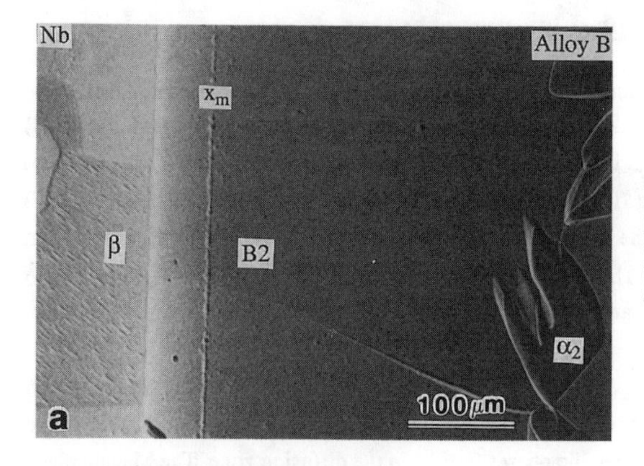

Fig.3.(a)SE image; (b)SAD pattern for B2 phase; (c)concentration profiles for the couple B/Nb/B annealed at 1100°C for 5 days.

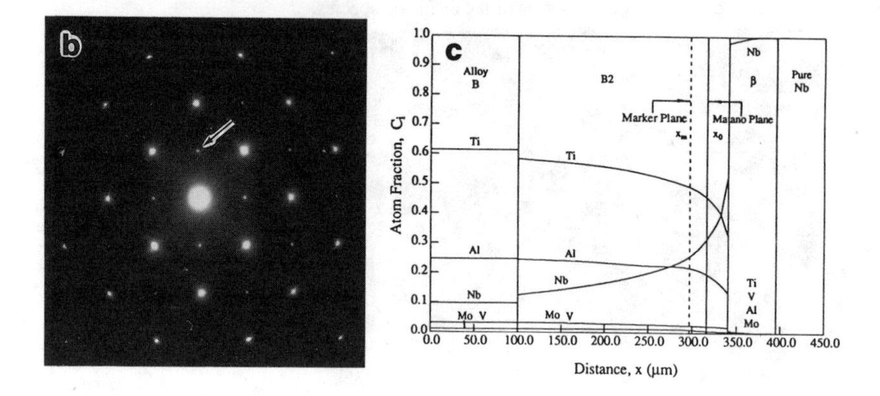

Diffusion Structures for Couple B/Nb/B after Diffusion Annealing

The interfacial structures evolved at the interface for couple B/Nb/B annealed at 1100 °C for 5 days is shown in Fig.3(a). The B2 single phase layer (will be identified later) grew by consuming both Nb and α_2 precipitates, which is evidenced by the visible marker plane (x_m). More interesting is that a sharp planar interface formed between B2 and β phases. The concentration profiles presented in Fig.3(c) also show the steps at the B2/β interface. This suggests that B2 phase is stable and in equilibrium with β at 1100 °C.

Identification of B2 phase was carried out by using transmission electron microscopy. The selected area diffraction (SAD) pattern taken from [001] zone axis of B2 single phase area with composition of Ti-20Al-31Nb-2.2V-0.8Mo is shown in Fig.3(b). The superlattice spots due to ordering of Ti atoms are clearly visible, indicating that B2 is an ordered β phase and similar to the previously reported B2 (nominal Ti_2NbAl), which was believed to be formed during quenching[4]. No antiphase boundary (APB) was observed under the dark field conditions. This also indicates that B2 is a stable phase at our annealing temperature. The streaks around the diffraction spots imply subsequent decomposition of the B2 phase[5].

Diffusion Paths on a Ternary Isotherm at 1100 °C

The concentration profiles for these two composite structures developed in the diffusion zone were mapped onto a Gibbs triangle to form the so-called diffusion paths. These are presented in Fig.4. The concentrations of Nb, V, and Mo were grouped together as one concentration variable in the plotting of diffusion paths. A schematic Ti-Al-Nb ternary isotherm at 1100 °C is also presented in the figure. Boundaries for selected phase regions are schematically

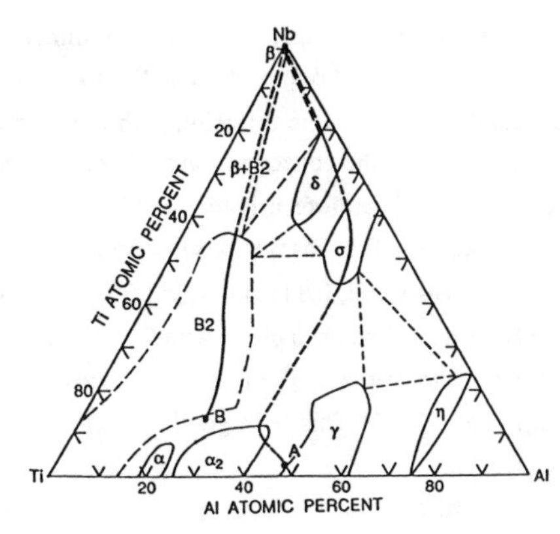

Fig.4.Schematic Ti-Al-Nb ternary isotherm showing experimental diffusion paths at 1100°C. The concentrations of Nb, V, and Mo are grouped together as one concentration variable in plotting diffusion paths.

drawn on the isotherm in terms of rules of diffusion paths[6] on the basis of the concentration profiles measured and diffusion paths observed in this work. In addition, available phase diagram information[7] for Ti-Al, Ti-Nb, and Nb-Al binary systems, and the isotherms[3,8,9] for Ti-Al-Nb ternary system were used to draw the phase boundaries.

DISCUSSION

The composite structures examined in this study displayed several interesting features in the development of diffusion structure. A wide variation in the intermetallic layers formed at the interfaces depends upon not only the annealing temperature but also the compositions of the terminal materials.

For structure A/Nb/A, the β/δ planar interface indicates that a two-phase equilibrium exists between each other. This corresponds to a diffusion path parallel to a tie-line showing the (β+δ) two-phase equilibrium[6], as shown in Fig.4. The δ/σ planar interface also established a two-phase equilibrium and corresponds to a diffusion path segment parallel to a tie-line representing the (δ+σ) two-phase equilibrium. The formation of δ and σ single phase layers corresponds to diffusion path segments within their single phase regions respectively. The interfaces of $\sigma/(\sigma+\alpha_2)$ and $(\sigma+\alpha_2)/\alpha_2$ are nonplanar. The protrusion of the σ phase towards $(\sigma+\alpha_2)$ two-phase region showing a columnar two-phase zone is seen in Fig.2(b). It corresponds to a diffusion path that passes into a $(\sigma+\alpha_2)$ two-phase region from a σ single-phase region at an angle to a tie-line but exits into an α_2 single-phase region. This is also clearly indicated by the large variations in the concentration profiles for Ti and Nb components presented in Fig.2(c). An α_2 single phase layer formed in the diffusion zone again represents a diffusion path segment within its single phase region. The volume fraction of the α_2 phase decreases towards the terminal alloy A. Such a two-phase region corresponds to a diffusion path crossing of the $(\alpha_2+\gamma)$ two-phase region. The diffusion structure observed in this couple is in good agreement with the diffusion path representation as shown in Fig.4.

The structure B/Nb/B developed a large layer of B2 phase between the terminal alloy B and pure Nb. A planar interface that formed between B2 and β indicates that these two phases is under equilibrium at 1100 °C, thus corresponding to a tie-line diffusion path crossing of the (B2+β) two-phase region on the isotherm. The steps in the concentration profiles also indicate the existence of the (B2+β) two-phase equilibrium. Therefore it is reasonable to suggest the possibility of a three-phase equilibrium among β, δ, and B2 phases at this moment.

From the SAD patterns taken from zone axes [001], [011], and [111] (the latter two are not shown here), one can conclude that the B2 is an ordered bcc phase with CsCl type structure due to the ordering of Ti atoms at the corners of the unit cell. The lattice parameter calculated from X-ray diffraction measurement is equal to 3.25 Å. This is in good agreement with the published value (a=3.22 Å) for the B2 phase[2].

Based upon an analysis of the diffusion structures and the corresponding diffusion paths for these diffusion couples, a schematic Ti-Al-Nb ternary isotherm at 1100 °C has been proposed. In spite of the limited terminal compositions employed in this study, the schematic isotherm does have several new features not reported in the literature. From a theoretical viewpoint, the present study establishes a B2 single phase field as well as a (B2+β) two-phase region. An equilibrium between B2 and β follows from the planar B2/β interface developed in the B/Nb/B composite structure isothermally annealed at 1100 °C. So we concluded that B2 is a stable phase at 1100 °C and is not necessarily quench-induced[4].

SUMMARY

Interdiffusion and phase relations have been studied in the titanium aluminide/Nb/titanium aluminide composite structures using solid state diffusion couple technique. Several important results have been obtained in this study. They are summarized as follows: (1) a large B2 single phase field and a (B2+β) two-phase region have been established and we experimentally confirmed that B2 phase is stable at 1100 °C; (2) a possibility of β, δ, and B2 three-phase equilibrium is suggested based upon the two-phase equilibria between β and δ, and between B2 and β; (3) a schematic Ti-Al-Nb isotherm at 1100 °C has been proposed.

ACKNOWLEDGEMENTS

We would like to express our thanks to Rockwell International Corporation for providing us the materials used in this study. Thanks are also due to School of Materials Engineering of Purdue University for supporting this work. Some of the materials characterization were carried out in the Center for Microanalysis of Materials at University of Illinois, which is supported by the Department of Energy.

REFERENCES

1. J.D. Destefani, Adv. Mater. Proc., **2**, 37(1989).
2. J.A. Peters and C. Bassi, Scripta Metall., **24**, 915(1990).
3. J.H. Perepezko, Y.A. Chang, L.E. Seitzman, J.C. Lin, N.R. Bonda, T.J. Jewett and J.C. Mishurda, in High Temperature Aluminides and Intermetallics, ASM/TMS-AIME, Indianapolis, Oct.1989.
4. R. Strychor, J.C. Williams and W.A. Soffa, Metall. Trans. A., **19**, 225(1988).
5. L.A. Bendersky, W.J. Boettinger, B.P. Burton and F.S. Biancaniello, Acta Metall., **38**, 931(1990).
6. J.S. Kirkaldy and L.C. Brown, Canadian Metall. Quart., **2**, 89(1963).
7. Binary Alloy Phase Diagrams, ed. T.B. Massalski, ASM, Ohio, 1986.
8. R.G. Rowe and E.L. Hall, Mat. Res. Soc. Symp. Proc., **213**, 449(1991).
9. L.A. Bendersky and W.J. Boettinger, Mat. Res. Soc. Symp. Proc., **133**, 45(1989).

FATIGUE CRACK GROWTH IN A TiAl ALLOY WITH LAMELLAR MICROSTRUCTURE

DAVID L. DAVIDSON
Southwest Research Institute, P.O. Box 28510, San Antonio, TX 78228

ABSTRACT

The mechanisms of fatigue crack advance are examined for a lamellar $\alpha_2 + \gamma$ alloy. Crack growth rates are highly dependent on the orientation of the loading axis to the lamellae direction. Thus, the material has some of the characteristics of a composite. For crack growth perpendicular to the lamellae, the mechanisms of crack advance are similar to those of other titanium alloys, while crack growth parallel to lamellae has other characteristics.

INTRODUCTION

The mechanisms and micromechanics of fatigue crack growth through the $\alpha+\beta$ titanium alloys, Ti-6Al-4V (RA) [1] and CORONA-5, [2] and the $\alpha_2+\beta$ titanium aluminide alloys, Super Alpha 2, [3] and 2411, have been examined and compared [4]. The growth rates for large fatigue cracks through these alloys is different, but the mechanisms of growth were similar. For all these alloys, crack growth near threshold required a large number of cycles (ΔN) before crack advance (Δa). The sequence of events accompanying crack extension was observed to be similar to that found for aluminum alloys; a sharp crack blunted as the number of cycles increased, followed by crack extension and resharpening. For the $\alpha+\beta$ alloys, slip lines were observed to form at the crack tip during the blunting process, and crack advance occurred by breakdown of these slip lines. For the $\alpha_2+\beta$ alloy 2411, crack blunting was observed also, but often an α_2 particle near the crack tip broke, and crack advance occurred by linking of this broken particle with the main crack tip.

This paper reports on fatigue crack growth through a titanium base alloy with a microstructure fundamentally different than any previously examined. The alloy is based on the intermetallic compound TiAl (γ) mixed with a substantial volume of α_2 phase. The composition tested was Ti-47Al-0.9Cr-0.8V-2.6Nb (in at.%). Processing resulted in a microstructure of $\alpha_2 + \gamma$ lamellae with some regions of equiaxed γ. Dissolved oxygen was measured as 700 ppm (weight). This alloy composition and microstructure has demonstrated higher ductility and

Mat. Res. Soc. Symp. Proc. Vol. 273.

toughness than other gamma-based alloys [5,6].

MATERIAL AND MICROSTRUCTURE

The microstructure of this alloy, supplied by Dr. Kim of the Metcut Materials Research Group, is complex. The large, approximately equiaxed regions seen in **Fig. 1**(a) are colonies of lamellae. Average size of these colonies was 1.2 mm. When a mixed acid etch was used [7], some of these platelets etched and some did not. Transmission electron microscopy (TEM), Fig. 1(b) revealed a very complex structure consisting of many wide and narrow lamellae.

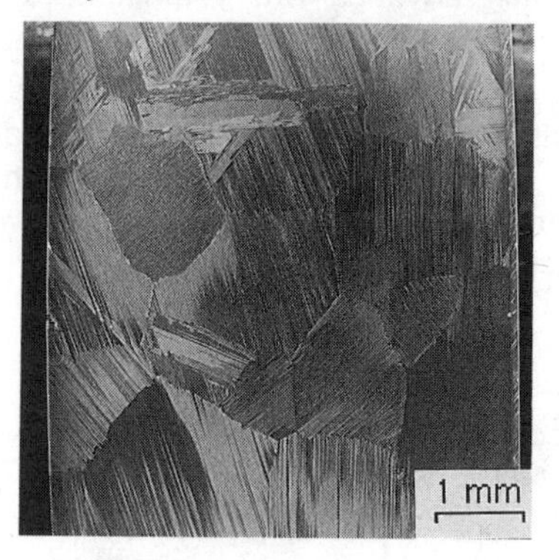

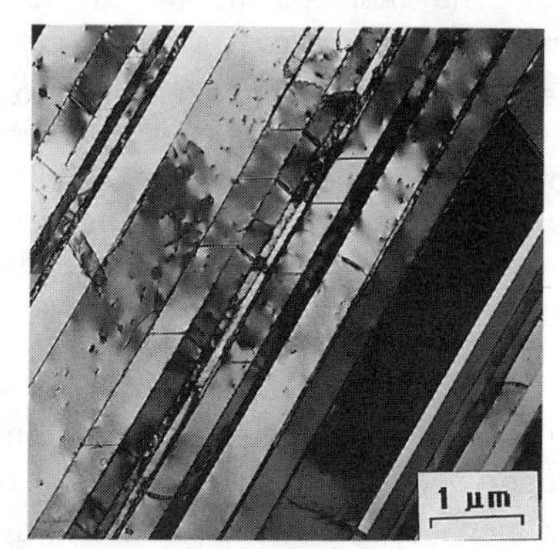

Fig. 1. Microstructure of the material showing (a) SEM (secondary image) of the colony structure, and (b) TEM of lamellae.

From the alloy composition and the phase diagram [8], the microstructure should consist of a mixture of Ti_3Al (α_2, DO_{19}) and TiAl (γ, $L1_0$); thus, each of the lamella is one of these two phases, but it is difficult to determine the phase of individual lamella. The alloy composition requires more γ than α_2, suggesting that wide lamellae are γ and some of the narrow lamellae are α_2. However, identification of individual phases by this means is complicated by the fact that boundary on the phase diagram are not vertical, which means additional lamellae are formed as the material cools. The work of Shong, et al. [9], on a Ti-43 at%Al alloy indicates that α, the high temperature phase, first transforms to an ordered α_2. Continued cooling causes precipitation of γ as lamellae, but Umakoshi, et al. [10], describe the transformation for a Ti-48 at.% Al process as γ forming first, followed by precipitation of α_2 lamellae. These studies indicate that the route to final microstructure very sensitive to aluminum content; cooling rate is also known to be important.

A number of TEM studies have been made of these two phase alloys to examine of the orientation relationships between the phases. The most recent report [11] indicates that four different orientation variants can form both between TiAl and adjacent Ti_3Al lamella as well as between two adjacent TiAl lamellae. The reason for this complexity is hypothesized as being due to the ability of different variants to accommodate the strain induced by transformation. Boundaries between lamellae may separate two crystallographic orientations of the same phase, or two different phases. The interfaces between adjacent lamellae, studied by Zhao and Tangri [12] using TEM, were found to contain many misfit dislocations. Twinning is one mechanism by which two adjacent γ lamellae may exist. In interfaces of this type, three types of defect may exist.

For the material used in this study, the lamellar microstructure has been characterized from secondary electron SEM images at a magnification of 2000X. Indications were that α_2 etched, and γ did not. The volume fractions measured were approximately 0.50±0.15 for each phase. The width of α_2 lamellae were measured between 0.3 and 2.8 μm, with most plates being 1 to 1.4 μm wide. Most of the γ plates were between 0.6 and 1.4 μm, but a few plates up to 3.5 μm wide were found. These data should be interpreted within the context of the limited resolution of secondary electron images from etched samples, which does not allow counting of plates less than about 0.3 μm. Thus, the bulk of the lamella are less than 2 μm wide and there is a large variation in lamellae widths. The lamellae dimensions of our alloy may be compared with those of the Soboyejo, Deffeyes and Aswath [13] alloy which had a colony size of approximately 18 μm, with a typical lamellae width of 0.26 μm.

At ambient temperature, tensile modulus for this material is 1.75×10^5 MPa and tensile yield was measured by Chan [14] as approximately 400 MPa (at 0.2% strain), which is typical of other measurements made for this material [5]. Strain to fracture was 0.9% (at $\approx 10^{-3}$/sec). Strain to fracture ranged from 1 to 9%.

EXPERIMENTAL METHODS

Fatigue cracks were grown at ambient temperatures from notches in two specimen designs: 3 mm thick single edge notched (SEN) specimens with a gauge section 20 mm wide, and 4.6 mm thick compact tension (CT) specimens with a distance of 16 mm between loading hole and back surface. Even with these fairly thick specimens, there were only 4-6 colonies of lamellae through the thickness specimen.

The notches approximately 0.5 mm wide and 3 mm deep for the SEN and 6 mm deep for the CT specimen were made using a low speed diamond saw. Cracks were initiated from notches by compression-compression loading at $\Delta K = 30$ MPa$\sqrt{m}$, R = 10 (R = minimum/maximum stress intensity). After crack initiation and some growth, cycling was changed to tension-tension loading, R = 0.1, at 10Hz. Cracks were grown mainly in laboratory fatigue machines under ambient conditions, but transferred to a special loading stage for the SEM [14] for detailed studies. Fatigue cracks were grown in air (relative humidity approximately 50%), except for cycles applied using the SEM stage.

RESULTS

Fatigue Crack Growth

Fatigue crack growth in the lamellar structure is highly dependent on the orientation of the crack to the direction of the lamellae and loading axis. Near to the threshold stress intensity factor for fatigue crack growth (ΔK_{th}), the application of many (thousands) of loading cycles often did not result in any crack growth. Therefore, the rates of crack growth were difficult to determine and depended strongly on the averaging process used. Furthermore, colony sizes were so large that observations of crack growth in individual colonies over small distances must be considered as essentially tests from single crystals.

All the fatigue crack growth rates, as measured from three specimens, regardless of the orientation between the crack and lamellae directions, are shown in **Fig. 2**, and compared to results obtained by Soboyejo, et al. [13] under similar conditions.

These data were fit to the correlation

$$da/dN = B\, \Delta K^{s} \tag{1}$$

where s = 5.7 and B = 1.9×10^{-14} m/cy.

Detailed information on fatigue crack tip mechanics at the lowest growth rates ($4 < \Delta K < 8$ MPa$\sqrt{m}$) have been confined to only one lamellar orientation to date because of the difficult task of growing a crack at very low rates at various orientations relative to the lamellae. Only a crack growing approximately parallel to the lamellae direction in air at 10 Hz has been observed in detail.

Fatigue crack growth near the threshold has been found to be intermittent for all of the materials studied thus far (Al, Ti, and Fe alloys) [15] using similar techniques, and this alloy behaved similarly.

However, for the lamellar microstructure, the intermittent nature of crack advance was more exaggerate than has been observed for any other material. Coaxing the crack to grow at near-threshold values of ΔK required extreme patience and was very sensitive to the microstructure and level of ΔK used.

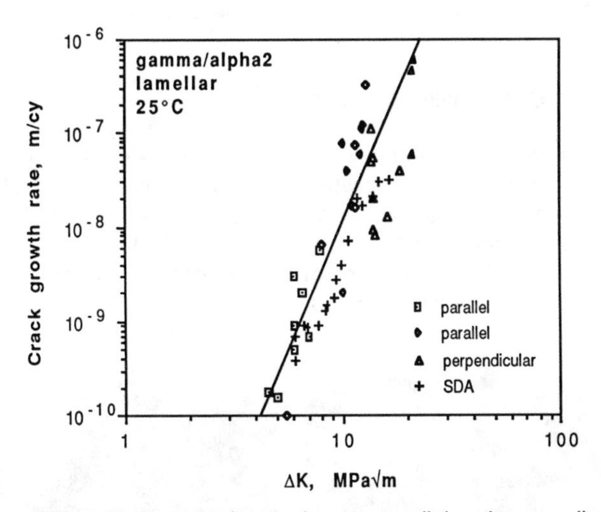

Fig. 2 Fatigue crack growth rates parallel and perpendicular to the lamella direction. SDA [Ref. 13]

Crack growth at intermediate ΔK ($8 < \Delta K < 12$ MPa√m), although well above the threshold, still required many cycles for crack advance and was very intermittent in growth behavior. Cracks growing approximately parallel to lamellae were mostly observed in this range.

At the high crack growth rates generated by stress intensity factors above about $\Delta K \approx 14$ MPa√m, it was possible to advance the crack into a grain where the lamellae were perpendicular to the crack plane on one side and thus obtain data for this crack/lamellae orientation unavailable at lower ΔK.

The primary knowledge desired was a detailed account of crack interaction with the microstructure for this orientation. Thus, the crack tip region was photographed on each successive loading cycle. In order for this procedure to yield useful data, the cyclic stress intensity factor was periodically increased until crack growth was achieved in a relatively few cycles. This was accomplished by loading a fixed number of cycles in the laboratory loading frame and then moving the specimen to the loading stage of the scanning electron microscope to continue cycling under high resolution conditions. In this process, it was discovered that the rate of crack growth was unusually sensitive to both the frequency of cyclic loading and the environment. The crack growing at a loading frequency of 10 Hz in air in the laboratory machine would not grow when cycled in the vacuum of the SEM at 0.3 Hz. This was confirmed by breaking the SEM vacuum and cycling the specimen in the SEM stage in air, followed by evacuation and measurement of the crack length with the SEM.

Crack growth perpendicular to lamellae

The cyclic stress intensity factor was increased in small increments while cycling in a laboratory loading frame at 0.3 Hz in vacuum (≈ 1 mPa) until crack advance was being achieved in only tens of cycles. Then the specimen was transferred to the SEM loading stage for detailed observation. A description of crack growth observed and photographed under this condition follows and is illustrated in **Fig. 3**.

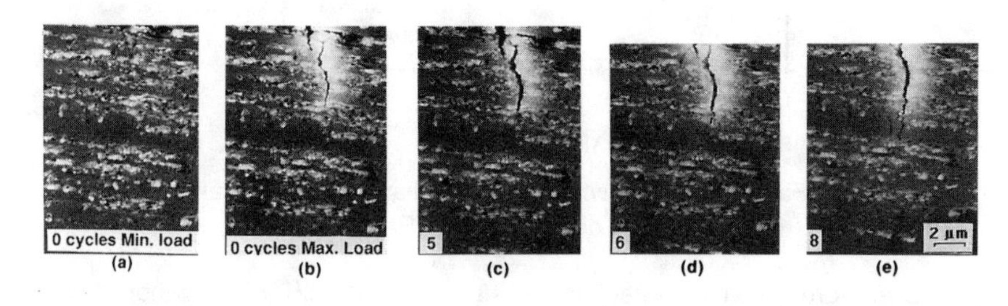

Fig. 3 Fatigue crack growth perpendicular to lamellae at $\Delta K = 23$ MPa$\sqrt{m}$. The sharp-blunt-sharp crack opening sequence is illustrated.

At the beginning of the sequence, the crack at minimum load was tightly shut and the loaded crack was quite sharp, Fig. 3(a). During the next 5 cycles, the crack tip blunted, Fig. 3(b), although not much. The crack also extended approximately 0.2 µm during these 5 cycles.

On the next loading (cycle 6), Fig. 3(c) the crack grew 0.8 µm and again became sharp. The next 2 cycles caused the crack to grow an additional 1.35 µm across a wide lamella (possibly gamma), resulting in a sharp crack tip, Fig. 3(d). From cycles 8 to 15, (photographs not shown) the crack tip grew 0.6 µm and blunted, but from cycle 15 to 23, the crack grew 2.4 µm across several narrow lamellae. Similar observations were made as this crack grew another 10 µm and indicated that the sequence described in Fig. 3 was typical. Crack length is shown as a function of loading cycles in Fig. 4, where the approximate location of the visible lamellae boundaries are also shown. There may be additional lamellae in the regions where lamellae are narrow, but there is less likelihood of additional boundaries within the wide lamellae.

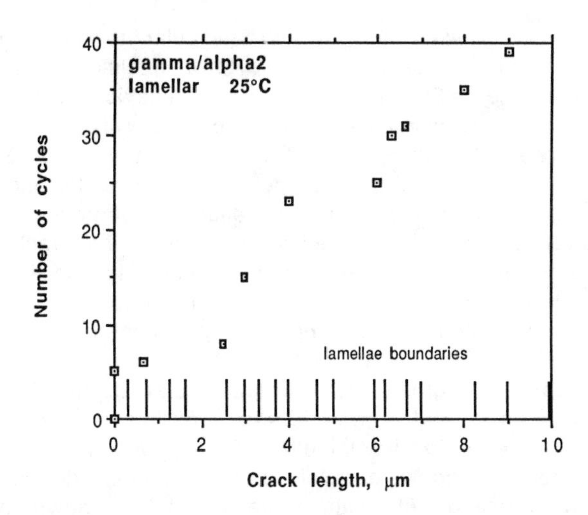

Fig. 4 Crack length as a function of loading cycles for the growth sequence shown in Fig. 3.

This crack growth sequence indicates that crack growth is still intermittent at high ΔK, approaching K_c, and that the crack alternates between being sharp and blunt in different parts of the growth sequence. This same behavior has been observed for a number of other materials, including titanium and titanium aluminide alloys, [1,4,15] but the crack tip blunting exhibited by the present material is smaller than for those alloys. Measurements of this parameter will be presented elsewhere.

The crack growth sequence of Fig. 3 and the crack length measurements in Fig. 4 suggest that lamella width may be an important factor in controlling the rate of fatigue crack growth in this microstructure. The crack was observed to traverse wide plates (most likely gamma) in one cycle, or at most a few cycles, while crack advance appeared to grow slowly or stop during blunting of the crack tip when the lamella spacing was small (probably a mixture of gamma and alpha 2 plates). However, there are other factors besides lamellae width which could also influence this behavior. The crystallographic orientation between adjacent lamellae may be important in controlling fracture resistance, as may be the sequence of γ and α_2 platelets. The lack of delamination along interfaces is notable, especially when the ease of crack growth in the lamellar direction is considered.

Crack growth parallel to lamellae

As the fatigue crack was growing perpendicular to the lamellae direction on one side of the specimen, it was growing almost parallel to the lamellae on the other side. Converse to the behavior illustrated in Fig. 3, the fatigue crack grew on nearly every cycle. This is shown in **Fig. 5** which illustrates the elongation of the crack in just 10 loading cycles. The crack was trying to grow perpendicular to the loading axis, but grew easier in the direction of the lamellae. Orientation of the lamellae was nearly parallel to the loading axis; thus, the crack grew alternately between parallel and nearly perpendicular to the lamellae direction. Since ΔK was the same as for Fig. 3, the effects of lamellae direction on crack growth behavior may be directly compared.

Average fatigue crack growth rate perpendicular to the lamellae direction, shown in Figs. 3 and 4, was 2.7×10^{-7} m/cy, while the growth rate approximately parallel to the lamella, Fig. 5, was 7×10^{-7} m/cy. However, crack growth roughly parallel to the lamellae direction is complicated by the formation of multiple crack tips, as shown in the figure. Multiple cracking and the extra energy dissipation mechanisms associated with it slows the rate of crack growth down to approximately that perpendicular to the lamellae direction.

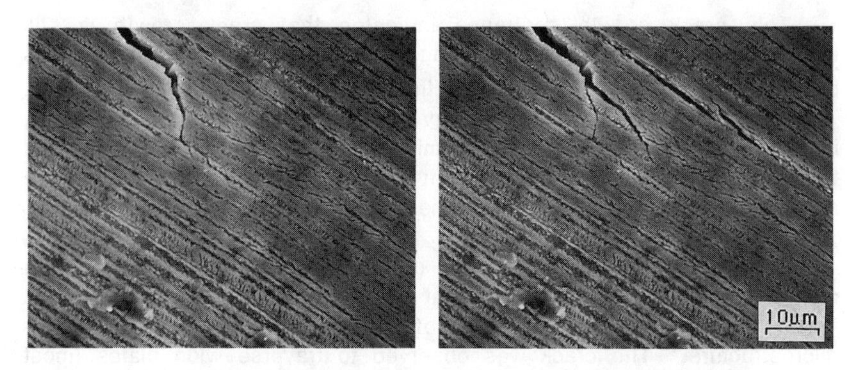

Fig. 5 Crack growth approximately parallel to the lamellae direction.
Left: 0 cycles, Right: 10 cycles. $\Delta K = 23$ MPa$\sqrt{m}$

Crack Closure

Measurements of crack opening load were made during these experiments by directly observing the crack tip in the SEM. Load application caused the crack to open in Mode I near the crack mouth and quickly open in a progressive manner (peeling open) with increasing load until the crack was open near to the tip (10-20 μm). Increasing load then resulted in the crack opening the remainder of the distance to the tip, but more slowly than during initial load application. Similar behavior has been observed many times in other alloys [16,17], and nothing unique to this alloy was observed in the Mode I crack opening process. The parameter $U = \Delta K_{eff}/\Delta K$ was computed from maximum, minimum and opening loads. Previously, it was found that U correlated with $1/\Delta K$ [17]. The measured data for this alloy are not extensive, but fit the concept, previously determined for five other metallic alloys and partially stabilized zirconia [19], that

$$U = 1 - K_{th}/\Delta K \tag{2}$$

which is equivalent to

$$\Delta K_{eff} = \Delta K - \Delta K_{th} \tag{3}$$

Since ΔK_{eff} can be determined, eq. (1) can be rewritten as

$$da/dN = B' \, \Delta K_{eff}^{s'} \tag{4}$$

FRACTOGRAPHY

The fracture surfaces of the specimen from which detailed measurements were made were carefully examined by secondary electron contrast in the SEM. A low magnification view of the fracture surface is given in **Fig. 6**, which shows a large variety of features, directly related to colony size. The micrograph of Figs. 6(b) was taken directly below the surface observations of Fig. 3. The appearance of the fracture surface is dependent on the orientation of the lamellae relative to the direction of crack growth, just as were the details of crack growth observed directly. The lamellae were perpendicular to the crack growth direction, but they were at an angle of 60° the surface; thus, lamellae widths shown in Fig. 3 are enlarged by about 15%. The same surface features appear to cross the entire width of a lamella, which correlates well with the surface observation of cracking across a whole lamella width in only one or two cycles. The direction of local features varies from lath to lath, probably indicating that the local direction of crack growth was variable and depended at least partially on the crystallography of each lamella. These fractographs, together with Fig. 3, are evidence that the growth increment in fatigue is related to the lamella width.

Fig. 6 Left: Overall view of fracture surface. Right: Detail of fracture surface formed by the crack of Fig. 3. Crack grew from bottom to top.

Fracture features similar to those found in Fig. 6(b) were attributed by Pao, et al. [18], to "following TiAl twin interfaces and Ti_3Al interface" to produce the step-like features seen. The lamellae widths measured from the fractographs of Fig. 5 indicate two peaks, one between 0 and 0.25 µm and another broad peak between 0.75 and 1.25 µm. These lamella widths agree well with those determined from surface measurements. Stereopairs of the fracture surface, made by tilting, show that it is relatively flat on a microscopic scale.

DISCUSSION

It has been shown for this and similar alloy compositions that a lamellar microstructure maximizes crack growth resistance [5]. The main purpose of the research reported here has been to understand how this microstructural form controls fatigue crack advance. The direction of crack growth in this material is highly dependent on the orientation of the loading axis relative to that of the lamellae. Crack growth, which usually occurs perpendicular to the loading axis for homogeneous materials, was altered as much as 45° by lamellae orientation. This crack growth behavior is similar to that observed in continuous fiber composites; thus, this material could be considered as a class of composites formed by microstructural manipulation (an in-situ composite).

High resolution observations have shown that crack growth is not a continuous process for this material, much as has been observed for a number of much more homogeneous microstructures, such as aluminum and titanium alloys and steels. Crack opening displacements are similar to those found for other alloys, as are the threshold for fatigue crack growth and crack closure measurements. Thus, the growth of fatigue cracks in this highly anisotropic microstructure has many similarities to crack growth in much more homogeneous materials, and it is logical to extend the description of crack growth physics developed for those alloys to this material, and that will be done elsewhere.

A model has been advanced [19,20] for predicting the threshold stress intensity for crack growth, ΔK_{th}, from microstructural parameters.

$$\Delta K_{th} = \sigma_y \sqrt{2\pi r_s} \quad (5)$$

where σ_y = the yield stress and r_s is the length of a slip line extending from the crack tip at threshold. The slip line length for this material is limited by the microstructure to approximately the lamella width when the crack is growing perpendicular to the lamellar structure, but what limits slip line length for cracks growing parallel to lamella is unknown. Using yield stresses and lamellae widths in eq. (5) gives (for ambient temperature) $0.8 < \Delta K_{th} < 1.2$ MPa$\sqrt{m}$. Comparison with measured values of ΔK_{th} indicates that this value is too small by at least a factor of 5. The inhomogeniety of crack growth in this material is thought to be the reason for this difference. The value calculated might be close to the measured value for a single colony of material, but the presence of multiple colonies having different orientations increases ΔK_{th} due to the anisotropy of crack growth. For crack growth parallel to lamellae, slip line lengths are expected to be larger because of the lack of an inherent microstructural barrier, such as the lamellae boundaries provide. A larger slip distance raises the estimate of ΔK_{th} using eq. (5). However slip is limited by other than colony boundaries; estimating ΔK_{th} from colony size gives $\approx$ 24 MPa$\sqrt{m}$.

Fatigue crack growth rates at ambient temperature, Fig. 2, are somewhat higher than those measured by Soboyejo, et al., [13], mainly at $\Delta K > 10$ MPa$\sqrt{m}$, but also near threshold. Their material had a smaller colony size and the lamellae width was smaller than for the material tested in this work. The difference in crack growth rates could be due also to the differences in technique for measurement of crack length (they used potential drop), but the most likely causes are: a difference in strain rate experienced by the material (cyclic frequency), and differences in colony size and in the dimensions of the lamellar microstructure. These factors may be a good indication of just how sensitive crack growth rates are to microstructure.

When the direction of crack growth is approximately the same as the lamellae, has not been possible to discern whether the crack was in the interface between two phases or within one of the phases. From studying photographs made from 2000 to 8000X, it appears that crack growth was within one phase in some places and along phase boundaries in others. Both fractography and dynamic observation on the specimen surface showed some secondary cracking. Conversely, for crack growth perpendicular to the lamellae direction, neither fractography nor surface observation indicated that cracking occurred either in interlamellae boundaries or within specific lamellae. These observations indicate that crack bridging could have had a limited effect on crack growth in the lamellae direction, but no effect for crack growth perpendicular to lamellae.

The detailed TEM studies of this microstructure by Tangri, et al. [12] has found large numbers of dislocations in the interlamellar boundaries, and these investigators believe that many of these dislocations would be glissile when stressed. They believe that this might lead to extensive cracking in the interlamellar boundaries. This work on fatigue crack growth has not found much evidence of this type of cracking. Tangri, et al., attribute the ductility exhibited by this material to limitations in slip caused by the lamellae widths; thus, cleavage failure is suppressed preventing extensive slip on a few planes. Meanwhile, the existing dislocation structure is activated. Strains measured in the region surrounding the crack tip, presented elsewhere, is relatively homogeneous and this homogeneity is not strongly influenced by the presence of lamellae, regardless of the relative orientation of the crack to the lamellae, which is surprising. This is indicative of multiple active slip systems within the high constraint region of the crack tip and seems to verify Tangri's view that the existing dislocation network can be activated without cleavage.

CONCLUSIONS

1. Fatigue crack growth was intermittent, as has also been observed for alloys of iron and aluminum and other titanium alloys.

2. At low ΔK, the direction of crack growth was greatly influenced by the direction of lamellae boundaries relative to the loading axis, and crack growth was strongly influenced by colony boundaries. Cracks appeared to grow both within lamella boundaries and inside lamellae.

3. At high ΔK, colony boundaries had much less effect on crack growth rate, and the lamellae orientation was less of an influence on the direction of crack growth. Cracks readily grew across lamellae boundaries.

4. Crack tip opening alternated between blunt and sharp. As for other alloys, blunt crack tips had the largest strains.

5. Some secondary cracking was observed and was found to occur generally when the crack was growing in the lamella direction at ≈ 45° to the loading axis.

6. For crack growth perpendicular to the lamella direction, the increment of crack advance was linked to the width of the lamellae.

7. It is concluded from these results and other published work, that lamellae width and colony size, which are determined by hot working and heat treatment, are likely to have a large influence on fatigue crack growth behavior.

8. Fatigue crack closure had about the same characteristics as for other alloys. The model used to estimate ΔK_{th} from microstructure was found to be inadequate because of the influences of other microstructural features such as colony boundaries and lamella direction.

ACKNOWLEDGEMENT

This work was funded by AFOSR, Contract F49620-89-C-0032, Dr. Alan Rosenstein technical monitor. Thanks to J.-W. Kim for furnishing the materials, John Campbell and Jim Spencer for experimental assistance, Dr. Pan and H. Saldano for the TEM work, and to Dr. Chan for useful discussions and help in interpreting the microstructures.

REFERENCES

1. D.L. Davidson and J. Lankford, Metall. Trans. A, v.15A, pp. 1931-1940 (1984).
2. D.L. Davidson, D. Eylon, and F.H. Froes in Microstructure, Fracture Toughness and Fatigue Crack Growth Rates in Titanium Alloys, edited by A.K. Chakrabarti and J.C. Chesnutt (TMS, Warrendale, PA, 1987) pp. 19-37.
3. D.L. Davidson, J.B. Campbell, and R.A. Page, Metall. Trans. A, v. 22A, pp. 377-391 (1991).
4. D.L. Davidson in Microstructure/Property Relationships in Titanium Aluminides and Alloys, edited by Y-W. Kim and R.R. Boyer (TMS, Warrendale, PA, 1991) pp. 447-461.
5. Y-W. Kim and D.M. Dimiduk, J. of Metals, v.43, pp. 40-47 (1991).
6. K.S. Chan and Y-W. Kim, ibid 4, pp. 179-196.
7. J.B. Campbell, Metallography, v. 18, pp. 413-420 (1985).

8. J.A. Graves, L.A. Bendersky, F.S. Biancaniello, J. H. Perepezko, and W.J. Boettinger, Mater. Sci. and Engineering, v. 98, pp. 265-268 (1988).
9. D.S. Shong, A.G. Jackson, and Y-W. Kim in Titanium Materials, Surfaces and Interfaces, TMS, Warrendale, PA, 1991 (in press).
10. Y. Umakoshi, T. Nakano, and T. Yamane, Scripta Met. et Met., v. 25, pp. 1525-1528 (1991).
11. Y.S. Yang and S.K. Wu, Scripta Met. et Met., v. 24, pp. 1801-1806 (1990).
12. L. Zhao and K. Tangri, Phil. Mag., v. 64, pp. 361-386 (1991).
13. W.O. Soboyejo, J.E. Deffeyes and P.B. Aswath Mat. Sci. and Eng., v. A138, 95-101 (1991).
14. Andrew Nagy, John B. Campbell and D.L. Davidson, Rev. of Sci. Instruments, v. 55, 778-782 (1984).
15. D.L. Davidson and J. Lankford, Int. Mater. Revs., v. 37, pp. 45-76 (1992).
16. D.L. Davidson, Eng. Fracture Mech., v. 38, pp. 393-402 (1991).
17. S.J. Hudak, Jr. and D.L. Davidson in Mechanics of Fatigue Crack Closure, edited by J.C. Newman and W. Elber, ASTM STP-982 (ASTM, Philidelphia, PA, 1988) pp. 121-138.
18. P.S. Pao, A. Pattmaik, S.J. Gill, D.J. Michel, C.R. Feng and C.R. Crowe Scripta Met. et Met., v. 24, pp. 1895-1900 (1990).
19. D.L. Davidson, Acta Met., v. 36, pp. 2275-2282 (1988).
20. D.L. Davidson, Eng. Fracture Mech., v. 33, pp. 965-977 (1989).

INTERNAL STRESSES IN TiAl BASED LAMELLAR COMPOSITES

P. M. HAZZLEDINE, B. K. KAD*, H. L. FRASER* and D. M. DIMIDUK**
UES Inc. 4401 Dayton-Xenia Road, Dayton, OH 45432, USA.
*Ohio State University, 2041 College Road, Columbus, OH 43210,
**Wright Laboratory, Materials Directorate, Wright-Patterson AFB, OH 45433.

ABSTRACT

In the in situ lamellar γ-TiAl based composites, very large elastic stresses ($\approx$1GPa) would be generated at coherent interfaces between close packed planes for three reasons: the tetragonality of TiAl, the larger atomic spacing in α_2-Ti_3Al than in γ and the differences between thermal expansion coefficients in α_2 and γ. These stresses appear to be partially relaxed by the creation of van der Merwe dislocations, diffusion across α_2/γ interfaces and cracking along interfaces. Measurements of lattice parameters by Convergent Beam Electron Diffraction (CBED) reveal stresses of the order of 100MPa. The presence of these stresses and the very specific form of the resulting stress tensor are used to discuss the hard/soft mode deformation of TiAl composites.

INTRODUCTION

The mechanical properties of TiAl-based alloys are sensitive to composition and microstructure. A promising two phase microstructure with good strength and ductility consists of laminates of flat slabs of γ/γ and/or α_2/γ depending upon the exact composition [1]. This microstructure is derived from solid state transformations $\alpha \rightarrow \gamma$ and $\alpha_2 \rightarrow \gamma$, with the orientation relationship of $(0001)\alpha,\alpha_2//(111)\gamma$, $<11\bar{2}0>\alpha,\alpha_2//<1\bar{1}0]\gamma$. Tetragonal γ lacks a triad axis along (111) whereas α_2 does have a hexad along (0001). Hence a total of six orientation variants exist for the α_2/γ and γ/γ adjacent slabs [2]. The resulting α_2/γ and γ/γ interfaces can all be described in terms of rotations of one grain relative to its neighbour by multiples of 60° about (111) or (0001). Lattice parameter mismatches give rise to stresses at α_2/γ and γ/γ interfaces as described below.

MISMATCH STRESSES

α_2/γ interface

On account of the six fold symmetry of the DO_{19} (α_2) phase, the atomic spacings in the $<11\bar{2}0>$ are all equivalent , but in the $L1_0$ (γ) phase they are $a/\sqrt{2}$ and $(1/2)\sqrt{(a^2+c^2)}$ in the <110] and <011] directions respectively. In order to maintain the atomic continuity across an interface, as observed in HREM micrographs [3], both plates must be strained. Since the atomic spacings in α_2 are larger than in γ, there is a biaxial tension in γ and a biaxial compression in α_2. In addition, the tetragonality of γ induces shear strains of opposite signs in the two phases. The mismatch strains e_i (which are shared unequally between the two phases), principal directions P_i, crystallography and Mohr circle are shown in Fig 1(a).

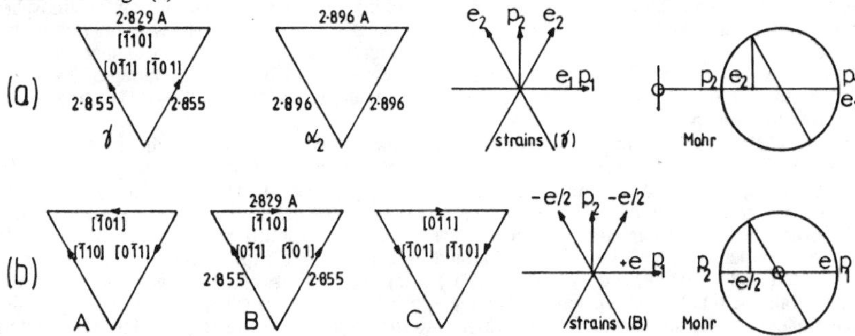

Figure 1. Elastic strains e resulting from lattice mismatches on close packed planes. (a) α_2/γ interface. (b) γ/γ interface. Atomic spacings are mean 300K values from [4,5].

The principal direction P_1 is along [$\bar{1}$10] and in this direction the close packed rows in each phase are parallel, but the <$\bar{1}$01] directions would be rotated by $(e_1-e_2)/\sqrt{3}$ relative to the close packed rows in α_2 in a fully relaxed lamella.

γ/γ interface

When a lamella B is sandwiched between two lamellae A and C, which are rotated by ±120° from B, the mismatches caused by tetragonality are pure shear strains, Fig 1(b). If the lamellae are all equally thick they would be expected to shear by equal amounts. ([$\bar{1}$10] is a principal direction in each lamella but this direction rotates by 120° across an interface). The strains in any lamella, when refered to standard <100> axes form a tensor:

$$e/3 \begin{pmatrix} 1 & \bar{2} & 1 \\ \bar{2} & 1 & 1 \\ 1 & 1 & \bar{2} \end{pmatrix} \qquad \text{where } e = (c/a-1)/3$$

Magnitude of internal stresses, coherent interface case

Assuming, initially, that all the mismatches are taken up elastically in the lamellae: In a γ/γ lamella, the principal strains in each lamella are ±e and consequently the principal stresses are $\pm Ee/(1+\upsilon)$, where E and υ are Young's modulus and Poisson's ratio in TiAl. This is a shear stress with magnitude ≈ E/200 ≈ 0.9GPa.at 300K Clearly such a large stress will be relaxed, but this estimate represents the largest possible internal stress caused by the mismatches at the γ/γ interface. In the α_2/γ lamellar interfaces, the shear stresses generated are of the same order of magnitude as above, but superimposed on this stress there is a biaxial tension of $E(e_1+2e_2)/6(1-\upsilon) \approx E/100 \approx 1.8$ GPa.

In making these estimates, two effects which in principle could significantly alter the magnitude of estimated internal stresses, have been neglected: a) Short range diffusion of Ti from α_2 to γ and a reverse flow of Al to α_2 would reduce the shear stresses (by decreasing the c/a ratio of γ) and also reduce the biaxial stresses (by decreasing a in α_2 and increasing a in γ).
b) The fact that the three thermal expansion coefficients i) of a in α_2, ii) of a in γ and iii) of c in γ are all different [6], ensures that the internal stresses depend strongly on temperature. For example, in the absence of any relaxation processes, the shear stresses in γ/γ and γ/α_2 interfaces roughly double between 300K and 1300K. Such uncertainities imply that the estimates of internal stresses have large margins of error, but they remain high, of the order of 1 GPa. This estimate is compared with experimental measurements below and is discussed in relation to the observed relaxation processes later. These processes destroy the coherency of the interfaces.

EXPERIMENTAL PROCEDURE

Master ingots of Ti-50at%Al alloy were drop cast into a water cooled copper mold 25mm in diameter and 150mm long. The cast ingots were levitation zone melted in a static helium atmosphere, yielding large grains of lamellar microstructure. Basically this is a crucibleless induction melting technique where a highly turbulent liquid freezes directionally. TEM disks were spark machined from the as-processed lamellar material. Thin foils were electropolished in a solution of 7 volume% sulphuric acid in methanol at ≈240K. Thin foils were examined using a Phillips CM12 transmission electron microscope operating at 100kV.

The CBED pattern with pole [334] was selected for lattice parameter determinations based on its high sensitivity to small variations in lattice parameters a, b, and c simultaneously and its simple array of high order Laue zone (Holz) lines. CBED provides a method for measuring lattice parameters to an accuracy of ≈0.1% from small volumes of material [7]. The method is used to determine the lattice parameters of adjacent slabs of material, their differences giving the strains. The lattice parameters in each lamella are obtained by matching the experimental CBED pattern to a simulated pattern (generated by 'DIFFRACT') using trial and error lattice parameters. For a full determination, 6 lattice parameters, unit cell lengths a,b,c and angles α,β,γ are required to give the 6 components of the strain tensor. The stresses are then found by using Hooke's law. The material used in the experiment contains only γ lamellae for which the unstrained values of α,β,γ are 90° and a=b.

CONVERGENT BEAM ELECTRON DIFFRACTION RESULTS

Figure 2 is a micrograph of γ TiAl showing six lamellae and five interfaces. The <334> CBED patterns corresponding to each lamella contain Holz lines which are displaced from lamella to lamella. These displacements provide visual proof of different strains in adjacent lamellae. The experimental lattice parameters are derived from simulations which can always be matched to within the line width of the experimental pattern. This is possible even by varying only three parameters (a,b,c keeping $\alpha=\beta=\gamma=90^o$). Hence unique values of the strains have not yet been obtained. However the orders of magnitude of the normal strains can be calculated by comparing the measured values of a, b and c (these have been normalised so that the volume abc is constant, in order to eliminate experimental errors such as voltage drift in the microscope) with the standard values given in figure 1. The results are shown in the table; stresses are related to strains by $\sigma=E\varepsilon/(1+\upsilon)$ with E=180GPa and υ=1/3. The accuracy of the stresses, corresponding to the line width of the Holz lines, is ±50MPa.

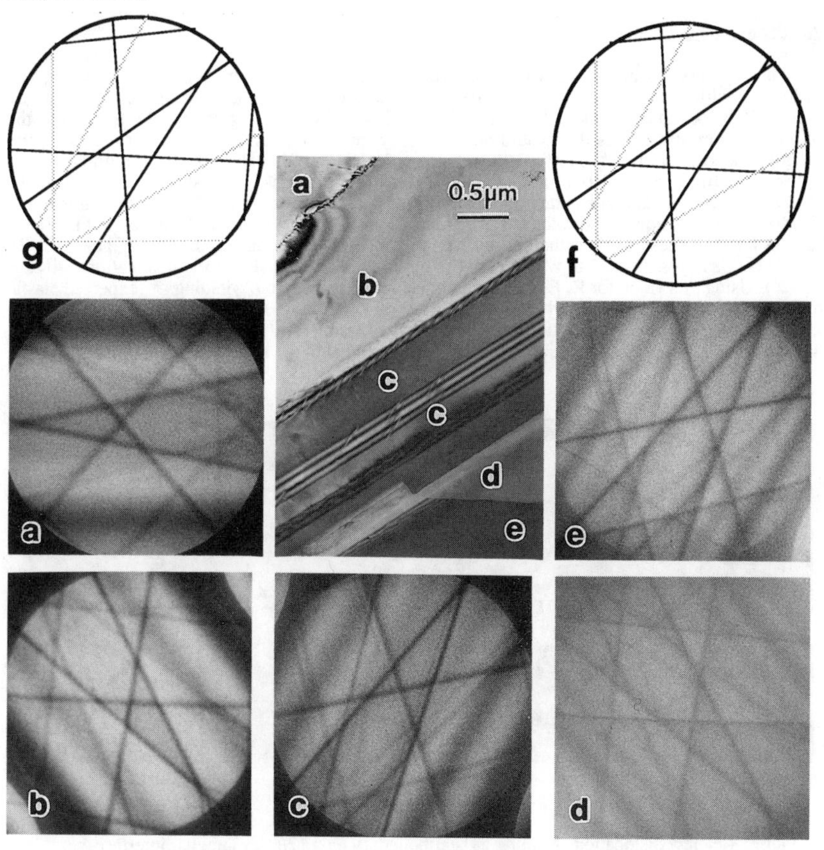

Figure 2. Lamellar microstructure showing five distinct lamellae (a-e) with corresponding <334] CBED patterns. All the interfaces are of the 120° type. Strains make the Holz line patterns differ slightly e.g. in the size of corresponding triangles. (f,g) simulated patterns from (f) unstrained TiAl (a=b=4.00A, c=4.06A) and (g) strained TiAl (a=4.02A, b=4.00A, c=4.06A)

Table: CBED measurements of strains and stresses in lamellae.

Lamella	a/Å	b/Å	c/Å	$10^4\varepsilon_1$	$10^4\varepsilon_2$ strains	$10^4\varepsilon_3$	σ_1/MPa	σ_2/MPa stresses	σ_3/MPa
a	3.999	4.004	4.073	-3.5	7.5	-3.5	47	101	-47
b	4.009	4.004	4.063	20.0	7.0	-27.0	270	95	-364
c	3.999	3.989	4.088	-2.5	-31.0	34.0	34	-419	459
d	3.995	4.000	4.080	-14.0	-2.5	16.0	-189	-34	216
e	3.995	4.005	4.075	-15.0	11.0	3.5	-203	149	47

RELAXATION OF THE INTERNAL STRESSES

Van der Merwe interfacial dislocations.

When two crystals are joined with mismatched lattice parameters semi coherent boundaries may be formed with some of the mismatch taken up by elastic strains, and with the rest taken up by mismatch dislocations. Frank and van der Merwe [8] showed how to calculate the fraction of the homogenous elastic strains that would be relaxed by interfacial dislocations to form a minimum energy structure. For the γ/γ interface, the total mismatch strain f is a shear, so the minimum energy dislocation structure consists of a single set of screw dislocations [9] as shown in Fig. 3. If the spacing of the dislocations is s then the dislocations account for a strain of b/s and the remaining $(f-b/s)=2\varepsilon$ is shared equally between the two lamellae, each having an elastic shear strain of ε. If each lamella has thickness h, then the energy per unit area of the boundary is $E = 2Gh\varepsilon^2 + Gb^2\ln(R/b)/4\pi s$. Usually $s<<h$, in which case the outer cutoff radius (R) of the dislocations is $\approx s \approx b/(f-2\varepsilon)$. Using this value for R, E may be minimised with respect to ε, yielding ε^* (expected elastic strain) $= -b\ln((f-2\varepsilon^*)/e)/8\pi h$, where e is the base of logarithms. In most cases $f>>2\varepsilon^*$ so, to an adequate approximation, the elastic strain in each lamella is $\varepsilon^* = (b/8\pi h)\ln(e/f)$. The important point is that ε^*, the residual elastic strain, is inversely proportional to the lamellar thickness.

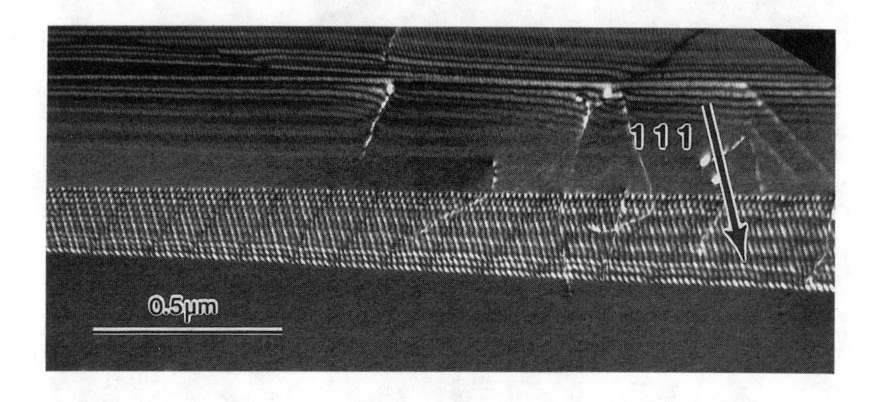

Figure 3. Screw dislocations forming a shear boundary on a 120° γ/γ lamellar interface

In TiAl, f is of the order of 0.01 in which case ε^* becomes a significant fraction of this, more than 0.1f, in lamellae of thickness h<250b. In lamellae much thicker than this the interfacial dislocations are expected to relax the mismatch strains even more effectively. In figure 3, the dislocations relax nearly all the interfacial stress.

Crystal Yield

In thin lamellae the elastic stresses may be sufficient to cause yield of the material. The forms of the stress tensors are such that no dislocations gliding on planes parallel to either γ/γ or α_2/γ interfaces are activated. At yield, therefore, 'hard mode' dislocations travel across the lamellae and are immobilized in the boundaries. There they act as further misfit dislocations which relax the elastic stress. Yield of the lamellae reduces the stress to the yield stress but this itself may be high, especially in thin lamellae, because of Hall-Petch hardening.

Cracking of γ/γ interfaces

When lamellar specimens are subject to tensile stresses they frequently fail by cracks which open up at γ/γ interfaces. Fig. 4 shows the tensile side of a bend specimen which has failed by multiple crack propagation. {111} fracture is common [10] and can be understood either because {111} is a weak cleavage plane or else on the grounds that by propagating down the γ/γ interface, the crack relieves all the mismatch energy of the boundary (dislocations and elastic) and consequently it propagates with substantial strain energy release rate.

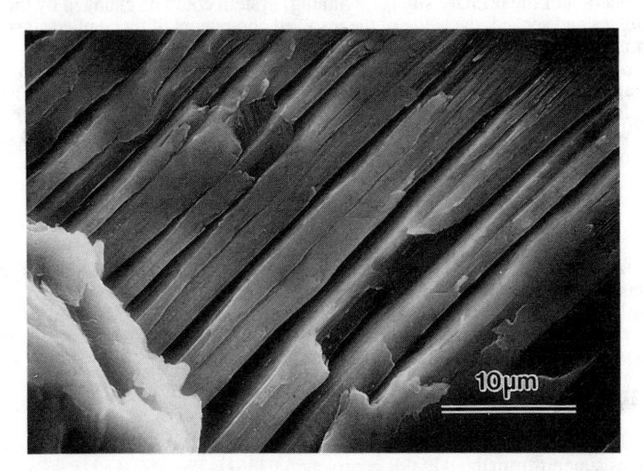

Figure 4. Cracked tensile face of a bend specimen of lamellar TiAl showing cracks separating {111} lamellar planes.

DISCUSSION

There is an apparent disparity between the HREM (foils ≈50Å thick) [3] and conventional TEM (foils ≈2000Å thick) experimental observations. In the HREM foils misfit dislocations are not observed at the interfaces whereas in conventional TEM foils they are. On the face of it, the HREM observations imply the presence of very high elastic stresses whereas conventional TEM observations (in γ/γ interfaces see Fig. 3 and in α_2/γ interfaces see [11]) imply that the stresses are largely relaxed. The CBED measurements, which reveal internal stresses in the order of one tenth the maximum possible, agree with the conventional microscopy.

For the case of the γ/γ interfaces, one possible explanation for the disparity is that the non dilatational misfits are relieved by screw dislocations (see Fig. 3) which are not easily detected in the HREM. However, for the case of the α_2/γ interface the dilatational misfit has to be relieved by edge dislocations which should be observable in the HREM. The fact that they are not observed might be because atomic scale migration of Ti and Al across the α_2/γ interface reduces the mismatch at the interface. This suggestion is supported by compositional measurements [12] reported on α_2/γ lamellar material (the α_2 composition is Ti-40at%Al instead of Ti-25at%Al and hence the atomic spacing of α_2 is closer to that of γ than it would be without diffusion) . The reduction in the lattice parameter mismatch consequently increases the separation of mismatch dislocations so that they are less likely to be observed in the small HREM specimens.

The remaining internal stresses are sufficient, especially in very thin lamellae to influence the mechanical properties of TiAl. Three effects may be considered:

i) The lamellar interfaces (γ/γ and γ/α_2) have been observed to crack in tensile and bend tested specimens. The elastic energy stored in the material and the energy of any interfacial dislocations are both available for conversion to surface energy when a crack propagates in this way. Thus crack propagation along lamellar interfaces is favored over other crack paths.

ii) The internal stresses are expected to have an effect on the general yielding behavior of the lamellar composites. Since the internal stresses are tied into the crystallography of the material, their contribution depends on the orientation of the lamellar plane relative to the compression direction. Yamaguchi and Umakoshi [13] report the orientation dependence of the yield stress and explain it by means of two different critical resolved shear stresses, one for soft modes and one for hard modes (the soft and hard modes are those with shear deformations parallel and across the lamellar boundaries respectively). The presence of internal stresses affects the argument in two distinct ways. First, the selection of the primary slip (or twinning) system could be changed by the presence of the internal stresses and second, the value of the resolved component of the applied stress required to operate the primary system would also be changed.

iii) The internal stresses act on all the hard mode dislocations and have similar resolved Schmid factors for the hard mode 1/2<110] and 1/2<112] twinning Shockley dislocations. The internal stresses exert no forces on soft mode dislocations . This may change the plastic anisotropy of the lamellar material which is already marked because of the Hall Petch effect acting on hard slip modes and because hard mode sources are necessarily short and therefore difficult to operate.

SUMMARY

In lamellar TiAl-based composites the lattice mismatches on the lamellar planes would, if unrelaxed, generate large biaxial tensions in γ at α_2/γ interfaces and large shear stresses at all interfaces. These stresses are mostly relaxed by diffusion across α/γ interfaces, by the formation of mismatch dislocations, by yield and by cracking. However,CBED measurements reveal residual stresses which are sufficient to modify the yield behavior and the plastic anisotropy significantly.

ACKNOWLEDGEMENTS

The authors would like to thank Dr. Ben F. Oliver for providing the material used in this study. PMH acknowledges support from the AFOSR contract no.F33615-39-C-5604.

REFERENCES

1. Y-W. Kim and D.M. Dimiduk, J. Metals, 43, 40, (1990).
2. Y.S. Yang, and S.K. Wu, Philos. Mag. A65 , 15, (1992).
3. H. Inui, H. Nakamura, M. H Oh and M. Yamaguchi, Ultramicroscopy, 39, 268, (1991).
4. J.L. Murray, Phase Diagrams of Binary Titanium Alloys, ed. J. L. Murray, (Cleveland, OH ASM monograph) p.12, (1986).
5. S.C. Huang and E.L. Hall, Acta Metall. et. Mater.,39, 1053,(1991).
6. R.D. Shull and J.P. Cline, High Temp. Sci., 28, (1989), private communication.
7. J. W. Steeds, Introduction to Analytical Electron Microscopy. eds. J.J. Hren, J.I. Goldstein and D.C. Joy. (New York: Plenum Press) p387, (1979).
8. F.C. Frank and J. H. van derMerwe, Proc. Roy. Soc. A, 198, 216 (1949).
9. J.W. Matthews, Philos. Mag. 29, 797, (1974).
10. B.K. Kad, B.F. Oliver and P.M. Hazzledine, Microstructural Science, 18, 431, (1990).
11. L. Zhao and K. Tangri: Philos. Mag. A64, 361 (1991).
12. Y. Umakoshi, T. Nakano and T. Yamane, Mat. Sci. & Eng., in press, (1992).
13. M. Yamaguchi and Y. Umakoshi, Prog. Mater. Sci., 34, 1, (1990).

INTERMETALLIC/METALLIC POLYPHASE IN-SITU COMPOSITES

D.R. JOHNSON, S.M. JOSLIN, B.F. OLIVER
The University of Tennessee, Knoxville, TN 37996-2200
Materials Science and Engineering Dept.

R.D. NOEBE AND J.D. WHITTENBERGER
NASA-Lewis Research Center, Cleveland, OH 44135

ABSTRACT

To evaluate various in-*situ* reinforcement schemes, a computer controlled containerless directional solidification system has been used to produce NiAl-based polyphase composites containing up to two intermetallic phases and at least one ductile phase. Systems evaluated include Ni-Al-Cr, Ni-Al-Mo, Ni-Al-V ternary systems that form NiAl/α-refractory metal eutectics and a three phase eutectic in the Ni-Al-Cr-Nb system. Initial screening of these in-*situ* composites has included morphological characterization, four point bend testing, temperature dependent yield strength evaluation and compressive creep testing. Occasional growth defects termed "banding" currently interrupt the continuity of these composite structures and limit the attainment of optimum properties. However, both the creep strength and toughness of NiAl were improved by in-*situ* reinforcement.

INTRODUCTION

While in-*situ* composites have been studied for some time [1,2], interest in them has been expressed for possible high temperature applications. The same is true of the B2 compound NiAl. However, limited low temperature toughness and poor elevated temperature strength render NiAl-based alloys inadequate as structural materials. These deficient properties may be simultaneously improved by reinforcing NiAl with "appropriate" second phases to form a composite material. This paper is part of an ongoing program to evaluate NiAl-based eutectics. These in-*situ* composites have been generated in a computer-automated, containerless levitation zone processor utilizing ultra-pure atmospheres. Thus, the possibility of contamination of the ingot from crucible materials is eliminated.

This preliminary report includes data for NiAl based eutectics whose compositions are given in Table 1. Hyper-eutectic alloys NiAl-12Mo and NiAl-15Mo were also investigated with the intent to increase the volume fraction of metallic reinforcement. A number of these alloys were deliberately doped with Zr to improve the high temperature strength of the NiAl matrix. Ternary additions of Zr to binary NiAl have been shown to greatly improve the strength of the resulting alloy [3] as well as to getter impurities from the alloy. The existence and composition of a three phase eutectic was determined from a series of arc-melted buttons. The matrix phase of this eutectic, (Cr,Al)NbNi, was found to have a Laves or C14 crystal structure by X-ray analysis. Mechanical properties are reported for all of these systems and compared to more conventionally processed NiAl alloys.

Table 1: NiAl-Based Eutectics Investigated

Alloy Composition (atomic %)	Morphology	Melting Point (K)	(ref)
NiAl-9Mo-0.1Zr	Mo rods	1778	(1)
NiAl-12Mo-0.1Zr	Mo rods	---	
NiAl-15Mo-0.1Zr	Mo rods	---	
NiAl-33Cr-0.1Zr	Cr rods	1718	(1)
NiAl-40V	V lamellae	1633	(2)
NiAl-33.3Cr-11.1Nb	(Cr,Al)NbNi matrix NiAl lamellae containing Cr rods	1633	

EXPERIMENTAL

The containerless levitation zone processor (CLZP) is described in detail elsewhere [4]. However, the system has been upgraded to record the following variables as a function of run time: the growth rate, liquid zone diameter, interface position, relative temperature of the interface region, and the resulting ingot diameter. Thus, the entire processing run can be documented and easily correlated with the ingot's morphological characteristics. The zone diameter at the solidifying interface and the height of the zone are controlled through digitized-image processing, independent of any temperature measurement. The liquid zone in this technique is intentionally mixed by the rotation of the freezing interface. The upper section of the ingot is movable relative to the lower section, but does not rotate. The rotation of the solidifying interface is intended to increase the thermal gradient in the liquid promoting stability of a planar interface and coupled growth.

Four-point bend specimens approximately 6.5mm x 4.5mm x 40mm were electrical discharge machined and notched 2.0mm x 0.35mm using a slow speed diamond impregnated saw. A fatigue crack was not initiated at the notch tip prior to testing. The length of each bend specimen was oriented parallel to the growth direction of the directionally solidified ingot with the notch perpendicular to the growth direction. A servo-hydraulic test frame under cross-head displacement control was used for testing. The displacement rate was $7.6x10^{-4}$ mm/s. The outer/inner spans were 30mm and 15 mm respectively. The load-displacement data were corrected for machine stiffness and stress versus strain curves for the outer fiber of the bend specimens were calculated from linear elastic beam equations [5]. Fracture toughness values were calculated using the K calibration for pure bending [6].

Cylindrical compression samples, 5mm diameter by 10mm long were EDM'ed so that the length of each specimen was parallel to the growth direction of the directionally solidified ingot. Compressive yield as a function of temperature and 1300 K compressive creep data were generated in a screw driven load frame. Tests were conducted in air under constant velocity conditions and the load displacement data were converted into true stress-true strain rate results [3,7].

RESULTS AND DISCUSSION

The longitudinal microstructures of directionally solidified ingots are shown in Figure 1. Composite growth is interrupted at random intervals, preventing the formation of continuous second phase reinforcement. This problem is under investigation and does not correlate with changes in rotational velocity or the computer control of the zone diameter during processing. Banding is also present in ingots processed with and without computer control. The nature of the banding is different than that previously reported by Quenisset et al. [8] which appeared to be periodic in nature and was eliminated by rotation at 200 rpm. Rotational velocities from zero to 220 rpm have not eliminated the banding problem in this study.

While the banding problem is not evident in NiAl-based eutectics processed using the more quiescent Bridgeman technique [1,2], the reactive nature of the liquid can cause contamination of the ingot by the crucible material. The purity of the processed ingots is a significant advantage of the CLZP processing technique. For example, a 25mm nominal diameter single crystal ingot of NiAl was analyzed for C, O, N, and S; all were individually less than 10ppm. Further investigation of this high purity ingot is in progress. Similar purity levels are expected in the eutectic alloys.

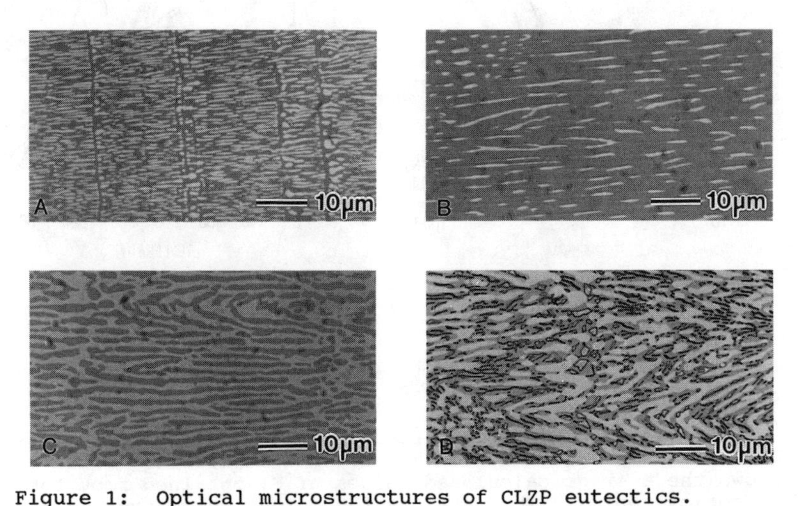

Figure 1: Optical microstructures of CLZP eutectics.
A) NiAl-34Cr (rod) showing typical banding
B) NiAl-9Mo (rod) typical 4-pt bend specimen
C) NiAl-40V (lamellar) typical cellular structure
D) 3-phase Light=(Cr,Al)NbNi, Dark=NiAl, Cr rods

Figure 2 compares the 1300 K true compressive creep behavior of the NiAl based eutectics evaluated to binary NiAl. Data from a directionally solidified NiAl-NiAlNb eutectic which has previously shown promising high temperature properties [7] are included for comparison. The strengths of all the NiAl based eutectics investigated are substantially better than binary NiAl but are less than that of the NiAl/NiAlNb alloy. However, the eutectics are expected to display even greater strengths when the banding problem is eliminated and the second phase is continuous.

Figure 3 illustrates the compressive yield strength data from 300 to 1300 K for the NiAl/refractory metal eutectics and [001] single crystal NiAl at a strain rate of $1.4x10^{-4}$ in air. The filled symbols refer to data taken from a single specimen tested at several temperatures (m-T). Note that there is good correlation between these multi-T tests and individual tests run at each temperature. All the eutectics show improved strength at elevated temperatures (>650 K) over stoichiometric NiAl. The anomalous peak in the NiAl-40V system tested in air is not understood at this time. However, it does occur near the melting point of V_2O_5 (960 K). Near 1000 K the strength of the NiAl-40V alloy drops off precipitously possibly due to the rapid loss of vanadium. The sharp decrease in yield strength between 600 and 800 K for binary NiAl single crystals, associated with the ductile/brittle transition, is more gradual for the eutectics.

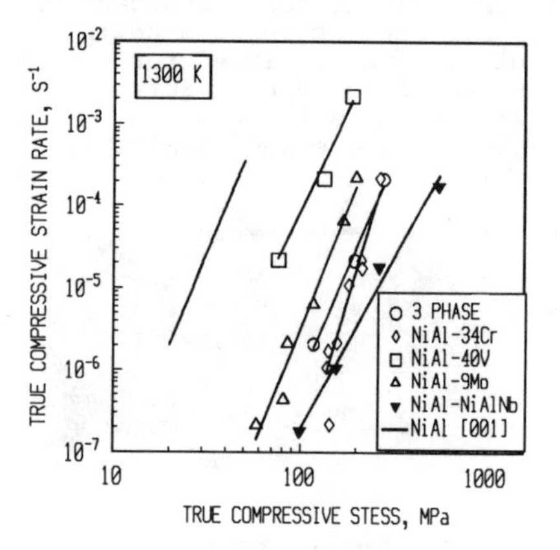

Figure 2: Compressive flow stress-compressive strain rate behavior for several NiAl-based eutectics at 1300 K.

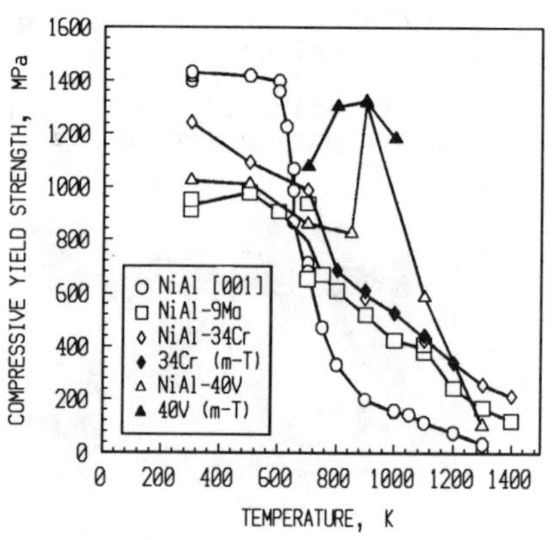

Figure 3: Compressive yield strength as a function of temperature for [001] oriented NiAl single crystal and CLZP eutectics.

Figure 4 shows typical results of the 4-point bend tests. Data for the NiAl-Cr system are not yet available. Table 2 shows the average calculated values of K_1 obtained from the eutectic samples. The data for the Mo eutectic agree well with that of Subramian et al. [9]. All of the NiAl/refractory metal eutectics have approximately twice the toughness of single crystal NiAl [10] except for the NiAl-40V alloy which has 4-5 times the toughness of monolithic NiAl. While the

NiAl-V system shows the most promising room temperature strength and toughness, this alloy also exhibits the poorest creep strength and would require a protective coating for prolonged elevated temperature exposure to prevent catastrophic oxidation of the vanadium. Figure 5 shows a fracture surface of the NiAl-9Mo eutectic and is an example of toughening by a crack bridging mechanism as indicated by the necking of the Mo fibers.

TABLE 2: CALCULATED K_1 VALUES NiAl EUTECTICS

Alloy	Modulus of Rupture (MPa)	K_1 (MPa√m)
[001] NiAl (ref [10])	---	4-8
NiAl-9Mo	164 ± 5	15.2 ± 0.7
NiAl-12Mo	164 ± 9	15.5 ± 0.7
NiAl-15Mo	156 ± 17	13.2 ± 0.7
NiAl-40V	336 ± 37	30.7 ± 3.5
NiAl-Cr-(Cr,Al)NbNi	137 ± 36	11.6 ± 3.3

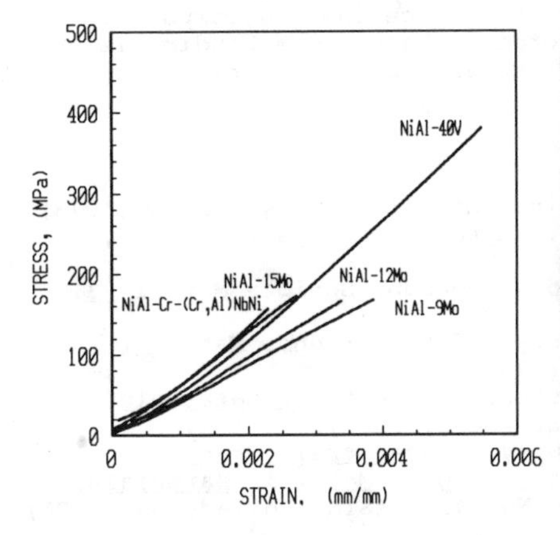

Figure 4: Summary of 4-point bend outer fiber stress-strain curves at room temperature.

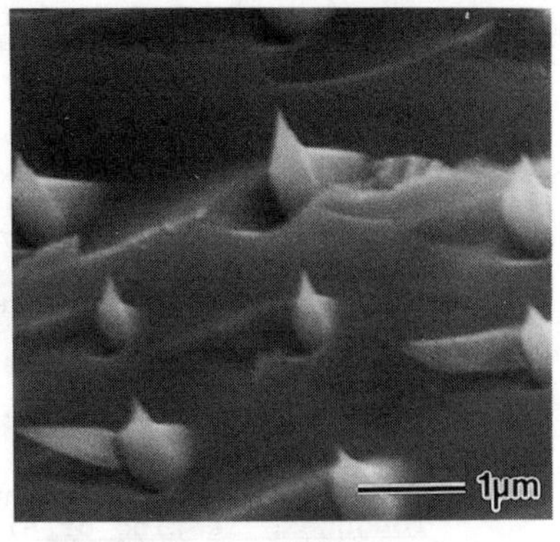

Figure 5: Fracture surface of NiAl-9Mo 4-point bend specimen showing ductile behavior of the reinforcement. SEM

While the properties reported here for the NiAl-based in-*situ* composites are promising and are dramatically better than monolithic NiAl, they can be improved. The first strategy for improving both low and elevated temperature properties of the NiAl-based eutectics investigated is to produce composite microstructures with a more continuous reinforcement phase uninterrupted by banding. More work is required to determine the composition region of absolute stability for both the eutectic and off-eutectic alloys. The same growth rate which produced uninterrupted coupled growth in the Bridgeman technique [11] results in some banding in the CLZP technique. The thermal gradients in the CLZP currently can only be estimated by pyrometry. Work is currently in progress to increase the thermal gradient in the solid and obtain better estimates of the existing gradients, especially in the higher thermal conductivity NiAl-based alloys.

SUMMARY AND CONCLUSIONS

In-*situ* reinforcement with refractory metal phases was found to increase both the low temperature toughness and creep strength of NiAl. The 3-phase eutectic exhibits creep resistance nearly equal to the higher melting NiAl-Cr eutectic. The melting temperature of the NiAl-V and 3-phase eutectics are equal, but the creep resistance of the tougher NiAl-V eutectic is much lower than the 3-phase eutectic. However, no single reinforcing scheme was optimum since the system with the best toughness also had the worst creep resistance. The compromise in creep strength for additional toughening depends upon the alloy system and will require further study to quantify.

While the CLZP technique produces extremely pure NiAl and NiAl-based eutectics, additional work is required to overcome the banding problem in order to produce in-*situ* composites with a more continuous reinforcing phase. The banding is not simply related to any single processing parameter.

REFERENCES

1. J.L. Walter and H.E. Cline, Metall Trans 4, 33-38 (1973).
2. P.W. Pellegrini and J.J. Hutta, J of Crys Growth 42, 536-539 (1977).
3. S.V. Raj, R.D. Noebe and R. Bowman, Scripta Metall 23, 2049-2054 (1989)
4. R.D. Reviere, B.F. Oliver, and D.D. Bruns, Mat & Mfg Proc,4(1), 103-131 (1989).
5. F.P. Beer and E.R. Johnston, Mechanics of Materials, (McGraw-Hill Book Co., St.Louis, 1981), P. 438.
6. W.F. Brown and J.E. Srawley, Plane Strain Crack Toughness Testing of High Strength Metallic Materials, ASTM Special Publication No. 410 (ASTM, Philadelphia, PA) pp. 13-14.
7. J. Daniel Whittenberger, R. Reviere, R.D. Noebe, B.F Oliver, Scripta Metall et Mater 26, 987-992 (1992).
8. J.M. Quessinet, J. Girol, and R. Naslain, in Conference on In Situ Composites-III, edited by . J.L. Walter, M.F. Gigliotti, B.F. Oliver, and H. Bibring (Ginn Custom Publishing, Lexington, Mass. 1979), pp. 78-85.
9. P.R. Subramanian, M.G. Mendiratta, D.B. Miracle and D.M. Dimiduk in Intermetallic Matrix Composites, edited by D.L. Anton et al. (MRS Symp. Proc. 194, Pittsburgh, PA 1990) pp. 147-352
10. K-M Chang, R. Darolia, and H. Lipsitt in High Temperature Oredered Intermetallic Alloys-IV, eds. L.A. Johnson et al. (MRS Symp. Proc. 213, Boston, Mass. 1991)597-602.
11. Harvey E. Cline and John L. Walter, Metall Trans 1, 2907-2917 (1970)

MICROSTRUCTURAL CHARACTERIZATION OF Al_2O_3/GAMMA TITANIUM ALUMINIDE COMPOSITES#

G.DAS * AND S. KRISHNAMURTHY **
* Metcut-Materials Research Group, Wright-Patterson AFB, OH. Now at Pratt & Whitney, P.O. Box 109600, West Palm Beach, Fl 33410-9600
** Metcut-Materials Research Group, Wright-Patterson AFB, OH. Now at Universal Energy System, 4401 Dayton-Xenia Road, Dayton, Ohio 45432

ABSTRACT

Alumina (Al_2O_3) fibers were incorporated into gamma titanium aluminide(TiAl) based powders by hot isostatic pressing (HIP'ing). The microstructure of as-HIP'd and heat treated composite specimens were characterized by scanning electron microscopy (SEM) and transmission electron microscopy (TEM). TEM studies reveal the presence of an amorphous reaction zone at the fiber/matrix interface. Numerous dislocations, dipoles and loops as well as twins are observed in Al_2O_3 fibers. In addition, it is determined that the fiber/matrix interface stability is significantly affected by the matrix microstructure.

INTRODUCTION

There is considerable interest in the development of fiber-reinforced gamma titanium aluminide-based composites because of their potential for high stiffness and strength at elevated temperature. Studies involving silicon carbide (SiC) fibers in gamma titanium aluminide matrices showed that SiC fibers readily react with gamma matrices (1,2). In addition, the coefficient of thermal expansion (CTE) of gamma matrices is nearly twice that of SiC fibers. This may lead to the development of residual stresses during cooling of the composite. However, based on CTE match and chemical compatibility, the Al_2O_3 fibers are considered to be a better candidate as reinforcement for gamma titanium aluminide matrices (2, 3).

This study is a continuation of earlier work (2) and aims at characterizing the nature of the reaction zone and the fiber/matrix interface stability in Al_2O_3/gamma composites as well as the microstructure of alumina fibers and gamma matrices.

MATERIALS AND PROCEDURES

Three gamma titanium aluminide alloy powders were used as starting matrices. The chemical composition of these alloy powders is shown in Table I. Two of the powders were produced by the rapid solidification rate (RSR) process while the third was made by the plasma rotating electrode process (PREP). According to chemical analysis, the two RSR powders correspond to two-phase gamma + alpha-2 microstructure while the PREP powder corresponds to single-phase gamma. Typical features of the as-received powders are shown in Figure 1. One of the RSR powders (Ti-48Al-1Ta) contains shattered particles with a high level of oxygen and carbon interstitials. The other RSR powder (Ti-48Al-3V) consists of spherical particles with somewhat high oxygen content. The PREP powder (Ti-48Al-2Nb-0.3Ta) contains coarser spherical particles with a lower interstitial content. The Al_2O_3 fiber (Saphikon) was in the form of single crystal monofilament with the 'c' axis parallel to the fiber axis. The fiber was characterized by a diameter of ~175 μm and had a smooth surface.

Work performed at Wright-Patterson AFB, OH

Table I. Chemical Analysis of Matrix Alloy Powders (at %)

Nominal Composition	Analysis						
	Al	Ta	V	Nb	C*	O*	Ti
Ti-48Al-1Ta (RSR)	44.7	0.34	-	-	1300	2300	Bal
Ti-48Al-3V (RSR)	47.2	-	1.3	-	120	1200	Bal
Ti-48Al-2Nb-O.3Ta (PREP)	50.5	0.08	-	3.9	140	500	Bal

* ppm (wt. basis)

The alloy powders and alumina fibers were mixed in steel cans, sealed under vacuum and subjected to HIP'ing at 1200°C/275 MPa/6 hr. Samples of HIP'd composites were encapsulated in vacuum and heat treated at 1000°C and 1150°C for different periods of time to determine the stability of the fiber/matrix interface. The microstructural characterization of various components of the composites was conducted by a combination of SEM and TEM. Thin foils of composites were prepared using ion-milling (4). TEM studies were conducted in a JEOL 2000FX transmission electron microscope equipped with a Tracor-Northern 5550 EDS system and operated at 200 keV.

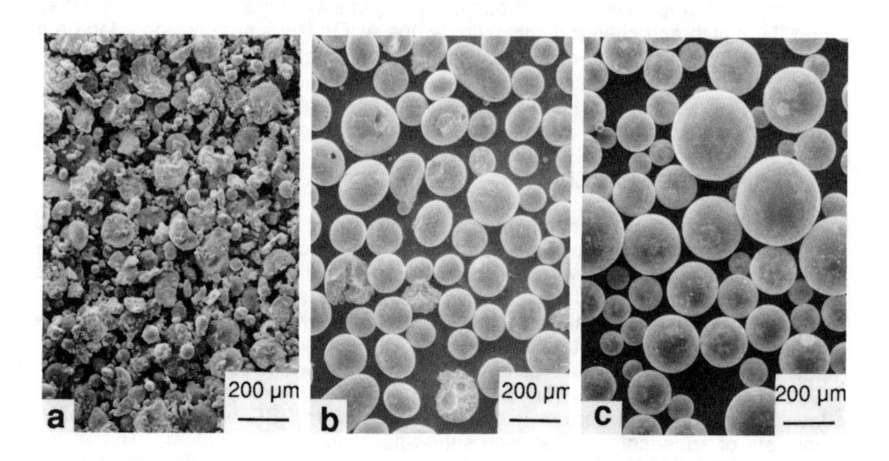

Figure 1 - SEM micrographs of gamma alloy powders: (a) Ti-48Al-1Ta (RSR), (b) Ti-48Al-3V (RSR), and (c) Ti-48Al-2Nb-0.3Ta (PREP).

RESULTS AND DISCUSSIONS

SEM CHARACTERIZATION

Figure 2 shows the fiber/matrix interface of Al_2O_3/Ti-48Al-3V and Al_2O_3/Ti-48Al-2Nb-0.3Ta composites in the as-HIP'd condition. No reaction zone formation is indicated at the fiber/matrix interface of these composites. However, there are important effects of matrix alloy chemistry on the fiber/matrix interface. The Al_2O_3/Ti-48Al-3V interface reveals cracking and void formation (indicated by arrows) while the Al_2O_3/Ti-48Al-2Nb-0.3Ta interface is free from these defects. Heat treatment of the Al_2O_3/Ti-48Al-3V composite at 1000°C/10 hr has caused waviness and serrations at the interface and an increase in void formation (shown by arrows in Figure 3a). At still higher heat treatment temperature, 1150°C/100 hr, void coalescence has led to debonding of the fiber from matrix (Figure 3b). Essentially similar results are obtained for Al_2O_3/Ti-48Al-1Ta composites. On the other hand, heat treatment of Al_2O_3/Ti-48Al-2Nb-0.3Ta composites has yielded markedly different results (Figure 4). The composite exhibits a greater thermal stability with only a few cracks forming at the fiber/matrix interface after heat treatment at 1150°C/100hr (indicated by arrows in Figure 4b).

It has been observed that the Al_2O_3 fiber reinforced two-phase gamma + alpha-2 matrix composites revealed the fiber/matrix instability in the form of cracks, serrations and voids at the fiber/matrix interface (Figure 2a and Figure 3). In contrast, when the matrix composition corresponds to a near single phase gamma microstructure, the fiber/matrix interface was stable and relatively free from these defects (Figure 2b and Figure 4). It would appear that the matrix microstructure was responsible for the observed instability of the fiber/matrix interface. However, it should be pointed out that the two-phase gamma + alpha-2 alloy powders also contained a high concentration of oxygen which is thought to stabilize the alpha-2 phase. Thus, contributions to the interface instability resulting from different matrix composition, microstructure, and interstitial content could not be isolated from each other.

TEM CHARACTERIZATION

Al_2O_3/Ti-48Al-3V Composite

Figure 5 shows the fiber/matrix interface in an Al_2O_3/Ti-48Al-3V composite heat treated at 1000°C/200 hr as revealed by TEM. A distinct reaction zone is observed which is attached to the fiber and the matrix while the middle of the reaction zone is lost once the perforation was achieved by ion-thinning at this location. The thickness of the reaction zone is estimated to be ~ 350nm. Selected area diffraction pattern (SAD) from the reaction zone shows diffused rings superimposed on the spot pattern of crystalline Al_2O_3. The diffused ring pattern is suggestive of the reaction zone being amorphous in nature. Figure 6 shows the concentration profiles of Ti and Al across the fiber/matrix reaction zone as determined from energy dispersive spectra (EDS). Unfortunately, the determination of oxygen was beyond the scope of the spectrometer used. Thus, a complete determination of the chemical composition of the reaction zone was not possible. It has been established that in composites involving alumina fibers and gamma titanium aluminide matrices, the chemical reaction occurred by diffusion of Al and atomic oxygen from alumina fibers into the matrix (5). It may therefore be speculated that the reaction zone consists of a mixture of Ti, Al, and oxygen.

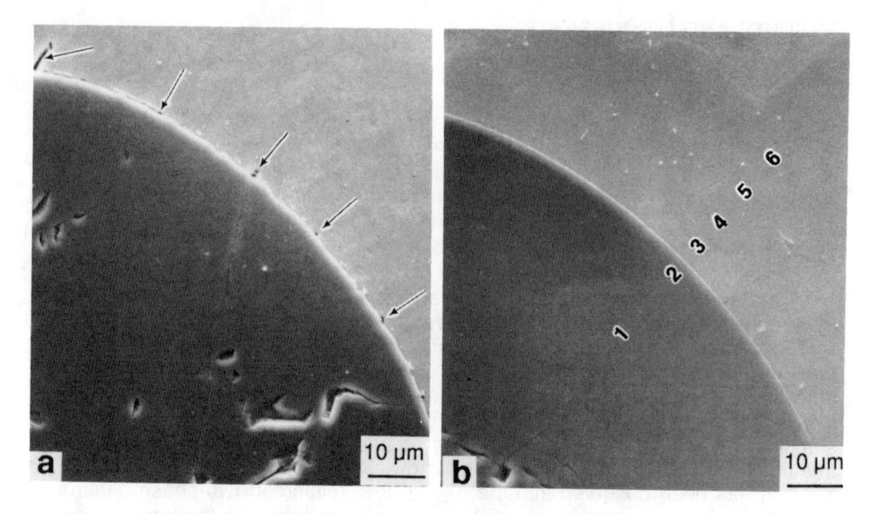

Figure 2 - SEM micrographs showing fiber/matrix interface in as-HIP'd composites: (a) Al_2O_3/Ti-48Al-3V and (b) Al_2O_3/Ti-48Al-2Nb-0.3Ta. Arrows in "a" indicate cracking and void formation at the fiber/matrix interface.

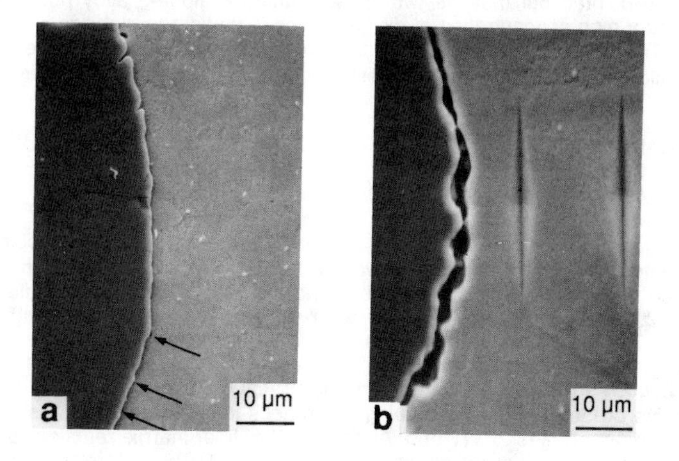

Figure 3 - SEM micrographs showing the effect of heat treatment on the fiber/matrix interface of Al_2O_3/Ti-48Al-3V composite: (a) 1000°C/10 hr, and (b) 1150°C/100 hr. The interface is marked by arrows.

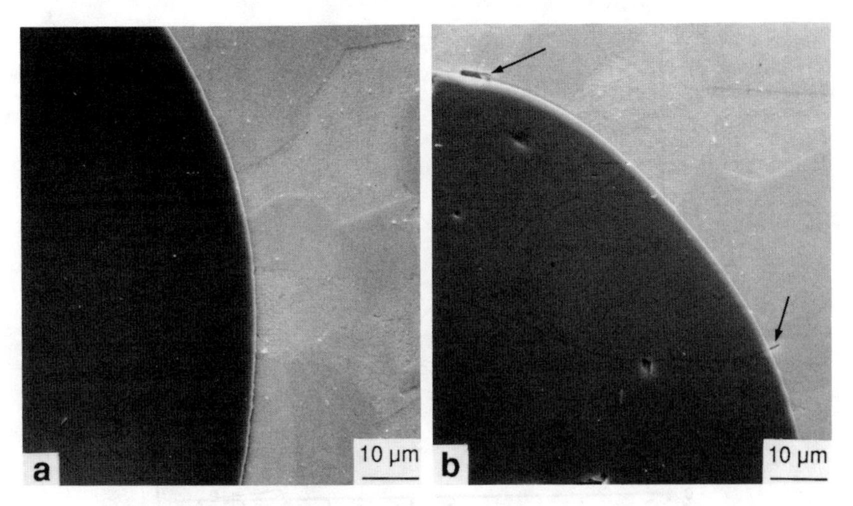

Figure 4 - SEM micrographs showing the effect of heat treatment on the fiber/matrix interface of an Al_2O_3/Ti-48Al-2Nb-0.3Ta composite: (a) 1000°C/10 hr, and (b) 1150°C/100 hr, with arrows indicating cracking near the interface.

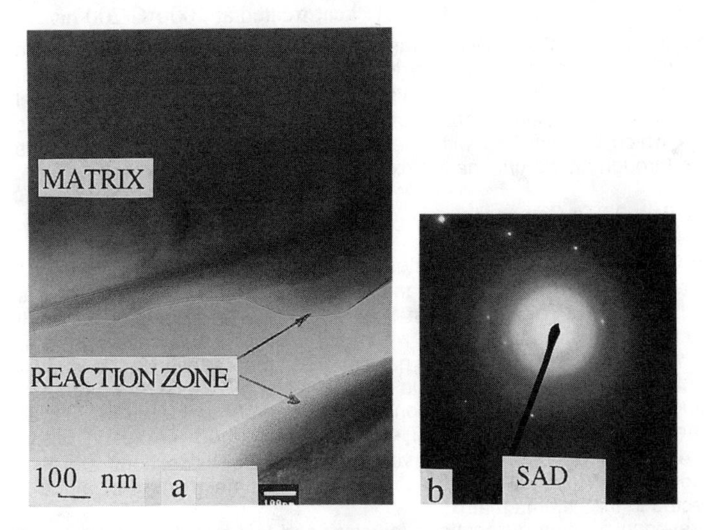

Figure 5 - (a) TEM micrograph showing the reaction zone at the fiber/matrix interface of an Al_2O_3/Ti-48Al-3V composite heat treated at 1000°C/200 hr. (b) SAD pattern from the reaction zone and part of the Al_2O_3 fiber. The diffused rings suggest that the reaction zone is amorphous in nature. The spot pattern is from the crystalline Al_2O_3 fiber.

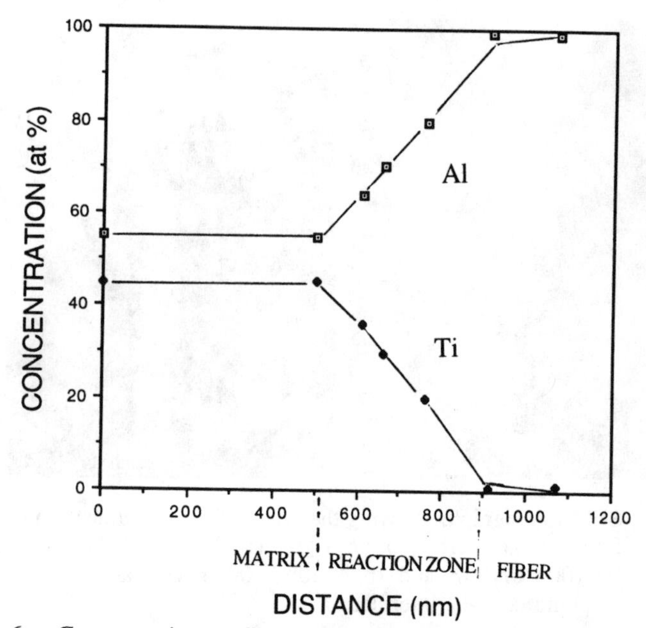

Figure 6 - Concentration profiles of Ti and Al across a fiber/matrix interface of an Al_2O_3/Ti-48Al-3V composite heat treated at 1000°C/200 hr.

A high density of glide dislocations, dipoles and loops are observed in Al_2O_3 fibers as shown in Figure 7. In addition, twins are found in the fibers. Similar observation of twins in deformed Al_2O_3 single crystals was made by Stofel and Conrad (6). A small volume fraction of alpha-2 phase is also observed in the gamma matrix which is consistent with chemical analysis (figure 8). Dislocations are also found throughout the gamma matrix.

Al_2O_3/Ti-48Al-2Nb-0.3Ta Composite

Figure 9 shows the fiber/matrix interface in an Al_2O_3/Ti-48Al-2Nb-0.3Ta composite heat treated at 1000°C/500 hr and 1150°C/100 hr. No reaction zone is observed for this composite heat treated at 1000°C/100 hr (9 a). However, a small reaction zone ~ 160 nm thick is revealed when the composite was heat treated at 1150°C/100 hr (Figure 9b). The SAD pattern from the reaction zone shows diffused rings too weak to be reproduced in the positive print. Again, the formation of diffused rings suggests that the reaction zone is amorphous in nature. This is similar to the observation in Al_2O_3/Ti-48Al-3V described earlier. Glide dislocations, dipoles and loops are observed in the fibers. Dislocations are also present in the matrix. In addition, a relatively small volume fraction of alpha-2 phase is detected in the gamma matrix.

The presence of glide dislocations, dipoles, and loops in the Al_2O_3 fibers may suggest that these fibers have undergone plastic deformation during HIP'ing at elevated temperature. Pletka et al (7) have observed that edge dislocation dipoles formed in Al_2O_3 crystals during basal glide at elevated temperatures break up into strings of loops by self-climb. It is believed that heat treatment of the composite at elevated temperatures following consolidation may have accelerated the process of breaking up of edge dislocation dipoles into dislocation loops by self-climb.

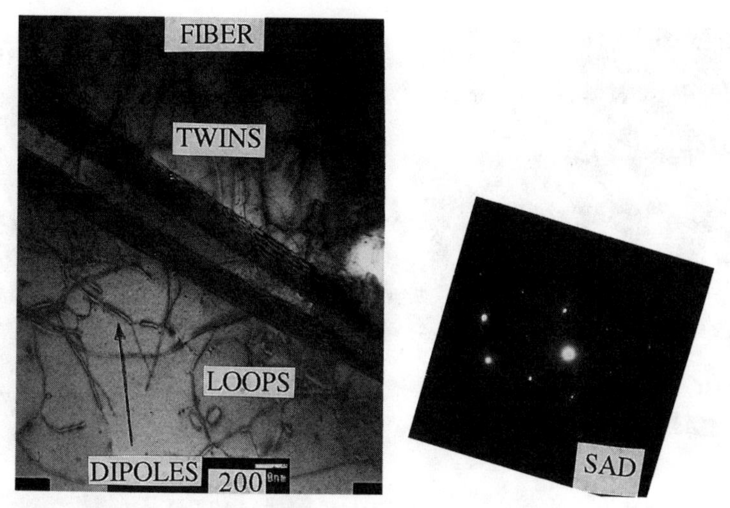

Figure 7 - Observation of glide dislocations, dipoles and loops as well as twins in Al_2O_3 fibers of an Al_2O_3/Ti-48Al-3V composite heat treated at 1000°C/200 hr.

SUMMARY AND CONCLUSION

(1) The presence of an amorphous reaction zone at the fiber/matrix interface in Al_2O_3/gamma titanium aluminide composites is revealed by TEM.

(2) Numerous glide dislocations, dipoles, and loops are observed in Al_2O_3 fibers. This may suggest that the Al_2O_3 fibers have undergone plastic deformation during HIP'ing at elevated temperature.

(3) Al_2O_3 fiber-reinforced single-phase gamma titanium aluminide composite exhibits a greater fiber/matrix interface stability compared to Al_2O_3-fiber reinforced two-phase gamma + alpha-2 titanium aluminide composites.

ACKNOWLEGEMENT

The authors would like to thank Dr. D. B. Miracle of Wright Laboratory, Wright-Patterson AFB,OH for reviewing the manuscript.

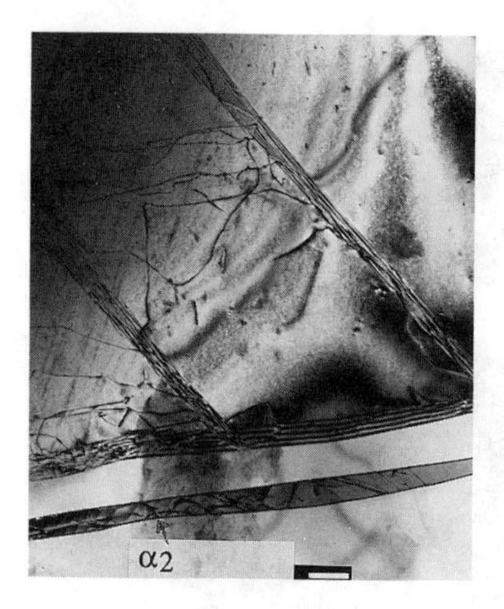

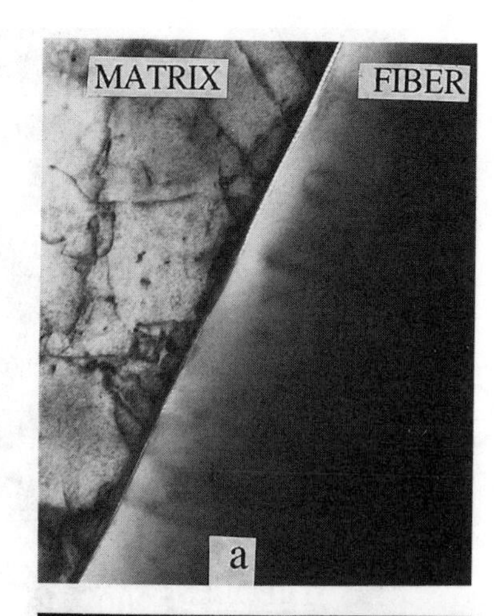

Figure 8 - Microstructure of an Al_2O_3/Ti-48Al-3V composite heat treated at 1000°C/200 hr showing the presence of alpha-2 in gamma matrix. Also seen are dislocations and loops in the matrix.

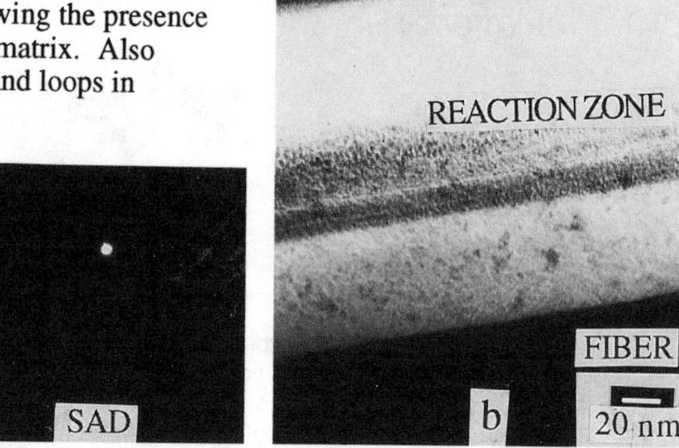

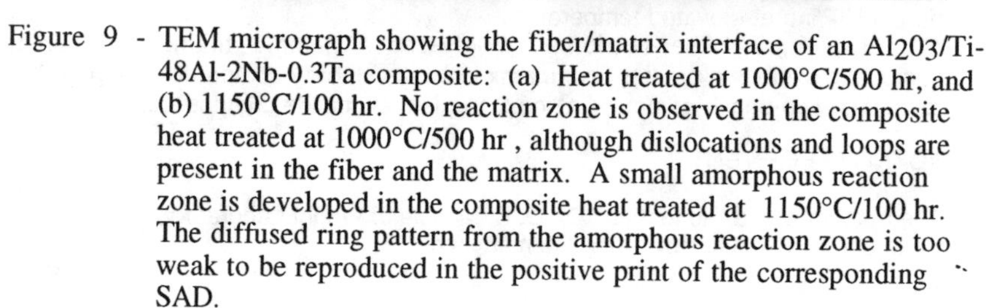

Figure 9 - TEM micrograph showing the fiber/matrix interface of an Al_2O_3/Ti-48Al-2Nb-0.3Ta composite: (a) Heat treated at 1000°C/500 hr, and (b) 1150°C/100 hr. No reaction zone is observed in the composite heat treated at 1000°C/500 hr , although dislocations and loops are present in the fiber and the matrix. A small amorphous reaction zone is developed in the composite heat treated at 1150°C/100 hr. The diffused ring pattern from the amorphous reaction zone is too weak to be reproduced in the positive print of the corresponding SAD.

REFERENCES

1) G.Das in Advances in Powder Metallurgy, (Metal Powder Industries Federation, Princeton, N.J., 1989), p. 491.
2) S. Krishnamurthy in Titanium Aluminide Composites, WL-TR-91-4020, Edited by P.R. Smith, S.J. Balsone, and T.D. Nicholas (Wright Laboratory, WPAFB, OH 45433, 1991), p.135.
3) G.G. Reynolds and P.W. Powell in Titanium Aluminide Composites, WL-TR-91-4020, Edited by P.R. Smith, S.J. Balsone, and T.D. Nicholas (Wright Laboratory, WPAFB, OH 45433, 1991), p. 217.
4) G. Das, E. Harper, R. Omler and R. Lewis, Practical Metallography, 26, 443 (1989).
5) A.K. Misra, Met. Trans. A, 22A, 715 (1991).
6) E. Stofel and Hans Conrad, Trans. AIME, 227 [5], 1053 (1963).
7) B.J. Pletka, T.E. Mitchell, and A.H. Heuer, J. Amer. Ceram. Soc. 57[9], 388 (1974).

INTERDIFFUSION BETWEEN Mo AND TiAl

Min-Xian Zhang, Ker-Chang Hsieh* and Y. Austin Chang
Department of Materials Materials Science and Engineering, University of Wisconsin, Madison, WI 53706
*On leave from Institute of Materials Science and Engineering, Sun Yat-Sen University, Kaohsiung, Taiwan, ROC

ABSTRACT

Diffusion couple experiments were carried out for Mo/γ-TiAl at 900, 1000 and 1100°C for periods of time ranging from 121 to 553 hrs. Using the Boltzmann-Matano analysis, the two diagonal interdiffusion coefficients for the two phases formed in these couples were obtained assuming the cross diagonal terms to be negligible. These two phases are δ-$(Mo,Ti)_3Al$ and β-(Mo,Al)Ti. The three intrinsic diffusion coefficients were also obtained using the Darken-type relationships between the interdiffusion and intrinsic diffusion coefficients. The calculated interdiffusion coefficients from the three intrinsic diffusion coefficients are in reasonable accord with those obtained directly from the Boltzmann-Matano analysis.

INTRODUCTION

Composite materials to be used as high-temperature structural components are becoming an important class of engineered materials. One such system which has been proposed consists of an intermetallic compound as the matrix and an oxide fiber as the reinforcement. An example is a composite of γ-TiAl and sapphire, which are in thermodynamic equilibrium with the exception of oxygen solubility in γ. A recent study on the fracture resistance of metal/ceramic/intermetallic interfaces by Evans, Bartlett, Davis, Flinn, Turner and Reimanis[1] reported that γ-TiAl/Al_2O_3 interfaces are resistant to debonding[2]. According to Evans et al.[1], brittle debonding of metal/ceramic interfaces has been observed only for Mo/Al_2O_3[3] and W/Al_2O_3[1]. On the basis of these observations, if one were to fabricate a γ-TiAl/Al_2O_3 composite, the sapphire fibers need to be coated with Mo in order to enhance interfacial debonding during mechanical deformation. But on the other hand, there is a need to know the chemical compatibility between the Mo coating and the γ-TiAl matrix.

The primary objective of the present study is to determine the extent of interactions between Mo and γ-TiAl at 900, 1000 and 1100°C.

Experimental Method

Diffusion couples of Mo/γ-TiAl were made by sputter deposition of ~300 μm Mo on well-characterized γ-TiAl. The γ-TiAl substrate was under constant rotation during the deposition process with the substrate temperature kept at about 870°C. The total sputter time was approximately 55 hrs.

The diffusion-couple samples sealed in evacuated quartz capsules were annealed at 900, 1000 and 1100°C. The samples were sealed under a vacuum of ~10^{-3} Torr. However, they were evacuated and then back-filled with high purity argon several times before sealing. In order to prevent interaction between the samples and the quartz capsules, the surfaces of the samples as well as the inside surfaces of the capsules were coated with Y_2O_3. This method was effective in preventing interaction between the samples and the capsules. After the completion of each annealing experiment the quartz capsule was removed from the furnace and quenched into water. Although the furnace temperature was controlled to within ±1°C, the estimated uncertainties of the samples temperatures are ±5°C. The quenched samples were subsequently characterized by optical microscopy, scanning electron microscopy (SEM), and electron probe microanalysis (EPMA).

The annealed samples, mounted in an epoxy resin were ground sequentially on 240, 320, 400 and 600 grit silicon carbide papers. The samples were then polished with 3 and then 1 micron alumina. The metallographic samples were examined in both the polished and etched conditions. The etchant used was Kroll's etch, ASTM 192, 1 part HF, 2 parts HNO_3 and 30 parts H_2O. The samples were swabbed for 30 to 60 seconds, rinsed in water, methanol, and then dried with air.

EXPERIMENTAL RESULTS

Two phases were found to form between Mo and γ-TiAl. They are: δ-$(Mo,Ti)_3Al$, and β-(Mo,Al)Ti. These results are consistent with the proposed 1000°C isotherm of Mo-Al-Ti[5] on the basis of the data reported in the literature[5-10]. The thickness of δ and β determined at 900, 1000 and 1100°C are summarized in Table I, as a function of annealing time. Typical results for the growth of δ and β at 900, 1000 and 1100°C are plotted in Fig. 1 in terms of layer thickness in μm, versus the square root of time in hrs. As shown in the figure, the layer thicknesses for both δ and β vary linearly with $t^{1/2}$, indicating diffusion-controlled kinetics. Figure 2 shows the Arrhenius-type plots for the parabolic growth constant k with $k^{1/2}$ being the slope obtained from Fig. 1.

Table I
Thickness of the Phases Formed
in Mo/γ-TiAl Diffusion Couples

T,°C	900	1000	1100
Annealing time, hrs.	121	138	137
δ-phase, μm	2	6	13
β-phase, μm	4	11	36
Annealing time, hrs.	285	306	308
δ-phase, μm	3	8	18
β-phase, μm	6	14	48
Annealing time, hrs.	553	551	548
δ-phase, μm	4	10	24
β-phase, μm	8	19	76

Concentration profiles for all samples were determined starting with Mo, through δ and β, and ending with γ-TiAl. Using the Boltzmann-Matano analysis[11] we can obtain the interdiffusional fluxes, $\tilde{J}_i$, where the subscript i represents Mo, Al or Ti. Once we have the values of $\tilde{J}_i$, we can derive values of the interdiffusion coefficients of the various phases. According to the Boltzmann-Matano analysis, the interdiffusion flux, $\tilde{J}_i$, is related to the

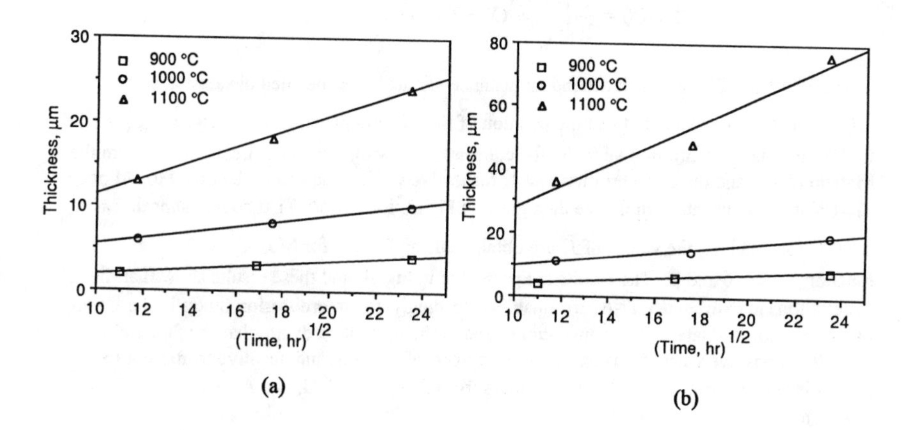

Fig. 1 Parabolic plots for the growth of the δ phase (a) and the β phase (b).

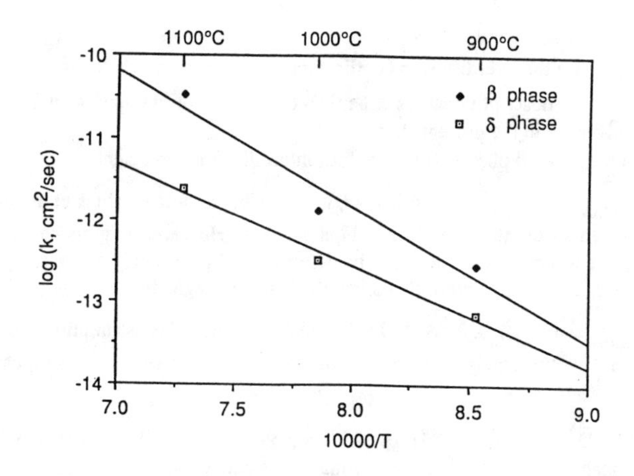

Fig. 2 An Arrhenius plot of the parabolic rate constants for the δ and β phases.

concentration profiles in a semi-infinite diffusion couple and the annealing time t, as follows:

$$\tilde{J}_i(X^*) = \frac{1}{2t}\int_{C_i^+}^{C_i(x^*)} (X - X_o)\, d\, C_i \qquad [1]$$

where X, X* and X_o are the independent distance variable, the specified distance at which $\tilde{J}_i(X^*)$ needs to be evaluated and the position of the Matano plane, respectively; and C_i, C_i^+, $C_i(X^*)$ are the concentration variable, the concentration within the bulk (i.e. far away from the Matano plane) and the concentration at X*, respectively. The subscript i denotes Mo, Al or Ti. Using the concentration profile data given in Fig. 3, for a Mo/γ-TiAl couple annealed at 1100°C for 308 hrs., the values of $\tilde{J}_i$ are obtained using Eq. [1] for Mo, Al and Ti as a function of X in δ and β. The results are presented in Fig. 4, and these results show that the penetrations in (Mo) and γ-Ti-Al are small. Accordingly, the interdiffusion fluxes in these two phases are not evaluated. The conventions used in the present study are that the fluxes are considered positive when diffusion takes place from left to right and negative from right to left. It is evident from Fig. 4 that Mo diffuses from (Mo) to γ-TiAl, and Al and Ti from γ-TiAl to (Mo).

Interdiffusion coefficient values for δ and β can be obtained from knowledge of their fluxes, $\tilde{J}_i$, and their concentration gradients, $\partial C_i^\delta / \partial X$ and $\partial C_i^\beta / \partial X$, via Fick's first law. The relationship is:

$$\tilde{J}_i^\phi = -\tilde{D}_{iMo}^\phi \frac{\partial C_{Mo}^\phi}{\partial X} - \tilde{D}_{iAl}^\phi \frac{\partial C_{Al}^\phi}{\partial X} \qquad [2]$$

where $\tilde{D}_{iMo}^\phi$ and $\tilde{D}_{iAl}^\phi$ are the interdiffusion coefficients. The superscript ϕ denotes either the δ or the β phases and the subscript i denotes either Mo or Al. The third component Ti is being used as the solvent[12] in the present study.

For each of the δ and β phases, there are four interdiffusion coefficients, $\tilde{D}_{MoMo}^\phi$, $\tilde{D}_{AlAl}^\phi$, $\tilde{D}_{MoAl}^\phi$ and $\tilde{D}_{AlMo}^\phi$. In principle, it is possible to obtain values of all the interdiffusion coefficients from the data given in Figs. 3 and 4. However, in practice this is difficult since there are too many parameters to be determined. Accordingly, the assumption is made that the cross terms of the interdiffusion coefficients are negligible, i.e. $\tilde{D}_{MoAl}^\delta, \tilde{D}_{AlMo}^\delta, \tilde{D}_{MoAl}^\beta$ and $\tilde{D}_{AlMo}^\beta \cong 0$. As to be demonstrated later, this assumption is quite reasonable. A second assumption is made that the interdiffusion coefficients are composition-independent.

The values of $\tilde{D}_{MoMo}^\delta$, $\tilde{D}_{AlAl}^\delta$, $\tilde{D}_{MoMo}^\beta$ and $\tilde{D}_{AlAl}^\beta$ obtained are given in Table II. They are obtained numerically by optimization considering all the data points. The uncertainties given in this table are the standard deviations from the average values. The activation energies obtained from the Arrhenius-type plots for the four coefficients are 251±50, 297±30, 192±30 and 228±20 kJ/mol respectively. The equations describing the temperature dependence of these coefficients are given in Table II.

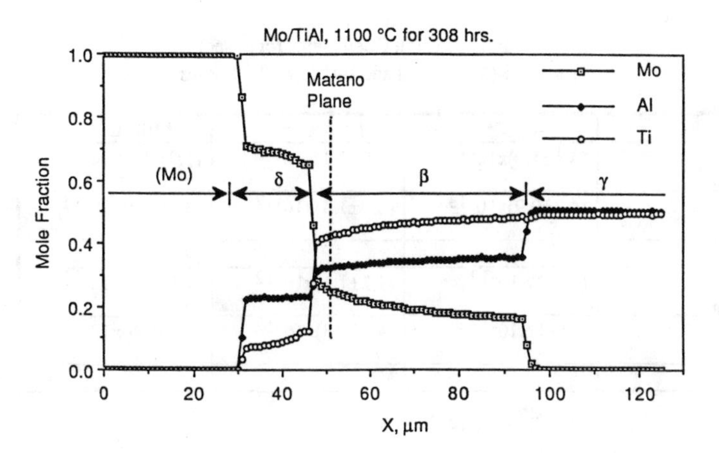

Fig. 3 Concentration profiles of Mo, Al and Ti in a semi-infinite diffusion couple of Mo/γ-TiAl annealed at 1100°C for 308 hrs.

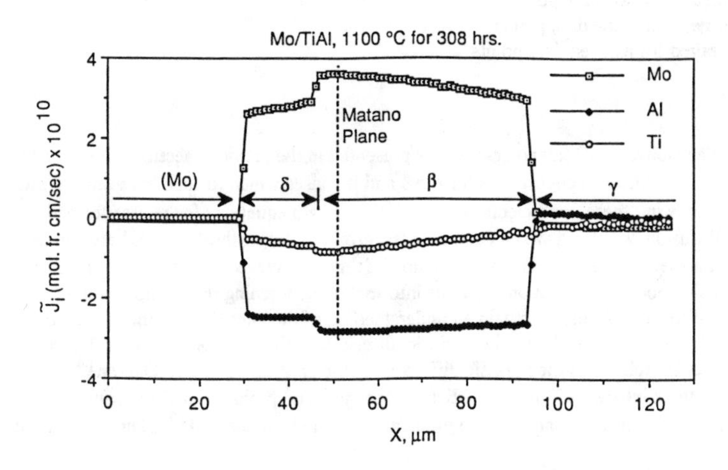

Fig. 4 Interdiffusion fluxes of Mo, Al and Ti for a semi-infinite diffusion couple of Mo/γ-TiAl annealed at 1100°C for 308 hrs. The fluxes are considered positive when the species diffuses from left to right and negative from right to left.

Table II
Interdiffusion Coefficients (cm^2/sec)
for the δ-$(Mo,Ti)_3Al$ and β-(Mo,Al,Ti) Phases

	900°C*	1000°C**	1100°C***
$\tilde{D}^{\delta}_{MoMo}$	$(3.1\pm1.0)x10^{-13}$	$(3.0\pm1.1)x10^{-12}$	$(1.3\pm0.5)x10^{-11}$
$\tilde{D}^{\delta}_{AlAl}$	$(4.7\pm0.9)x10^{-13}$	$(2.9\pm1.4)x10^{-12}$	$(3.8\pm0.7)x10^{-11}$
$\tilde{D}^{\beta}_{MoMo}$	$(10\pm2)x10^{-13}$	$(3.2\pm1.1)x10^{-12}$	$(1.8\pm0.3)x10^{-11}$
$\tilde{D}^{\beta}_{AlAl}$	$(11\pm1)x10^{-13}$	$(4.0\pm0.8)x10^{-12}$	$(3.4\pm0.6)x10^{-11}$
$\tilde{D}^{\delta}_{MoMo}$	$5.1x10^{-2}$ exp[-(251±50 kJ)/RT]		
$\tilde{D}^{\delta}_{AlAl}$	4.1 exp[-(297±30 kJ)/RT]		
$\tilde{D}^{\beta}_{MoMo}$	$3.2x10^{-4}$ exp[-(192±30 kJ)/RT]		
$\tilde{D}^{\beta}_{AlAl}$	$1.3x10^{-2}$ exp[-(228±20 kJ)/RT]		

*Averaged from two data points
**Averaged from nine data points
***Averaged from seven data points

DISCUSSION

The Boltzmann-Matano analysis, as presented in the previous section, allows us to obtain the interdiffusion coefficients for the δ and β phases which formed between Mo and γ-TiAl. Knowing these coefficients, and using the growth equations for phases formed in a ternary diffusion couple[13], we can compute the growth rates of these phases between Mo and γ-TiAl at a specified temperature as a function of time. However, diffusion analysis using interdiffusion coefficients can only give us information concerning the phenomenological aspect of diffusion. In order to gain an understanding of the microscopic phenomena of diffusion it is necessary to know the intrinsic diffusion coefficients. Since interdiffusion coefficients are related to the intrinsic diffusion coefficients, according to Darken[14], we can obtain the three intrinsic diffusion coefficients for the δ and β phases directly from the interdiffusion fluxes and concentration gradients given in Figs. 3 and 4[13]. These equations are

$$\tilde{D}_{11} = D_1 + C_1(D_3 - D_1)U_1 \quad [3]$$

$$\tilde{D}_{12} = C_1(D_3 - D_2)U_1 \quad [4]$$

$$\tilde{D}_{21} = C_2(D_3 - D_1)U_2 \quad [5]$$

$$\tilde{D}_{22} = D_2 + C_2(D_3 - D_2)U_2 \qquad [6]$$

with

$$U_i = 1 + \frac{2}{M}\left(1 - \frac{D_i}{\sum_{n=1}^{3} C_n D_n}\right) \qquad [7]$$

where the quantities D_i are the intrinsic diffusion coefficients for each of the three components, and M is a parameter dependent on the crystal structure[15].

Values of D_{Mo}, D_{Al} and D_{Ti} are obtained for δ and β at 900, 1000 and 1100°C. Values of D_{Mo}, D_{Al} and D_{Ti} for the β-phase at 1100°C are given in Table III. Knowing these values, the four interdiffusion coefficients are determined using Eqs. [3] to [7] and are given in Table III. The values for $\tilde{D}^{\beta}_{MoMo}$ and $\tilde{D}^{\beta}_{AlAl}$ obtained directly from the interdiffusion flux data assuming the cross terms to be negligible are also given in Table III. Table III shows that the values of $\tilde{D}^{\beta}_{MoMo}$ and $\tilde{D}^{\beta}_{AlAl}$ do not differ appreciably from the values calculated using the intrinsic diffusion coefficients D^{β}_{Mo}, D^{β}_{Al} and D^{β}_{Ti}. This is understandable since the three intrinsic diffusion coefficients do not differ significantly from each other.

Table III
Diffusion Coefficients for the β-phase at 1100°C

D_{Mo}	$= (1.1 \pm 0.3)10^{-11}$ cm^2/sec
D_{Al}	$= (6.0 \pm 0.7)10^{-11}$ cm^2/sec
D_{Ti}	$= (2.9 \pm 0.6)10^{-11}$ cm^2/sec
$\tilde{D}_{MoMo}$	$= (1.5 \pm 0.3)10^{-11}$ cm^2/sec
$\tilde{D}_{MoAl}$	$= (-0.75 \pm 0.08)10^{-11}$ cm^2/sec
$\tilde{D}_{AlMo}$	$= (0.48 \pm 0.14)10^{-11}$ cm^2/sec
$\tilde{D}_{AlAl}$	$= (5.2 \pm 0.7)10^{-11}$ cm^2/sec
$\tilde{D}_{MoMo}$ (from Table II)	$= (1.8 \pm 0.3)10^{-11}$ cm^2/sec
$\tilde{D}_{AlAl}$ (from Table II)	$= (3.4 \pm 0.6)10^{-11}$ cm^2/sec

Knowing the intrinsic diffusivities, we can calculate the lattice velocity relative to the Matano plane using the following relationship relating the interdiffusion and intrinsic diffusion fluxes,

$$\tilde{J}_i = J_i + vC_i \qquad [8]$$

where J_i denotes the intrinsic diffusion flux and v denotes the lattice velocity. A rearrangement of Eq. [8] yields,

$$v = \frac{1}{C_i}\left[\tilde{J}_i + D_i \frac{\Delta C_i}{\Delta X}\right] \qquad [8A]$$

Using the concentration gradient data for the Mo/γ-TiAl couple annealed at 1100°C for 308 hrs. and the intrinsic diffusion coefficients, the lattice velocity is calculated for each of the three components at the Matano plane. These three values of v, and the average, are given in Table IV.

Table IV
Lattice Velocity Relative to the Matano Plane

	v(cm/sec)
Mo	6.0×10^{-10}
Al	3.6×10^{-10}
Ti	5.5×10^{-10}

Avg.	$(5.0\pm1.5)10^{-10}$

The above calculation shows that the experimental data is reliable and that the method used to evaluate the data appropriate.

According to Fig. 4, the two dominating interdiffusion fluxes between Mo and γ-TiAl are those of Mo and Al. The fluxes of Ti are close to zero. One can not tell from this data whether or not the diffusion of Mo and Al is due primarily to the intrinsic movement of Mo and Al on the lattice or the movement of the lattice. Knowing the intrinsic diffusivities, we can estimate these two contributions. Again, using the data of Mo/γ-TiAl annealed at 1100°C for 308 hrs., we can calculate J_i, vC_i and $\tilde{J}_i$ (mol fr. cm/sec) at the Matano plane. The calculated values are given in Table V.

Table V
Fluxes of Mo, Al and Ti

	J_i	vC_i	$\tilde{J}_i$
Mo	2.1×10^{-10}	1.5×10^{-10}	3.6×10^{-10}
Al	-4.0×10^{-10}	1.2×10^{-10}	-2.8×10^{-10}
Ti	-3.2×10^{-10}	2.4×10^{-10}	-0.8×10^{-10}

The data given in Table V show the absolute values of the intrinsic diffusion fluxes are the highest for Al and lowest for Mo, i.e. J_{Al} and J_{Mo}. However, the fluxes due to lattice movement are positive for all three elements. Since the intrinsic diffusion flux of Mo is positive, this yields an even larger interdiffusion flux of Mo. On the other hand, the positive flux due to lattice movement for Ti almost compensates for its negative intrinsic diffusion flux. This results in nearly zero interdiffusion flux for Ti. These results are quite reasonable since we would expect the diffusivity for Al to be the highest and that for Mo the lowest on the basis of their melting points.

CONCLUSIONS

Both interdiffusion and intrinsic diffusion coefficients for δ-$(Mo,Ti)_3Al$ and β-(Mo,Al)Ti can be obtained from the Mo/γ-TiAl diffusion couples annealed at 900, 1000 and 1100°C for periods of time varying from 121 to 553 hrs. These data provide the needed information to predict quantitatively the extent of interaction between Mo and γ-TiAl. Such information is essential in the development of composite materials involving γ-TiAl and Mo.

The intrinsic diffusivities obtained could be used to evaluate the lattice movement in Mo/γ-TiAl diffusion couples. The analysis of the data for Mo/γ-TiAl couple annealed at 1100°C for 308 hrs. in terms of the two flux contributions, i.e. intrinsic and lattice movements, shows that the high interdiffusion fluxes of Mo are the results of the fact that the fluxes due to intrinsic diffusion and lattice movements take place toward γ-TiAl with respect to the Matano plane.

ACKNOWLEDGMENT

We wish to thank Paul Allard and Ralph Hecht of Pratt-Whitney, W. Palm Beach, Florida for fabricating the diffusion couple specimens by sputtering Mo on γ-TiAl, Alton Romig, Jr., at Sandia National Laboratories, Albuquerque, NM and E. Glover of the Geology Department of UW-Madison for getting some of the EPMA measurements and DARPA for partial financial support through ONR Contract No.
0114-86-K-0753 as a part of the URI program at University of California-Santa Barbara.

References

1. A. G. Evans, A. Bartlett, J. B. Davis, B. D. Flinn, M. Turner and I. E. Reimanis, Scr. Metall. Mater. 25, 1003 (1991).
2. M. L. Emiliani, H. J. Hecht, E. E. Dève, J. B. Davis and A. G. Evans, to be published. This study was cited as Ref. 9 of Evans et al.[1].
3. J. B. Davis, G. Bao, H. C. Cao and A. G. Evans, Acta Metall. Mater., 1991, in press.
4. Y. A. Chang, J. P. Neumann and S.-L. Chen, in Alloy Phase Stabilities and Design edited by G. M. Stocks, D. P. Pope and A. F. Giamei (Mat. Res. Soc. Symp. Proc. 186, Pittsburgh, PA, 1991), pp. 131-140.
5. H. Böhm and K. Löhberg, Z. Metallk. 49, 173 (1958).
6. T. Hamajma and S. Weissmann, Metall. Trans. 6A, 1535 (1975).
7. A. Zangvil, K. Osamura and Y. Murakami, Metal Sci. 9, 27 (1975).
8. R. C. Hansen and A. Raman, Z. Metallk. 61, 115 (1970).
9. T. B. Massalski, J. L. Murray, L. H. Bennett and H. Baker, Binary Phase Diagrams, (ASM, Metals Park, Ohio 44073, 1986).
10. U. kattner, J.-C. Lin and Y. A. Chang, "Thermodynamic Assessment and Calculation of the Ti-Al System", Metall. Trans. A, to be published (1992).
11. R. J. Borg and G. J. Dienes, Introduction to Solid State Diffusion (Academic Press, Inc., Boston, 1988) pp. 173-177.
12. J. S. Kirkaldy and D. J. Young, Diffusion in the Condensed State (The Institute of Metals, London, SWIY 5DB, U.K., 1987).
13. C.-H. Jan, D. Swenson, X.-Y . Zheng, J.-C. Lin and Y. A. Chang, Acta Metall. Mater., 39, 303 (1991).
14. L. S. Darken, Trans. AIME, 180, 430 (1949).
15. L. S. Castleman, Metall. Trans. 14A, 45 (1983).

WORKABILITY OF A DUAL PHASE TITANIUM ALUMINIDE-TiB_2 XD™ COMPOSITE

D. ZHAO, K.G. ANAND, J.J. VALENCIA, AND S.J. WOLFF
Metalworking Technology, Inc., 1450 Scalp Avenue, Johnstown, PA 15904, U.S.A.

ABSTRACT

High temperature compression tests have been performed on a Ti-44^{a}/o Al-3^{a}/o V-7.5^{v}/o TiB_2 XD composite over the temperature range 1000 to 1300°C and the strain rate range 10^{-3} to $10s^{-1}$. The workability of this material for metalforming processes was determined using both dynamic material modeling and workability testing approaches to cover both internal and surface cracking. At higher temperatures and strain rates internal fracture occurred in the material, while at lower temperatures and higher strain rates surface cracks occurred. Use of lower temperatures and strain rates inhibited internal instabilities and surface cracking in the composite. A finite element model (FEM) was developed to describe the stress and strain states during the deformation process. Mechanical flow behavior obtained from the compression tests was used as input to the model. A fracture criterion developed from Kuhn's surface crack equation was coupled with the FEM model to predict bulk formability in a forging process.

INTRODUCTION

Titanium aluminide matrix composites have attracted much attention recently for their potential use as structural materials in the aerospace industry [1,2]. While the materials are known for their high specific strength and modulus at elevated temperatures [3], their low ductility and fracture toughness may limit their workability. The workability of the composites must be determined before any successful metal forming process is designed. Prior work has indicated that the grain refining effect of the TiB_2 reinforcement increases the hot workability of the as-cast XD composite [4]. Other work studied the workability of a XD composite as a function of flow localization [5]. The objective of the present work is to determine the workability limits of a dual phase (α_2+γ) Ti-44Al-3V-7.5 ^{v}/o TiB_2 XD composite using both the dynamic material modeling (DMM) [6-8] and the workability testing approaches [9]. Processing and workability maps were developed to evaluate the most favorable conditions for processing in terms of temperature and strain rate. The workability data were combined with a fracture criterion [10] to predict crack formation in bulk forming processes. In this paper, a forging process is simulated for defect prediction.

EXPERIMENTAL PROCEDURE

The material used in this work was a Ti-44Al-3V-7.5 ^{v}/o TiB_2 XD composite. The as-cast material was HIP'ed to close porosity and to homogenize the ingot microstructure. Figure 1 shows the optical microstructure of the as-HIP'ed composite. This consists of an (α_2+γ) lamellar matrix with a uniform distribution of the TiB_2 reinforcement. The grain size of the matrix was found to be approximately 30 μm and the size of irregular shaped TiB_2 particulates varied from 1 to 15 μm.

Compression tests were conducted on cylindrical specimens with a diameter of 12.7 mm and a height of 15.9 mm. Prior to testing, circumferential parallel lines were inscribed on the workability specimens to enable the determination of the exact surface compressive strains after deformation. The tensile strain was calculated by measuring the change in diameter.

The dynamic material modeling tests were conducted over the temperature range 1050 to 1300°C at 50°C intervals, and at strain rates of 0.001, 0.01, 0.1 and $1s^{-1}$. The workability tests were conducted at temperatures of 1000, 1100, and 1200°C, and strain rates of 0.001, 0.1

TM XD is a trade mark of Martin Marietta Corporation, Baltimore, MD.

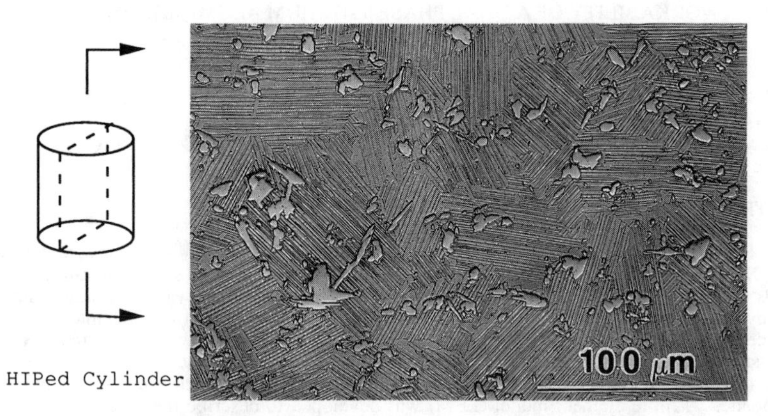

Figure 1. Optical microstructure of the as-HIP'ed XD composite.

and $10s^{-1}$. The range of testing temperatures and strain rates was determined by metalworking practices found in the literature [5,11]. All tests were done in vacuum with boron nitride lubricant. Load and stroke data from the test were acquired by a microcomputer and later converted into true stress-true strain curves. Immediately after the compression test was finished, the specimen was quenched with helium in an attempt to retain the microstructure. The microstructural analysis was performed on selected specimens using optical microscopy.

RESULTS AND DISCUSSION

Figure 2 shows the processing map developed from dynamic material modeling [6]. The information in this diagram is presented as an isoefficiency map in temperature and strain rate space. The unstable regions as predicted by DMM are shown by the hatched areas in the diagram, and the predicted stable region for processing is at lower temperatures (below 1150°C) and lower strain rates (below 0.1 s^{-1}).

Figures 3 to 6 illustrate typical microstructures found in the stable and unstable regions of the processing map. Figure 3 shows the microstructure of a specimen deformed in the stable region at 1100°C and at a strain rate of 0.01 s^{-1}. Large deformation of the lamellae and flow localization were not sufficient to produce failure of the composite. Figures 4 to 6 show the microstructures of specimens processed in the unstable regions, where internal cracking was observed.

At processing temperatures below the eutectoid transformation (~1100°C) [12], microstructural damage was observed to occur by both inter- and translamellar fracture (Figure 4). At temperatures above the eutectoid transformation, intergranular fracture was predominant, as is shown in figure 5 for a specimen deformed at 1250°C and $1.0s^{-1}$, and figure 6 for a specimen deformed at 1300°C and $10^{-3}s^{-1}$. The internal fracture found in these specimens affirms the processing map.

Figure 7 shows the workability map, which is constructed from the experimental observations of surface cracking of specimens subjected to a true compressive strain of 0.75. This map is also presented in terms of strain rate and temperature, and provides the safe and unsafe conditions for processing. The unsafe region is found at strain rates above 1.0 s^{-1} and temperatures below the eutectoid transformation (1100°C). The hoop tensile stress is critical in the surface cracking of this material, which is in agreement with earlier work by Bryant et al [5]. A typical example of the surface cracking found is illustrated in Figure 8. The remaining portion of the diagram represents the safe region for processing.

In brittle materials such as titanium aluminides it is expected that both internal and surface cracks would quickly propagate. At this time it is difficult to ascertain whether the cracks in the XD composite originated within the lamellae due to an unfavorable orientation

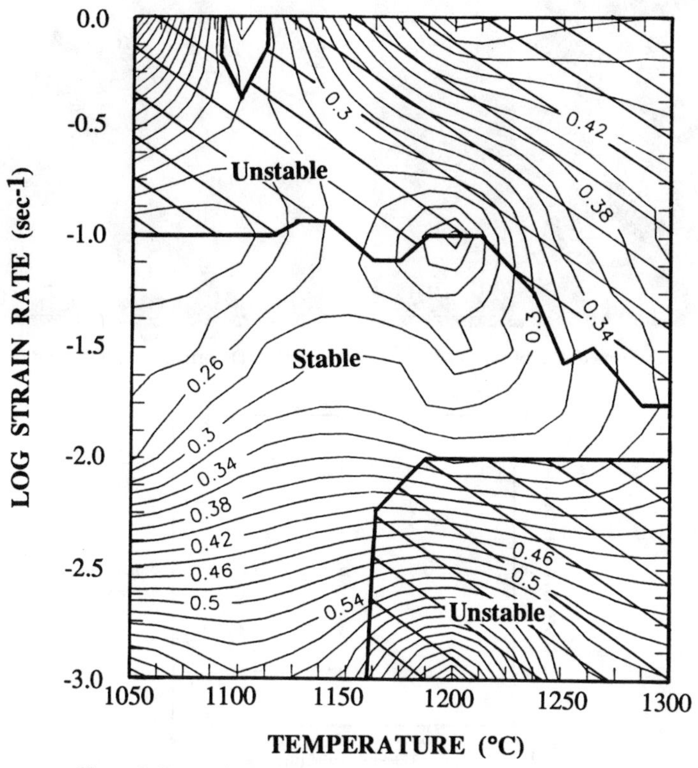

Figure 2. Processing map showing stable and unstable regions.

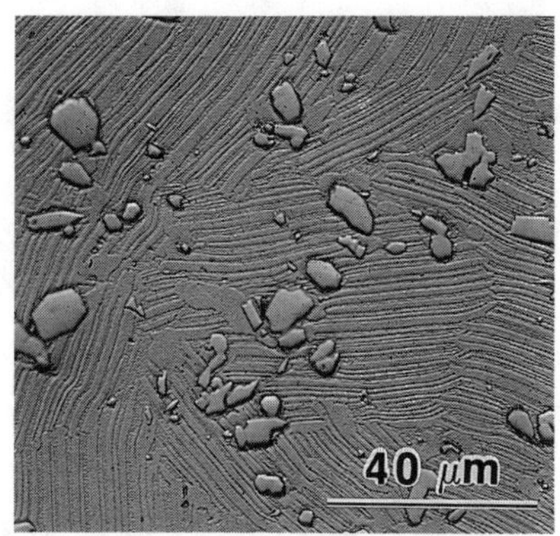

Figure 3. Deformed at 1100°C and 0.01s-1.

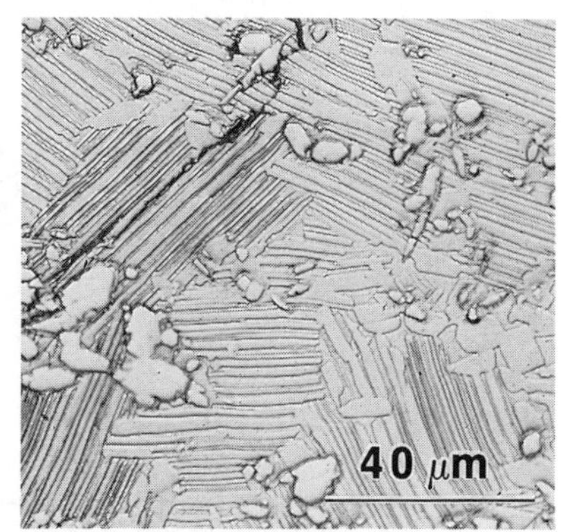

Figure 4. Deformed at 1100°C and 1.0s^{-1}.

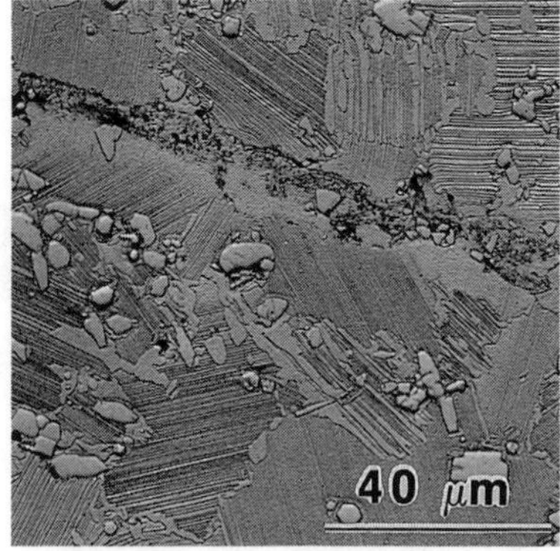

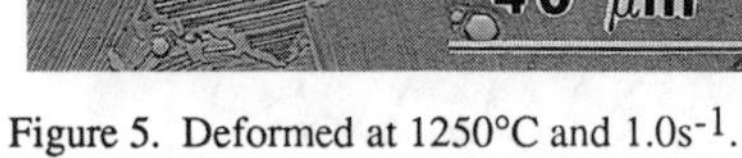

Figure 5. Deformed at 1250°C and 1.0s^{-1}.

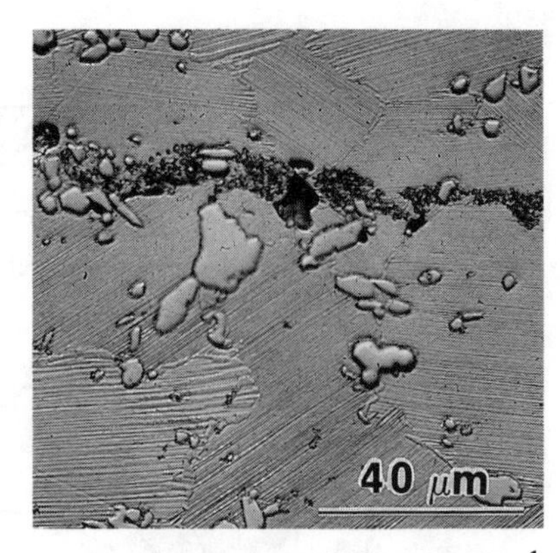

Figure 6. Deformed at 1300°C and 0.001s^{-1}.

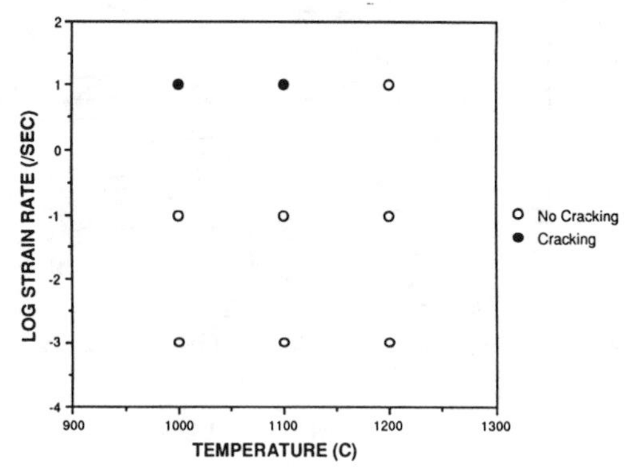

Figure 7. Workability map for XD composite

Figure 8. Compressed specimen with surface cracking.

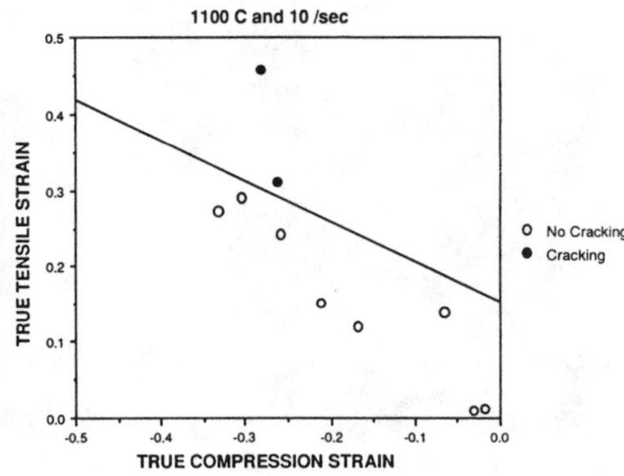

Figure 9. True tensile strain and compression strain at the surface of the specimens.

with respect to the maximum applied stress [13],or by decohesion between the matrix and the TiB2.

Figure 9 is a forming limit diagram showing the measured tensile and compressive strains obtained from the surface of both cracked and uncracked workability specimens. The results presented here are not inconsistent with the findings of Kuhn for many materials [9], which defines a fracture locus by a straight line with a slope of -1/2.

The straight line in Figure 9 can be represented by [9]:

$$\varepsilon_{1f} = \mathbf{C} - \frac{1}{2}\varepsilon_{2f} \tag{1}$$

where ε_{1f} and ε_{2f} are the tensile and the compression strains at fracture, respectively, and **C** is a constant.

It was suggested recently [10], that the straight line be described as:

$$\sigma_{1f}\frac{\bar{\varepsilon}}{\bar{\sigma}} \geq \mathbf{C'} \tag{2}$$

where $\mathbf{C'} = \mathbf{C}/(1-\nu^2)$, ν is Poisson's ratio, $\bar{\sigma}$ is the equivalent stress, and $\bar{\varepsilon}$ is the equivalent strain.

Equation (2) predicts cracking by utilizing the data from Figure 9, and an example for a forging process is given in Figure 10. The FEM simulation of the forging process was performed using the public domain software NIKE^{TM}. Figure 10a depicts the original mesh, and 10b shows the fracture criterion contours predicted by equation (2) in a workpiece being deformed at 1100°C and a stroke rate of $10s^{-1}$. Note that the material is very sensitive to tensile stress under this condition. At 10% reduction, the sites in contact with the curved part of the dies have already exceeded the value of **C'** from the test data. This predicts that fracture should occur at these sites in this forging process.

SUMMARY

The workability of a Ti-44Al-3V-7.5 %TiB_2 XD composite has been determined with both

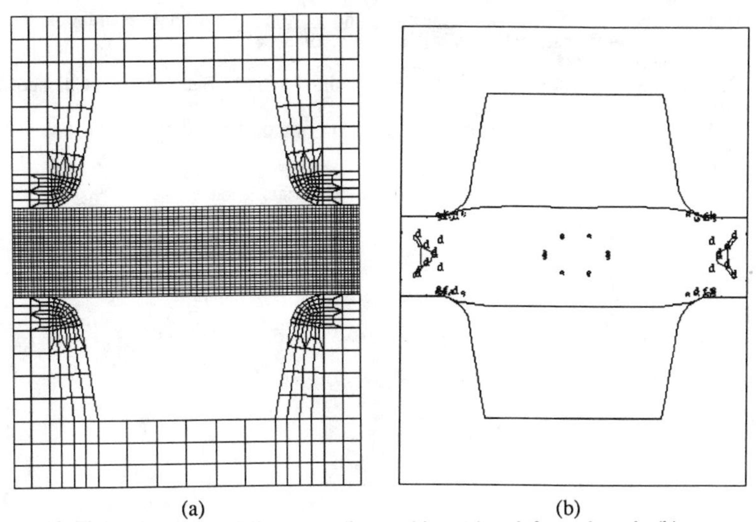

(a) (b)

Figure 10. Finite element modeling to predict cracking, (a) undeformed mesh, (b) contours of Zhao and Kuhn's fracture criteria at 10% deformation.

TM Registered trademark of Lawrence Livermore National Laboratories, Livermore, CA.

the dynamic material modeling approach and bulk workability tests. The safe region to process this material is between the temperatures of 1000 to 1150°C, and strain rates of 10^{-3} to $10^{-1}s^{-1}$. The material fails internally at the higher temperatures and cracks at the surface at the lower temperatures. In both cases, lower strain rate should be used to avoid cracking. Microstructural observation revealed that the internal fracture was both inter- and trans lamellar at temperatures below the eutectoid transus, and intergranular at temperatures above. The crack propagation has no preference through the matrix or particle-matrix interface. No particle cracking from the deformation was observed.

ACKNOWLEDGEMENTS

This work was conducted by the National Center for Excellence in Metalworking Technology, operated by Metalworking Technology, Inc., under contract to the U.S. Navy as a part of the U.S. Navy Manufacturing Technology Program.

The authors wish to thank Jim Bandstra for his help with the finite element model.

REFERENCES

1. M. Saqib, I. Weiss, G.M. Mehrotra, E. Clevenger, A.G. Jackson, and H.A. Lipsitt, Met. Trans A, 22A, 1721 (1991).
2. S.L. Kampe, J.D. Bryant, and L. Christodoulou, Met Trans A, 22A, 447 (1991).
3. D.E. Larsen, M.L. Adams, S.L. Kampe, Scripta Met., 24, 851 (1990).
4. L. Rothenflue, A. Szaruga, and H. A. Lipsitt, in Physical Metallurgy of Intermetallic Compounds (TMS Warrendale, PA 1991).
5. J.D. Bryant, M.L. Adams, A.R.H. Barrett, J.A. Clarke, G.B. Gaskin, L. Christodoulou, and J. Brupbacher, NASP Contractor Report 1095, Martin Marietta Laboratories, (1990).
6. K.G. Anand, D. Zhao, J.J. Valencia, and S.J. Wolff, in Application of Mechanics and Material Models to Design and Processing III, edited by E.S. Russell, (TMS, Warrendale, PA, 1992).
7. Y. V. R. K. Prasad, H. L. Gegel, S. M. Doraivelu, J. C. Malas, J. T. Morgan, K. A. Lark, and D. R. Barker, Met. Trans. A, 15A, 1883 (1984).
8. H. L. Gegel and J. C. Malas, in Forming and Forging, 14, Metals Handbook, Ninth Edition (ASM International, Metals Park, Ohio, 1988), 417.
9. H. A. Kuhn, ibid, 388.
10. D. Zhao, J.P. Bandstra, and H.A. Kuhn, submitted to TMS Fall Meeting, Chicago, IL, (1992).
11. D. Popoola et al, in Interfaces in Metal-Ceramic Composites, ed. R. Y. Lin et al, (TMS, Warrendale, PA, 1989) 465.
12. P. Chaudhury, M. Long, and H. J. Rack, Mater. Sci. Eng., (in press).
13. M. Yamaguchi and Y. Umakoshi, Progress in Materials Science, 34, 1 (1990).

RESIDUAL STRESSES AND RESULTING DAMAGE WITHIN FIBERS INTERSECTING A FREE SURFACE

J. M. GALBRAITH, M. N. KALLAS, D. A. KOSS, AND J. R. HELLMANN
Center for Advanced Materials, The Pennsylvania State University, University Park, PA 16802

ABSTRACT

Finite element as well as indentation fracture mechanics modeling have been used to analyze the evolution of fiber damage that was observed at the ends of fibers intersecting a free surface in sapphire-reinforced TiAl matrix composites. Experimental observations indicate that, under certain conditions, surface cracks introduced during cutting will propagate along the fiber axis due to thermally-induced residual stresses. Finite element computations predict that significant thermally-induced residual tensile stresses exist near the ends of sapphire fibers which are embedded within TiAl-based matrices and are oriented normal to a free surface. Crack growth behavior induced by microhardness indentations is used to experimentally verify the FEM predictions. The results indicate that a biaxial tensile residual stress state exists near the fiber ends due to a thermal expansion mismatch. The magnitude of the residual stresses are a sensitive function of interfacial bond strength and elastic/plastic properties of the interfacial region and may be sufficient to propagate pre-existing cracks.

INTRODUCTION

Since most intermetallic matrix composites (IMCs) are reinforced with brittle ceramic fibers, it is important to understand the implications of exposing the fiber ends as a result of cutting and machining operations which may induce damage. It is well known that, due to the large mismatch in coefficients of thermal expansion (CTE) between constituents and the high stress-free temperature in these systems, significant thermally-induced residual stresses develop within IMCs during post-consolidation cool-down. The purpose of this study is to illustrate that the combination of brittle fibers, significant tensile residual stress, and machining damage can result in longitudinal cracks (i.e., fiber splitting) when a fiber intersects a free surface. Assuming a well-bonded interface, finite element modeling (FEM) calculations of the residual thermoelastic stress distribution within a sapphire fiber in a TiAl matrix are presented for the case of a fiber intersecting a free surface and with its axis normal to the surface. The computations predict a gradient of tensile stresses within the fiber such that, given the relatively low fracture toughness of single crystal sapphire, fiber splitting will occur along the fiber axis as matrix material is removed.

Due to the size of the sapphire fibers and the distribution of stresses within the composites, experimental verification of the predicted residual stress distribution by conventional experimental methods is extremely difficult at best. However, it has been shown in several studies of brittle materials that indentation cracks can be used to estimate the magnitude of pre-existing residual stresses [1-4]. This method offers the advantages of being simple and able to probe residual stresses on a small scale. In addition, an indentation crack represents a model surface flaw that can aid in understanding multiple contact processes such as cutting and machining [5]. Thus, this study also presents an experimental validation of the FEM predicted stress state using indentation crack behavior within the sapphire fibers.

EXPERIMENTAL PROCEDURE

Materials

This study is based on low volume fraction fiber-reinforced matrices consisting of sapphire fibers (Saphikon) embedded in three different matrices: Ti-48Al-1V (hereafter referred to as Ti-48-1), Ti-48Al-2Cr-2Nb (i.e., Ti-48-2-2), and pure niobium (compositions are given in at. pct.). It should be noted that the first two matrices are based on the intermetallic TiAl. These unidirectionally reinforced fiber composites were prepared via vacuum hot-pressing powder and ~160 μm diameter fibers. The average values of CTE over the range of hot pressing temperatures used for TiAl and Nb are 12.7 and 8.8×10^{-6}/°C respectively. For sapphire, the CTE along the c-axis, 9.0×10^{-6}/°C, is slightly higher than in the transverse direction, 8.3×10^{-6}/°C. The Ti-48-2-2 matrix material was consolidated by G E Aircraft Engines, Evendale, OH, containing approximately 9 vol. pct. sapphire fibers with roughly uniform fiber spacing on the order of three fiber diameters. A much lower volume fraction of sapphire reinforcement (~0.04 vol. pct.) was produced for the Ti-48-1 matrix. Both TiAl-based matrices reacted chemically with the sapphire to produce a chemical bond with high average interfacial shear strength. The sapphire/niobium composite also contained a low volume fraction of reinforcement (~0.04 vol. pct.) and was hot-pressed, using a foil-fiber-foil technique, under conditions that resulted in a chemical bond between the fiber and matrix. Since the CTE mismatch between niobium and sapphire is small, the predicted thermoelastic residual stress upon cooling from the fabrication temperature is also very small ($\sigma_{rr} \approx \sigma_{\theta\theta} \approx -22$ MPa).

The single crystal sapphire fibers, with the c-axis of sapphire parallel to the fiber axis, were fabricated by Saphikon, Inc. and have a diameter of 150-175 μm. The protective organic coating present on as-received fibers was removed prior to hot-pressing. Both uncoated and tantalum-coated fibers were consolidated in the Ti-48-1 matrix. The tantalum coating (CTE = 7.0×10^{-6}/°C)was deposited via sputtering at the Naval Research Lab with a coating thickness of 0.2-0.3 μm. No cleaning or other preparation was performed on the as-received fibers prior to sputtering.

Specimens were prepared in the form of thin (i.e., 0.3 to 0.6 mm) sections cut with a diamond wafering saw perpendicular to the fiber direction. Care was taken to align the fibers normal to the specimen surface, i.e., the (0001) basal plane parallel to the specimen surface. The specimen surface was initially ground flat on a glass plate with 3 μm and subsequently polished with 1 μm diamond paste on nylon cloth.

Indentation of the fibers was performed using a Vickers diamond pyramid indenter and a standard microhardness tester at a load of 5 N such that well developed radial crack patterns were obtained (i.e., $c \geq 2a$, where a is the hardness indent diagonal [1]) and the overall crack size, 2c, was maintained less than the fiber diameter. These indentations were made in air with a load dwell time of 10 sec. Mean crack sizes were determined from 2 to 9 indentations for each fiber/matrix system.

Indentation Crack Growth Analysis

The cracks formed during the elastic/plastic contact between a sharp indenter and a brittle material have been classified into two systems: (1) half-penny shaped, radial/median cracks (hereafter referred to as radial cracks) that form on symmetry planes containing the load axis and principal indenter diagonals and (2) lateral cracks that form on shallow sub-surface planes approximately normal to the load axis [6-7]. Of these two crack systems, the radial cracks are of interest here since they closely model the type of damage experimentally observed in as-cut sapphire fibers. The lengths of the radial cracks should be a sensitive measure of in-plane residual stresses [1].

The driving force for the radial cracks that emanate from the impression corners in a residual stress-free substrate comes from the residual-contact field [8]. For an indentation load, P, the radius, c, of the radial half-penny shaped cracks may be determined from the residual stress intensity factor K_i [8-9]:

$$K_i = \frac{\chi P}{c^{3/2}} \tag{1}$$

where χ is a dimensionless parameter dependent on the indenter cone half angle and the stiffness-to-hardness ratio of the substrate. For the present study, χ has a value of 0.0687 based on calibration tests on the presumably stress-free sapphire fibers embedded in niobium. It should be noted that the form of K_i is such that it decreases with increasing crack length. This will result in crack propagation to a length dictated by the fracture toughness and subsequent arrest.

If the radial crack system is also acted upon by thermally-induced residual stresses, the net mode I stress intensity factor, K, can be obtained by use of the superposition principle:

$$K = K_i + K_r \tag{2}$$

where K_r is the stress intensity factor for a half-penny shaped crack subject to a thermally-induced residual stress field. For equilibrium fracture conditions, the crack extends until K reaches the fracture toughness of the material K_{IC}.

By generalizing an earlier approach by Lawn and Fuller [1], K_r may be obtained for any depth profile of stress. Taking the fracture mechanics solutions for a concentrated force acting on an elemental area within the crack perimeter [10] and integrating this solution over the entire crack area, we obtain for the stress intensity due to the residual stresses:

$$K_r = \frac{\psi}{\sqrt{c}} \int_0^c \left(\sqrt{\frac{c}{z}} - 1\right)\sigma(z)dz \tag{3}$$

where $\sigma(z)$ defines the residual stress field profile as a function of depth, z, into the fiber from the free surface, and ψ is a crack geometry term approximately equal to unity. We assume a fit of the residual stress profiles within the fiber (calculated by FEM) to 4th-order polynomials according to:

$$\sigma(z) = \sigma_R(1 + az + bz^2 + dz^3 + ez^4) \tag{4}$$

where σ_R is the maximum tensile stress in the free surface of the fiber and a, b, d, and e are constants. Note that $\sigma(z)$ refers to either the radial, σ_{rr}, or hoop stress, $\sigma_{\theta\theta}$, component, whichever is greater. Substituting equation (4) into (3) and combining the result with equations (1) and (2), we obtain an expression that allows us to calculate the net mode I stress intensity:

$$K = \frac{\chi P}{c^{3/2}} + [\psi\sigma_R\sqrt{c}(1 + \frac{a}{6}c + \frac{b}{15}c^2 + \frac{d}{28}c^3 + \frac{e}{45}c^4)] \tag{5}$$

Note that the presence of tensile surface stresses is equivalent to reducing the intrinsic resistance to fracture.

Equation (5) permits us to calculate a K-value for a measured indentation load and crack length, as well as the predicted residual stress state. Comparison of K to the K_{IC} value of sapphire then serves to test the validity of the predicted stress state.

RESULTS AND DISCUSSION

This study was initiated by the observation that thin-slice specimens ($\leq$ 0.6 mm thick) of sapphire-reinforced Ti-48-1 contained cracked fibers, regardless of several different methods used for sectioning the specimens. In short, we observed that damage to the fiber ends during sectioning could not be removed by subsequent polishing; it was "incurable". The extent of fiber fracture and crack orientation with respect to the fiber axis is shown in Figure 1 in which the matrix was chemically removed by etching with a concentrated Kroll's reagent. It appears that the zigzag crack path follows the rhombohedral planes, the preferred cleavage planes in sapphire at room temperature. On the other hand, Ta-coated fibers hot-pressed in the same matrix and under the same consolidation conditions as the uncoated fibers contained no cracks. Likewise, uncoated sapphire fibers in the multi-filament reinforced Ti-48-2-2 sample were not cracked during specimen preparation. Quite significant is the fact that metallographic evidence of voids at the sapphire/Ti-48-2-2 interface may indicate a weaker fiber/matrix bond in this system than in sapphire/Ti-48-1. In addition, the Ti-48-2-2 matrix hardness is 17% lower.

A possible contributor to the fiber damage is the presence of residual tensile stresses within a fiber near a free surface. In order to characterize the stress state in a fiber along its axis, an axisymmetric finite element model (FEM) was used to emulate the actual thermal loading during post-consolidation cool-down [11]. The bi-linear elastic-plastic model assumes that the fiber intersects a free surface at right angles and is bonded to the matrix at a well-defined interface. Since the CTE of the matrix is greater than that of the fiber, the fiber is subject to a compressive triaxial stress state when end effects are ignored (i.e., as in an analytical concentric cylinder model). However, as illustrated in Figure 2a, thermally induced displacements of the matrix and fiber surfaces during cross-sectioning create a "dome-shaped" fiber end. Within it, thermoelastic residual stresses occur resulting in a tensile triaxial stress state within the fiber.

The profiles of radial and hoop stresses (i.e., stresses parallel to the free surface plane) within the fibers at depths up to 20 μm from the free surface are illustrated in Figure 2 for a sapphire-reinforced TiAl composite with a 0.1 fiber volume fraction. Figure 2 shows that: (1) at or near the free surface, a state of nearly equal biaxial tension exists, (2) the magnitude of the tensile stresses decreases rapidly with increasing depth, z, below the free surface, and (3) the hoop stress is somewhat larger than the radial stress near the fiber/matrix interface at $z>0$. Experimental support of the presence of biaxial tension in the fiber's free surface is shown in Figure 3 of a cross-sectioned, uncoated fiber in Ti-48-1. This "mud cake" fracture pattern is characteristic of cracking in an equal biaxial tensile stress state.

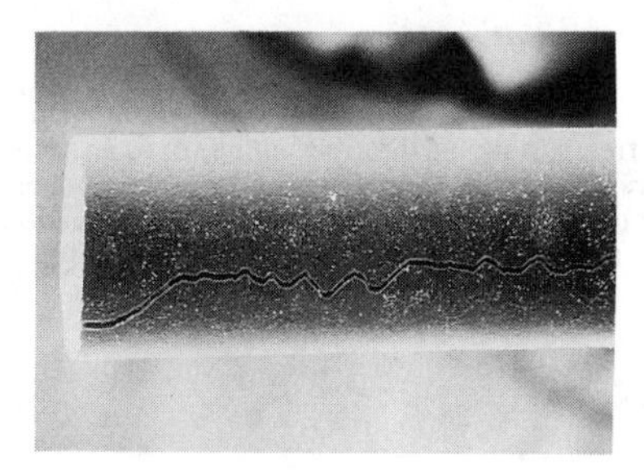

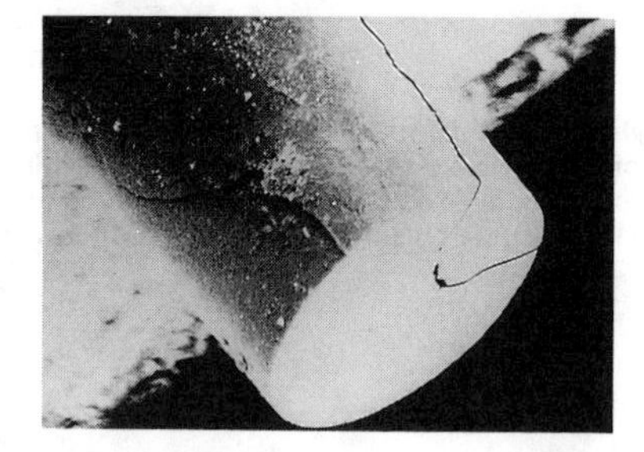

Figure 1. SEM micrographs showing the extent of fiber fracture and crack orientation with respect to the fiber axis in sapphire-reinforced Ti-48Al-1V.

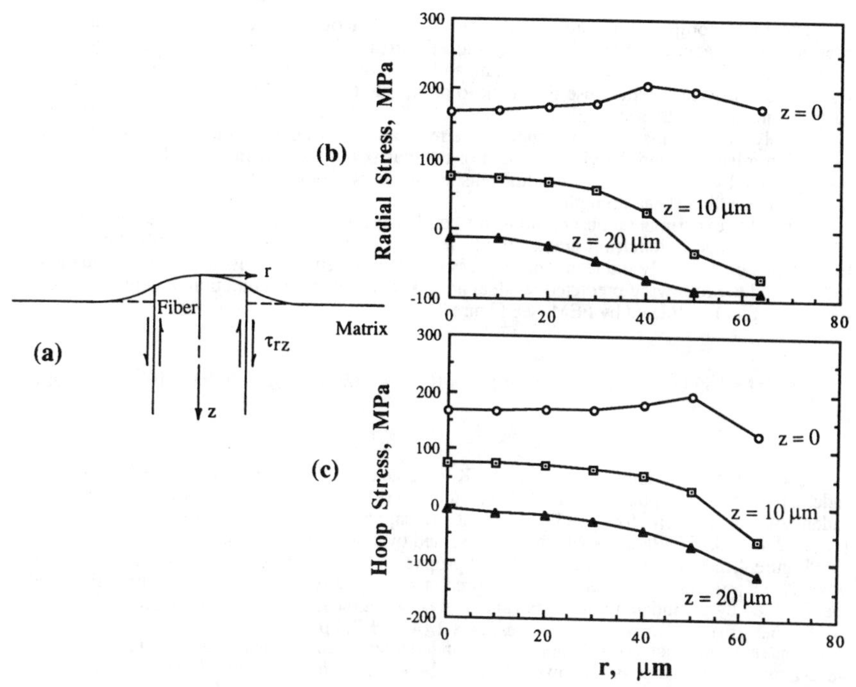

Figure 2. Free and sub-surface thermoelastic residual stress profiles in the fiber for a sapphire-reinforced TiAl composite with a 0.1 fiber volume fraction (predicted by FEM): (a) fiber orientation showing the important role of the interfacial shear strength and resulting displacements of the matrix and fiber surfaces, (b) radial stress (σ_{rr}), and (c) hoop stress ($\sigma_{\theta\theta}$).

Figure 3. SEM micrographs showing the characteristic "mud cake" fracture pattern on the cross-sectioned fiber surface in sapphire-reinforced Ti-48Al-1V due to cracking in an equal biaxial tensile stress state: (a) cross-sectioned fiber and (b) high magnification micrograph of upper right quadrant of fiber end.

The FEM computations suggest that, given a fiber of low fracture toughness and the presence of a crack, the thermally-induced tensile stresses within the fiber may propagate the crack to a length dictated by the stress state and fracture toughness. Furthermore, subsequent removal of the matrix will cause the crack to propagate along the fiber, splitting it as the free surface and stress state is spatially displaced. A "split" fiber, such as shown in Figure 1, results.

Finally, we note that the magnitude of the residual stresses depend on not only the thermal expansion mismatch and the yield strength of the matrix but also the interfacial shear strength. As suggested by Figure 2a, large residual tensile stresses will develop within the fiber only if the interface has a high shear strength.

While the experimental observations of fiber damage are qualitatively in agreement with the FEM predictions, indentation crack growth results provide experimental verification of the magnitudes of thermally induced tensile stresses within the fiber near its end. Using equations (3), (5), and the following predicted residual hoop stress profile (as a function of depth, z, below the free surface) calculated by FEM (see Figure 2):

$$\sigma_{\theta\theta}(z) = 166.6(1 - 6.114\times10^{4}z + 4.520\times10^{8}z^{2} + 3.675\times10^{11}z^{3} - 9.760\times10^{15}z^{4})\ \text{MPa}, \quad (6)$$

the dependence of the total stress intensity, K, on crack length is shown in Figure 4 for a 4.9 N indentation load in Sapphire/TiAl. These results are representative of similar data for the other indentation loads. K_{IC} for the $\{11\bar{2}0\}$ planes in sapphire was measured by Iswsa and Bradt [12] as 2.43 ± 0.26 MPa-m$^{1/2}$ which is represented by a horizontal line on the figure.

Figure 4 indicates that the crack should propagate to a length of 33 µm (i.e., the value of c when $K = K_{IC}$) and arrest, given the FEM-predicted, thermally-induced, residual stress field and that induced by the indentation. This value lies between those measured for sapphire/Ti-48-2-2 (29.5 ± 6.5 µm) and Ta-coated sapphire/Ti-48-1 (44.7 ± 7.4 µm).

This range of crack arrest lengths can be understood given the lower yield strength of the Ti-48-2-2 matrix and the possible lower chemical bond strength in the sapphire/Ti-48-2-2 system

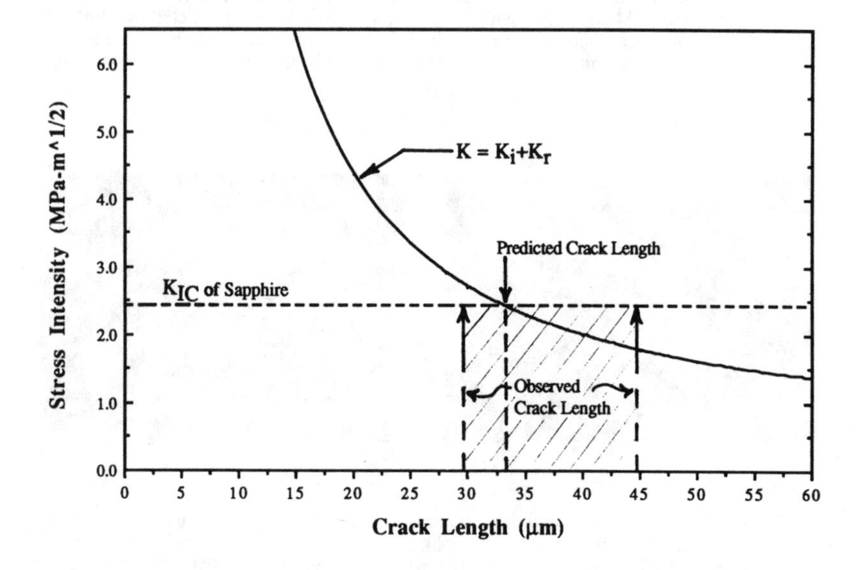

Figure 4. The predicted total stress intensity (from equations (5) and (6)) as a function of crack size in the fiber in sapphire-reinforced TiAl for a 4.9 N indentation load.

which suggest smaller thermally-induced residual stresses are present. Thus, the smaller crack length at arrest. This is consistent with our hypothesis that the in-plane residual tensile stresses in the fiber's free surface are a function of: (1) CTE mismatch between constituents, (2) fiber/matrix chemical bond strength, and (3) elastic/plastic properties of the interfacial region (namely yield stress).

SUMMARY

This study was initiated by the observation that sapphire-reinforced Ti-48Al-1V contained fibers which were cracked near their ends, regardless of the cross-sectioning or specimen preparation technique used. Finite element computations predict that significant residual tensile stresses exist near the ends of sapphire fibers which are embedded within TiAl and are oriented normal to a free surface. The FEM computations suggest that, given a fiber of low fracture toughness, the presence of a crack created during cross-sectioning or machining, and a high interfacial shear strength, the thermally-induced tensile stresses within the fiber may propagate the crack even if the surface is polished or the matrix is removed. Indentation crack behavior within the sapphire fibers provided experimental validation of the FEM predicted stress state. The results of this study suggest that the magnitude of these stresses can be decreased through the use of ductile and/or compliant interphases and by tailoring the interfacial bond strength to decrease the shear strength of the interface.

ACKNOWLEDGEMENTS

This research was supported by NASA under HITEMP Grant #NAGW-1381 and by the US Air Force.

REFERENCES

1. B. R. Lawn and E. R. Fuller, Jr., J. Mater. Sci. 19, 4061-4067 (1984).

2. D. B. Marshall and B. R. Lawn, J. Am. Ceram. Soc. 60 (1-2), 86-87 (1977).

3. R. Tandon and D. J. Green, J. Am. Ceram. Soc. 73 (4), 970-977 (1990).

4. M. F. Gruninger, B. R. Lawn, E. N. Farabaugh, and J. B. Wachtman, Jr., J. Am. Ceram. Soc. 70 (5), 344-348 (1987).

5. B. R. Lawn, in Fracture Mechanics of Ceramics, Vol. 5, edited by R. C. Bradt, A. G. Evans, D. P. H. Hasselman, and F. F. Lange (Plenum Press, New York, 1983) pp. 1-25; H. P. Kirchner and E. D. Isaacson, ibid., pp. 57-70.

6. B. R. Lawn and T. R. Wilshaw, J. Mater. Sci. 10, 1049 (1975).

7. B. R. Lawn and M. V. Swain, J. Mater. Sci. 10, 113 (1975).

8. D. B. Marshall and B. R. Lawn, J. Mater. Sci. 14, 2001-2012 (1979).

9. B. R. Lawn, D. B. Marshall, and A. G. Evans, J. Am. Ceram. Soc. 63, 574-581 (1980).

10. H. Tada, P. C. Paris, and G. R. Irwin, The Stress Analysis of Cracks Handbook (Dell Research Corp., Hellertown, PA, 1973), p. 24.2.

11. M. N. Kallas, D. A. Koss, H. T. Hahn, and J. R. Hellmann, J. Mater. Sci. 27 (to be published in 1992).

12. M. Iwasa and R. C. Bradt, Bull. Am. Ceram. Soc. 60, 374 (1981).

MICROSTRUCTURAL EFFECTS ON FATIGUE-CRACK GROWTH BEHAVIOR IN γ-TiAl/β-TiNb INTERMETALLIC COMPOSITES

K. T. VENKATESWARA RAO AND R. O. RITCHIE
Department of Materials Science and Mineral Engineering, University of California at Berkeley
Berkeley, CA 94720

ABSTRACT

Cyclic crack-propagation behavior is examined in a series of γ-TiAl intermetallic alloys reinforced with pancake-shaped, ductile β-TiNb particles as a function of microstructure and specimen orientation. In contrast to results under monotonic loading, TiNb reinforcements are found to be far less effective in impeding crack extension under cyclic loading due to their susceptibility to premature fatigue failure, and consequently to the diminished role of shielding from crack-bridging mechanisms. Modest improvements in fatigue-crack growth resistance are observed in TiAl/TiNb composites compared to monolithic γ-TiAl, provided the particle faces are oriented perpendicular to the crack plane; however, properties are compromised in orientations where the particle edges are stacked normal to the crack plane. Microstructural effects on cyclic crack growth are less prominent in the composites, with crack-growth rates exhibiting a strong dependence on the applied ΔK level; measured exponents for the da/dN-ΔK relationship range between 10 and 20, and are found to decrease with increasing ductile phase content, yet are independent of particle thickness.

INTRODUCTION

Previous studies on ductile-particle toughening of brittle intermetallic alloys at ambient temperature [1-5] have clearly demonstrated the contrasting role of the reinforcement phase in influencing crack advance under monotonic vs. cyclic loading conditions. Specifically, in the case of γ-TiAl intermetallic reinforced with ductile β-TiNb, a substantial improvement in toughness is observed under monotonic loading (Fig 1a); compared to a K_{Ic} value of ~8 MPa√m for monolithic TiAl, the composite containing 10 vol.% TiNb phase exhibits higher crack-initiation toughness (K_o = 16 MPa√m) from crack trapping and renucleation effects [2-5]. The fracture resistance increases with crack extension (resistance-curve or R-curve behavior) as the crack is progressively bridged by unbroken TiNb ligaments in the wake (Fig. 1c); such crack-bridging zone dimensions approach ~4 mm corresponding to steady-state toughness (K_{SSB}) values of ~25 MPa√m. In contrast, the TiNb ligaments fail prematurely under cyclic loading (Fig. 1d), such that crack velocities in the TiAl/TiNb composite can be marginally faster compared to unreinforced TiAl (Fig. 1b).

It is objective of the present study to further examine the cyclic crack-growth behavior in TiAl/TiNb composites, particularly with respect to variations in microstructure and reinforcement orientation. These intermetallic-matrix composites are being developed as advanced structural materials for potential use in future hypersonic and transatmospheric vehicles.

MATERIALS AND EXPERIMENTAL PROCEDURES

The γ-TiAl (Ti-50.5 at.%Al) intermetallic powder was blended with 5, 10 and 20 vol.% of β-TiNb (Ti-33 at.%Nb) particles, consolidated to near net density at 1025°C (235 MPa pressure) using vacuum hot pressing, and then hot forged at 1025°C. As a result, the TiNb particles are

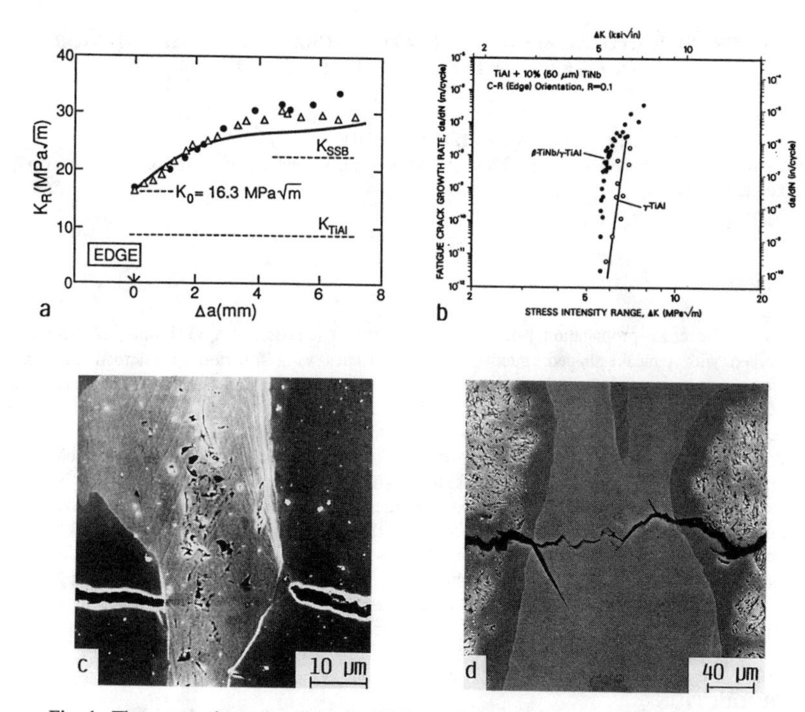

Fig. 1: The contrasting role of ductile TiNb reinforcements during monotonic vs. cyclic crack growth in a TiAl/TiNb intermetallic composite, showing improved toughness compared to unreinforced TiAl (a) from crack bridging by intact TiNb ligaments (c), yet faster growth rates under cyclic loading (b) due to premature TiNb failure (d).

deformed into irregularly dispersed, thin pan cakes (~100-300 μm wide, and 50 μm thick normal to the forging direction) in a TiAl matrix (Fig. 2a). In addition, the thickness of TiNb phase in one of the microstructures was modified to 25 μm, at a constant volume fraction of 20 vol.%. At room temperature, pure γ-TiAl has an elastic modulus of 150 GPa, yield and ultimate strengths between 300-350 MPa with limited ductility (tensile elongation ≤1.5%) and toughness (K_{Ic} ~8 MPa√m). The TiNb solid solution has an yield and fracture strength of 430 MPa with no ability to strain harden because of localized, planar-slip deformation [4]; fracture strains are thus dependent on sample geometry with estimated toughness values in excess of 60 MPa√m.

Cyclic crack-growth behavior under tension-tension loading was examined using compact tension C(T) samples (25 mm wide, 2.5 mm thick) in the edge or C-R orientation (Fig. 2b), where the crack plane is normal to the particle edges, and single edge-notched bend SEN(B) specimens (2.5 mm thick, 50 mm span) in the face or C-L orientation, where the crack plane is perpendicular to the TiNb particle faces [6]. Experiments were conducted in a room-temperature air environment (22°C, 45% relative humidity), using automated servohydraulic testing machines operating under stress-intensity (K) control, at a load ratio ($R = K_{max}/K_{min}$) of 0.1 and a frequency of 50 Hz (sine wave). Crack lengths were continuously monitored using indirect electrical-potential measurements on thin metal foils bonded to the specimen surface, similar to those used for fatigue testing of ceramics [7]. Load-shedding schemes with a normalized K-gradient of -0.1 mm^{-1} were used to obtain growth rates (da/dN) ranging between 10^{-6} and 10^{-12} m/cycle, approaching a threshold stress-intensity range (ΔK_{TH}) below which no appreciable crack growth is seen.

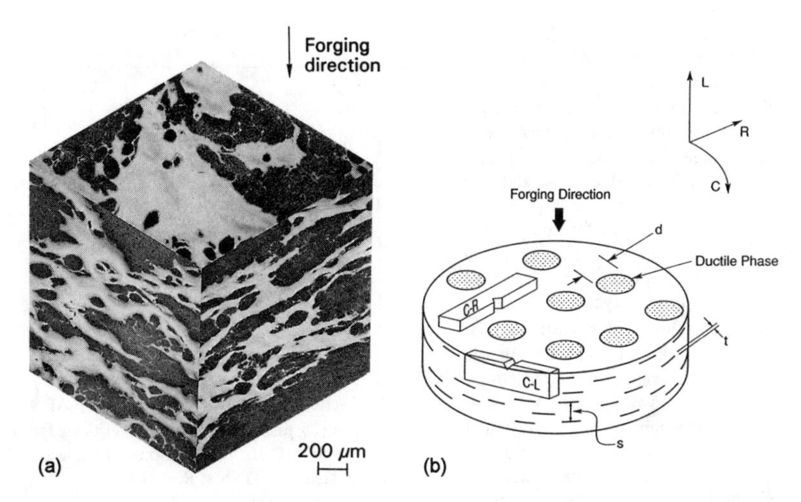

Fig. 2: (a) Three-dimensional optical micrograph of a typical TiAl/TiNb composite microstructure (TiAl + 25 mm-thick 20 vol.% TiNb), and (b) the nomenclature for specimen orientation in the composite; L, C and R refer to the forging, radial and circumferential directions, respectively. C-R and C-L are also referred to as the edge and face orientations.

RESULTS AND DISCUSSION

Reinforcement Orientation Effects

Under monotonic loading, the incorporation of TiNb particles in γ-TiAl leads to remarkable improvements in toughness (Table I), with properties being somewhat better in the face orientation compared to the edge [2-5]. For example, the plateau toughness values measured on resistance curves of γ-TiAl composites reinforced with 20 vol.% TiNb phase are about 36 and 40 MPa√m, in the edge and face orientations, respectively, concurrent with the formation of large (up to 5 mm long), uncracked TiNb bridging zones in the crack wake. With an initiation toughness of 20 MPa√m, the steady-state contribution to toughening from the intact TiNb ligaments (under small-scale bridging conditions) is estimated to be ~15 MPa√m [5].

TABLE I: Summary of Monotonic and Cyclic Crack-Growth Data in TiAl+TiNb Composites

Material	Toughness (MPa√m)	ΔK_{TH} (MPa√m)	Exponent m*	Constant C**
Monolithic γ-TiAl	8	5.8	29.4	9.7×10^{-31}
TiAl + 5% TiNb (50 μm)	--	4.5	17.6	5.3×10^{-18}
TiAl + 10% TiNb (50 μm)	24	5.6	14.1	1.1×10^{-16}
TiAl + 20% TiNb (50 μm)	40	5.0	9.6	2.0×10^{-13}
TiAl + 20% TiNb (25 μm)	--	5.3	9.7	2.5×10^{-13}
Monolithic β-TiNb	60	2.9	3.9	7.4×10^{-9}

* Taken for crack-growth rates between 10^{-9} and 10^{-6} m/cycle

** Given in units of $(MPa\sqrt{m})^{-m}$.m/cycle

Corresponding cyclic crack-growth results in the TiAl +20 vol.% (25 μm thick) TiNb omposite are plotted in Fig. 3 in terms of the growth rate per cycle as a function of the applied stress-intensity range, ΔK (= K_{max} - K_{min}), at R = 0.1, for both the edge (C-R) and face (C-L) orientations; behavior in pure TiAl and TiNb is also shown for comparison. In accordance with previous studies [2], it is clear that cracks can propagate subcritically under cyclic loading at stress-intensity levels of 5-8 MPa√m, far below the crack-initiation and plateau (K_{SSB}) toughness values of the composite. In addition, growth rates are very sensitive to the applied ΔK level similar to behavior in other brittle material systems [8,9]; nearly four orders of magnitude increase in da/dN is observed for ~2 MPa√m increase in ΔK. Expressed in terms of the empirical Paris power-law relationship:

$$da/dN = C\,\Delta K^m$$

the exponent m ~10 for the composite and ~30 for unreinforced γ-TiAl; in contrast, m is on the order of 4 for pure TiNb, similar to values reported for metallic materials (Table I).

Such cyclic crack-growth response is consistent with observations of minimal ductile-ligament bridging by TiNb particles in the crack wake. As shown in Fig. 4, fracture paths in the plane of loading are fully continuous, both in TiAl and TiNb, as the crack traverses the particle without significant blunting and renucleation effects. Similarly, the crack front across the specimen thickness is uniform with TiNb ligaments being fractured to within 150 μm behind the crack tip. In essence, the role of ductile-phase toughening via crack-bridging mechanisms, which is the principal cause for the enhanced monotonic crack growth resistance in these composites, is severely diminished under cyclic loading.

These results clearly demonstrate that the ductile-phase bridging approach

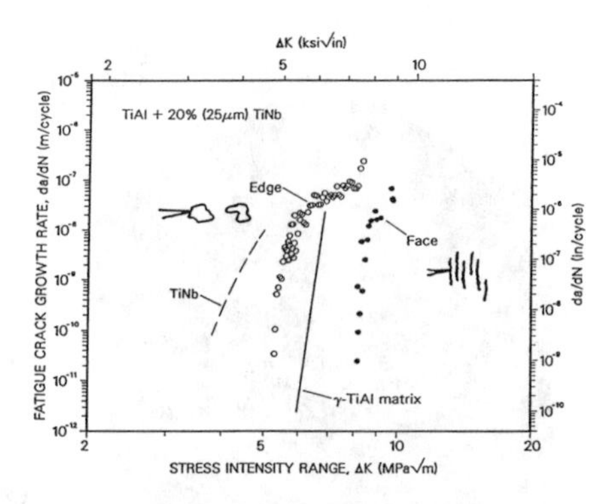

Fig. 3: Cyclic crack-growth behavior in TiAl + 20 vol.% TiNb (25 μm thick) composite in the edge (C-R) and face (C-L) orientations compared to monolithic TiAl and TiNb at R = 0.1.

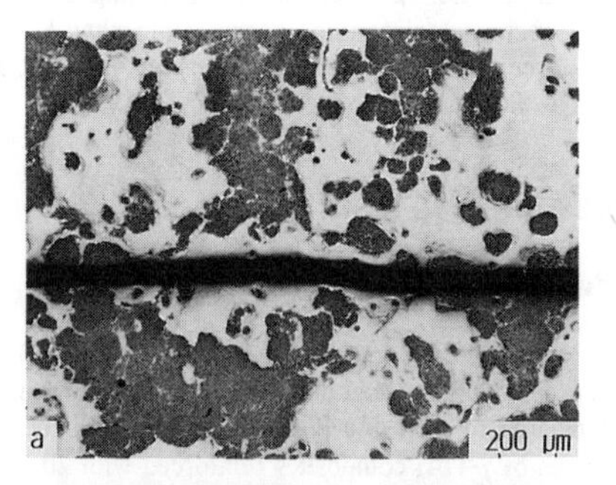

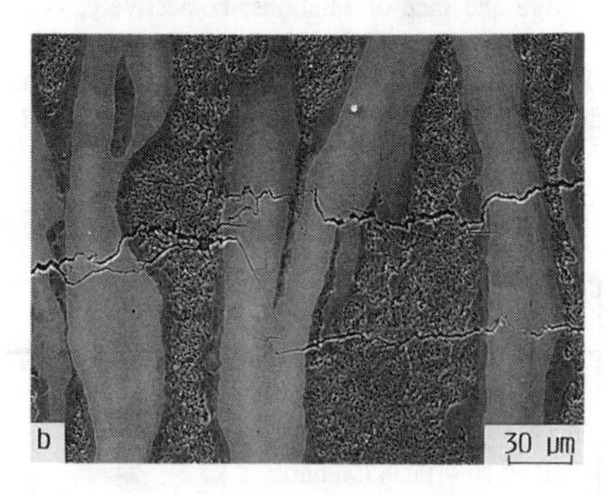

Fig. 4: Scanning electron microscope images cyclic crack-path profiles in TiAl + 20 vol.% TiNb (25 μm thick) composite in the (a) edge (C-R) and (b) face (C-L) orientations, at ΔK levels of ~6 MPa√M. Arrow indicates the crack growth direction.

to toughening intermetallic alloys is far less effective in impeding cyclic crack propagation. Fatigue thresholds are slightly higher and crack-growth rates marginally slower in the TiAl + 20 vol.% TiNb composite, specifically in the face (C-L) orientation, compared to pure TiAl, as fatigue cracks tend to be more deflected and show evidence of coplanar bridging due to multiple cracking and crack branching in the matrix (Fig. 4b). Moreover, there is evidence of crack renucleation in the matrix ahead of the crack tip at high stress-intensity levels above 7-8 MPa√m; yet, bridging-zone dimensions from such cracking phenomena are limited to a few hundred micrometers, compared to several millimeters under monotonic loading. Accordingly, this results in only a modest improvement in cyclic crack-growth resistance, observed by the ~2 MPa√m shift in the da/dN-ΔK curve between 10^{-11} to 10^{-7} m/cycle to the right, compared to the fivefold increase in toughness of TiAl from ~8 to 40 MPa√m. Moreover, cyclic fracture resistance of the composite in the edge (C-R) orientation is compromised by the presence of TiNb particles, particularly at near-threshold ΔK levels (da/dN below ~10^{-10} m/cycle) where crack velocities in the TiAl/TiNb composite are an order of magnitude faster compared to TiAl; crack deflection and bridging effects here are relatively small (Fig. 4a).

Several factors are responsible for the lack of a significant effect on the cyclic fracture resistance of γ-TiAl by the addition of ductile TiNb particles. Typical of most metallic materials, ductile TiNb is susceptible to subcritical cracking by fatigue; properties are inferior to TiAl as shown by the lower threshold values and faster growth rates below ~6 MPa√m (Fig. 3). In other words, fatigue cracks can propagate in pure TiNb at ΔK levels of 3 MPa√m; the high degree of constraint experienced by TiNb particles from lack of debonding at the relatively strong TiAl/TiNb interface [4,5] is further expected to accelerate fatigue cracking in the composite. As a result, crack-tip shielding contributions from ductile-phase bridging during cyclic crack growth are virtually non-existent. Moreover, continuum models for fatigue-crack growth, based on the notion that crack extension is the net result of accumulated damage at the crack tip with number of cycles (or time), which is proportional to the cyclic crack-tip opening displacement (ΔCTOD), yield:

$$da/dN \propto \Delta CTOD \propto \Delta K^2/\sigma_y E$$

where E and σ_y refer to the elastic modulus and the yield stress respectively. In the case of TiAl/TiNb composites, the lower modulus of ductile TiNb reinforcements increases the CTOD and leads to faster crack velocities for a given applied driving force (ΔK). Shielding effects due to crack deflection and coplanar bridging, from multiple cracking and crack branching in the matrix (Fig. 4a) are thought to be the principal reasons for better fatigue resistance in the face orientation compared to the edge.

Microstructural Effects

The influence of microstructure on fatigue-crack propagation in TiAl/TiNb composites is summarized in Fig. 5. The general features of da/dN-ΔK curves are quite similar for all composite microstructures characterized by low fatigue thresholds (compared to steady-state K_R values) and high *m* values (Table 1). Although crack-growth rate behavior is clearly unaffected by particle thickness, the exponent is found to decrease with increasing TiNb volume fraction, consistent with the reduced influence of brittle γ-TiAl cracking on cyclic crack advance. Fatigue thresholds, do not exhibit any clear trend with microstructural variations presumably since the crack-tip damage zone dimensions (e.g., cyclic-plastic zone size) are ~30-40 μm at ΔK_{TH}, typically below the interparticle spacing. In contrast, monotonic toughness and ductility are expected to show a greater dependence on ductile-phase content and reinforcement size [3-5].

In view of these contrasting results on cyclic vs. monotonic crack growth in TiAl/TiNb composites, subcritical fatigue-crack propagation must be considered an important aspect in damage tolerance of ductile/brittle composites. The design of microstructures to date has been to optimize toughness through the development of bridging zones, which may prove counterproductive to cyclic fracture resistance since ductile ligaments are observed to fail prematurely by fatigue at low stress intensity levels. In order to resist fatigue fracture via crack-bridging mechanisms, reinforcements must avoid failure and remain unbroken; this may involve the substitution of aligned and stiffer ductile particles with greater work- hardening capability (compared to TiNb) and weaker interfaces to promote bridging at low stress intensity levels without compromising the transverse properties.

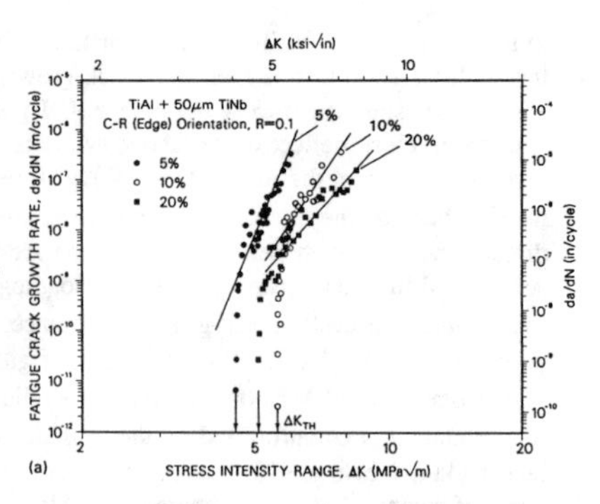

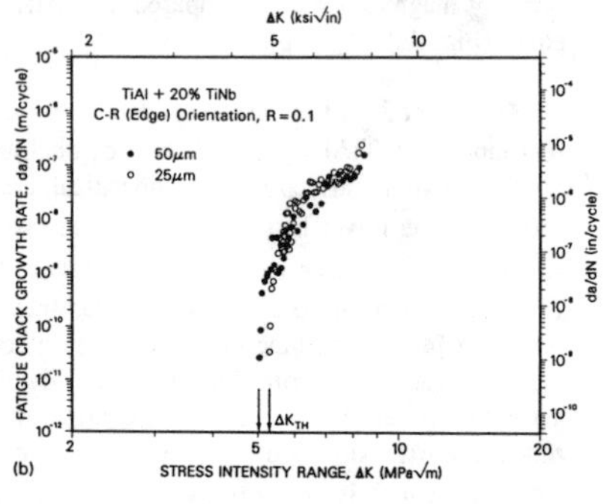

Fig. 5: Microstructural effects of (a) volume fraction and (b) particle thickness on fatigue-crack growth rates in TiAl/TiNb intermetallic composites tested in the edge (C-R) orientation at R = 0.1.

CONCLUSIONS

From an experimental study on the microstructural effects influencing cyclic crack-growth resistance in γ-TiAl intermetallic composites reinforced with ductile TiNb particles, the following conclusions can be made:

1. Ductile particle toughening is less effective in retarding crack advance under cyclic loading compared to results under monotonic loading, as ductile particles are unable to bridge the crack because of their susceptibility to premature fatigue failure; cracks propagate subcritically at ΔK levels of 5-7 MPa√m far below their monotonic toughness levels of ~25-40 MPa√m.

2. Moderate influence of reinforcement orientation on cyclic crack growth is observed. For crack orientations normal to the faces of disc-shaped TiNb particles, crack velocities are slower in TiAl+TiNb compared to pure TiAl; in the edge orientation, however, the composite properties are marginally worse than the unreinforced TiAl matrix.

3. Microstructural effects on cyclic crack growth are relatively minor; measured values for the exponent *m* in the Paris power-law relationship between da/dN and ΔK range between 10-20 and decrease with increasing volume fraction of ductile phase (independent of particle thickness).

Acknowledgements

This work was supported by the U. S. Air Force Office of Scientific Research under Grant No. AFOSR-90-0167, with Dr. A. H. Rosenstein as contract monitor. Our thanks to Prof. G. R. Odette (University of California, Santa Barbara) and Dr. R. H. Dauskardt for helpful discussions.

References

1. M. F. Ashby, F. J. Blunt and M. Bannister, Acta Metall. 37, 1847 (1989).
2. K. T. Venkateswara Rao, G. R. Odette and R. O. Ritchie, Acta Metall. Mater. 40, 353 (1992).
3. C. K. Elliott, G. R. Odette, G. E. Lucas and J. W. Sheckherd, in High-Temperature /High-Performance Composites, edited by F. D. Lemkey, A. G. Evans, S. G. Fishman and J. R. Strife (Mater. Res. Soc. Proc. 120, Pittsburgh, PA, 1988) pp.95-101.
4. H. E. Dève, A. G. Evans, G. R. Odette, R. Mehrabian, M. L. Emiliani and R. J. Hecht, Acta Metall. Mater. 38, 1491 (1990).
5. G. R. Odette, H. E. Dève, C. K. Elliott, A. Harigowa and G. E. Lucas, in Interfaces in Ceramic Metal Interfaces, edited by R. Y. Lin, R. J. Arsenault, G. P. Martins and S. G. Fishman (TMS-AIME Symp. Proc., Warrendale, PA, 1990) pp. 443-63.
6. Anon, American Society for Testing and Materials Standard E399-83, 3.01, 487 (1989).
7. R. H. Dauskardt and R. O. Ritchie, Closed Loop 27, 7 (1989).
8. R. O. Ritchie and R. H. Dauskardt, J. Ceram. Soc. Japan 99, 1047 (1991).
9. R. H. Dauskardt, M. R. James, J. R. Porter and R. O. Ritchie, J. Amer. Ceram. Soc 75, 759 (1992).

THE EFFECT OF FATIGUE LOADING ON THE INTERFACIAL SHEAR PROPERTIES OF SCS-6/Ti-BASED MMCS

PETE KANTZOS,[1] J. ELDRIDGE,[2] D.A. KOSS,[3] AND L.J. GHOSN[4]
[1] Ohio Aerospace Institute, Brook Park, OH 44135
[2] NASA Lewis Research Center, Cleveland, OH 44135
[3] Pennsylvania State University, University Park, PA 16802
[4] Sverdrup, Brook Park, OH 44142

ABSTRACT

Fractographic analysis of SCS-6/Ti-24Al-11Nb(a/o) (Ti-24-11 hereafter) and SCS-6/Ti-15V-3Cr-3Al-3Sn(w/o) (Ti-15-3 hereafter) composites subjected to fatigue crack growth conditions indicates that the interface is prone to wear damage as a result of fiber/matrix sliding. In this study, the effect of fatigue loading on the integrity of the interface was studied by using fiber pushout testing to compare the interfacial shear strength of composite specimens in the as-received condition with specimens that were previously subjected to fatigue loading. Fatigue loading was also simulated by pushing fibers back and forth (multiple reverse pushouts). It was concluded that interfacial sliding during fatigue loading results in interfacial damage and degradation of the interfacial shear strength. Tensile testing of extracted fibers exposed to fatigue-induced interfacial damage was also performed to determine the effect of interface damage on the fiber strength. Interfacial damage also resulted in decreased fiber strength of the SCS-6 fiber. Fracture and wear of the outer carbon coatings on the SCS-6 fiber is the main contributing factor in the deterioration of these interfaces.

INTRODUCTION

Interfacial debonding and fiber/matrix sliding are important mechanisms which influence fatigue crack growth and fracture toughness of composites. These mechanisms play a particularly important role in the crack bridging behavior of composites. Fiber bridging of cracks occurs when fatigue cracks propagate in the matrix leaving the fibers intact in the wake of the crack. Under these conditions the bridging fibers carry some of the applied load and thus shield the crack tip. In the process, the interfaces of the bridging fibers undergo relative fiber/matrix sliding. This phenomenon results in improved fatigue crack growth resistance, toughness, and enhanced fatigue life in composites. Composite systems with interfaces that accommodate interfacial debonding and sliding during fatigue thus promote fiber bridging.

Many of the titanium-based metal matrix composite systems that show promise for advanced aerospace applications are reinforced with SCS-6 (SiC) fibers. These fibers have several carbon-rich outer coatings which are instrumental in accommodating debonding and fiber/matrix sliding. Consequently, these systems also display extensive crack bridging under fatigue loading conditions [1-3]. Recently, however, it has been observed that during the bridging process, the interfaces in these composites exhibit fretting damage [1,3].

The purpose of this study is to investigate the extent of fatigue-induced interfacial damage and its effect on the mechanical properties of the interface and the reinforcing fiber. These issues will be addressed by employing the fiber pushout response to compare the interfacial properties of composites which have been exposed to fatigue crack growth conditions with the interfacial properties of composites in the as-received condition. In addition, tensile testing of fibers extracted from the fatigue crack region was also performed in order to determine the effect of interfacial damage on the fiber strength. The materials used in this study were SCS-6/Ti-24-11 and SCS-6/Ti-15-3 both of which display crack bridging during fatigue crack growth.

EXPERIMENTAL PROCEDURE

Thin composite slices (≈0.5 mm) were cut from bulk composite panels or specimens and polished to final thicknesses of 0.2-0.3 mm. Individual fibers (143μ dia.) were pushed out (on a support plate with a 0.36 mm groove) using flat bottomed tungsten carbide indenters. A displacement rate of .05 mm/min was used. The pushout data was recorded in the form of load versus time in conjunction with acoustic emissions to help identify the debond load. A more detailed description of the equipment and procedure can be found elsewhere [4].

Pushout testing was performed on samples which were previously subjected to fatigue crack growth conditions. As shown in Figure 1, these samples were obtained by cutting slices near the fracture surface of single-edge notched specimens that exhibited crack bridging and failed during fatigue loading. The sample shown in Figure 1 is a failed Ti-24-11 specimen which was tested at room temperature in vacuum at $\Delta\sigma$ = 276 MPa for over 750,000 cycles. Pushout testing near the fracture surface was also performed on a Ti-15-3 specimen which was previously tested at $\Delta\sigma$=200 Mpa for over 420,000 cycles. Generally 25-50 fibers were pushed out over the entire width of the specimen (x=5 mm). In Figure 1, the region of interest, where stable fatigue crack growth occurred, is between x=1.0-2.5 mm. The initial machined notch was 1.0 mm deep and the tensile overload occurred at x ≈2.5 mm. Pushout testing of material in the as-received condition was also performed for use as a reference when comparing and interpreting the effects of fatigue loading on the interface.

In order to simulate the relative fiber/matrix sliding that occurs at the interface during crack bridging conditions, multiple reversed pushouts were performed. These pushouts were performed on material in the as-received condition and are shown schematically in Figure 2. Fibers were typically pushed back and forth 3 or 4 times.

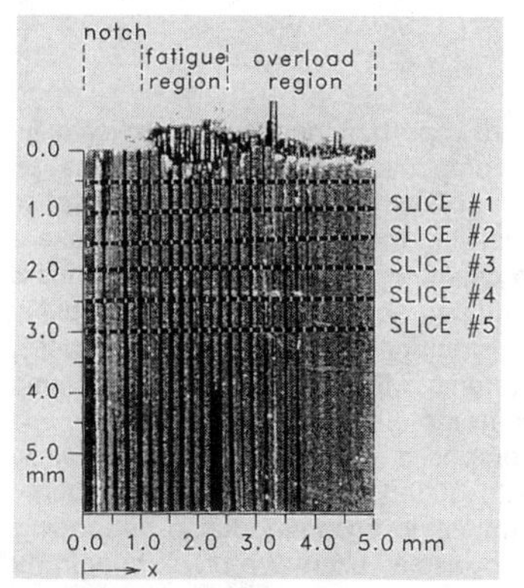

Figure 1. Failed single edge notch specimen (SCS-6/Ti-24-11) showing the approximate location of the slices taken for pushout testing.

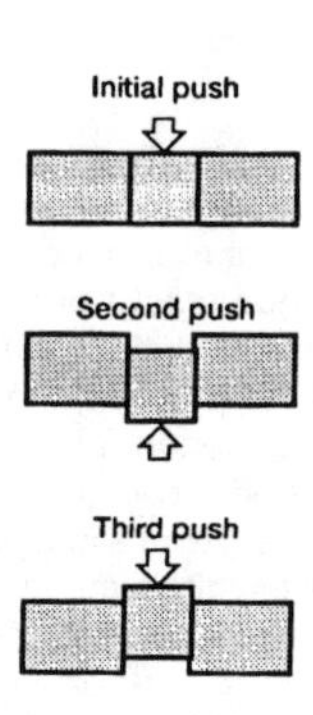

Figure 2. Schematic of the multiple reversed fiber pushout method.

In order to determine the effect of fatigue damaged interfaces on fiber strength, a double-edge notched specimen was mechanically fatigued (at $\Delta\sigma$ = 248 Mpa) in vacuum and room temperature for over a million cycles to produce a bridged crack. The specimen was then chemically etched in a 10%HF/5%HNO_3 bath (for several hours) to dissolve the matrix. Fibers within the fatigue crack growth region (bridging fibers) and fibers ahead of the crack tip (non-bridging fibers) were carefully removed and tensile tested. Fibers were tested using a 12.7 mm gage length and a strain rate of 1.27 mm/min.

RESULTS

Interface properties of composites in the as-received condition

Typical pushout data of both composites in the as-received condition are shown in Figure 3. The load versus time curves shown in Figure 3a and b represent two different types of behavior that was commonly observed in both composites. In both cases the debonding event (P_d) coincides with a load drop and pronounced acoustic emissions. In the first case, Figure 3a, debonding is followed by a gradual decrease in load as the fiber is displaced. This represents the frictional resistance of the interface (P_f). Examination of the pushout fiber (Figure 4a) in this case indicates failure along a single interfacial plane between the carbon coating. In the second case, Figure 3b, debonding is followed by a load increase. This behavior is a result of multiple failures within the outer-carbon coatings of the SCS-6 fibers resulting in a zigzag fracture path. This failure mode results in mechanical interlocking of the carbon coatings as shown in Figure 4. The load needed to break down the interlocking is typically greater than the debond load (P_d). In the case where failure occurs solely within one interface (Figure 3a) limited interlocking occurs since the interface between the carbon coatings is very smooth (roughness $<<1\mu$). On the other hand, when multiple failures between the carbon coatings occur, interlocking can be on the scale of the carbon coating thickness $\approx 3\mu$. In the as-received condition, both composites exhibited chemical bonding.

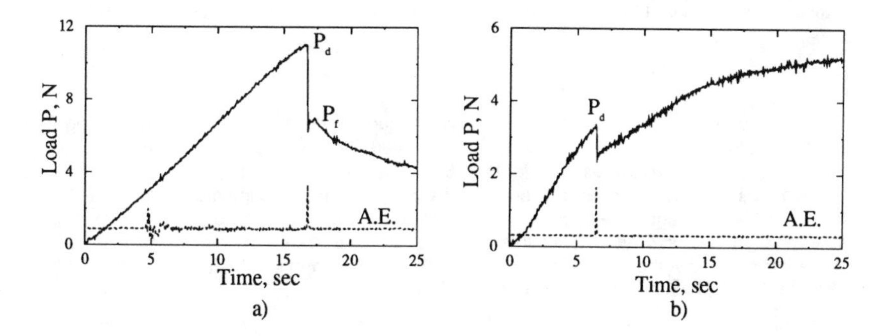

Figure 3. Load versus time response of fibers in the as-received condition; a) Typical behavior observed when the debonding failure occurred along a single interface; b) Typical behavior observed when failure occurred along multiple interfaces.

The interfacial shear strength, calculated by a simple force balance and assuming an average value τ_d along the interface, was determined to be $\tau_d \approx 125$ MPa for SCS-6/Ti-24-11 and $\tau_d \approx 115$ MPa for SCS-6/Ti-15-3. Likewise, the as-received frictional shear strength was $\tau_f \approx 75$ MPa and $\tau_f \approx 50$ MPa for SCS-6/Ti-24-11 and SCS-6/Ti-15-3 respectively. Interfacial failure occurred primarily between the carbon coatings of the SCS-6 fiber in both composites. The SCS-6/Ti-24-11 composite also displayed some interfacial failure between the outer carbon coating and the reaction zone.

Pushouts on previously fatigued specimens

The influence of fatigue damage on the interfacial shear response is shown in Figure 5a along with the as-received response for comparison. The response of the fatigued specimen in Figure 5a clearly suggests that these interfaces were already debonded since no load drop or acoustic emission event was observed. Furthermore, the load at which fiber/matrix displacement

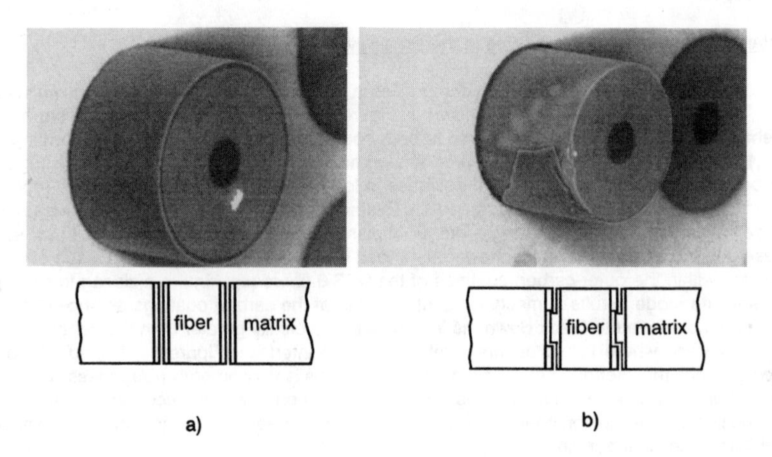

Figure 4. Fractographs of the interface after pushout; a) When failure occurs within one interface, the interface is very smooth; b) When multiple failures occur the interface is interlocked by the carbon coatings.

occurs is much lower than that of the as-received material even after fiber matrix debonding. This suggests fatigue-induced damage occurs at the interface, which is evident in Figure 5b. This behavior was observed consistently for both SCS-6/Ti-24-1 and SCS-6/Ti-15-3 composites. The corresponding average interfacial strength of the previously fatigued interfaces was 32 $\pm$10 MPa for SCS-6/Ti-24-11 and 38 $\pm$11 MPa for SCS-6/Ti-15-3. When compared to the frictional shear strength of the as-received material, this represents about a 57 percent loss (from 75 to 32 MPa) in the frictional resistance of the interface for SCS-`'Ti-24-11 and a 24% decrease (from 50 to 38 MPa) in SCS-6/Ti-15-3.

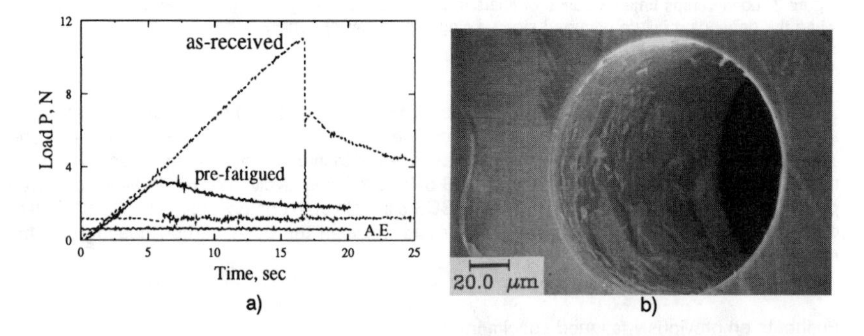

Figure 5. Typical response of previously fatigued interfaces; a) The load versus time behavior of the fatigued interface shows lower debond load and no load drop or acoustic emissions event in comparison to the as-received condition; b) Fractograph of the interface shows significant damage.

As stated earlier, in order to determine the extent of fatigue-induced interfacial damage, several slices were tested near the fracture surface. As shown in Figure 1, for SCS-6/Ti-24-11, slice #1 was taken at 0.5-1.0 mm from the fracture surface, while slice #5 was taken at a depth of about 2.5-3.0 mm. A complete summary of the pushout results for all the slices is shown in Figure 6a and 6b for SCS-6/Ti-24-11 and SCS-6/Ti-15-3 respectively. For the SCS-6/Ti-24-11 composite, the results from slices #1 and #2 were basically identical. All interfaces were debonded and displayed a much lower interfacial shear strength. This was the case even for the fibers in the overload region. This is somewhat disturbing since these interfaces were not subjected to relative fatigue-induced fiber/matrix displacements. This is probably related to deformation during fracture which would result in relaxation of the residual stresses. Subsequent slices (#3 and #5) typically display well bonded interfaces with exception of the fibers in the fatigue crack growth region which again displayed debonded interfaces. However, at this distance from the fracture surface (> 2.0 mm) the strength of the interfaces in the fatigue region were in the order of the as-received friction strength. This implies that the interfaces at this depth are debonded but not significantly damaged; which is consistent with the fact that the relative fiber/matrix sliding (during bridging) would be greater near the crack surface and lower at higher depths. Another interesting phenomenon was observed in the region of the tensile overload, at x ≈2.5 mm for SCS-6/Ti-24-11 and x ≈3.5 mm for SCS-6/Ti-15-3. The debond strength of these fibers were much lower than those typically observed in the overload region implying a larger debond length. This is probably related to large bending stresses which can sometimes result in longitudinal failure (i.e. failure along the interface) in single edge notch and compact tension specimens [1].

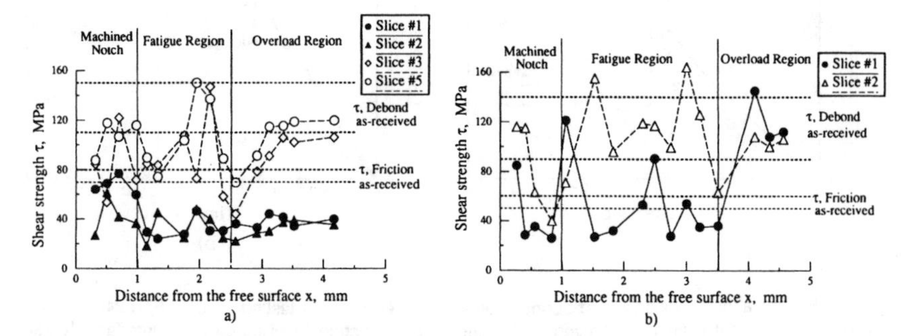

Figure 6. Summary of the pushout results for the various slices tested near the fracture surface; a) SCS-6/Ti-24-11; b) SCS-6/Ti-15-3; (for SCS-6/Ti-15-3, slices #1 and #2 were approximately 0.5-1.0 mm and 1.5-2.0 mm from the fracture surface.) Also shown for comparison are the ranges for debond and frictional strength of the as-received interfaces.

Similar trends were also observed for the SCS-6/Ti-15-3 composite. The main difference in the two composites is that in general the damage and debond zones observed for SCS-6/Ti-15-3 did not extend as deep as observed in the SCS-6/Ti-24-11 system. Likewise, interfacial damage was less extensive in SCS-6/Ti-15-3. As stated earlier, the SCS-6/Ti-24-11 specimen was tested at $\Delta\sigma$= 276 MPa for over 750,000 cycles in comparison to $\Delta\sigma$= 200 MPa and ≈420,000 cycles for the SCS-6/Ti-15-3 specimen. Therefore it seems likely that more interfacial damage can be expected in the SCS-6/Ti-24-11 composite. A schematic profile of the debond and damage zones constructed from the pushout results in Figure 6 are given in Figure 7 for both composites.

<u>Multiple reversed pushouts</u>

Multiple reversed pushouts were performed on thin-slice specimens in the as-received

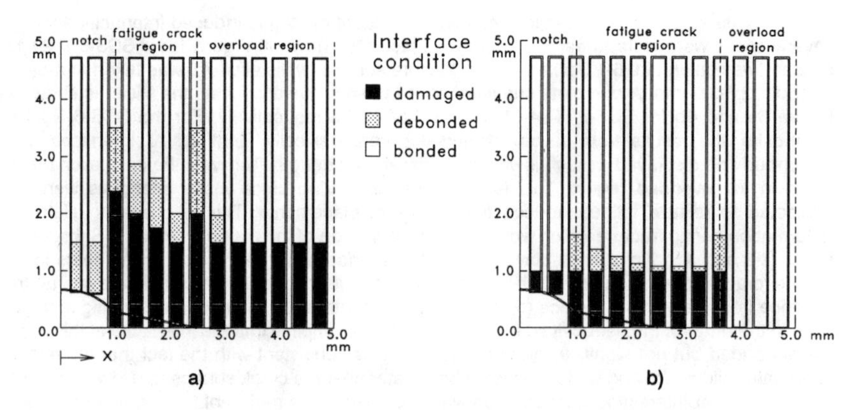

Figure 7. Schematic of the extent of damage and debonding observed during crack bridging; a) SCS-6/Ti-24-11; b) SCS-6/Ti-15-3.

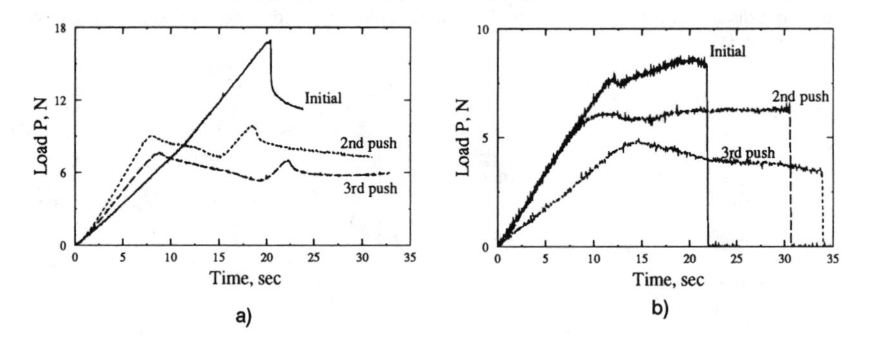

Figure 8. Typical behavior observed during multiple reversed pushout testing; a) SCS-6/Ti-24-11; b) SCS-6/Ti-15-3 (Note that even in the presence of interlocking the frictional resistance of the interface degrades very rapidly).

condition in order to simulate the effect of fatigue and to determine how rapidly interfacial degradation occurs. The fibers were typically displaced between 5-10μ. The results are shown in Figure 8. The initial pushout shows behavior typical of the as-received material with a bonded interface. Subsequent pushouts show a steady decrease in the frictional resistance of the interface. Also as shown in Figure 8b, (for SCS-6/Ti-15-3), the interlocking of the carbon coatings which initially increases the sliding resistance of the interface (after debonding) also degrades significantly on subsequent pushouts. The corresponding interfacial friction stresses for each pushout are given in Table I. The results indicate that the interfaces in these composites degrade very quickly. For example after only 4 pushouts the frictional strength of the interfaces in SCS-6/Ti-24-11 has decreased to 39 MPa which is only slightly higher than the interfacial strength observed near the fracture surface ($\tau \approx 32$ MPa). However, it needs to be noted that the same testing conditions during pushout testing do not prevail during the fatigue crack growth process. For one, the stress state will be different due to the Poisson's effect. Also the relative magnitude of the fiber/matrix displacements during bridging ($\approx 2\mu$) are smaller than those imposed during pushout testing (5-10μ). Likewise, it was also observed that for multiple reverse pushouts over larger distances ($\approx 60\mu$) the interfacial damage and loss of friction strength was more severe.

TABLE I. Multiple reverse pushout results

INTERFACIAL FRICTION STRENGTH (MPa)		
	SCS-6/Ti-24-11	SCS-6/Ti-15-3
INITIAL	75 ± 6	83 ± 7[1]
SECOND PUSH	55 ± 6	64 ± 7[1]
THIRD PUSH	48 ± 8	55 ± 7[1]
FOURTH PUSH	39 ± 7	-
FRACTURE SURFACE	32 ± 10	38 ± 11

[1] Refers to interfaces with multiple failures and extensive interlocking

Fiber tensile testing

Since extensive damage was observed for interfaces that were subjected to cyclic loading, tensile testing was performed on bridging fibers to determine the effect of this damage on the fiber strength. Fibers extracted from the fatigue crack growth region, (bridging fibers) displayed damaged interfaces (primarily in the carbon coatings) while fibers outside the fatigue zone (ahead of the crack tip) did not exhibit surface damage. Fibers whose interfaces were subjected to sliding during fatigue (bridging fibers) had an average tensile strength of 2664 ±514 MPa (10 tests). Fibers near the crack tip, which were subjected to lower fiber/matrix displacements and fewer fatigue cycles had and average tensile strength of 3430 ±286 MPa (5 tests). Fibers ahead of the crack tip (outside the bridged region), whose interfaces were not subjected to relative fiber/matrix sliding, were significantly stronger with an average tensile strength of 4415 ±684 MPa (5 tests). Typically, virgin SCS-6 fibers have strengths greater than 3490 MPa (Textron). Therefore, damage incurred by the interface during fatigue has a detrimental effect on the fiber strength.

DISCUSSION

It is evident that interfacial damage incurred during fatigue results in degradation of the mechanical properties of both the interface and the SCS-6 fiber. The interfacial damage clearly depends on the nature of the carbon-rich outer coatings on the SCS-6 fiber. These coatings contain weak turbostratic (graphitic) phases [5] which are very susceptible to fretting damage during cyclic loading. However, it should be noted that the presence of the carbon coatings in these systems is beneficial both in protecting the fiber and in accommodating crack bridging. Chan [6] has shown that degradation of the interface during crack bridging will reduce the overall effectiveness of the bridging process. Interfacial degradation in combination with loss of fiber strength can also be expected to reduce crack bridging even more. On the other hand, a composite system with a strong fiber/matrix interface (that does not favor crack bridging) would benefit from the use of a fiber coating that degrades during fatigue reducing the interfacial shear strength and thus promoting fiber bridging.

Other factors that influence the extent of interfacial damage include the magnitude of the relative fiber/matrix displacements, the number of fatigue cycles, and the residual stress state. The relative magnitude of fiber/matrix displacements in combination with the number of fatigue cycles seems to have the most drastic effect. As shown by the multiple reverse pushout tests, large displacements ($\approx 10\mu$) will cause significant damage even after very few cycles (3 or 4). Marshall [7] has also shown that for greater displacements ($\approx 200\mu$) some degradation occurs even during monotonic loading. On the other hand, for smaller displacements, as those observed during crack bridging ($\approx 2\mu$), many thousands of fatigue cycles may be needed to produce similar damage.

The extent of fiber/matrix sliding that occurs at the interface during bridging is dictated by the residual radial clamping stresses and the size of the asperities. For example, large clamping stresses will limit the relative fiber/matrix displacements but the higher stresses on the interface will also make it easier to cause damage in the carbon coatings. Eldridge [8] has shown, that the reduction of residual clamping stresses (at higher temperatures) reduces multiple failures and the interlocking between the carbon coatings resulting in less interfacial damage. Likewise, large asperities will also limit fiber/matrix displacements due to greater interlocking. However, in these interfaces the asperities are comprised of carbon coatings which break down very readily, as shown by multiple reverse pushout testing. In other composite systems interfacial asperities may not breakdown as rapidly.

There are many factors other than the inherent properties and characteristics of the interface which influence the magnitude of fiber/matrix sliding and consequently the integrity of the interface. These include the remote applied stresses, R-ratio, crack geometry, testing conditions such as thermal-mechanical fatigue, thermal-cycling and environmental effects. For example, Revelos and Smith [9] have observed that thermal-cycling in combination with oxidation will also result in interfacial damage and degradation in fiber strength.

Finally, it should be noted that the fatigue-induced interfacial damage observed in the present study, is a result of damage to the carbon-rich outer coatings present on the SCS-6 fiber. While it is likely that other fiber/matrix combinations will also display fatigue-induced degradations of the interface, the extent and nature of these effects will likely differ.

CONCLUSIONS

1. Relative fiber/matrix sliding during fatigue crack growth results in interfacial damage and degradation of the interfacial properties and fiber strength.

2. Fracture and wear of the weak outer carbon coatings on the SCS-6 fiber is the main contributing factor in the deterioration of the interfaces in these composites.

3. The relative fiber/matrix sliding distance in combination with the number of cycles greatly influences the extent of interfacial degradation.

REFERENCES

1. P.T. Kantzos, MS Thesis, The Pennsylvania State University, Metals Science and Engineering (1991).
2. M.D. Sensmeier, and P.K. Wright, "The Effect of Fiber Bridging on Fatigue Crack Growth in Titanium Matrix Composites," Fundamental Relationships between Microstructure & Mechanical Properties of Metal-Matrix composites, P.K. Liaw and M.N. Gungor, eds., The Minerals, Metals & Materials Society, 1990.
3. D.L. Davidson, K.S. Chan, and J. Lankford, "Crack Growth Processes at Elevated Temperatures in Advanced Materials, AFSOR ANNUAL REPORT FOR 1989, F4962ac9-89-C-0032, Jan. 1990.
4. J.I. Eldridge, R.T. Bhatt, and J.D. Kiser, Ceram. Eng. Sci. Proc. 12, 1152 (1991).
5. X.J. Ning, P. Pirouz, K.P.D. Lagerlof, and J. DiCarlo, "The Structure of Carbon in Chemically Vapor Deposited SiC Monofilaments," J. Mater., Vol. 5, No. 12, Dec. 1990.
6. K.S. Chan, "Effects of Interface Degradation on Fiber Bridging of Composite Fatigue Cracks," (To be Published).
7. D.B. Marshall, M.C. Shaw, and W.L. Morris, "Measurement of Interfacial Properties in Intermetallic Composites," Titanium Aluminide Composite Proceedings, P.R. Smith, S.J. Balsome, and T. Nicholas, eds., WL-TR-91-4020, Feb. 1991.
8. J.I. Eldridge, "Fiber Push-out Testing of Intermetallic Matrix Composites at Elevated Temperatures," MRS Spring Meeting, San Francisco CA. Apr. 1992.
9. W.C. Revelos and P.R. Smith, Metall. Trans. A, Volume 23A, 1992.

PART II

Nickel Aluminide Composites

INFLUENCE OF INTERFACIAL CHARACTERISTICS ON THE MECHANICAL PROPERTIES OF CONTINUOUS FIBER REINFORCED NiAl COMPOSITES

RANDY R. BOWMAN
NASA Lewis Research Center,
Mail Stop 49-3, 21000 Brookpark Road, Cleveland, Ohio, 44135.

ABSTRACT

As part of a study to assess NiAl-based composites as potential high-temperature structural materials, the mechanical properties of polycrystalline NiAl reinforced with 30 vol.% continuous single crystal Al_2O_3 fibers were investigated. Composites were fabricated with either a strong or weak bond between the NiAl matrix and Al_2O_3 fibers. The effect of interfacial bond strength on bending and tensile properties, thermal cycling response, and cyclic oxidation resistance was examined. Weakly-bonded fibers increased room-temperature toughness of the composite over that of the matrix material but provided no strengthening at high temperatures. With effective load transfer, either by the presence of a strong interfacial bond or by remotely applied clamping loads, Al_2O_3 fibers increased the high-temperature strength of NiAl but reduced the strain to failure of the composite compared to the monolithic material. Thermal cycling of the weakly-bonded material had no adverse effect on the mechanical properties of the composite. Conversely, because of the thermal expansion mismatch between the matrix and fibers, the presence of a strong interfacial bond generated residual stresses in the composite that lead to matrix cracking. Although undesirable under thermal cycling conditions, a strong interfacial bond was a requirement for achieving good cyclic oxidation resistance in the composite. In addition to the interfacial characterization, compression creep and room temperature fatigue tests were conducted on weakly-bonded NiAl/Al_2O_3 composites to further evaluate the potential of this system. These results demonstrated that the use of Al_2O_3 fibers was successful in improving both creep and fatigue resistance.

INTRODUCTION

There is currently considerable interest in developing intermetallics for use in high-temperature applications, where high strength and low density are absolute requirements. Specifically, NiAl is considered an especially attractive candidate due to its combination of high melting point, low density, and excellent oxidation resistance. Although NiAl boasts several advantages over present high-temperature materials, it is currently unsuitable for structural applications due to a lack of high-temperature strength and low toughness below about 500 K. The use of continuous reinforcing fibers is being explored as a means to overcome many of NiAl's inherent deficiencies.

The choice of the reinforcing fiber is limited by requirements of high-temperature strength as well as compatibility with the matrix. The compatibility requirement refers to both mechanical compatibility (similar coefficients of thermal expansion (CTE)) as well as chemical compatibility with the matrix. Additionally, the fiber must be readily available in quantities sufficient for developmental studies. At present, Al_2O_3 fibers are the best choice for meeting these requirements. Interfacial properties are considered key to the overall response of a composite. It is as yet unclear to what degree interfacial control, such as through the use of fiber coatings, will be required to realize the full potential of $NiAl/Al_2O_3$. Before a systematic study of fiber coatings should be undertaken, it will be necessary to clearly understand what effect a weak *vs* strong interfacial bond has on specific composite properties. Only then can the characteristics of the interface be prescribed, and perhaps tailored by the use of coatings, for specific requirements. This paper describes the results of such a study in which composites with either a strong or weak interfacial bond were evaluated over a temperature range of 300 to 1300 K in bending and tension, thermal cycling, and cyclic oxidation. In addition, creep and fatigue testing was conducted to further judge the viability of Al_2O_3 fiber reinforcements for improving NiAl's properties.

EXPERIMENTAL PROCEDURES

All composites in this study were fabricated from prealloyed stoichiometric NiAl powders (50-150 μm) and reinforced with either continuous single crystal 125 μm diameter Al_2O_3 or 200 μm TZM Mo fibers. Molybdenum fibers were included in the study to compare the properties of a ductile reinforcement to those obtained in the brittle Al_2O_3 system. The yield strength of NiAl ranges from 120 to 300 MPa at room temperature to approximately 50 MPa at 1300 K. Below about 500 K, ductility varies from 0 to 2% depending on composition and processing [1]. The Al_2O_3 fibers had as-received strengths of approximately 3200 MPa at room temperature and 600 MPa at 1300 K. No appreciable ductility was measurable in the fibers even at 1300 K. Mo fibers had tensile strengths of 300 MPa at 300 K and 220 MPa at 1000 K. The Mo fibers had a 50% reduction in area at room temperature and over 80% at 1000 K.

Most of the composites were fabricated using the powder-cloth technique [2]. In this process, the matrix material was processed into flexible cloth-like sheets by combining matrix powders with a fugitive organic binder. Likewise, fiber mats were produced by winding the fibers on a drum and applying another organic binder. The composite panel was assembled by stacking alternate layers of matrix cloth and fiber mats. This assemblage was consolidated by hot pressing followed by HIPing to ensure complete densification of the composite. As an alternative to the powder-cloth technique, composite plates were fabricated without the use of fugitive binders by directly hot-pressing fibers in a die of matrix powders, followed by HIPing. Eliminating the use of binders can reduce contamination in the composite, although a disadvantage is that this technique is more labor intensive and can result in a less uniform distribution of fibers.

Four-point bending, uniaxial tension, and fatigue specimens were waterjet machined from the 150 cm x 5 cm x 0.2 cm thick as-fabricated plates. Four-point bend samples were taken from a single composite plate while the thermal cycling and oxidation specimens were machined from a different plate. Multiple plates were needed to fabricate the required number of tensile, fatigue, and creep specimens. Typical dimensions of the bend samples were 5.08 cm long and 1 cm wide. The span of the lower supports was 4 cm and the upper span was 2 cm. The flexure tests were performed in a universal testing machine at a cross-head speed of 0.0508 cm/min over a temperature range of 300 to 1300 K. Tensile specimens, with reduced gage dimensions of 1.524 cm long and 0.635 cm wide, were tested at a cross-head speed of 0.508 cm/min at 300 and 1200 K. Fatigue specimen gage dimensions were 2.54 cm long by 1 cm wide. Fatigue tests were conducted at 0.3 Hz, R = 0, under displacement-controlled conditions between prescribed stress ranges. Tensile and fatigue specimen gripping was achieved using water-cooled hydraulic grips with a 225 kg load. Strain was measured in these specimens using a 1.27 cm gage length extensometer. All high temperature testing was performed in a resistance furnace in air with a temperature gradient of ±3 K.

Thermal cycling was performed in air using the bend sample geometry. The thermal cycle consisted of specimen insertion in a resistance-heated furnace for 10 minutes at 1400 K followed by removal for 15 minutes. The 15 minute hold outside the furnace allowed the samples to cool to 400 K. Oxidation coupons (2.5 cm x 1.25 cm) were cycled between 1473 K and near room temperature with a 1-hour hold at temperature. Rectangular cross-section compression creep specimens were waterjet machined from a 36-ply composite plate. The specimens were 1 cm long x 0.5 x 0.6 cm with the fibers oriented along the compression axis. Compression creep testing was conducted in air under constant load conditions for temperatures between 1200 and 1400 K. Creep rates ranging from 10^{-11} to 10^{-7} were obtained.

For each temperature, four bend tests and three tensile tests were performed. Three fatigue and one creep test was performed at each load-temperature condition. All of the fatigue tests were interrupted at 100,000 cycles if failure had not occurred.

RESULTS AND DISCUSSION

As-Fabricated Properties

Metallography of the as-fabricated composites revealed a fully consolidated matrix with no evidence of fiber or matrix cracking. Fiber volume fractions were approximately 30-35% in all composites although the fiber spacing in powder-cloth processed composites was more uniform than that obtained in the binderless composites. Even though no obvious damage was visible in optical or electron microscopy, a degradation in the tensile strength of fibers etched from consolidated composites has been noted [3]. As-received fibers had average tensile strengths of approximately 3200 MPa whereas those etched from the composite plates fabricated by the powder-cloth technique averaged around 1500 MPa. The reason for the decrease in fiber strength resulting from the fabrication is unknown and is under further study. Fibers have not been etched from plates fabricated without binders so it is unknown whether the strength degradation occurs in these composites as well.

Fiber push-out experiments revealed that bonding between the matrix and the Al_2O_3 fibers was inhibited by the presence of binders. NiAl/Al_2O_3 specimens fabricated by the powder-cloth technique had interfacial bond strengths ranging from 50-150 MPa. The bond was frictional in nature and was a consequence of the matrix/fiber CTE mismatch which caused mechanical clamping of the fibers by the matrix during cooling from the consolidation temperature. Conversely, NiAl/Al_2O_3 composites fabricated without the use of binders had bond strengths in excess of 280 MPa. Only a lower limit can be established since matrix cracking generally occurred prior to breaking the matrix/fiber bond. In the binderless material, the matrix/fiber bond remained in excess of 280 MPa after more than 1000 thermal cycles between 400 and 1300 K. In the few instances when fiber push-out was achieved in the binderless composite, the shape of the load-displacement curve was indicative of a chemical bond between the fiber and matrix. The nature of the chemical bond is unclear since no reaction was visible in either optical or electron microscopy. While the NiAl/Al_2O_3 interfacial bond strength was sensitive to the presence of binders, the bond strength of the NiAl/Mo composites was greater than 280 MPa regardless of the processing technique.

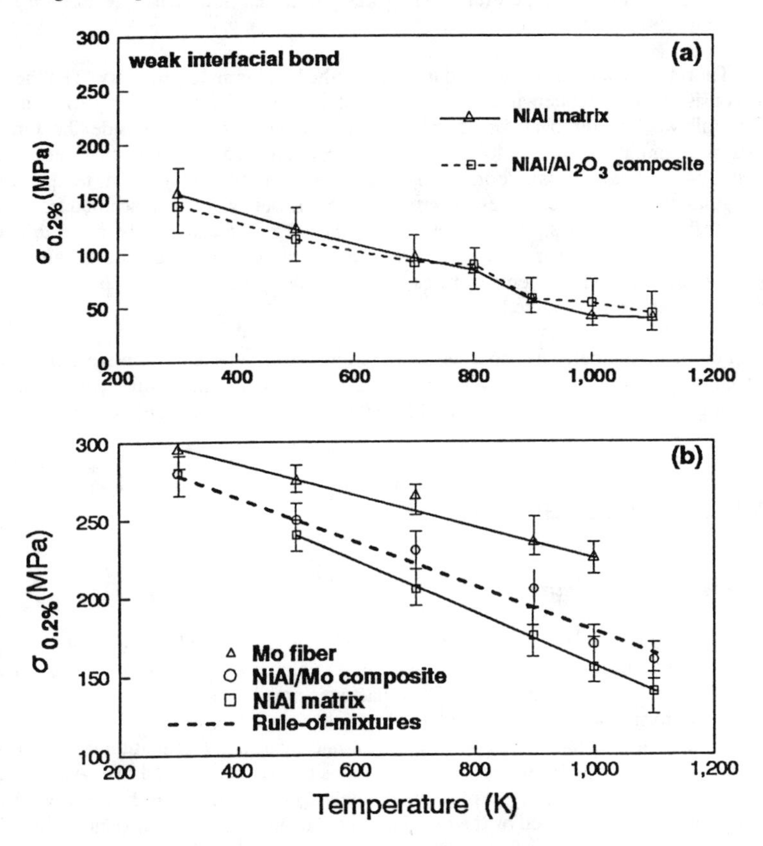

Figure 1. Bend strength at 0.2% strain for (a) NiAl/Al_2O_3 composites with weakly bonded fibers and (b) strongly-bonded NiAl/Mo.

Bend Properties

Bend strength as a function of temperature for weakly-bonded NiAl/Al_2O_3 and strongly-bonded NiAl/Mo composites are shown in Figures 1(a) and (b) respectively. The lack of strengthening in the NiAl/Al_2O_3 composite was a direct result of the weak interfacial bond. It was occasionally observed that the composite strength was actually slightly lower than that of the matrix material. With the weak bond, only a small fraction of the load could be transferred to the fibers, thus reducing the effective load carrying cross-sectional area of the composite, resulting in lower composite strengths. However, while providing no strengthening, a weak interfacial bond is beneficial to composite toughness [4]. Because of the weak interface, debonding occurs in the wake of an advancing crack thereby reducing the driving force for continued propagation. Binderless NiAl/Al_2O_3 specimens have been fabricated and tested in bending at various temperatures. Preliminary results have shown that the binderless specimens have flow strengths which exceeded those of the weakly-bonded composites and approached ROM predictions. This result is not unexpected, but is as yet incomplete. Fiber push-out experiments have not as yet been performed on these specimens to confirm the presence of a strong bond.

Although data on strongly bonded NiAl/Al_2O_3 is relatively scarce, significantly more information has been generated on strongly-bonded NiAl/Mo. As seen in Figure 1(b) the presence of a strong bond resulted in ROM composite strengths. In addition to strengthening, because of the ductility of the fibers, composite toughness was also increased. The properties of NiAl, and in particular the ductility, are highly dependent on processing. In the Mo reinforced material, because of the particular processing conditions, the matrix had no ductility below 500 K. However, with the addition of the Mo fibers, elongations of greater than 0.2% were possible.

Tensile Properties

Tensile results confirmed much of the bend data but also differed in may respects. Most notably, because of differences in fabrication conditions along with NiAl's sensitivity to processing conditions, the NiAl matrix material used for fabricating the composite tensile specimens had much greater ductility than the NiAl matrix of the bend specimens. Room temperature and 1200 K tensile curves for weakly-bonded NiAl/Al_2O_3 and strongly-bonded NiAl/Mo composites are shown in Figure 2. No tensile data is currently available for the strongly-bonded NiAl/Al_2O_3. Strengthening was observed at 300 and 1200 K for both the weakly-bonded NiAl/Al_2O_3 and the strongly-bonded NiAl/Mo. By mechanically gripping the specimen, as required in a tensile test, load was transferred to the fibers even in the absence of a strong interfacial bond, as is the case for the NiAl/Al_2O_3. Several features of the low-temperature tensile curves, in addition to the strengthening, are worthy of mention. The first is the change in composite modulus noted at 60 MPa in the NiAl/Al_2O_3 system and at 50 MPa in the NiAl/Mo. The composite modulus in the initial loading regime (E_1) agreed with ROM predictions. Assuming iso-strain conditions at the point where composite stiffness is reduced the stress in the matrix was calculated to be 41 MPa in the NiAl/Al_2O_3 and 33 MPa for the NiAl/Mo. Both of these values are less than the monolithic yield stress of 120 MPa. Metallographic examinations of specimens interrupted after the change in modulus failed to show any matrix or fiber cracking that could account for the change in stiffness. Therefore, the reduction in modulus is likely to be the result of matrix plasticity rather than fiber and/or matrix damage. Further evidence for this contention will be presented in the fatigue section. The reason for the lower yield stress of the composite can be explained in terms of the residual tensile stresses present in the matrix [5]. These stresses are generated during cooling from the consolidation temperature (1350 K) due to the CTE mismatch of the matrix (16×10^{-6} /K) and fibers ($Al_2O_3 = 9 \times 10^{-6}$ /K, Mo = 4.8×10^{-6} /K). The presence of residual tensile stresses therefore reduces the applied yield stress of the composite. NiAl/Mo composites yielded at even lower stresses than the NiAl/Al_2O_3 because of the larger CTE mismatch (and hence larger residual tensile stresses in the matrix) between NiAl and Mo. At 1200 K (Fig. 2(b) and (d)), the residual stresses are relaxed and the calculated stress in the matrix corresponding to the change in composite modulus was 91 MPa which is in close agreement to the 82 MPa yield stress of monolithic NiAl.

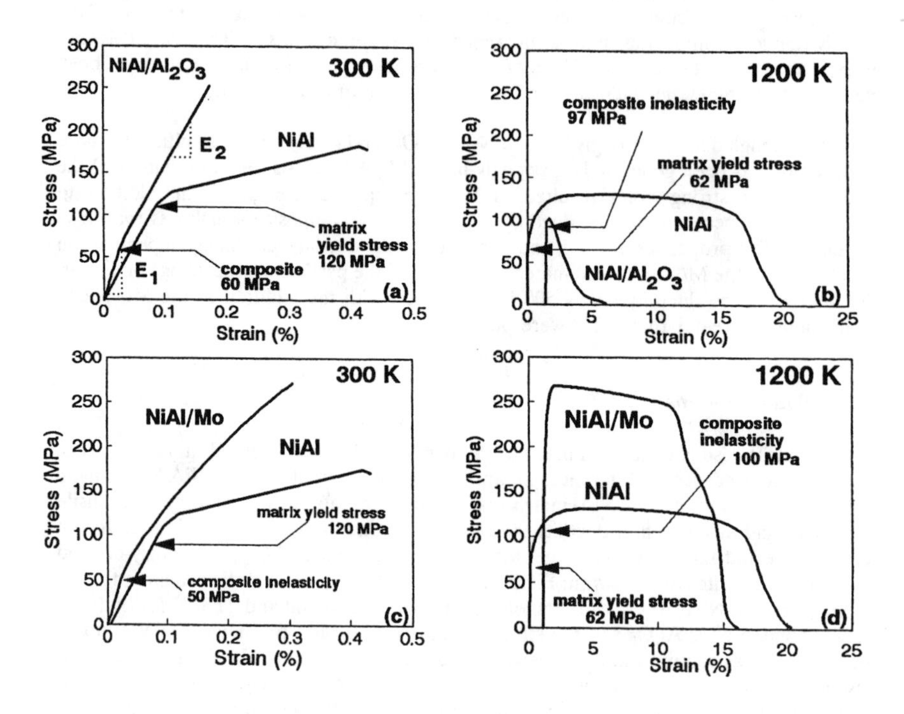

Figure 2. 300 and 1200 K tensile curves of (a)-(b) NiAl and weakly-bonded NiAl/Al_2O_3 and (c)-(d) NiAl and NiAl/Mo.

Another point to be made about the tensile curves in Figures 2(a) and (b) is that the toughness of NiAl/Al_2O_3 is less than monolithic NiAl. The reduced toughness (or failure strain) of the NiAl/Al_2O_3 composite can be attributed to the brittle nature of the fibers. Previous work has shown that failure of Al_2O_3 fibers in a FeAl matrix occur at approximately 0.17% strain in tension [6]. In the case of the NiAl/Al_2O_3 composites, the relatively low matrix ductility is incapable of tolerating flaws which are created from the fiber failures. Consequently, the composite fails at the onset of fiber failure. As for the NiAl/Mo composites, toughness is not severely reduced because of the ductility of the fibers which tend to neck rather than fracture. Although the strong bond and ductility of the Mo fibers results in a good balance of properties, this system may not be a practical high-temperature structural candidate due to the severe oxidation and high density of Mo.

Thermal Cycling

Thermal cycling experiments were undertaken on $NiAl/Al_2O_3$ composites to determine if the CTE mismatch would cause measurable composite damage. If so, coatings may be required to serve as a compliant layer between the fiber and matrix. In the weakly-bonded specimens no evidence of either fiber or matrix cracking was observed. Bend testing of thermally cycled weakly-bonded specimens also failed to show any degradation of properties, either in terms of strength or failure strain, as a result of the thermal cycles. Strongly-bonded composites when subjected to the same thermal cycling conditions developed extensive cracking in both the matrix and fibers. This is a result of the large internal stresses generated by the CTE mismatch which are not alleviated by interfacial debonding and fiber sliding due to the strong matrix/fiber bond.

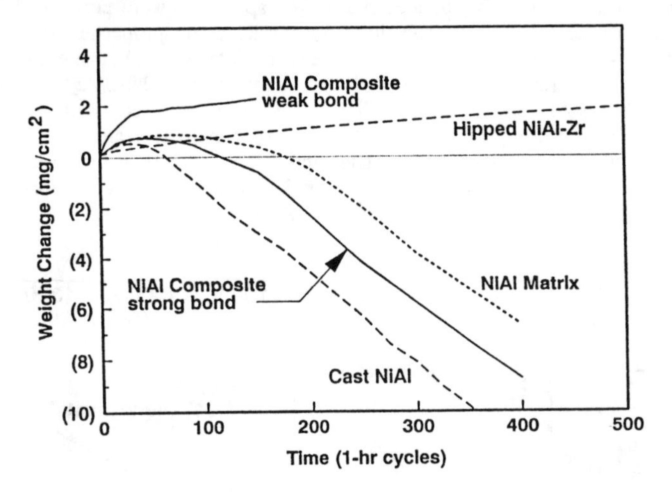

Figure 3. Cyclic oxidation behavior of $NiAl/Al_2O_3$ composites containing either strongly or weakly bonded fibers.

Cyclic Oxidation

While the thermal cycling experiments provided data on the effect of CTE, cyclic oxidation tests were conducted to determine the effect of longer hold times at temperature where significant oxidation could occur. The weight change data in Figure 3 shows that in the weakly-bonded material the oxidation resistance has been reduced from that of the matrix. Metallography revealed that with a weak interfacial bond, oxidation occurred along the matrix/fiber interface into the interior of the composite [7]. Protective oxide scales that formed in the interface were damaged during cool-down due to the thermal clamping of the matrix. On subsequent heat-up, fresh matrix material was exposed to the atmosphere and thus provided a regenerating source for oxidation. Although a detailed metallographic examination of the strongly-bonded composite is not available, it is thought that since matrix/fiber debonding did not occur, the strong bond may prevent oxidation from occurring along the fiber length. As a result, the properties of the strongly-bonded composite were equivalent to that of the monolithic material [8]. Although oxidation resistance of the fiber and matrix individually are excellent, this in itself does not necessarily yield an oxidation resistant composite. An additional requirement is that a strong chemical bond which remains intact during cycling, forms between the fiber and matrix . Alternatively, the fast diffusion path provided by the interface also may be eliminated by processing in such a way that fiber ends are not exposed to the oxidizing atmosphere. Or if fiber ends are exposed, they could be protected by applying an oxidation resistant coating.

Compression Creep

To further judge the merits of continuous fiber Al_2O_3 reinforced composites, compression creep of weakly-bonded NiAl/Al_2O_3 specimens were performed. Although compression creep data is no substitute for tensile creep properties, it does provide information which can compliment the tensile creep data when it becomes available. The steady state creep rates in Figure 4 show that in compression, NiAl/Al_2O_3 is far superior to monolithic NiAl and approaches that of the most advanced NiAl alloys currently available. The low creep rates were obtained in specimens subjected to strains of up to 10%. Beyond this point, specimen buckling and delamination occurred. Although tensile creep data is not yet available, the compression results are very encouraging and in themselves justify continued optimization of this system.

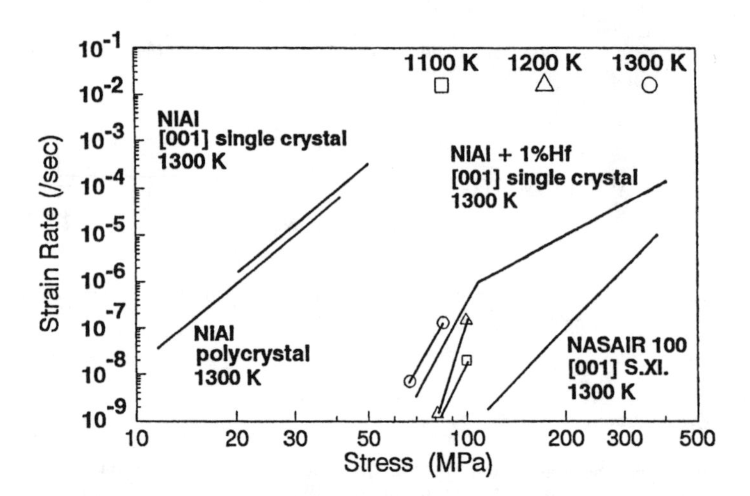

Figure 4. Compression creep rates of weakly bonded NiAl/Al_2O_3 composites compared to advanced NiAl single crystals. Single crystal NiAl and NiAl(Hf) data taken from refs. [9] and [10] respectively.

Fatigue

Due to the low ductility of the matrix, there was some concern that the matrix/fiber interface might act as an initiation site for cracks. Such cracks would severely degrade the fatigue resistance of the composite. To explore this possibility, room temperature fatigue tests were conducted on NiAl/Al_2O_3 specimens at various stress levels. Initial loading was limited to maximum strains less than approximately 0.2% which is the tensile fracture stress of the composite (Fig. 2). In Figure 5, typical stress-strain curves for NiAl and NiAl/Al_2O_3 composites are shown. In both cases, strain ratchetting occurred because of the inability to impose compressive loads on the flat panel specimens.

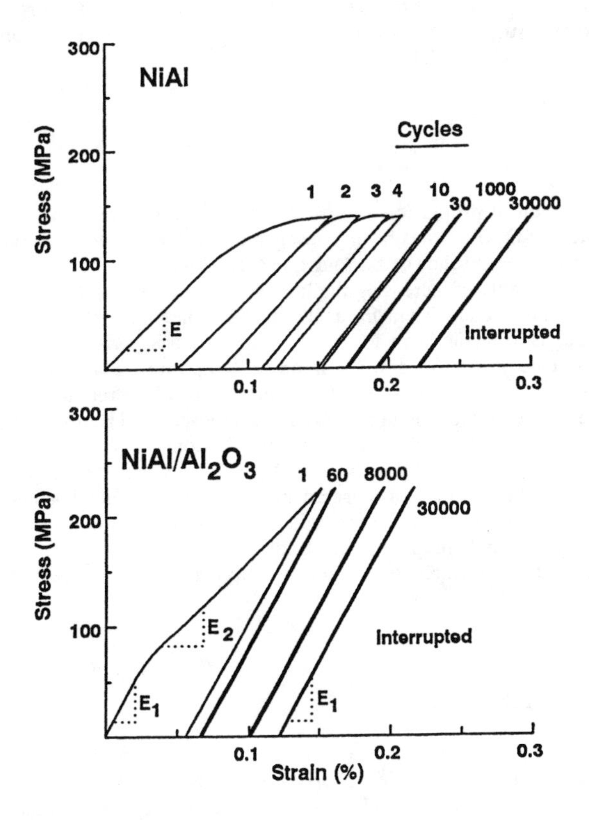

Figure 5. Interrupted stress-strain curves for (a) NiAl and (b) weakly bonded NiAl/Al_2O_3 specimens fatigued at room temperature.

In the preliminary test thus far conducted at various initial strain levels, all tests were interrupted if the lives exceeded 100,000 cycles. For an initial strain of 0.15% (which approaches the tensile fracture strain of the composite of 0.18%) both the monolithic and composite remained intact after 100,000 cycles. Although both materials survived cycling, accumulated strain in NiAl was continually increasing while the composite had reached equilibrium by 30,000 cycles. As shown in Figure 5, after 30,000 cycles the monolithic material had rachetted to an accumulated strain of 0.22% and was continually increasing while the NiAl/Al_2O_3 composite had saturated at 0.12% strain. Also of importance, it is noted in Figure 5 that the unloading modulus for the composite was equivalent to that of the initial loading regime. This is further evidence, in addition to the metallography discussed earlier, that the change in modulus is due to matrix yielding and not fiber and/or matrix damage. Metallography of the fatigue specimens after 100,000 cycles, which corresponds to an accumulated strain of 0.12% in the composite and 0.35% in the monolithic, revealed no cracking in either the matrix or fiber. Therefore, for loads below the fracture stress of the composite, the presence of the fibers did not degrade fatigue resistance, at least for lives up to 100,000 cycles. Incorporation of fibers does, however, limit the maximum strain that can be initially imposed without failure to approximately half that of the monolithic material.

CONCLUSIONS

The mechanical properties of a NiAl composite reinforced with either weakly or strongly-bonded Al_2O_3 fibers has been investigated. With effective load transfer, which is possible either by the presence of a strong interfacial bond or by externally applied loads, Al_2O_3 fibers are capable of increasing the high-temperature strength of NiAl. The use of fiber coatings is not a requirement for achieving well bonded fibers although they may be necessary to act as a ductile layer for blunting matrix cracks. However, even the use of coatings will not be sufficient to vastly improve composite failure strains owing to the limited ductility of the Al_2O_3 fibers. A strong interfacial bond is a requirement for oxidation resistance and can be obtained in uncoated fiber composites. CTE mismatch does not lead to composite damage in the weakly-bonded material. Conversely, the high CTE mismatch combined with a strong interfacial bond did result in severe crack formation in the composite due to the thermally generated internal stresses. Barring improvements in matrix ductility, fiber coatings will be required as a compliant layer between the matrix and Al_2O_3 fibers. Fatigue and creep data on weakly-bonded $NiAl/Al_2O_3$ also indicate that the use of Al_2O_3 fibers is a viable technique for improving these properties over that of monolithic NiAl.

ACKNOWLEDGMENTS

I would like to thank Dr. Jeff Eldridge for allowing access to the fiber push-out equipment and for his many helpful discussions and suggestions. I would also like to acknowledge Dr. Jim Nesbitt's efforts in leading the cyclic oxidation study and greatly appreciate his permission to include previously unpublished results in this paper.

REFERENCES

1. R.R. Bowman, R.D. Noebe, S.V. Raj, and I.E. Noebe, *Metall. Trans.*, Vol. 23A, No. 5, 1992, pp. 1493-1508.

2. J.W. Pickens, R.D. Noebe, G.K. Watson, P.K. Brindley, and S.L. Draper, NASA Technical Memorandum 102060, (1989).

3. J.P. Pickens, NASA Lewis research Center, unpublished research, 1992.

4. R.R. Bowman, R.D. Noebe, J. Doychak, and K.S. Crandall, *in 4th Annual HITEMP Review - 1991*, NASA CP-10082, pp. 43-1 to 43-14.

5. P.K. Brindley, S.L. Draper, J.I. Eldridge, M.V. Nathal, and S.M. Arnold, accepted for publication, *Metall. Trans*, 1992.

6. S. L. Draper, P.K. Brindley, and M.V. Nathal, in *4th Annual HITEMP Review - 1991*, NASA CP-10082, pp. 44-1 to 44-12.

7. J. Doychak, J.A. Nesbitt, R.D. Noebe, and R.R. Bowman, "Oxidation of Al_2O_3 Continuous Fiber Reinforced NiAl Composites": *in 15th Conference on Metal Matrix, Carbon, and Ceramic Matrix Composites*, cosponsored by NASA and the Department of Defense, Cocoa Beach, Florida, January 16-18, 1991.

8. J. Nesbitt and R. Bowman, NASA Lewis Research Center, unpublished research, 1992.

9. R.D. Noebe, NASA Lewis Research Center, unpublished research, 1991.

10. R.R. Bowman, I.E. Locci, and J.D. Whittenberger, NASA Lewis Research Center, and R. Darolia, GE Aircraft Engines, unpublished research, 1992.

MECHANICAL EVALUATION OF FP ALUMINA REINFORCED NiAl COMPOSITES

D.L. Anton* and D.M. Shah**
* United Technologies Research Center, E. Hartford, CT 06084
** Pratt & Whitney, E. Hartford, CT 06084

ABSTRACT

NiAl is well known for its low density, superb oxidation resistance and low ductile to brittle transition temperature. It is equally renowned for its low room temperature fracture strength and poor high temperature creep strength. A compositing approach has been used to introduce chopped and aligned FP Al_2O_3 into a matrix of fine grain NiAl via a powder metallurgical approach. This resulted in composites with approximately 40 vol. per cent undamaged alumina reinforcement. Fiber orientation effects on strength have also been characterized. Room temperature tests resulted in yield strength increases of 425% for chopped FP reinforcements and 800% for aligned composites. Elevated temperature tests conducted at 1200°C were even more dramatic in their strength increment with 200% and 3600% increases respectively. Fractographic results show matrix ductility, fiber-matrix decohesion and minimal fiber pull-out.

INTRODUCTION

For high melting point intermetallics to become commercially viable as structural materials, their well known lack of low temperature ductility and fracture toughness will need to be addressed. One method of imparting damage tolerance into brittle ceramics is by incorporating strong fibers into them. These reinforcements, typically SiC, Si_3N_4 or Al_2O_3, impart damage tolerance through a fiber pull-out mechanism [1,2]. In addition to damage tolerance, NiAl which has excellent oxidation resistance and a leading candidate for gas turbine application, also will need to be strengthened at high temperatures where its creep resistance has been shown to be very poor [3]. Thus the introduction of strong fibers into an NiAl matrix can lead to both enhanced damage tolerance as well as increased high temperature strength. NiAl was also chosen as an ideal matrix candidate because a great deal of information is available in the literature on its ambient and high temperature properties.

A recent study [4] of W(218) and Saphicon® fiber reinforced NiAl illustrated the effects of composite ply geometry on the strength of large fiber diameter, 250μm, composites. Much of this effect stemmed from inter-fiber defects in the matrix.

Alumina fibers were chose as the preferred reinforcing phase because of their chemical stability [5] and commercial availability. Specifically, DuPont's FP® alumina fibers were used because of their fine diameter, approximately 12μm, which would allow them to be incorporated easily into thin wall structures. As such, two reinforcement geometries were studied in this system. Specifically, aligned FP fibers and chopped and randomly aligned FP fibers. In this study we will conform to the following convention for referring to the composites: Matrix/Reinforcement, where the reinforcement will be designated AFP for Aligned FP and CFP for Chopped FP.

EXPERIMENTAL PROCEDURES

The NiAl powder was received from Homogeneous Metals Inc. in the form of +100 mesh powder. The as-received material was dry ball milled in an argon atmosphere with ultrasonic classification, also in argon, first to -400 mesh and finally to -10μm. Particle size distribution measurements were made to verify largest, average and mean particle sizes to a resolution of 0.1μm.

Mat. Res. Soc. Symp. Proc. Vol. 273. ©1992 Materials Research Society

Monolithic NiAl specimens were prepared by vacuum hot pressing -400 mesh NiAl powder at 1400°C, 34.5 MPa (5 ksi) pressure for 1.5 hours. NiAl/CFP composites were prepared by high energy dry blending -10μm NiAl powder with 14.4 wt./o FP fibers chopped to approximately 1 cm lengths then hot presses at conditions identical to those given above. Continuous tows of FP fiber were immersed in an agitated solution of -10μm NiAl powders with a volatile binder. After fiber tow infiltration, they were aligned and stacked, the binder driven off at 200°C, and the composite hot pressed at 1400°C for 30 minutes under 3.4 MPa (0.5 ksi) pressure.

The mechanical evaluation of the composites consisted of four basic tests, (i) bending, (ii) creep, (iii) fracture toughness and (iv) thermal cycling. All of these tests were carried out on specimens that were nominally 3.17 x 0.63 x 0.32 cm. Bend testing was performed using a 2.54 cm lower span and 1.27 cm upper span at a constant cross head displacement of 0.025 cm/min. Testing was conducted at temperatures to 1200°C, with the elevated temperature tests conducted under an argon atmosphere. Strain was measured using a center point deflectometer and recorded as a function of load. By using the procedure outlined in ASTM D-790, stress-strain curves representative of the outside tensile ligament of the specimens could be calculated. Data observed after tensile ligament cracking is not indicative of true strength, but rather is of more engineering significance in that pseudo plastic strain and fiber pull-out could be identified. Additionally, signs of graceful failure and damage tolerance could also be distinguished. Strain to failure, as reported in this study, is defined as the total strain to 50% load drop from the maximum stress. In many instances where extensive fiber pull-out caused pseudo plastic deformation or general matrix yielding occurred, this failure strain was significantly higher than the strain to peak stress.

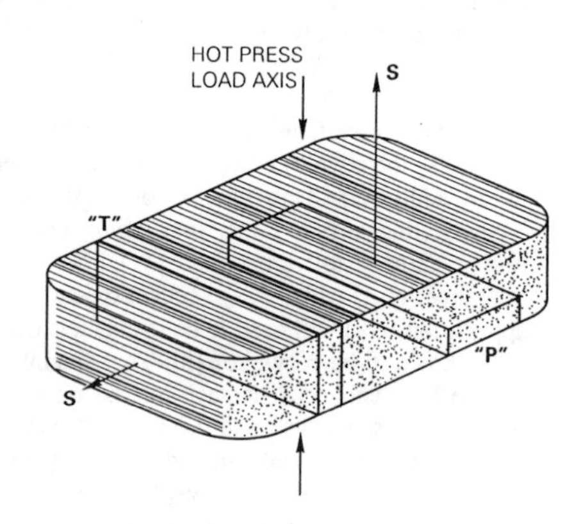

Figure 1 Specimen Orientation with Respect to Hot Press Axis.

Bend test specimens were cut in two orientations with respect to the hot press load axis. These two orientations are designated "P" and "T" as shown in Fig. 1. Orientation "P" refers to the specimen plane normal, **S**, being parallel to the hot press load axis. In the case of orientation "T", the specimen plane normal, **S**, is perpendicular to the hot press load axis. Due to geometrical constraints, it was found that these two composite orientations result in different properties and thus the orientation is imbedded in the specimen number. In most cases the specimens were cut in the "P" orientation, and thus no special addition was made to the specimen number (i.e. IMC-021-1) while those of the "T" orientation always have the letter "T" following the specimen number (i.e. IMC-038-1T).

Creep testing was also carried out under bending conditions to test the composites in their weakest mode, that of tensile loading. The specimen dimensions were nominally identical to those described above, with all testing being conducted at 1200°C under an argon atmosphere. All other test details were also identical to those of the bend testing, but load was maintained constant at 1 to 10 ksi, and the tests conducted until 3% bend strain deformation was achieved.

Fracture toughness testing was conducted in bending as per ASTM E-819 and ASTM E-399. These tests were all conducted under ambient conditions on notched specimens having nominal dimensions identical to those cited above. The notch consisted of a 45° cut with a 0.08" maximum radius at its root. Valid K_{IC} results were obtained from this test configuration.

An evaluation of the stability of the composite under cyclic thermal conditions was gained using either one of two test sequences. The first placed the specimen in a 1200°C furnace for 55 min. followed by removal to ambient conditions for 5 min. and is referred to as 55/5., while the second test utilized a 20 min. in, 10 min. out cycle at 1200°C, referred to 20/10. This was carried out for 100 cycles after which the specimen was studied through optical examination.

RESULTS AND DISCUSSION

Optical microstructural characterization of the monolithic NiAl showed the material to be essentially pore free with prior particle boundaries clearly delineated. A very uniform and high degree of matrix density was, obtained from hot pressing NiAl/AFP composites using -10μm NiAl powder as shown in Fig. 2. The microstructure of the NiAl/CFP is given in Fig. 3 in the transverse orientation, viewing perpendicular to the hot press axis. Because of the axial nature of hot press consolidation, a true random distribution of fibers could not be obtained. Instead, the fiber segments are random in the plane perpendicular to the applied load while they lie solely with in that plane. This planar isotropy is graphically depicted in Fig. 1. All three of these materials can be characterized as having a fully consolidated matrix while the two composite structures maintained a well dispersed reinforcement with minimal damage.

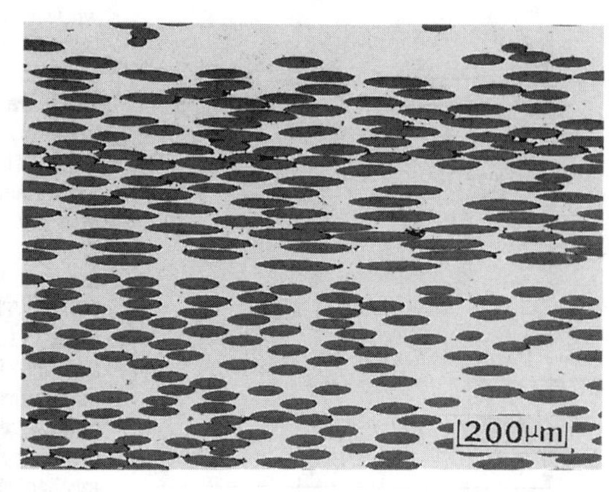

Figure 2 Optical Micrograph of NiAl/AFP Composite

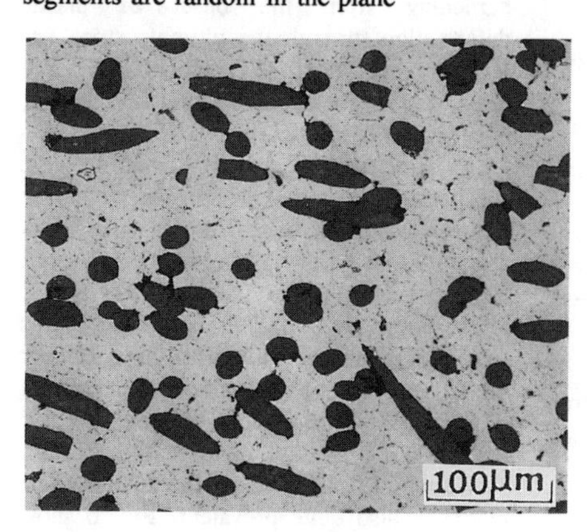

Figure 3 Optical Micrograph of NiAl/CFP in Transverse View

The results of bend test evaluation of NiAl/AFP, NiAl/CFP and the monolithic NiAl are given in Table I. This table gives the specimen number, test temperature, maximum stress, proportional limit yield strength, modulus, 50% load drop failure strain, and a qualitative assessment of the stress-strain curve.

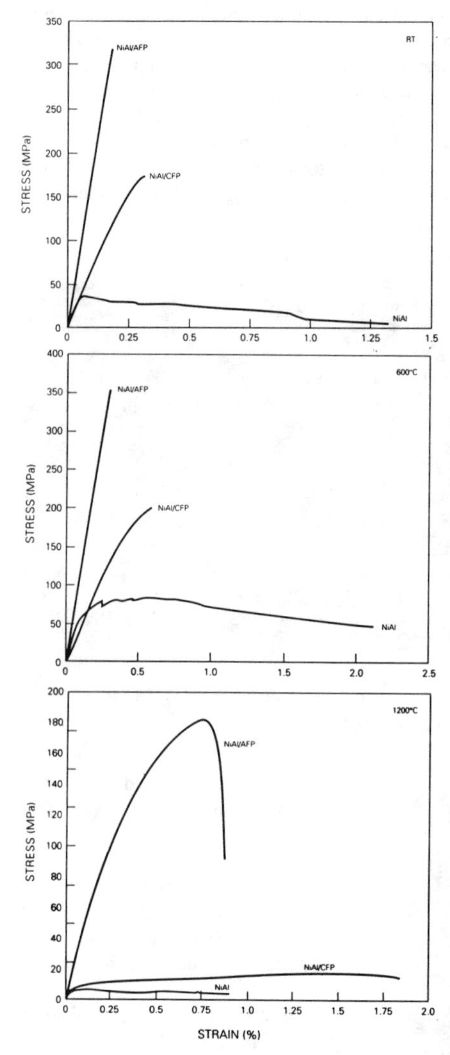

Figure 4 Comparison of Bend Test Stress Strain data for NiAl and its composites at (a) 25, (b) 600 and (c) 1200°C.

Bend Testing

The monolithic NiAl bend test results are in general agreement with the literature findings. Since this is a relatively fine grained material, starting from -400 mesh powder, a degree of plasticity can be achieved at room temperature. The doubling of strength from ambient to 600°C is due to an increase in fracture stress along with trebling of ductility. At 1200°C strength dropped dramatically while plasticity increased to the point where fracture strain exceeded the bend test fixture capabilities. Stress-strain plots for monolithic NiAl are given at 25, 600 and 1200°C in Figs. 4a, b and c respectively.

The aligned composites displayed more linear elastic behavior at 25 and 600°C than the monolithic. This is depicted in Fig. 4a and b where a comparison can be made for the NiAl/AFP composites at 25 and 600°C. The maximum strengths increased from 35 to 240 MPa at ambient temperatures, a 580% increase. At 600 and 900°C, the maximum strengths increased 320% and 2300% respectively. It is apparent that the brittle reinforcement is dominating the mechanical properties of the composite with vastly increased strength and much more brittle failure. The stress-strain curve at 1200°C, Fig. 4c shows significant strain hardening from the onset of plastic deformation through the ultimate stress, as the individual fibers are being loaded. Failure is rapid after the ultimate stress is achieved, but it is not catastrophic.

Fractography was conducted on these specimens to gain insight into the composite failure mechanism. Fig. 5 shows evidence of fiber pull-out , but on a very limited level. Regions where the fibers were closely spaced led to large shear walls. These shear walls were not numerous enough to greatly affect properties, but they do point out the necessity of a highly uniform microstructure and also the weak matrix/fiber bond in this system. Transgranular cleavage of the intermetallic matrix is also quite prevalent. Fig. 6 show the fracture surface of the 600 and 1200°C tested specimens, respectively. It can be noted that

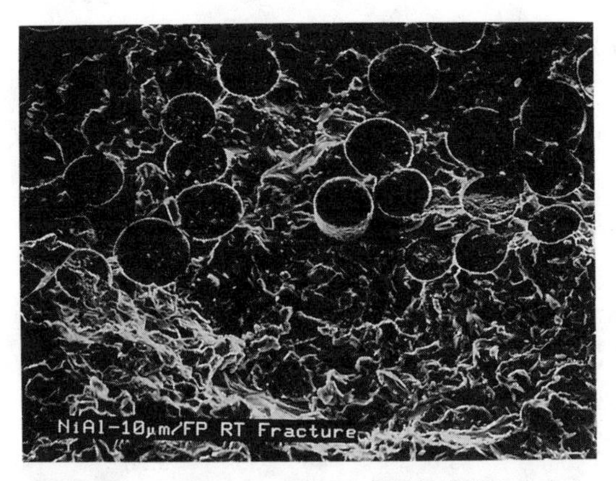

Figure 5 Fracture surfaces of NiAl/AFP tested at 25°C.

there is a transition to ductile tearing of the matrix as temperature increases. At 1200°C, tear lips resulting in extended fiber/matrix debonding are evident.

Bend tensile testing was carried out on specimens from two hot press runs in the "T" orientation. At all of the test temperatures, the specimens from IMC91-038 were by far the strongest. The former hot press run was carried out at a pressure of 34.5MPa (5.0 ksi), a factor of ten greater than for the latter. This increased pressure resulted in much better bonding between the

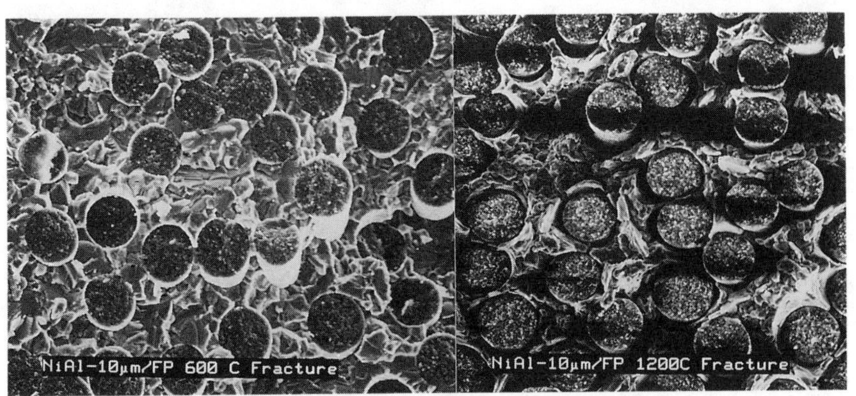

Figure 6 Fracture surface of NiAl/AFP tested at (a) 600°C and (b) 1200°C showing increasing ductile tearing with temperature.

prior powder particles, and a stronger composite. Specimens originating from the IMC91-038 plate also displayed significantly greater failure strains and plastic deformation at room temperature.

Fractography conducted on failed bend specimens at ambient temperature showed little evidence of fiber pull out. However, surfaces of exposed fibers lying parallel to the fracture plane appeared spotty. EDS analysis showed the spots to be the actual fiber surface but the continuous region consistently showed the presence of Ni. It was concluded that the continuous layer was NiAl and the Ni/Al ratio in the EDS spectrum was distorted by the presence of Al in both the substrate and the layer. Occasional evidence of a strongly adherent NiAl particle supported this conclusion. In general, the continuous layer appeared strongly bonded to the surface.

Table I NiAl Composite Bend Test Results

NiAl Spec. No.	Temp. (°C)	Max. Stress (MPa)	Prop. Limit (MPa)	Modulus (GPa)	Failure (%)
P167-1	25	35.1	25.4	87	0.88
P167-2	600	84.6	26.1	68	2.30
P167-3	1200	6.9	3.4	37	2.00+

NiAl/AFP Spec. No.	Temp. (°C)	Max. Stress (MPa)	Prop. Limit (MPa)	Modulus (GPa)	Failure (%)
10-1	25	242.7		107	0.26
10-2	25	242.9		126	0.19
10-3	600	347.8		112	0.31
10-4	1200	178.7	53.6	46	0.86
21-1	600	308.7	158.8	94	0.44
21-2	1200	141.6	21.3	71	0.76

NiAl/CFP Spec. No.	Temp. (°C)	Max. Stress (MPa)	Prop. Limit (MPa)	Modulus (GPa)	Failure (%)
22-1T	25	164.8	96.5	67	0.28
22-2T	25	171.7	102.7	65	0.31
38-12T	25	378.5	239.9	102	0.42
38-13T	25	254.4	124.8	64	0.46
38-1T	25	450.2	142.0	156	0.35
22-3T	600	200.6	86.2	52	0.54
22-4T	600	197.9	113.1	46	0.58
38-3T	600	424.7	231.7	113	0.47
38-4T	600	379.2	171.7	144	0.37
38-14T	900	356.5	106.9	73	2.29+
38-15T	900	311.6	86.9	63	2.67+

Bend Creep

One creep test was successfully completed. Two other tests resulted in specimen failure at 48 MPa (7ksi), substantially below the ultimate strength of 150MPa measured in bending. The one successfully completed test, conducted at a load of 69MPa (10ksi), resulted in a minimum creep rate of 4.20×10^{-6} sec^{-1}, comparable to monolithic Cr_3Si and V_3Si [6], much more refractory materials, under the same testing conditions. This is an extremely promising result, but it must be tempered with the fact that two specimens did not reach their testing load, pointing out the great variability in properties from one specimen to another. The resulting creep curve was typical, with a small primary region followed by extensive steady state creep.

The results of creep testing NiAl/CFP material are reported in Table II. In order to

Table II NiAl/CFP Creep Results

Spec. No.	Applied Stress (MPa)	Minimum Creep Rate ($sec.^{-1}$)
28-10	6.9	8.9×10^{-8}
38-10	13.8	6.2×10^{-8}
38-11	13.8	5.6×10^{-7}

insure the achievement of steady state creep, the applied load was dropped to 6.9 MPa (1 ksi). Primary creep was rather extensive in specimens from hot pressing IMC91-028 (the low pressure hot pressings), to approximately 1% strain. Otherwise the specimens displayed typical stain-time response with the attainment of steady state creep. Specimens from IMC91-038, pressed at 34.5MPa (5ksi), displayed significantly lower primary creep strains, on the order of 0.25%. Minimum creep rates were not greatly affected however, by better interparticle bonding. These creep rates are far better than one would expect from single phase unalloyed NiAl. The alumina fibers must be aiding creep strength greatly. If one assumes most of the creep strength is resulting from the fibers, then the integrity of the interparticle bonding would not greatly influence creep strength.

Fracture Toughness

One fracture toughness test was performed on the monolithic NiAl material at room temperature, and a valid K_{IC} of 12.02 MPa√m obtained. This toughness is a very encouraging finding, and is indicative of the plastic deformation reported above. Three valid K_{IC} measurements were made on NiAl/AFP specimens with toughness values of 3.43, 8.54 and 8.5 MPa√m achieved. The toughness variation was not detected in the bend test results, and is somewhat puzzling since the consolidation parameters were essentially identical.

Three valid K_{IC} measurements were made on specimens from the two hot pressed plates mentioned above. The measured fracture toughness, 2.1, 2.8 and 4.5MPa√m, did not very significantly from plate to plate. This was not in keeping with the results found in the NiAl/AFP specimens which maintained comparable fracture toughness to the monolithic material. The reduced toughness is attributed to the relatively large number of transverse, weakly bound fiber/matrix interfaces in the crack propagation direction. Again, the fracture toughness was not greatly affected by the hot press pressure.

Thermal Cycling

Two thermal cyclic tests were completed on NiAl/AFP specimens, both in the "T" orientation. The first test was carried out using the 55/5 cycle. In the first test, delamination occurred between the fiber tow plies within the first 30 cycles with macroscopic bending as

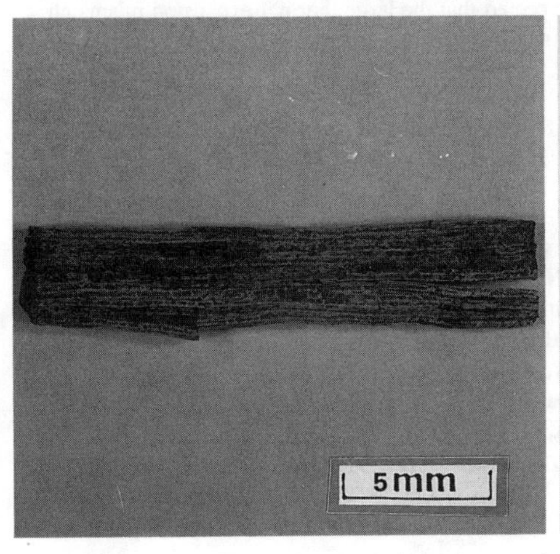

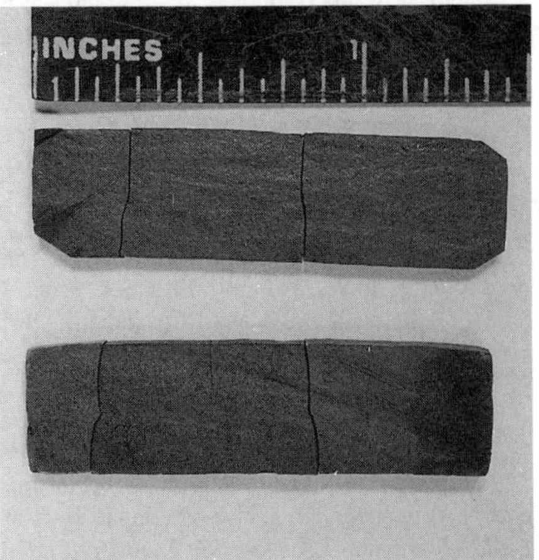

Figure 7 Thermally Cycled Specimens of (a) NiAl/AFP and (b) NiAl/CFP

shown in Fig. 7a. By contrast, in the second test, delamination was not observed. In both cases, however, microcracking in the plane normal to the fiber direction was noted on the surface.

Duplicate thermal cycling for hot pressed NiAl/CFP composites was carried out to 200 cycles utilizing the 20/10 cycle. In both samples, almost a 40% weight gain was observed after 100 cycles. In contrast to the NiAl/AFP composites, these specimens showed no signs of distortion or splitting. However, upon final withdrawal from the furnace, the specimens cracked transversely into two or three large pieces, approximately 1 cm^2 as shown in Fig. 7b. SEM examination of the fracture surface of these specimens showed the matrix to have undergone severe plastic deformation, most probably stemming from the high thermal expansion mismatch between the matrix and reinforcement.

CONCLUSIONS

Bend test results obtained in this study show the great enhancement of tensile strength with the incorporation of aligned FP fibers. The material, plastic at ambient temperatures, is transformed into a brittle system with tensile strength increased significantly. At 600°C, the strength increment is maintained for aligned FP reinforcement, however its degree is somewhat lower. At this temperature the aligned composite is totally brittle with slight plastic deflection found in chopped reinforcement composite. The strengthening increment of the aligned FP composite is quite striking at 1200°C. Where both the monolithic and CFP composite show very poor strength, the AFP material has substantial strength. Since the matrix is very plastic at these high temperatures, significant plasticity is also observed in the aligned FP composite.

The creep strength of both NiAl/AFP and NiAl/CFP composites was substantially increased through incorporation of the FP reinforcement. Monolithic NiAl data obtained at 1027°C can be used for comparison [7]. Here, minimum creep rate values of $4x10^{-7}$, $1x10^{-6}$ and $1x10^{-3}$ $sec.^{-1}$ were obtained at 8, 14 and 70 MPa respectively. Comparison of these numbers with composite data generated here reveals a one to two order of magnitude enhancement for the CFP composites and a three order of magnitude strengthening for the AFP composite, in addition to the 200°C advantage of the data.

These advantages come at the cost of enhanced brittle behavior at all temperatures leading to reduced ductilities and lowered fracture toughness. The fiber pull-out mechanism was not achieved in this composite system, and it is expected that the large thermal expansion mismatch between the matrix and fiber induce a clamping stress around the reinforcements.

ACKNOWLEDGEMENTS

This work was performed under Air Force Systems Command contract F33615-88-C-5405. The helpful discussions with Dr. D.B. Miracle, are acknowledged.

REFERENCES

1. Marshal, D.B. and A.G. Evans, J. Am. Ceramic Soc., 68(5), 225, (1985).
2. Rice, R.W., Ceramic Eng. Sci. Proc., 2(7-8), 661, (1981).
3. Polvani, R.S., A.W. Ruff and P.R. Strutt, J. Mat. Sci. Letters, 3, 287, (1984).
4. Noebe, R.D., R.R. Bowman and J.I. Eldridge, in *Intermetallic Matrix Composites*, D.L. Anton et. al. eds., MRS, Pittsburgh, PA, p. 323, 1990.
5. Shah, D.M., D.L. Anton and C.W. Musson, in *Intermetallic Matrix Composites*, D.L. Anton et. al. eds., MRS, Pittsburgh, PA, p. 333, 1990.
6. Anton, D.L. and D.M. Shah, in *Intermetallic Compounds-Structure and Mechanical Properties*, O. Izumi ed., Japan Inst. of Metals, Sendai, Japan, (1991) p. 379.
7. Stephens, J.R., in *High Temperature Ordered Intermetallic Alloys*, C.C. Koch et. al. eds., MRS, Pittsburgh, PA, (1985), p. 381.

EXTRUSION TEXTURES IN NiAl AND REACTION MILLED NiAl/AlN COMPOSITES

T.R. BIELER, R.D. NOEBE*, J.D. WHITTENBERGER*, and M.J. LUTON**
Department of Materials Science and Mechanics, Michigan State University, East Lansing, MI 48824
*NASA-Lewis Research Center, Cleveland OH 44135
** EXXON Research Center, Annandale, NJ 08801

ABSTRACT

Extrusion textures in monolithic NiAl and an NiAl/AlN particle composite are compared. The NiAl has a generally increasing grain size with extrusion temperature, and a corresponding transition from <110> to <111> fiber texture. The results suggests that <110> is more closely related to the deformation texture, while the <111> is a recrystallization orientation formed by preferential growth. The composite exhibits <311> fiber texture, and this is consistent with slip in <100> directions. The effect of the AlN particles is to prevent the orientation changes observed in the monolithic NiAl during recrystallization.

INTRODUCTION

Extrusions are known to have fiber textures. The components and strength of the texture provide information about the changes in the microstructure that occur during deformation at elevated temperature. The effects of the dominant slip systems, and the mechanisms of recovery and/or recrystallization can also be studied[1-3]. Particle strengthened materials often exhibit different textures than monolithic materials, due to alterations in slip, recovery, and recrystallization behavior[2,4]. This information is useful in understanding subsequent microstructures and material properties. The variation in extrusion conditions can affect the microstructure in a controllable way, and provide a means for choosing an optimum processing strategy.

Texture in extruded aluminides has not been extensively studied, however extrusion textures have been measured by Khadkikar, et. al.[5], and Stout and Crimp[6] in nickel and iron aluminides. Khadkikar found [111] fiber texture in both cast-extruded and powder-extruded samples of NiAl, but their characterization was done using normal diffractometer scans of the sample in three orientations (normal, longitudinal, and 45° to the axis), and inverse pole figures were made using the method by Harris[7]. Khadkikar mentioned a lack of agreement between measured and computed inverse pole figures, indicating that the fiber texture was not "pure".

The current work started as an exercise to determine differences in creep behavior arising HIP'ing or extrusion of the NiAl/AlN composite. Since the extrusion texture discovered was unusual, the work expanded into a study of extrusion parameters on monolithic NiAl to provide a basis for comparison. In the process, the effect of extrusion temperature on texture became significant and in the end provides some insight on the texture evolution in the composite. Also, the current work allows a comparison between texture measurements using standard pole figure techniques, with the technique used by Khadkikar et.al.

EXPERIMENTAL RESULTS

The monolithic NiAl samples were produced by extruding pre-alloyed NiAl powder in 76 mm evacuated steel cans at a 16:1 reduction ratio. The temperature of the samples prior to extrusion was between 1100 and 1505 K. The extrusion speed was rapid, about 2-3 seconds to make a circular rod about 2 m long. The extrusion cooled naturally in air. Two compositions were used, Ni-50.6Al at 1150 and 1300K, and Ni-48.25Al at 1100 and 1505K. The NiAl/AlN composite was prepared by ball milling a slurry of NiAl and Y_2O_3 powders in a liquid nitrogen medium (cryomilling) at Exxon[8,9]. Following cryomilling the powder was extruded at 16:1 and 1505K requiring the maximum stress (1.3 GPa) the press could impart, similar to the conditions needed for extrusion of NiAl powder at 1100K. Examination of the extruded composite showed that it contained about 1.3v% alumina and 10.3 v% AlN. A cursory TEM investigation indicated that the ceramic particles are about 10-50 nm in diameter making a net-like mantle about 300 nm thick around the original particle boundaries[8,10].

The grain structure following extrusion depends upon processing parameters. In monolithic NiAl, the grain shape is equiaxed for nearly all specimens extruded at temperatures at or above 1100K. The as-extruded grain size is linear with increasing extrusion temperature. Slope and intercepts (respectively) are .095μm/K and -102μm for Ni-50.6Al, and .032μm/K and -30 μm for Ni-48.25Al[11,12].

The only exception was a 1150K sample shown in Figure 1 that has a combination of equiaxed and elongated grain structures. Figure 1 suggests a grain aspect ratio of 2-3. On the other hand, the cryomilled material has a grain aspect ratio of 25:1[8], so extrusion can cause more strain in the center of the specimen than near the edges.

The cryomilled powder was also hot isostatically pressed (HIPed) in an extrusion can (207 MPa, 5 hr, 1623K). The surface studied was a plane with its normal parallel to the axis of the can and about 2 mm from the bottom of the can. The bottom of the can had bulged inward from the outer edge, so the deformation was not truly isostatic in this region.

(110), (200), and (211) pole figures were measured with the extrusion axis normal to the specimen. A Scintag XDS-2000 diffractometer with a 4-axis Huber ring goniometer system was used to make the measurements. Data were taken every 5° of azimuthal rotation through 360° at each angle of tilt. Tilt angles were incremented by 5° from 0° to 75° for (110) and up to 80° for (200) and (211). Counting times were 3-5 times longer for the weaker (200) and (211) peaks to provide a signal-to-noise ratio similar to the (110) peak. Background radiation was measured on each side of the peak for each increment of tilt. An experimental defocusing curve was obtained from a powder sample.

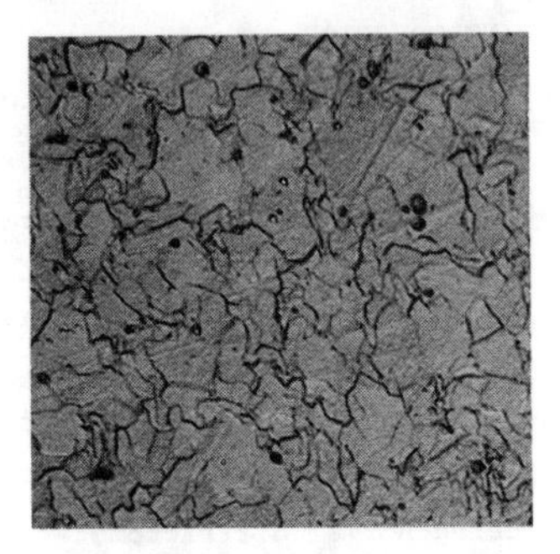

a. Extrusion axis is horizontal 10μm b. Extrusion axis is normal to the page.

Figure 1 Extrusion microstructure of the 1150K Ni-50.6Al specimen.

ANALYSIS

The data was reduced using "popLA" software (the preferred orientation package from Los Alamos[13]). The pole figures were corrected for defocusing, normalized, and a harmonic analysis was used to complete the rim of the pole figures. A WIMV algorithm was used to compute the sample orientation distribution (SOD). For a fiber texture normal to the sample, the SOD is essentially the same for each section. All sections averaged together provide the inverse pole figure. Inverse pole figures (equal area projection) for each sample are in Figures 2 and 3, where maximum values of the dominant fiber components are indicated. A plot of the intensity of certain components of texture as a function of temperature appears in Figure 4. Texture components in monolithic and the composite are compared in the histogram in Figure 5. It is apparent that with increasing temperature there is a shift from <110> to <111> fiber texture in NiAl. The components and intensities are similar to the work of Khadkikar[5]. In the NiAl/AlN composite, the <111> component is below random and a <311> peak has the highest intensity. The magnitude of the texture is less than NiAl, and the texture was nearly random in the HIPed sample.

Though the slip direction in NiAl is commonly understood to be <100>, independent of temperature[14-16], combined <100> and <110> slip has been observed under special conditions[17-19]. <110> dislocation segments have also been observed in the microstructure of as-extruded alloys[20,21], though they only compose a small fraction of the dislocations observed and are thought to be reaction products from gliding <100> dislocations[21,22], which constitute the bulk of the substructure in NiAl after extrusion.

The Schmid factor m has been computed in Table I for all four combinations of slip plane and slip vector, at orientations observed in the inverse pole figures ($m=\cos\phi\cos\lambda$, where ϕ and λ are the angle from the deformation axis to the

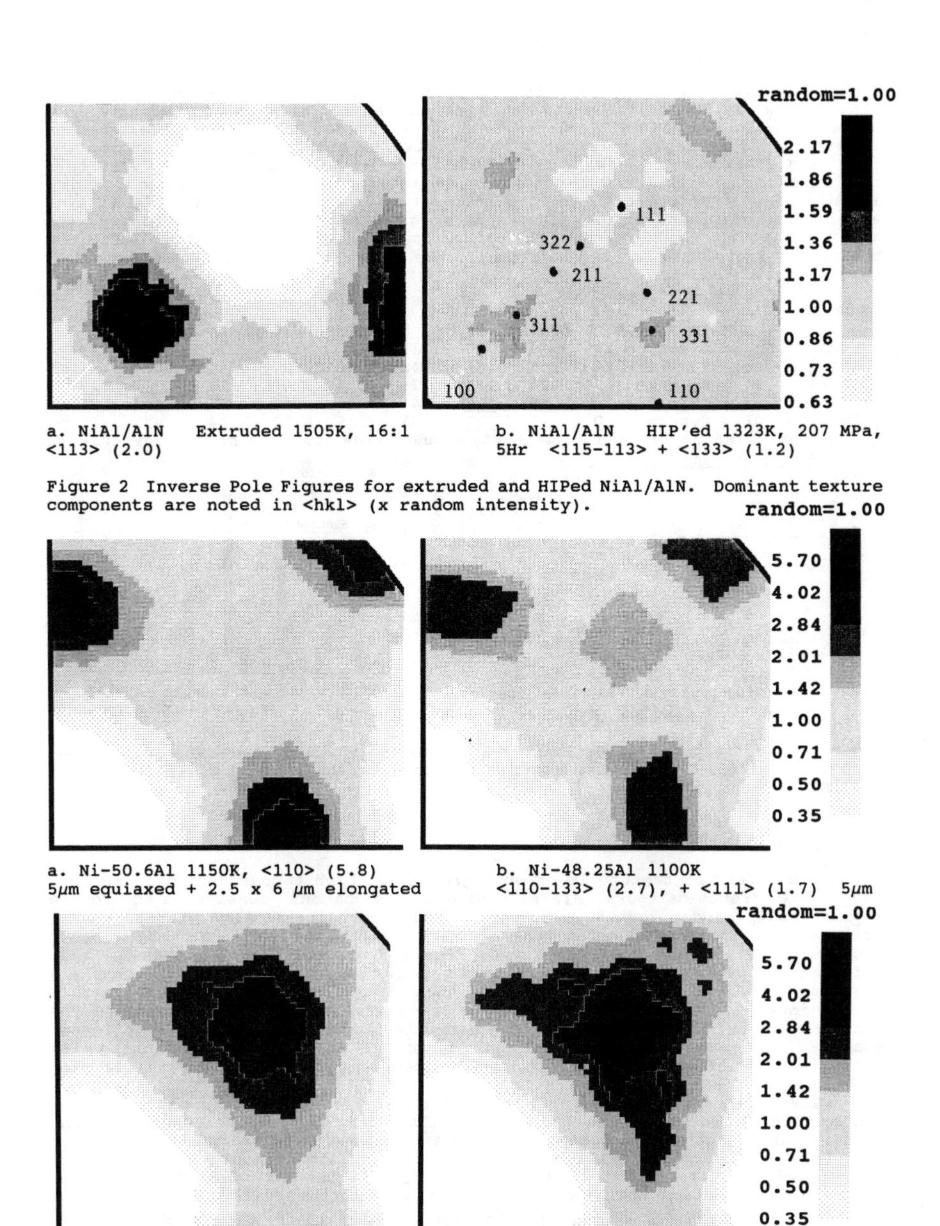

a. NiAl/AlN Extruded 1505K, 16:1
<113> (2.0)

b. NiAl/AlN HIP'ed 1323K, 207 MPa, 5Hr <115-113> + <133> (1.2)

Figure 2 Inverse Pole Figures for extruded and HIPed NiAl/AlN. Dominant texture components are noted in <hkl> (x random intensity).

a. Ni-50.6Al 1150K, <110> (5.8)
5μm equiaxed + 2.5 x 6 μm elongated

b. Ni-48.25Al 1100K
<110-133> (2.7), + <111> (1.7) 5μm

c. Ni-50.6Al 1300K
<111> (4.1) 22 μm

d. Ni-48.25Al 1500K
<111> (5.3) 18μm

Figure 3 Inverse Pole Figures for NiAl extruded 16:1 at various temperatures. Grain size in μm[11,12] and dominant texture components are noted in <hkl> (x random intensity).

slip plane normal and slip directions, respectively). During elongation of single crystals, the slip direction and plane rotate toward the elongation axis. This causes the Schmid factor to decrease, and an increase in applied stress is needed to maintain the critical resolved shear stress. Though the Schmid factor is strictly applicable only for single crystals, high temperature deformation permits diffusion processes (e.g. dislocation climb and grain boundary sliding), that relaxes the 5 slip system constraint needed for compatibility. For example, in superplastic deformation at comparable strain rates in mechanically alloyed aluminum, a component of texture due to slip on only two planes increased with strain in a context where several texture components were present[23]. Therefore the Schmid factor can be used to estimate the rotations and stresses observed in polycrystals.

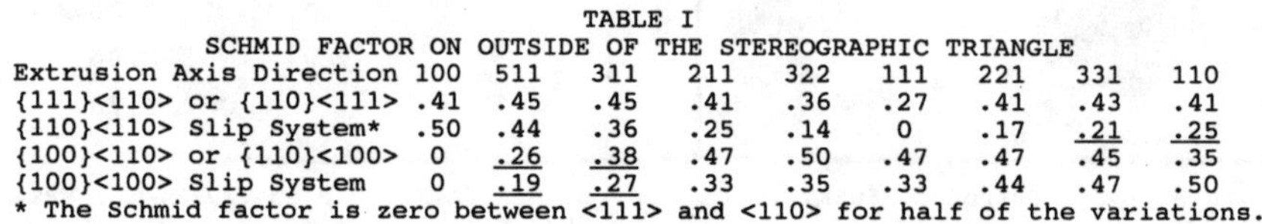

TABLE I
SCHMID FACTOR ON OUTSIDE OF THE STEREOGRAPHIC TRIANGLE

Extrusion Axis Direction	100	511	311	211	322	111	221	331	110
{111}<110> or {110}<111>	.41	.45	.45	.41	.36	.27	.41	.43	.41
{110}<110> Slip System*	.50	.44	.36	.25	.14	0	.17	.21	.25
{100}<110> or {110}<100>	0	.26	.38	.47	.50	.47	.47	.45	.35
{100}<100> Slip System	0	.19	.27	.33	.35	.33	.44	.47	.50

* The Schmid factor is zero between <111> and <110> for half of the variations.

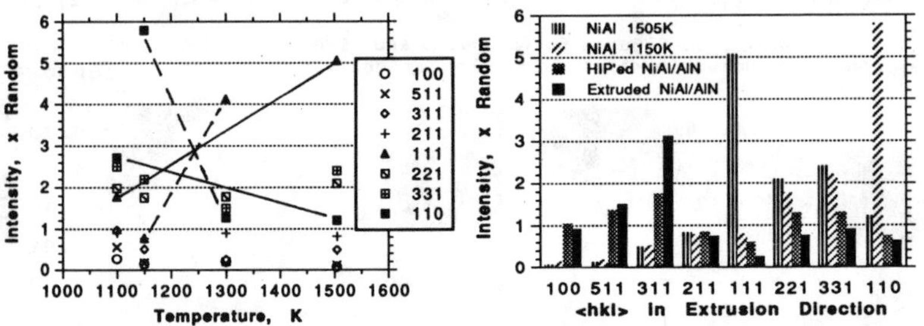

Figure 4 Variation of fiber texture components with extrusion temperature in NiAl

Figure 5 Comparison of texture components in NiAl and NiAl/AlN

DISCUSSION

The effect of texture is often neglected in high temperature deformation, due to the effects of recrystallization, which is commonly considered to provide a practically random texture. However, since recrystallization starts from an orientation already present in the microstructure[3], preferred orientations in recrystallization are commonly found, though they are generally weaker than deformation textures. Also, particles change the way deformation and recrystallization occurs[2,4]. Particles smaller than 1 μm tend to distribute strain homogeneously, but larger particles can stimulate new crystal growth from the local rotations. The non-uniform fine particle distribution is more complicated than the homogeneous particle distribution, but the effects of Zener pinning will resist motion of grain boundaries or dislocations through the particle-rich zones.

The microstructure and properties resulting from hot deformation depend highly upon the working temperature and cooling rate. In this study, the cooling rate is rapid, but not quenched. The presence of equiaxed grains in monolithic NiAl reflects recrystallization with significant grain growth. The grain size is reduced by extruding at lower temperatures (Figure 3), where the time for grain boundary motion is reduced, since the working temperature is lower. The Ni-50.6Al has a faster grain growth rate, and its effect shows in the different rates of growth of the <111> component in the two compositions (Figure 4).

In metals, extrusion textures tend to be <100> and/or <111> in FCC materials, and <110> in BCC materials. At the higher temperatures, new orientations from grain growth will often dominate the texture change. If boundary mobility is inhibited by a low temperature, the deformed texture can be retained[24]. Thus a low extrusion temperature can be expected to support retention of the deformation texture or an early stage of recrystallization.

Recrystallization in materials with well defined deformation textures exhibit particular new orientations, e.g. 40° rotations from a common <111> axis occur in FCC materials, and 25° rotations are common in recrystallized BCC materials[1]. The creations of new orientations can occur from either growth advantage, or a preferred nucleation orientation. The rise of the <111> component at the expense of the <110> component in NiAl (Figures 2,4) with increasing extrusion temperature is consistent with orientation changes that occur in recrystallization, in this case a rotation of 35°. Conversely, in the composite where particles inhibit grain growth, or in low temperature extrusion conditions, effects of the deformation texture remain (Figure 3).

Since the B2 structure of NiAl is like BCC, it is tempting to think that similar deformation processes occur, since there is a large fraction of grains with a <110> slip direction near the extrusion axis. However, the slip systems are not the same; <100> is more often observed in NiAl than <110>. <100> slip would support rotations toward <100>, and the observed <311> orientation has a low (underlined) Schmid factor. To justify a significant amount of <110> slip, a low Schmid factor is needed near <110>, and the {110}<110> system has a lower (underlined) Schmid factor than the other possible systems. Perhaps a better interpretation is that <110> represents a nucleation texture which is later consumed by a preferred growth of grains having <111> orientations. In this case <110> slip may contribute to the formation of the <110> component in a minority of the grains. A 16:1 reduction in area causes a 4 fold decrease in diameter of a spherical 40μm (typical) particle[8]. The smaller 2.5 μm grain size in Figure 1 supports the latter interpretation. The <110> nucleation component survives until growth above about 5 μm occurs.

The composite exhibits a different major texture component, <311>. This orientation has a low Schmid factor for the {110}<100> <u>and</u> the {100}<100> slip systems. The higher stress associated with the extrusion of the composite may result from <100> slip in a condition where recrystallization cannot occur, as evidenced by elongated grain diameters of about 2 μm. The Schmid factor for <100> slip near <111> is high, and few grains in this orientation could be expected to remain after large strains. This is consistent with the below random <111> intensity. A similar effect occurs for the same reasons near the <110> orientation, also. The low Schmid factor for <311> will cause the flow stress of <100> slip to be twice as high compared to <110> and <111> orientations. The composite exhibits a creep strength 8 times larger than NiAl[8], so part of the superior creep resistance in the composite could be due to texture strengthening.

This work provides information that will assist development of mechanistic models for deformation, recovery, and recrystallization during extrusion of NiAl and NiAl composites. Though further work is needed in determining the appropriate Taylor factor, these observations taken together suggest the following model for extrusion of NiAl and its composites: During extrusion, large strains are accommodated by slip and climb of <100> dislocations in the <100> directions. This causes rotations of crystals toward the <100> axis to a position near <311>, and a corresponding increase in applied stress. Perhaps with the assistance of <110> slip, further deformation assists in nucleation of <110> oriented recrystallized grains. As these grains grow up to a size of about 5 μm, they permit a lower stress for continued <100> slip. Following the deformation, grains oriented near <111> have a growth advantage over <110> and this component becomes dominant. However, due to the pinning efficiency of the mantle of AlN particles, boundary migration needed for recrystallization nuclei to form are precluded, maintaining the high stress needed for further deformation. Consequently, higher temperatures and/or pressures are needed to effect extrusion.

CONCLUSIONS

Texture measured in extruded samples of NiAl and a NiAl/AlN particle composite exhibit different deformation mechanisms and annealing phenomena that suggest a model accounting for deformation in both materials. When grain boundary mobility is reduced by particles that interfere with their motion, a <311> deformation texture remains. Without the particles, recrystallization produces new orientations in the <110> direction. With higher extrusion temperatures, time is available for grain growth, and a <111> texture forms.

ACKNOWLEDGEMENTS

T.R. Bieler acknowledges support from the ASEE Summer Faculty Fellowship Program administered by Case Western Reserve University for NASA-Lewis Research Center.

REFERENCES

1. K. Lücke, International Conference on Texture of Materials (ICOTOM 7), eds. C.M. Brakman, P. Jongenberger, E.G. Mittemeijer, (Netherlands Society for Materials Science, 1984), p. 195-210.

2. W.B. Hutchinson and B.A. Wilcox, Metal Science 7, 6, (1973).

3. Cahn... Physical Metallurgy 1983.

4. F.J. Humphreys, Metal Science 13, 136, (1979).

5. P.S. Khadkikar, G.M. Michal, and K. Vedula, Metall. Trans. 21A, 279, (1990).

6. J.J. Stout and M.A. Crimp, Mat. Sci. & Eng., in press.

7. G.B. Harris, Phil. Mag., 43, 113, (1952).

8. J.D. Whittenberger, E. Arzt, M.J. Luton, J. Mater. Res., 5, 2819, (1990).

9. J.D. Whittenberger, E. Arzt, M.J. Luton, in Intermetallic Matrix Composites, edited by D.L. Anton, R. McMeeking, D. Miracle and P. Martin, (Mater. Res. Soc. Proc. 194, (1990), 211-18

10. E. Arzt, unpublished TEM images

11. J.D. Whittenberger, J. Mat. Sci. 22, 394, (1987).

12. J.D. Whittenberger, J. Mat. Sci. 23, 235, (1988).

13. J.S. Kallend, U.F. Kocks, A.D. Rollett and H.-R. Wenk, Mat. Sci. and Eng. A132, 1, (1991).

14. A. Ball and R.E. Smallman, Acta Metall., 14, 1517, (1966).

15. R.J. Wasilewski, S.R. Butler and J.E. Hanlon, Trans. Metall. Soc. AIME 239, 1357, (1967).

16. M.H. Loretto and R.J. Wasilewski, Phil. Mag., 23, 1311 (1971).

17. J.T. Kim and R. Gibala, in High Temperature Ordered Intermetallic Alloys IV, edited by L. Johnson, et.al., (Mater. Res. Soc. Proc. Vol. 213, Pittsburgh, PA, 1991), pp. 261-66.

18. R.D. Field, D.F. Lahrman and R. Darolia, Acta Metall. Mater., 39, 2951, (1991).

19. D.B. Miracle, Acta Metall. Mater., 39, 1457, (1991).

20. C.H. Lloyd and M.H. Loretto, Phys. Stat. Sol., 39, 163, (1970).

21. I. Baker and E.M. Schulson, Metall. Trans., 15A, 1129, (1984).

22. N.J. Zaluzec and H.L. Fraser, Scripta Metall., 8, 1049, (1974).

23. Z. Jin and T.R. Bieler, Superplasticity in Advanced Materials (ICSAM-91), eds. S. Hori, M. Tokizane, and N. Furushiro, Japan Society for Research on Superplasticity, (1991), p. 587.

24. W.C. Leslie, Trans. AIME 221, 752, (1961).

IN-SITU SYNTHESIS OF PARTICLE DISPERSED NANOCRYSTALLINE NiAl BY CRYOMILLING

Benlih Huang†, J. Vallone††, C. F. Klein††, and M. J. Luton††
† Department of Materials Science and Eng., Rutgers University, Piscataway, NJ 08854.
†† Corporate Research, Exxon Research and Engineering Company, Annandale, NJ 08801.

ABSTRACT

The present research was carried out to study the cryogenic synthesis NiAl from elemental powders with the view of forming particle dispersed NiAl. It was found that nanocrystalline NiAl with a crystallite size in the range 5nm to 10nm is produced during cryomilling. Additionally, the nanocrystalline NiAl maintains the fine crystallite size when annealed at high temperatures. It is thought that the resistance of the microstructure to coarsening is due to the presence of nano-scale particles of AlN formed by reaction between milling medium, liquid nitrogen, and aluminum. The transformation of Ni and Al to ordered NiAl was studied using x-ray diffraction. This showed that cryomilling not only produces a solid solution but also can induce in-situ ordering of stoichiometric NiAl. The B_2 NiAl structure forms during milling after the elemental powder blend is cryomilled for about 40 hours. It also forms when the "mechanical" solid solution is annealed at room temperature for about twelve hours.

INTRODUCTION

Mechanical alloying is a ball milling process by which elemental or pre-alloyed powders are rolled, flattened, fractured and welded to synthesize novel materials. The process is carried out either in an attritor, a shaker or a conventional ball mill. Various intermetallics and amorphous materials have been synthesized. Recently, cryomilling, or cryogenic slurry mechanical alloying process was introduced to decrease the grain size of oxide dispersion strengthened alloys [1]. The rationale for cryogenic processing is that decreasing the milling temperature leads to an increase in the strain energy stored in the material, which, in turn, increases the density of recrystallization nuclei during subsequent annealing. In this way significant grain refinement is achieved. Cryomilling in liquid nitrogen has been shown to cause the in-situ formation of aluminum nitride or oxy-nitride particles in aluminum and dilute aluminum alloys[1]. Here the dispersoids are found to be typically 2 to 10 nm in diameter and with a mean spacing 50 to 100 nm. The consolidated alloys have an extremely fine grain size, typically of the order of 50 nm.

Due to its high melting temperature, 1638°C [2], relatively low density, 6 g/cm^3 [3] and resistance to high temperature oxidation, the intermetallic compound NiAl is of interest for aerospace applications. Considerable effort has been made to improve the mechanical properties. Recently, Whittenberger et al [4] employed cryomilling to disperse yttria a pre-alloyed NiAl matrix. Although 0.2 % of yttria was added, the predominant second phase present in the consolidated alloy was AlN. In these alloys about 10 vol.% of AlN is present. Compression test between 927°C and 1127°C of extruded powder indicate that the creep resistance is six times better than NiAl and twice that of a NiAl particulate composite containing 10 vol.% TiB_2.

The purpose of the present work was to study the synthesis of NiAl from elemental powders with a view to gaining an understanding of the interaction that occurs during mechanical alloying at cryogenic temperatures. In particular, the study was focused on the distinction between those events that are induced by the milling alone and those due to interaction of the powders with the milling medium. To this end, NiAl was synthesized in both liquid nitrogen and liquid argon.

EXPERIMENTAL

In the present study, the cryomilling was carried out in a Szegvari 10 liter attritor manufactured by Union Process Inc. The mill was cooled by flowing liquid nitrogen through its water cooling jacket. In the first set of experiments the milling medium was liquid nitrogen while in the second liquid argon was used. The milling medium was introduced to the mill to form a slurry of the metal powders. Throughout each run the level of the slurry was monitored by a suspended T-type thermocouple; additional liquid being added when necessary. Elemental Ni (99.99 wt.%) and Al (99.95 wt.%) powders were used that were supplied by Cerac Inc. with a particle size of 44 μm. The powders were blended in a 1:1 atomic ratio in a Patterson Kelly Co. twin shell dry blender before they were cryomilled.

As-milled powder specimens were discharged from the mill at selected intervals and examined by x-ray diffraction, optical and scanning electron microscopy and by transmission electron microscopy. X-ray diffraction analysis was performed using a Rigaku II-A horizontal diffractometer with Mo-Ka radiation and a nickel filter. In order to study the as-milled powder at the process temperature, a cryogenic sample holder for the x-ray diffractometer was designed and constructed. This device is illustrated in Figure 1. The sample holder is essentially a cold finger made from copper which is attached to a liquid nitrogen tank. The sample holder is equipped with cartridge heaters so that sample can be examined either at -190°C or at other temperatures up to room temperature. Temperature control is achieved by the thermal balance between the energy input from the heating elements and the cooling due to the evaporation of the liquid nitrogen.

Once the powder samples reached room temperature, they were blended with a Pacer RX-100 instant adhesive, mounted, ground and polished for optical and scanning electron microscopic examination. A JEOL JSM-840A scanning electron microscope, with a EDAX 9900 energy dispersive spectrometer, was employed. Samples for transmission electron microscopy were prepared by placing the powder in a 1.3 cm diameter stainless steel tube and cold swaging to 3 mm diameter. This procedure produced samples which were sufficiently strong to allow thinning and ion milling. These samples were examined using a Philips CM12 transmission electron microscope.

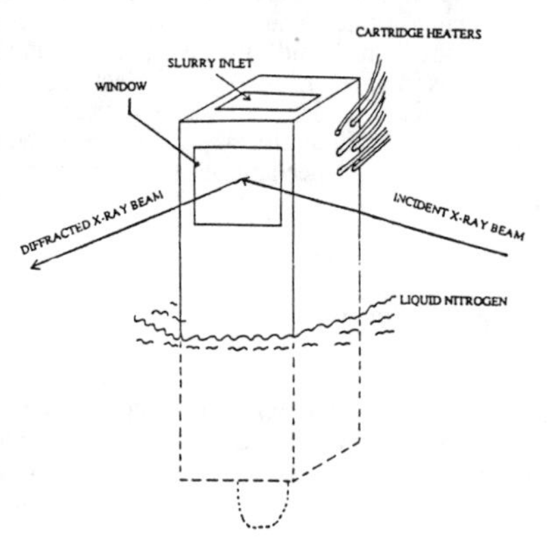

Figure 1
Illustration of the cryogenic difractometer sample holder used to examine milled samples at the temperature of milling.

RESULTS AND DISCUSSION

Synthesis NiAl in Liquid Nitrogen

In the early stages of cryomilling, agglomerates of Ni and Al form everywhere. After 21 hours of milling, few elemental particles remain. Instead, fine layered, powder agglomerates are observed in the optical and scanning electron microscope, see Figure 2. This microstructure is characteristic of what Gilman and Benjamin [5] described as the early stage of mechanical alloying. As the cryomilling proceeds, further mixing of Ni and Al occurs.

In order to follow these fine-scale changes, it was necessary to use x-ray diffraction analysis. Samples of the spectra obtained after 31 to 170 hours of milling are compared with that obtained from the original powder blend in Figure 3. It should be noted that in each case

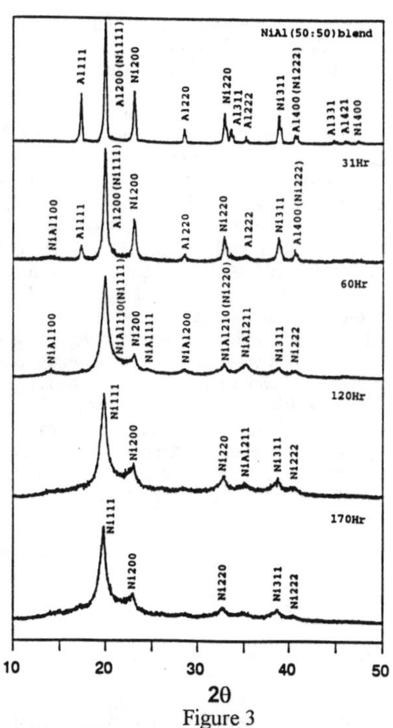

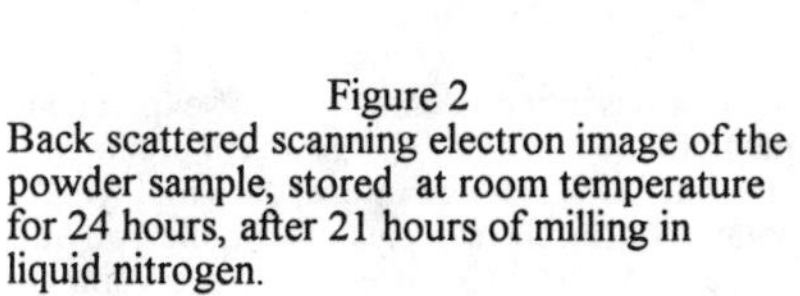

Figure 2
Back scattered scanning electron image of the powder sample, stored at room temperature for 24 hours, after 21 hours of milling in liquid nitrogen.

Figure 3
X-ray diffraction spectra obtained on powder samples, stored at room temperature for 24 hours, after milling in liquid nitrogen for 31 to 170 hours. The upper spectrum was obtained on the unprocessed powder blend.

the x-ray spectrum was obtained on powder samples that had been stored at room temperature. After 31 hours, the superlattice peaks of the B_2 structure begin to appear. The presence of the ordered compound in the processed powder could either be due to in-situ formation during milling or during the heating of the powders to room temperature. These possibilities will be discussed in more detail below. As the milling proceeds beyond 31 hours, the intensity of the NiAl {100} superlattice peak increases and reaches a maximum value after 60 hours. On further milling, the intensity of the NiAl {100} superlattice peak decreases and is eventually lost in the background. After 120 hours of milling the x-ray diffraction pattern consists predominantly of Ni or a Ni-rich solid solution.

Transmission electron microscopy shows that in the early stages of milling, less than 31 hours, the powders consist of a fine scale mixture of Ni, Al and the intermetallic phases, Ni_2Al, NiAl and Al_3Ni. These intermetallic phases tend to be less than 100 nm in diameter. In addition the microstructure contained many nanocrystalline areas with a "peppery" structure. These areas were examined by energy dispersive x-ray spectroscopy and found to contain Ni and Al near the stoichiometric ratio. The structure of these particles was examined using convergent beam electron diffraction (CBED), with a 20 nm spot size. Because of the fine crystallite size, however, it was generally not possible to obtain single crystal CBED patterns from these regions. Accordingly, the phases that were present could not be identified unambiguously . Nevertheless, a sample of powder that had been cryomilled for 170 hours and then annealed 1100°C for 1 hour, showed that the nanocrystalline regions consist of NiAl with a crystallite size in the range 10 nm to 40 nm. A dark field image of such a region is shown in Figure 4. It is thought that the extreme stability of the nanocrystalline material is due to the presence of nanoscale particles of AlN which impede growth of the NiAl.

It appears that there are two competing factors which dominate the cryomilling process in liquid nitrogen. One is the formation of the ordered NiAl and the other is the reaction between the liquid nitrogen and Al to form AlN. It is notable that the formation enthalpy of AlN is - 318.6 KJ/mol compared with. - 118.5 KJ/mol for NiAl. This suggests that, as milling proceeds, and the nitrogen content increases, the formation of AlN dominates the cryomilling process. In this way, the reacted Al tends to form AlN in stead of NiAl, thereby causing the Ni to persists in the x-ray spectra. The NiAl that was formed in the earlier stages of the process appears to be stable, but is present on a such a fine length scale that it does not contribute to the x-ray diffraction pattern.

Figure 4
Dark field transmission electron micrograph of a cold swage compacted sample of powder milled for 170 hours in liquid nitrogen and annealed for 1 hour at 1100°C. The image is typical of the nanocrystalline regions.

Synthesis of NiAl in Liquid Argon

In order to separate the two competing processes described, a similar set of experiments were performed using liquid argon as the milling medium at -190 °C. The powders prepared in this way were examined by x-ray diffraction, at the milling temperature by employing a cryogenic sample holder. The diffraction spectra obtained from the samples of as-milled powder processed between 10 and 50 hours at -190° C† are shown in Figure 5. As cryomilling proceeds, the intensity of the Al peaks decrease and eventually are lost in the background. At the same time Ni peaks appear to split. The new peaks, at lower angles on the 2θ scale, can be identified as being derived from a solid solution of aluminum in nickel. After 30 hours of cryomilling, the spectrum consists of that due to the solid solution and residual amounts of nickel and aluminum. Examination of the pattern obtained on the 40 hour sample shows only peaks due to ordered NiAl. However, the diffraction pattern obtained after 50 hours shows some evidence for the presence of the Ni-Al solid solution. It appears, therefore that ordered NiAl can form during milling at temperature as low at -190 °C. Further work is needed to understand the mechanism of ordering in the absence of long range diffusion.

Several of the cryomilled samples were scanned in the diffractometer at several intermediate temperatures as the temperature of the sample was increased to room temperature. The spectra that were obtained from the 30 hour milled sample are shown in Figure 6. At -190° C, the majority of the material is an fcc solid solution of Al in Ni. The broad peak at low angles can be attributed to scattering from the liquid argon. As the temperature increases, the residual aluminum peaks disappear indicating that this small amount of Al is dissolved in the Ni. At temperature above -170°C no further change in the spectrum occurs. After the sample had been stored for 12 hours at room temperature a massive change in the spectrum was observed, as shown in Figure 6. The solid solution spectrum is completely replaced by the diffraction peaks of the B_2 structure of NiAl along with pure Ni.

† The boiling temperature for Ar is -186°C, the temperature quoted here and in the figures is the temperature of the milling chamber or the copper block of the cryogenic sample holder.

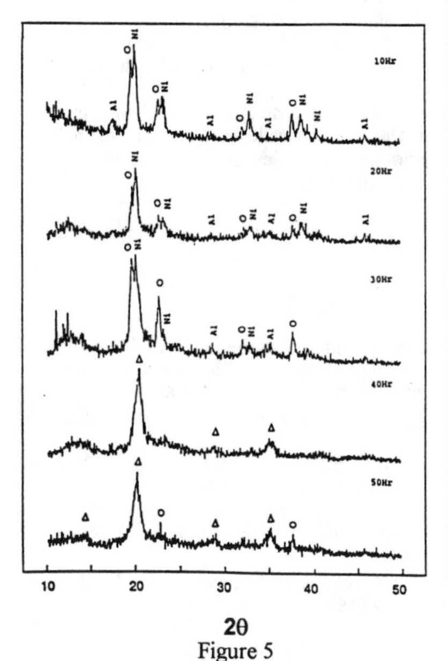

Figure 5
X-ray diffraction spectra obtained at - 190°C on powder samples immediately after milling and before being heated to room temperature. The spectra are for samples milled between 10 and 50 hours in liquid argon.
O - Solid Solution of Al in Ni
Δ - Ordered NiAl

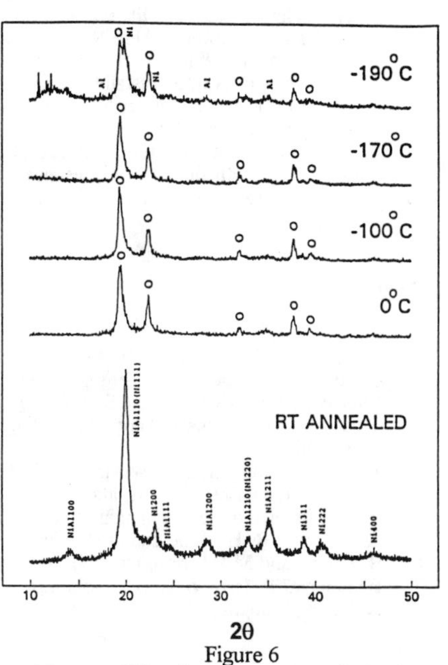

Figure 6
X-ray diffraction spectra obtained at temperatures between - 190°C and 0°C on a powder sample prepared by milling for 30 hours in liquid argon. The lower spectrum was obtained after the sample had been held for 12 hours at room temperature.

In previous investigations, using cryomilling in liquid nitrogen to synthesize Ni_3Al a similar transition to the ordered compound via the solid solution phase was observed [6]. Here, the evidence suggests that aluminum dissolves in the nickel to form an fcc solid solution which has the same structural form as a disordered $L1_2$ intermetallic Ni_3Al. On subsequent annealing of this "disordered Ni_3Al" $L1_2$ intermetallic Ni_3Al is formed. Ivanov et al [7] mechanically alloyed a mixture of pure Ni and Al powder in a conventional mill at room temperature and observed an extended fcc solid solution at the composition where the equilibrium Ni_3Al phase normally occurs. They also observed the formation of ordered Ni_3Al when this material was annealed. Since the free energy of formation of Ni_3Al is larger than that of NiAl, the higher annealing temperature required to transform to the ordered phase suggests that the structural changes that are required must be different in the two cases. Further work is required to develop a detailed understanding of these phenomena.

CONCLUSIONS

This first attempt to synthesize NiAl by cryomilling in both liquid nitrogen and liquid argon provides some valuable insights to the mechanisms of this particular process. They are:

1) Cryomilling is not only capable of forming a solid solution but can also produce ordered NiAl in-situ.

2) The formation of the B_2 NiAl intermetallic compound also occurs when the solid solution of Al in Ni is annealed for short times at room temperature.
3) Nanocrystalline NiAl forms after prolonged cryomilling in liquid nitrogen. Due to its fine crystallite size, further work is needed to establish whether or not the crystallites are ordered or disordered.
4) This nanocrystalline phase maintains its fine crystalline size even when it is annealed at high temperatures. This is thought to result from the inhibition of growth due to the presence of nano-scale particles of AlN formed by reaction between the liquid nitrogen and aluminum.

REFERENCES

1. M. J. Luton, C. S. Jayanth, M. M. Disko, S. Matras, and J. Vallone in Cryomilling of Nano-crystalline Dispersion Strengthened Aluminum in Multicomponent Ultrafine Microstructures (Mater. Res. Soc. Pro. 132, Pittsburgh PA 1989) pp.79-86
2. T. B. Massalski, Binary Alloy Phase Diagrams Vol.1 (American Society for Metals, Metal Park, Ohio 1987)
3. J. D. Whittenberger, L. J. Westfall, and M. V. Nathal, Scripta Metallurgica 23, 2127 (1989)
4. J. D. Whittenberger, E. Arzt, and M. J. Luton, J. Mater. Res. 5, 271 (1990)
5. P. S, Gilman and J. S. Benjamin in Annual Review on Materials Science Vol.1, edited by R. A. Huggins, R. H. Bube and D. A. Vermilyea (Annual Review Inc., California 1983) p. 279
6. B. Huang and M.J. Luton, Unpublished research
7. E. Ivanov, T. Grigorieva, G. Golubkova, V. Boldyrev, A.B. Fasman, S. D. Mikhailenko, and O. T. Kalinina, Materials Letters, 7 (1,2) 51-54 (1988)

FATIGUE OF EXTRUDED STEEL/NIAL COMPOSITES

M.K. BANNISTER, S.M. SPEARING, J.P.A. LÖFVANDER, M. DE GRAEF
High Performance Composite Center, Materials Department, College of Engineering, University of California, Santa Barbara CA 93106

ABSTRACT

Fatigue tests were performed on a novel, extruded, stainless steel/NiAl composite having good impact and tensile properties. A high fatigue limit was observed to occur at approximately 67% of the σ_{UTS}. The fracture surface showed a distinct change in morphology between the fatigued and fast fracture areas and the formation and growth of microcracks was postulated as the initial fatigue mechanism. The microcrack development was monitored by intermittent measurement of the elastic modulus and associated hysteresis. Microstructural characterization by means of SEM, TEM and EDS revealed the existence of approximately 100nm diameter Al_2O_3 particles decorating the interface between the NiAl and the stainless steel tubes.

INTRODUCTION

The B2 ordered material, NiAl, is one of a variety of systems under close examination for use as a high temperature, structural material [1, 2]. Its use is hampered by a lack of ductility and inadequate toughness at low temperatures. Previous work has shown the benefits of reinforcing brittle materials with ductile particles [3, 4]. Studies performed by Nardone *et al* [5, 6] provided evidence for a dramatic improvement in toughness and ductility through the use of a novel composite design. The material was formed by extrusion at 1066 °C and consisted of a NiAl matrix containing continuous, uniaxial tubes of 304 stainless steel and B_4C particulates. However, previous work [7] has raised doubts concerning the fatigue properties of brittle materials reinforced with ductile particles or fibers. The results presented here outline some research into the fatigue properties of two types of materials[1] : NiAl/stainless steel with Al_2O_3 (**MT-90-18**) or B_4C (**MT-90-19**) particulates.

ANALYTICAL TECHNIQUES

Electron transparent foils for analysis in a transmission electron microscope (TEM) were prepared by cutting thin slices of material with a low speed diamond saw perpendicular to the extrusion direction. These slices were ground to roughly 100 μm using diamond paste before discs 3 mm in diameter were ultrasonically drilled out. The samples were dimpled using 3 and 1 μm paste with 1/4 μm used for the final polish. Subsequent thinning was accomplished by ion milling with Ar at 5 kV, 1 mA and 14° incidence angle. Samples were examined in a JEOL 2000FX TEM equipped with a LINK eXL high take-off angle energy dispersive spectroscopy system; the microscope was operated at 200 kV. Indexing of diffraction patterns was aided by the Desktop Microscopist software[2]. Conventionally mounted and polished samples were also prepared and examined in a JEOL SM840A scanning electron microscope. The TEM samples were also characterized in the SEM.

[1]Material supplied by United Technologies Research Centre, East Hartford, Ct, 06108
[2]Virtual Laboratories, Ukiah, CA 95462

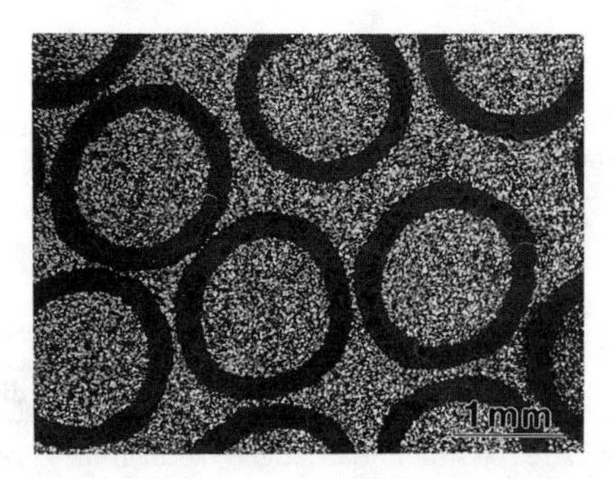

Figure 1: SEM micrograph of a polished **MT-90-18** cross-section showing stainless steel tubes with circular cross-section in NiAl-Al_2O_3 matrix.

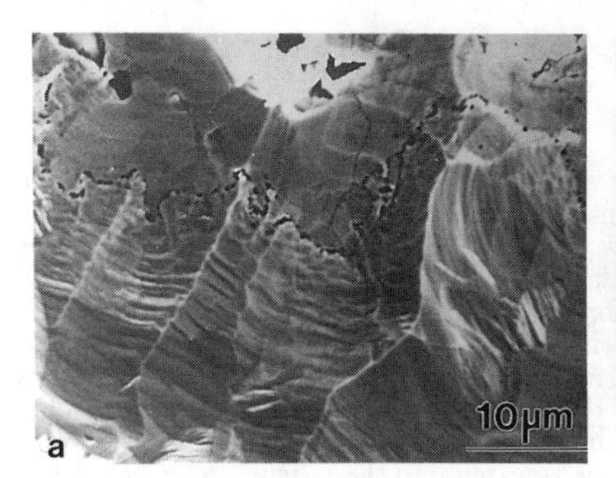

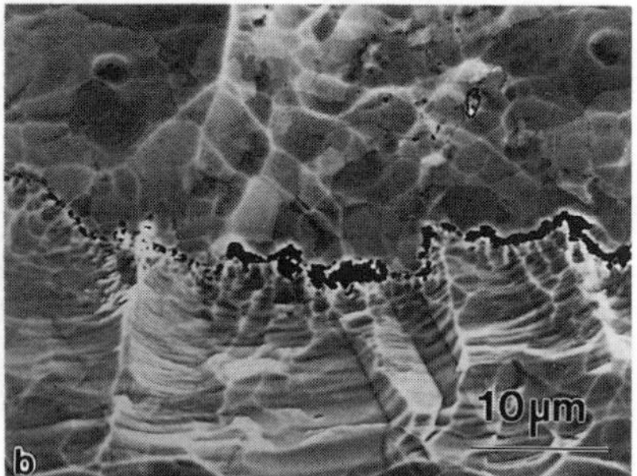

Figure 2: SEM micrograph of TEM foils showing particles decorating the interface of stainless steel (bottom) and NiAl (top) of a) **MT-90-18**, b) **MT-90-19**.

RESULTS

The uniform distribution of the stainless steel tubing is evident in the SEM micrograph in Fig. 1 which shows a typical area in the material containing Al_2O_3 platelets. The cross-section of the non-touching tubes was roughly circular with an inner diameter of approximately 1 mm and a macroscopically even wall thickness of ~200 μm. A higher magnification micrograph shows an uneven interface between the stainless steel reinforcement and the matrix NiAl of the same material (Fig. 2a). The stainless steel in the lower half of the micrograph is separated from the matrix by an almost continuous film of precipitates, which are in the order of 100 nm in diameter. The interface microstructure between the steel and the matrix containing B_4C particulate is similar in appearance albeit somewhat smoother, as shown in Fig. 2b. The area between the tubing and the matrix material is dominated by approximately 100 nm precipitates. The average grain size of the NiAl in the two materials, $\sim$ 6 μm in diameter, was determined by recording images with backscattered electrons which revealed that the particles decorating the interface of **MT-90-19** were comprised of lighter elements (Fig. 3).

The TEM micrograph in Fig. 4a shows such a particle at the interface between stainless steel on the left and NiAl on the right in **MT-90-18**. The diffraction pattern in the inset

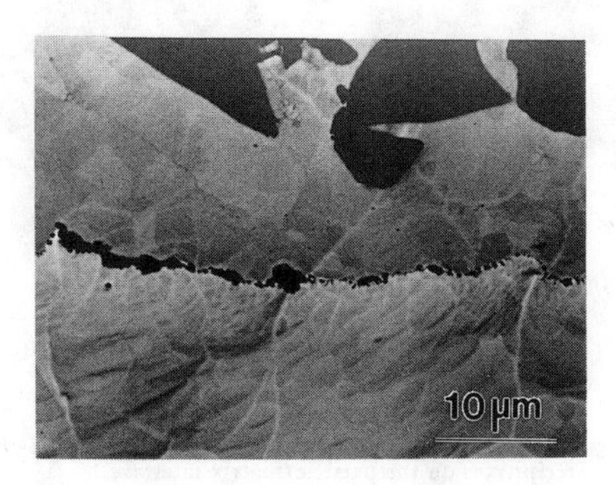

Figure 3: Backscattered image of **MT-90-19** TEM foil showing light elements decorating the interface and the uniformity in the grain size and composition of the matrix.

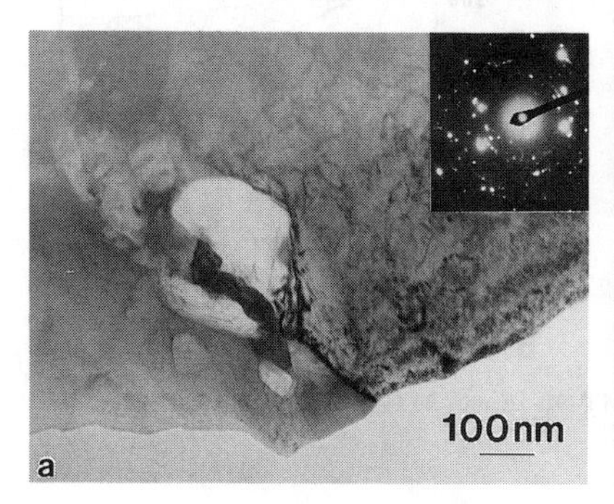

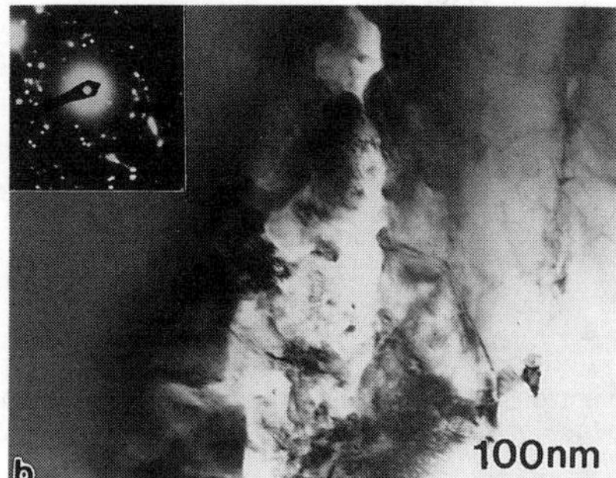

Figure 4: TEM micrographs of a) **MT-90-18** and b) **MT-90-19** showing Al_2O_3 particles decorating the matrix/stainless steel interface. Insets show the diffraction patterns from the interfacial region.

was recorded from the region in the center of the picture and the ring-like intensity maxima are consistent with randomly oriented polycrystalline alumina. This was supported by EDS analysis from the same region which yielded aluminum as the only metallic constituent. The **MT-90-19** material exhibited identical microstructural features as shown in Fig. 4b; the region of poly crystalline alumina between the NiAl matrix and the stainless steel reinforcement was identified by electron diffraction (see inset) and accompanied EDS analysis.

A detailed study of the interfaces between the NiAl and the filler materials revealed a difference. The **MT-90-18** material shows a continuous layer of particles separating the Al_2O_3 platelets from the matrix. The scale of the particles (Fig. 5a), ~ 20 nm, precluded electron diffraction analysis; EDS analysis of an approximately 100 nm region across the interface revealed only Ni and Al as constituents. No precipitates were found at the interface between the matrix and the B_4C in material **MT-90-19** (Fig. 5b). It can be noted that this interface was frequently observed to have debonded in the thin TEM sample, possibly caused by relaxation of residual stresses during specimen preparation. No debonding was observed in the **MT-90-18** sample.

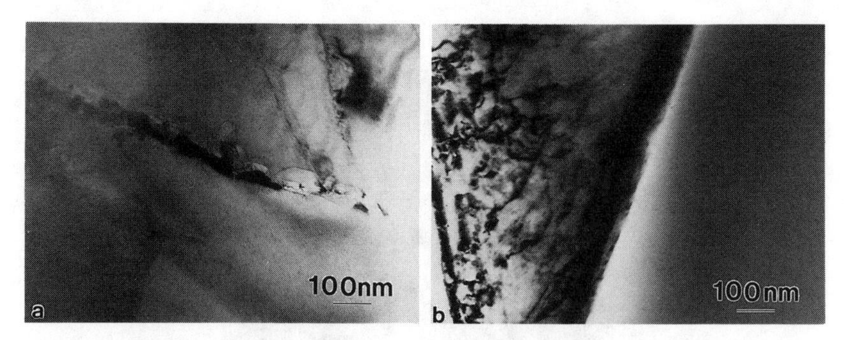

Figure 5: TEM micrographs of the particulate/matrix interface of a) **MT-90-18** and b) **MT-90-19**. The interface of a) shows precipitates on the particle/matrix interface.

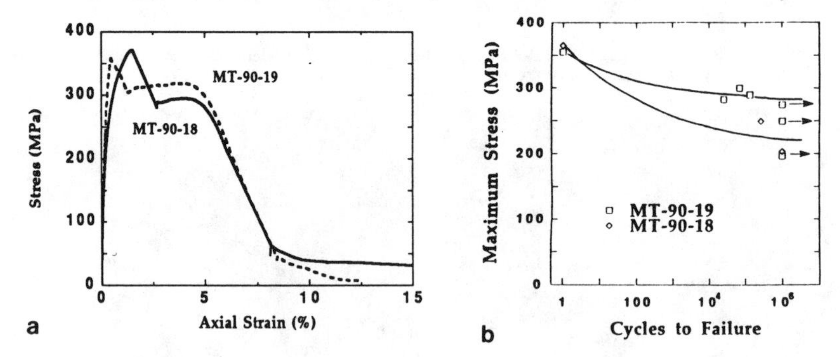

Figure 6: a) Monotonic stress-strain curves for **MT-90-18** and **MT-90-19**, b) S–N curve comparing fatigue behavior of **MT-90-18**and **MT-90-19**.

MECHANICAL TESTING

Tensile specimens of square cross section (9.5×9.5 mm) within a 20 mm gauge length were produced by EDM. Monotonic tensile tests and tensile fatigue tests were performed in a servohydraulic universal load frame, with hydraulic grips. Monotonic tests were run in displacement control at a strain rate of 2×10^{-3} s^{-1}. Fatigue tests were conducted at 10 Hz with a constant minimum stress of 5 MPa and strains were measured using a 10 mm clip gauge. The fatigue tests were monitored continuously. The modulus was chosen as a useful indirect measure of damage.

Flexural tests were conducted in situ within an optical microscope. 50×7×3 mm bars were loaded in four point bending with outer and inner spans of 40 mm and 20 mm, respectively. Examination of the tensile face was undertaken at a series of stress levels in order to observe the formation and propagation of damage, such as matrix cracks.

MECHANICAL BEHAVIOR

(i) Monotonic

Both materials had an initial peak stress (σ_{UTS}) of $\sim$ 365 MPa that was followed by a sudden drop in load due to the formation of a single large crack within the matrix (Fig. 6a). The σ_{UTS} occurred at a strain of 1.5% in material **MT-90-18** but at a strain of only 0.5% in material **MT-90-19**. The subsequent secondary peak in stress was due to work-hardening of the stainless steel tubes that were left bridging the specimen halves. The slight difference in the secondary peaks of the two materials was due to a greater number of tubes being present in material **MT-90-19** ($\sim$ 8%). In both samples the initial loading portion of the curve shows a slight "knee" at high loads (greater in material **MT-90-18**) due to the yield of the tubes ($\sigma_y \sim$ 200MPa) and the subsequent microcracking of the matrix.

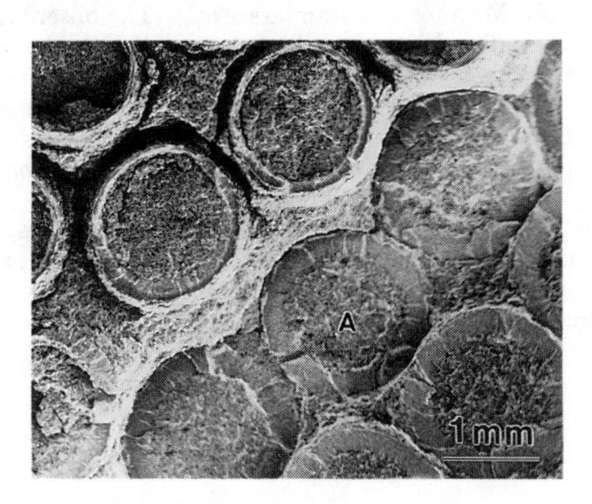

Figure 7: SEM micrograph of **MT-90-18** failure surface illustrating fatigue damage in area A.

(ii) Flexure Test

Matrix microcracks were first observed in material **MT-90-19** when the stress on the tensile face was $\sim$ 270 MPa. These microcracks were observed to initiate at the particulates and voids with the size of the microcracks dictated by the particle spacing. As the load was increased, large numbers of microcracks were generated in the matrix, both from observable surface flaws and without any noticeable initiation point. Finally, a single large crack was formed with further loading up to $\sim$ 390 MPa.

(iii) Fatigue

For both materials no fatigue behavior occurred at peak stress of 200 MPa or less, up to 10^6 cycles. At 250 MPa **MT-90-18** failed after 250000 cycles. Failure was preceded by a small reduction in modulus immediately prior to failure. **MT-90-19** survived 10^6 cycles at 250 MPa but failed after 50000 cycles at a peak stress of 300 MPa. The modulus reduced from the first cycle, the total modulus reduction at failure was 6%, similar to that observed in the **MT-90-18** specimen. The data from the fatigue tests was plotted on a S–N curve (Fig. 6b) which illustrates the high fatigue limits of both materials relative to their tensile strengths. Fractography clearly revealed the failure mechanism (Fig. 7). In each material several steel tubes failed on the same plane by fatigue crack growth. The onset of final fracture was the result of ductile rupture of the remaining (undamaged) tubes. Both specimens failed outside the gauge length of the clip gauge, therefore, the modulus reduction is not associated with the localized growth of fatigue cracks in the steel, which led to ultimate failure. The degradation of stiffness is caused by the growth of distributed microcracks in the matrix, which may contribute to the initiation of the dominant fatigue crack.

DISCUSSION

Both materials examined here showed fatigue limits that were substantial proportions of the composites' ultimate strength (67% σ_{UTS} for **MT-90-18** and 83% σ_{UTS} for **MT-90-19**). At stresses below these levels the specimens did not fail after $\sim 10^6$ cycles and, through monitoring of the composite modulus, showed no evidence of any accumulated damage. These fatigue limits are quite high when compared with other materials (e.g., 304 stainless steel ~40% σ_{UTS} [8]) and occur at stress levels in which the stainless steel tubes will have yielded ($\sigma_y \sim 200$ MPa for 304 stainless steel). The observation of a lower fatigue limit in the **MT-90-18** material appears consistent with the data from the monotonic tests in which more yielding was observed in **MT-90-18** during the initial loading than in **MT-90-19**. Thus a lower stress is needed to initiate damage necessary for fatigue crack growth in material **MT-90-18**. This is also consistent with the observation that the axial strain at the fatigue stress for both materials was 0.3%, implying that a certain amount of strain is needed before fatigue cracks can initiate and grow.

The presence of the Al_2O_3 precipitates at the stainless steel tubes/NiAl matrix interface is possibly due to reaction between the Al in the NiAl and surface oxide on the tubing. It is not clear at this moment what causes the occurrence of the nanoscale precipitates at the Al_2O_3 filler platelet interfaces nor their effect on the fatigue properties of the composite material.

CONCLUSIONS

The two materials tested (**MT-90-18** and **MT-90-19**) have fatigue limits that are significant proportions of their ultimate strengths; 67% for **MT-90-18** and 83% for **MT-90-19**. The fatigue data appears to be consistent with the monotonic test results suggesting that the mechanical properties are controlled by the growth of distributed microcracks in the matrix and the properties of the matrix/tube interface. Analysis of this interface shows the presence of 100nm Al_2O_3 particles but it is, as yet, uncertain what role these particles play in the composite properties

ACKNOWLEDGMENTS

We would like to thank K. Fields for the monotonic tensile data and V. Nardone at United Technologies Research Center for supplying the material. M. De Graef is a Research Associate with the Belgian National Fund for Research. This work was sponsored in part by the Defense Advanced Research Project Agency (DARPA) under a University Research Initiative Grant N00014-86-K-0753, supervised by Dr. W. Coblenz and monitored by Dr. S.G. Fishman of the Office of Naval Research.

REFERENCES

[1] J.R. Stephens, Mat. Res. Soc. Symp. Proc., 39, 381 (1985)

[2] P.R. Subramanian, M.G. Mendiratta, D.B. Miracle, D.M.Dimiduk, Mat. Res. Soc. Symp. Proc., 194, 147 (1990)

[3] M.K. Bannister, Ph.D Thesis, Cambridge 1990

[4] J. Besson, M. De Graef, J.P.A. Löfvander, S.M. Spearing, to be published in J. Mat. Sci. (1992)

[5] V.C. Nardone, J.R. Strife, K.M. Prewo, Mat. Res. Soc. Symp. Proc., 194, 205 (1990)

[6] V.C. Nardone, Met. Trans. A, 23A, 563 (1992)

[7] K.T. Venkateswara Rao, G.R. Odette, R.O. Ritchie, "Fatigue of Advanced Materials", Proc. Eng. Foundation Int. Conf., pp 429–436, MCEP Ltd. (Santa Barbara, 1991)

[8] Metals Handbook, Ninth Edition, 3, ASM (1980)

OXIDATION OF POWDER PROCESSED NiAl AND NiAl/TiB_2 COMPOSITES

P.S. Korinko, D.E. Alman, N.S. Stoloff, and D.J. Duquette, Materials Engineering Department, Rensselaer Polytechnic Institute, Troy, NY 12180

ABSTRACT

NiAl and NiAl/TiB_2 composites were tested in air at 800, 1000, and 1200°C. The oxidation resistance of the composites depends on the fabrication route, and subsequently on the reinforcement phase morphology and distribution. The oxidation resistance of NiAl reinforced with large TiB_2 particles was found to decrease with increasing TiB_2 content. NiAl reinforced with large TiB_2 particles was completely oxidized at 1200°C after a 100 hour exposure.

INTRODUCTION

Intermetallic matrix composites have emerged as potential next generation high temperature structural composites, as evidenced by this conference and the MRS conference of 1990 [1]. Features such as high melting temperatures, low densities, and excellent oxidation resistance make intermetallic compounds attractive for high temperature applications, especially for the aerospace industry. However, these materials have poor creep resistance and poor room temperature toughness. Compositing may improve these material properties and consequently make intermetallics useful structural materials.

The intermetallic compound NiAl, reinforced with TiB_2 particles, displays a number of properties that make it attractive for use at elevated temperatures. For example, NiAl has a high melting temperature, 1640°C, a low density, 5.86 g/cm^3, and excellent oxidation resistance [2]. In addition, the inclusion of TiB_2 particles has been shown to improve the mechanical behavior of NiAl. For instance, the creep resistance of NiAl/10 vol% TiB_2 is markedly improved compared to monolithic NiAl [3], and large increases in the room temperature strength of NiAl/TiB_2 composites have also been reported [4]. Finally, it has been experimentally determined that NiAl and TiB_2 are thermodynamically compatible at elevated temperatures [5].

Most research efforts have centered on determining the macromechanical properties of NiAl/TiB_2 composites [6],[7]. Little work has been reported on the chemical stability of NiAl/TiB_2 alloys when exposed to an oxidizing environment, such as air at elevated temperatures, although the oxidation behavior of binary nickel-aluminum alloys, as a function of aluminum content, has been extensively studied and is reasonably well understood [8],[9]. For example, it is well known that a minimum of 17 wt% Al is required to exclusively form a continuous alumina scale on binary nickel-aluminum alloy [10].

NiAl oxidized at 900°C has a minimum oxidation rate constant at 42 at% Al [11]. This minimum is attributed to the manner in which the activity of the dissolved Al varies with composition. More recent work [2] did not show the same compositional dependence of the rate constants, but this work was conducted at higher temperatures at which an alpha Al_2O_3 forms rather than the theta Al_2O_3 oxide that forms below 1000°C.

Further, the oxide formed on near stoichiometric NiAl has been observed to spall readily on cooling [12],[13]. Spalling can be diminished or prevented by appropriate alloying, [2] typically with reactive elements.

In the current work, the oxidation resistance of binary monolithic NiAl and NiAl/TiB_2 composites containing 10 and 20 vol% TiB_2 produced via powder metallurgy has been determined between 800 and 1200°C.

MATERIALS FABRICATION

For this study, NiAl/TiB_2 samples were produced by two powder metallurgical approaches: Reactive synthesis (RS) of Ni, Al and TiB_2 powders and conventional hot isostatic pressing (HIP) of XD™ synthesized NiAl/TiB_2 powder blends.

Reactive synthesis of NiAl/TiB_2

Reactive synthesis involves producing alloys or compounds from elemental constituents. Upon heating the mixture of elemental powders to the lowest liquidus temperature in the desired system (640°C in the Ni-Al system), a transient liquid forms. The liquid phase enhances both diffusion and densification via capillary forces. The high thermodynamic stability (i.e. large heat of formation) provides the driving force for the reaction. The high thermal stability of the forming compound manifests itself by the exothermic nature of the reaction. As a consequence, the reaction becomes self-sustaining. This type of process has been termed self-propagating high temperature synthesis (SHS), and has been used to synthesize NiAl, NiAl/TiB_2, as well as other alloys and compounds [4]. It was found that the reaction between Ni and Al to form NiAl was so exothermic that the addition of a dilutant phase (up to 15 wt% of either prealloyed NiAl or a high melting temperature second phase) was necessary to prevent melting of the container and sample during the synthesis and pressureless densification stages.

The procedure for the fabrication of reactive synthesized material is as follows: Elemental Ni and Al powders were mixed to a composition corresponding to 51 at% Ni (70 wt% Ni). To decrease the peak temperature and control the reaction, 10 wt% prealloyed NiAl was added to the elemental powder mixture. Either 0, 10, or 20 vol% TiB_2 was then blended with the Ni, Al, and NiAl mixture. Mixing was performed in a turbula type mixer for 60 minutes. Table I lists the characteristics of the powders used in this study. Cylindrical compacts (31.7 mm diameter by 75 mm long) were produced by cold isostatic pressing the powder to approximately 70% theoretical density. The compacts were encapsulated in 304 stainless steel HIP cans. While attached to a vacuum system the cans were heated to 700°C to allow the Ni and Al to react and densify. The samples were then sealed and HIPed at 1250°C, 172 MPa for either 1 or 2 hours depending on the sample chemistry, to remove residual porosity. Figures 1a-1c show the starting microstructures of the RS HIP materials.

Table I. Characteristics of the starting materials.

Powder	Size	Shape	Process
Al	3 μm	spherical	gas atomized
Ni	3-7 μm	spikey surface	carbonyl
NiAl	-325 mesh	angular	--
TiB_2	3 μm	--	--

HIP of XD™ Synthesized Powder

XD™ is a proprietary process developed by Martin Marietta Corporation, Baltimore, MD, in which fine (1-3 μm) TiB_2 particles are dispersed in a variety of matrices. Details of the process have been described in various patents [14],[15]. Composite NiAl/10 vol% TiB_2 powder obtained from Martin Marietta was pressed directly into 304 stainless steel HIP cans (12.5 mm diameter) and vacuum degassed at 300°C for several hours and then vacuum encapsulated. The HIP conditions were 1250°C, 172 MPa for 2 hours. Figure 1d shows the microstructure of the XD sample. Note the fine TiB_2 dispersoids compared to the coarser TiB_2 dispersoids found in the RS-10 vol% TiB_2 sample shown in Figure 1b.

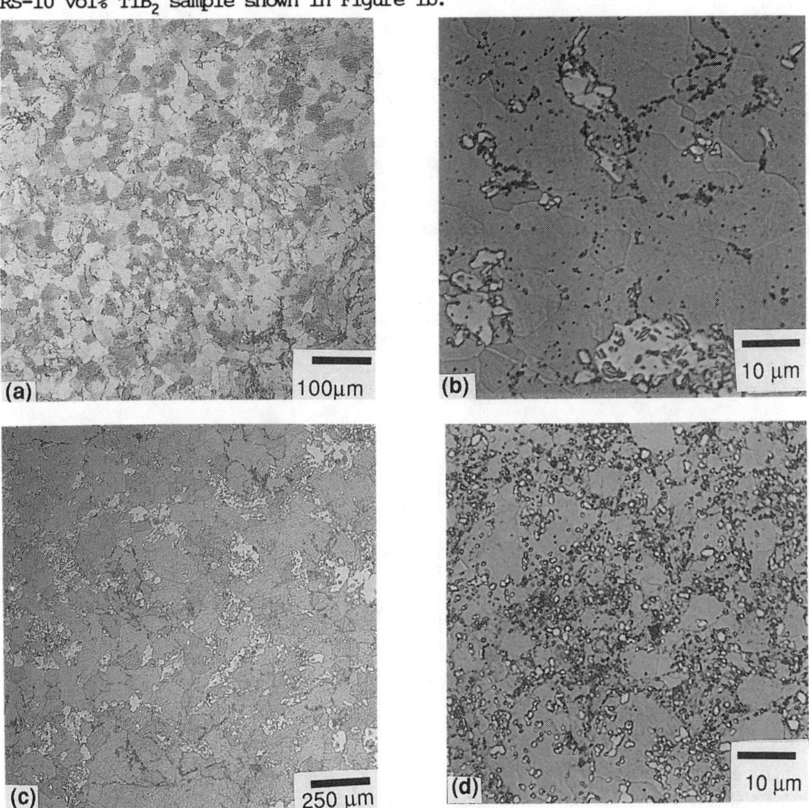

Figure 1 showing the initial etched microstructures of the samples used in this study. (a) RS-monolithic NiAl (b) RS-10TiB$_2$ (c) RS-20TiB$_2$ (d) XD-10 TiB$_2$. Etchant: Kallings Reagent.

OXIDATION SAMPLE PREPARATION

Oxidation samples were prepared by cutting disks of approximately 1 cm^2 of the HIPed material with a diamond saw. The samples were metallographically polished from 240 grit to 0.3 μm alumina on all sides. After polishing, the samples were ultrasonically cleaned in alcohol/50% acetone, measured, and weighed. The samples were placed in alumina combustion boats and inserted into a preheated furnace. Dry, compressed air was passed through the furnace tube at 50 cc/min. Thermal exposure was conducted by placing a single sample in the furnace and removing it after 6.25, 25, or 100 hours. The sample was cooled to room temperature, weighed, and then returned to the furnace until the next interval. The tests were conducted at 800, 1000, and 1200°C. Any loose or lightly adherent oxide was removed before the sample was weighed.

The samples were examined with optical microscopy, scanning electron microscopy (SEM), energy dispersive spectroscopy (EDS), and x-ray diffraction (XRD). Standardless semiquantitative (SSQ) analysis was used in conjunction with the EDS.

RESULTS AND DISCUSSION

The results of the isothermal exposure at 800°C are shown in Figure 2. These results show that the monolithic NiAl (RS-0) initially oxidized slightly and that the weight gain remained nearly constant after the first few hours. The oxide on the sample was so thin that it only showed a slight red-gold heat tint after the 100 hour exposure.

Figure 2 also shows that the inclusion of 10 vol% TiB_2 into the NiAl, either through the RS process, or the XD process, resulted in only a small reduction of the oxidation resistance compared to the monolithic alloy. However, increasing the TiB_2 content to 20 vol% resulted in an increased weight gain of approximately 250%.

The composites had a grayish surface when examined at 10X magnification, however, at 100X magnifications the surface showed islands of grey oxide with a heat tinted oxide on the balance of the surface.

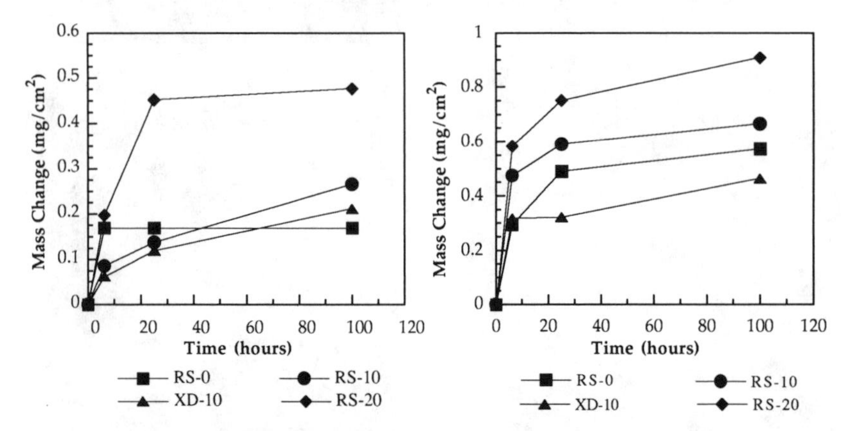

Figure 2 Isothermal oxidation data for NiAl and composites at 800°C.

Figure 3 Isothermal oxidation data for NiAl and composites at 1000°C.

X-ray diffraction of the composite samples' surfaces showed predominately NiAl and TiB_2, but also included peaks for Al_2O_3 and TiO_2. The X-ray diffraction pattern from the monolithic samples showed only NiAl, with the oxide peaks at about the same intensity as the background.

The results of the isothermal exposure at 1000°C are shown in Figure 3. The XD-10 sample shows improved oxidation resistance compared to the RS processed materials. This is possibly to due differences in alloy composition or morphology of the reinforcing phase. EDS of the oxides on the 10 TiB_2 samples showed that there is a small amount of Ti in the oxide on the XD material but not in the RS material. The RS-0 sample has better

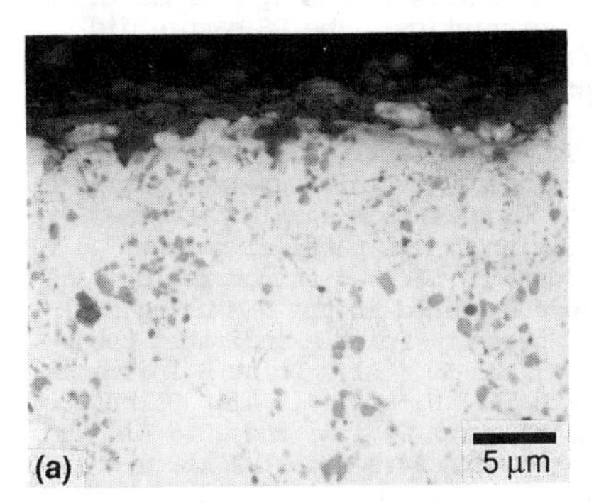

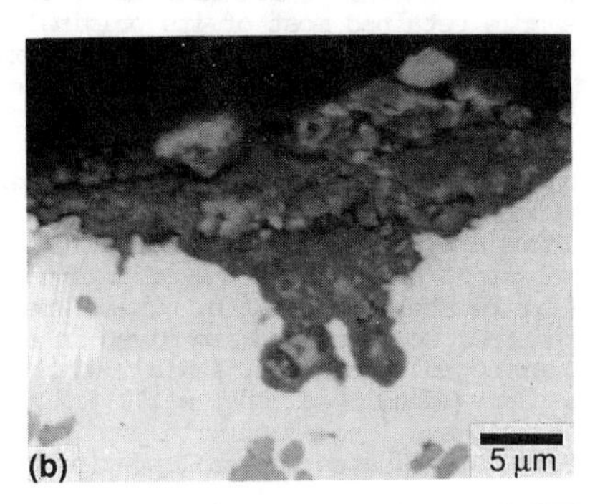

Figure 4 Oxides formed on NiAl/10TiB_2 (a) XD processed (b) RS processed SEM backscattered electron images (BEI).

oxidation resistance than any of the RS processed composites, and the weight gains increase with increasing volume fraction of TiB_2.

Figures 4a and 4b show the oxide scale that formed on the 10 vol% TiB_2 composites after a 100 hour exposure at 1000°C. Figure 4a shows the oxide on the XD processed sample. The oxide is generally planar and compact, compared to the oxide formed on the RS sample which tends to be thicker with some oxide intrusions along the TiB_2 particles. This difference in reaction product morphology can most likely be attributed to the difference in reinforcement morphology shown in Figures 1b and 1d. Although the difference in alloy composition could also play a role. Figure 5 shows the isothermal exposure data for the NiAl/TiB_2 composites exposed at 1200°C. These data show that the monolithic RS sample is more oxidation resistant than any of the composites, although it spalled after the 25 hour exposure

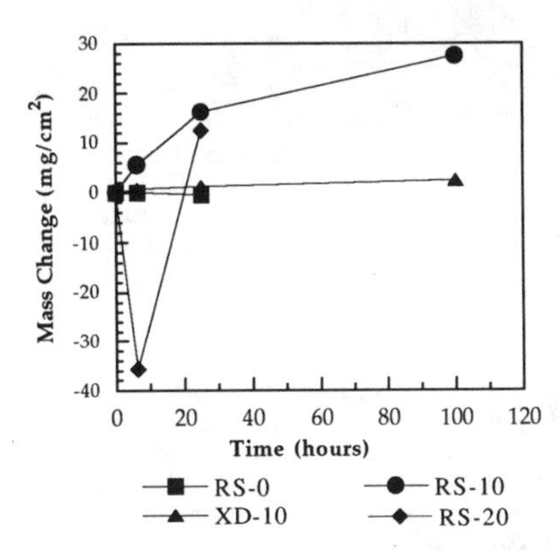

Figure 5 Isothermal oxidation data for NiAl and composites at 1200°C.

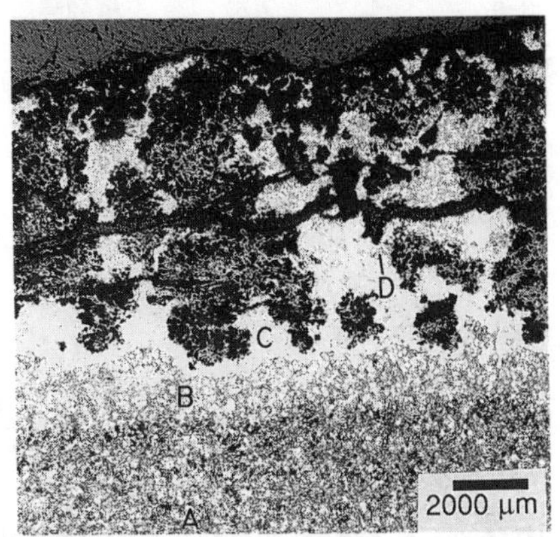

Figure 6 Photomicrograph of a cross-section of NiAl/20TiB_2 exposed at 1200°C for 25 hours; note the four distinct zones; Kalling's reagent.

and the test was terminated. The XD-10 sample is more oxidation resistant than the RS-10; in fact, the RS-10 sample was nearly consumed during the testing and only small islands of unoxidized matrix remained, whereas the XD sample retained most of its original characteristics. The RS sample did notspall despite the extent of the oxidation.

The RS-20 sample readily spalled after the first cycle. The weight gain for the RS-20 between 6.25 and 25 hours is due to the rapid oxidation and lack of spalling on this thermal cycle. This test was aborted after 25 hours because of the sample's poor performance. Several very interesting reactions occurred during the oxidation testing of this sample. Figure 6 shows a cross section of the RS-20 sample with cracks parallel to the substrate and large areas of internal oxidation. Four distinct regions can also be seen, points A-D. These areas were examined in the SEM using EDS and SSQ; compositions are given in atomic percent. Point A indicates the unaffected base alloy, β NiAl, with 49.6 Ni and 50.4 Al. Point B indicates a narrow dealloyed zone, still β NiAl, with 56.7 Ni and 43.3 Al. Point C indicates a large band with precipitates, γ', with 74.4 Ni and 25.6 Al. And Point D indicates internal precipitates, γ', with 74.8 Ni, 21.2 Al, and 4 Ti.

The oxide also shows variable compositions. These variations are indicated by points E-G in Figure 7, a backscattered electron image (BEI) of the RS-20 sample after a 25 hour exposure at 1200°C. Point E shows a dark continuous phase; presumably Al_2O_3 since only Al was detected. Point F shows a light grey precipitate in the Al_2O_3, most likely TiO_2, since only Ti was detected. Point G shows a white dispersed phase which consisted of 86.25 Ni, 9 Al, and 4.75 Ti.

Table II shows the weight changes after the 100 hour exposure. The specific mass change increases with increasing volume fraction of TiB_2 for the RS processed material.

Table II. Mass changes after 100 hour exposure (mg/cm^2).

Sample	800°C	1000°C	1200°C
RS-0	.17	.57	-.42*
XD-10	.21	.46	2.40
RS-10	.22	.66	27.55
RS-20	.48	.91	12.60*+

* test was terminated after 25 hours
+sample initially spalled

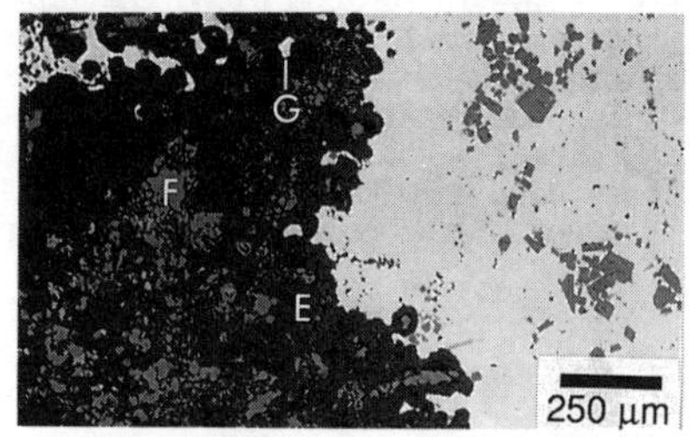

Figure 7 Backscattered electron image of the oxide on NiAl/20TiB_2 exposed at 1200°C for 25 hours; note the three distinct phases.

SUMMARY AND CONCLUSIONS

1. At 800, 1000, and 1200°C the oxidation resistance of RS NiAl reinforced with TiB_2 decreases with increasing volume fraction of reinforcement.
2. Apparently, the reinforcing phase morphology affects the oxidation resistance with small particles being less detrimental to the oxidation resistance than large ones, although processing and matrix composition differences may be partially responsible for this result.
3. RS NiAl reinforced with large TiB_2 particles is completely oxidized at temperatures in excess of 1000°C; fortunately, the temperature limit for tensile strength of NiAl with large TiB_2 particles also seems to be 1000°C [4],[16].
4. XD-NiAl/10TiB_2 has superior oxidation resistance to the RS-NiAl/10TiB_2 at 1000 and 1200°C.

ACKNOWLEDGEMENTS

The authors would like to thank ONR-DARPA (contract no. N00014-86-K-0770) for financial support and Dr. K.S. Kumar of Martin Marietta Research Laboratories for supplying the XD powder.

REFERENCES

1. Intermetallic Matrix Composites, Edited by D.L. Anton, P.L. Martin, D.B. Miracle, and R. McMeeking, Vol 194, Materials Research Society, Pittsburgh, PA 1990.

2. J. Doychak, J.L. Smialek, and C.A. Barrett, in Oxidation of High Temperature Intermetallics, Edited by T. Grobstein and J. Doychak, The Minerals, Metals and Materials Society, 1989.

3. J.D. Whittenburger, R.K. Viswanadham, S.K. Mannan, and B. Sprissler, Journal of Materials Sciences, Vol 25, pp 35-44, 1990.

4. D.E. Alman and N.S. Stoloff, International Journal of Powder Metallurgy, Vol 27, No. 1, pp. 29-41, 1991.

5. M. Saqib, G.M. Mehrotra, I. Weiss, H.Beck, and H.A. Lipsitt, Scripta Metallurgica, Vol 24, pp 1889-1894 1990.

6. R.K. Viswanadham, S.K. Mannan, K.S. Kumar, and A. Wolfenden, Journal of Materials Science Letters, Vol 8, p 409-410, 1989.

7. R.K. Viswanadham, J.D. Whittenberger, S.K. Kumar, and B. Sprissler, in High Temperature Ordered Intermetallic Alloys III, Edited by C.T. Liu, A.I. Taub, N.S. Stoloff, and C.C. Koch, Materials Research Society, Pittsburgh, PA 1989.

8. J.S. Wolf and E.B. Evans, Corrosion, Vol 18 pp 129-136, 1962.

9. F.H. Stott and G.C. Wood, Corrosion Science, Vol 17 pp 647-670, 1977.

10. F.S. Petit, Transactions of the Metallurgical Society of AIME, Vol 23, pp 1296-1305, 1967.

11. R.H. Hutchings and M.H. Loretto, Metal Science, pp 503-510, 1978.

12. H.M. Hindam and W.W. Smeltzer, Journal of the Electrochemical Society, Vol 127, pp 1630-1635, 1980.

13. J.L. Smialek, Metallurgical Transactions A, Vol 9A, pp 309-320, 1978.

14. D.C. Nagle, J.M. Brupbacher, L. Christodoulou, U.S. Patent No. 4,774,052 Sept. 27, 1988.

15. L. Christodoulou, D.C. Naglem, and J.M. Brupbacher, U.S. Patent No. 4,751,048, June 14, 1988.

16. D.E. Alman and N.S. Stoloff, in Advanced Composite Materials, Edited by M.D. Sacks, American Ceramics Society, Westerville, OH, 1991.

FRACTURE MECHANISMS IN NiAlCr EUTECTIC COMPOSITES

Keh-Minn Chang
GE Corporate Research and Development, P.O. Box 8, Schenectady, NY 12301

ABSTRACT

Selected eutectic compositions in Ni-Al-Cr ternary systems were processed by directional solidification (DS) with various growth rates. Fracture toughness tests were performed at room temperature and 400 °C; fracture surfaces of broken specimens were examined using SEM to investigate fracture behavior of each alloy. The alignment of eutectic phases was found to play an important role in composite toughening for the intermetallic matrix. Binary eutectic composites consisting of bcc α-Cr and B2 β-NiAl phases with a directional, well-aligned structure showed improved fracture properties over NiAl single crystals. Ternary eutectics, which contain an fcc γ-Ni phase, offered an excellent fracture resistance at room temperature.

INTRODUCTION

Microcomposite structures resulting from the eutectic reaction during directional solidification (DS) can offer many unique microstructural advantages such as thermodynamic stability, directional alignment, and fine dispersion of component phases. Extensive research efforts of directional eutectics in the past emphasized high temperature properties aimed at a better temperature capability than that of Ni-base superalloys [1-3]. Among many candidates, DS NiAl-Cr eutectics show a great promise as the high-temperature structural material for gas turbine applications [3]. This microcomposite material offers not only high-temperature strength but also several important physical properties, such as low density, superior oxidation resistance, and high thermal conductivity.

The purpose of this paper is to investigate the fracture behavior of DS NiAlCr eutectics at low temperatures. The matrix phase, NiAl intermetallic, has a low fracture resistance at room temperature even in single crystal form [4]. The ductile-to-brittle transition temperature (DBTT) for recrystallized Cr is known to be above 300 °C, and almost any alloying addition in Cr tends to raise DBTT [5]. However, composite toughening is expected in the microcomposites with a desired directional structure that contains uniformly distributed, well-aligned long fibers. The toughening mechanisms will be investigated by examining the fracture mode of individual eutectic phases. Another effective approach of composite toughening is the introduction of ductile fcc $\gamma-$NiCrAl phase in the ternary eutectics. The influences of alloy composition and DS growth rate on the composite microstructure are to be characterized; the improvement of fracture toughness is directly related to the alignment of eutectic phases.

EXPERIMENTAL

Based on the Ni-Al-Cr ternary phase diagram [6], a pseudobinary β(NiAl, B2 type) + α(Cr, bcc type) eutectic trough covers a wide composition range and allows a variation of alloy chemistry of eutectic component phases. An invariant ternary eutectic point, which consists of α(Cr, bcc type) + β(NiAl, B2 type) + γ(NiCrAl, fcc type) phases, is located at about Ni-17Al-38Cr (in at.%). Table I lists the four ternary alloy compositions that were chosen for this study. They represented: (1) a binary NiAl-Cr eutectic with a stoichiometic NiAl phase (CK-9); (2) a binary NiAl-Cr eutectic with a low-Al NiAl phase (CK-10); (3) a mixture of CK-10 and NiAl dendrites (CK-11); and (4) a ternary eutectic (CK-12).

Prealloyed ingots of alloy compositions in Table I were prepared by vacuum induction melting (VIM) using high purity laboratory raw materials. Each alloy melt was cast into two ingots. A Bridgman furnace was used for directional solidification of these eutectic alloys. A standard growth rate of 12.7 mm/h was applied to the first ingot of every alloy. A different DS rate was applied to the second ingots of alloy CK-9 and CK-12.

Table I Compositions and growth rates of DS NiAlCr eutectic composites.

Alloy Designation	Composition wt.%	at.%	Growth Rate mm/h	Remarks
CK-9	Ni-19.4Al-38.4Cr	33Ni-33Al-34Cr	254, 12.7	Stoichiometric NiAl
CK-10	Ni-12.7Al-42.7Cr	37Ni-23Al-40Cr	12.7	Al-lean NiAl
CK-11	Ni-15.7Al-27.0Cr	47Ni-28Al-25Cr	12.7	NiAl dendrites
CK-12	Ni-9.0Al-38.9Cr	45Ni-17Al-38Cr	12.7, 3.0	Ternary eutectic

Fracture toughness of as-DS eutectic composites was measured by four-point bending of notched rectangular bars. The specimens, which had a cross section of 7.6 mm by 15.2 mm and a length of 25.4 mm at least, were cut along the direction of crystal growth. A side notch of 40% of specimen width was cut at the center of the specimen by wire-EDM using a 0.05 mm diameter wire. Fracture tests were performed at room temperature and 400 °C in air, and the load speed was set at 0.5 mm/min.

There was a great composition difference among eutectic phases, which provided a convenient way to examine the DS eutectic structures by using the backscattered electron (BSE) image technique under a scanning electron microscope (SEM). The contrast of the image for each eutectic component phase depended on the phase chemistry, and a heavy element resulted in a bright image under a positive polarity. Fracture surfaces of broken specimens were also examined under a SEM using the normal (secondary electron) image technique. Special attention was concentrated on the fracture behavior of individual eutectic component phase and on the mechanisms for the improvement of fracture resistance through microcomposite structure.

RESULTS AND DISCUSSION

Microstructure of DS Eutectic Composites

Pseudobinay NiAl-Cr eutectic alloys were reported to develop a fibrous composite structure after directional solidification [3]. The matrix phase was a Cr-rich, B2 ordered NiAl phase, and the fibers were round Cr rods containing some Ni and Al. The similarity of crystal lattices between eutectic phases (both bcc) allowed the fibrous structure to exist even with a high volume fraction (> 40%) of the secondary eutectic phase. Figure 1 shows the transverse section of as-DS structures in CK-9 grown at 12.7 and 254 mm/h, respectively. At the standard growth rate, Cr rods of an average of 1 μm diameter were well aligned in the NiAl matrix. The cross section of Cr rods was relatively circular in a random pattern of their distribution and there was hardly any morphology discontinuity across cell or grain boundaries. The increase in the growth rate by 20X resulted in two major changes in microstructure. The diameter of Cr rods was dramatically reduced to 0.2 μm; an empirical relationship between the fiber diameter, d, and the growth rate, v, was verified, i.e., $d^2 * v$ = constant. The sizes of cells and grains were also reduced in a similar fashion. The most important change associated with the fast growth rate was the misalignment of Cr fibers near cell or grain boundaries. A cellular structure was developed by the drag of cell boundaries on the solidification front, and the growth direction of Cr fibers was deviated from the DS crystal growth direction toward the normal of cell or grain boundaries. The band of cellular structure was about 10 μm wide, developed along both sides of cell boundaries.

When the eutectic composition shifted toward the Al-lean NiAl matrix (CK-10 and CK-11), the DS composites still maintained the fibrous structure as seen in Figure 2. The Cr rods in these low-Al, pseudobinary eutectics were not as circular and as fine as those in CK-9; the change of Cr fiber morphology might be caused by the existence of dendrites. CK-11 was originally selected to form NiAl dendrites (65%) in addition to the eutectics. However, CK-10, which was expected to be a eutectic composition, developed a small fraction of Cr dendrites.

The dendrites formed ahead of the eutectic during directional solidification and consequently affected the local chemistry and solidification conditions.

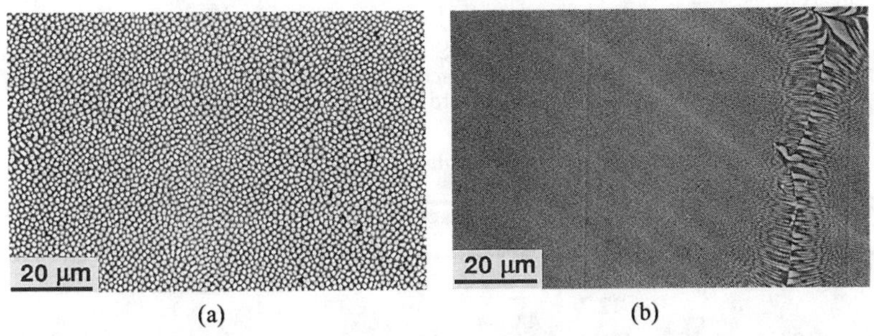

(a) (b)

Figure 1 Transverse sections of DS CK-9 eutectic composites grown at: (a) 12.7, (b) 254 mm/h.

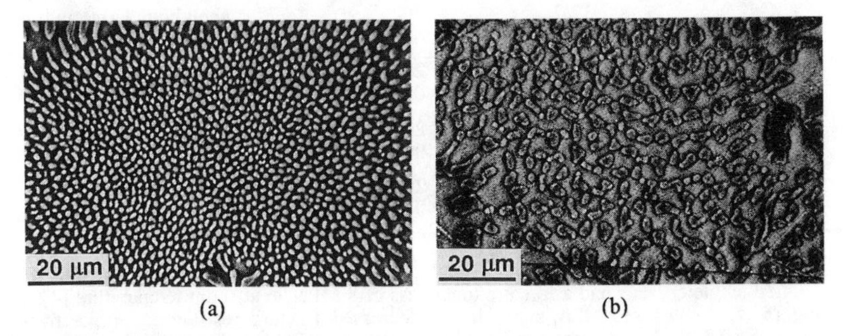

(a) (b)

Figure 2 Transverse section of DS eutectic composites: (a) CK-10, (b) CK-11.

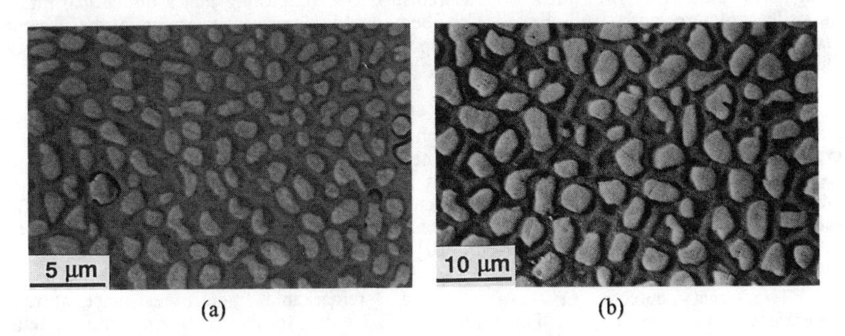

(a) (b)

Figure 3 Transverse section of DS CK-12 eutectic composites grown at: (a) 12.7, (b) 3.0 mm/h.

The ternary eutectic alloy, CK-12, developed an interesting microcomposite structure through DS. The standard growth rate (12.7 mm/h) generated the cellular structure, and therefore a slow growth rate of 3.0 mm/h was employed for the second ingot to form a fully aligned structure. Figure 3 shows the transverse sections of ternary eutectics after DS. A unique fibrous structure was developed: the matrix was the bcc Cr phase; two fiber phases, fcc NiCrAl (the light phase) and B2 NiAl (the dark phase), coalesced side by side in all cases. The irregular shape of fiber cross section was observed for each fiber phase, but the aspect ratio of an overall composite fiber was near unity. Jackson reported a similar ternary eutectic in a Ni-Cr-Fe-Al system, but the DS composite had a lamellar structure [7].

Fracture Toughness

Longitudinal fracture resistance of DS eutectic composites was measured at room temperature and 400 °C. The maximum fracture load was used to calculate the fracture toughness through the formula in ASTM standard E399. All results are listed in Table II.

Table II Low temperature fracture toughness of DS NiAlCr eutectic composites.

Alloy Designation	Growth Rate mm/h	Fracture Test Temp., °C	Kc, MPa$\sqrt{m}$	Remarks
CK-9	12.7	20	11.0	aligned Cr fibers
	12.7	400	51.4	
CK-9A	254.0	20	4.4	cell boundaries
	254.0	400	15.6	
CK-10	12.7	20	7.0	cell boundaries and few Cr dendrites
	12.7	400	15.6	
CK-11	12.7	20	8.1	NiAl dendrites and cell boundaries
	12.7	400	19.4	
CK-12	12.7	20	16.4	cell boundaries
	12.7	400	30.8	
CK-12A	3.0	20	25.4	aligned NiCrAl + NiAl fibers
	3.0	400	23.6	

NiAl single crystals had a fracture toughness of 8 MPa$\sqrt{m}$ at room temperature [4]. Cr-alloying in NiAl increased alloy strength but decreased fracture resistance; e.g., a fracture toughness of 6.7 MPa$\sqrt{m}$ was measured in a DS NiAl-5Cr crystal [8]. Eutectic alloy CK-9 had a Cr-saturated NiAl matrix, whose fracture toughness was expected to be below the above values. A toughness value of 11 MPa$\sqrt{m}$ was measured in CK-9, suggesting that a directional micro-composite structure with well-aligned Cr fibers could significantly improve the fracture resistance at room temperature. Such a composite toughening effect disappeared when the cellular structure was developed under an inappropriate solidification condition.

The results measured on CK-10 and CK-11 suggested that room temperature toughness of the eutectic component Cr phase was not higher than that of NiAl. Two alloy compositions were near on a two-phase (NiAl and Cr) tie-line but contain different dendrite phases. Alloy CK-11 with 65 vol.% NiAl had a somewhat higher fracture resistance than alloy CK-10 containing a few Cr dendrites. The crack deflected ~45 ° from the notch plane in CK-11. This fracture character resembled what was observed in NiAl <100> single crystals [4], i.e., the fracture path tended to follow the preferred {110} cleavage plane.

The ternary eutectic, CK-12A, exhibited a remarkable fracture resistance at room temperature as shown in Table II. There was no available data for the room temperature fracture toughness of Cr-Ni-Al alloys. The toughening effect associated with the composite structure could still be verified by comparing the results obtained from the same alloy with different growth rates. Both pseudobinary (CK-9) and ternary (CK-12) eutectics showed a better fracture resistance at room temperature when a slower growth rate was used during DS.

All alloys showed a remarkable increase in fracture toughness at 400 °C except CK-12A, which had a good fracture resistance at room temperature. Therefore, the ductile-to-brittle transition for this class of microcomposites occurs between 400 °C and room temperature. The ternary eutectic alloy CK-12A with a slow growth rate had a similar fracture toughness at both temperatures; its DBTT was expected to be below room temperature.

Fractography and Mechanisms

Directional composite structures can offer a brittle matrix the enhancement of fracture resistance through various toughening mechanisms, such as ductile phase toughening, phase transformation toughening, bridging effect, and microcracking effect, etc. In case of NiAlCr microcomposites, a new toughening mechanism associated with the preferred cleavage planes was observed.

Figure 4 shows some representive SEM fractographs of broken specimens. Alloy CK-9 tested at 400 °C revealed a ductile fiber fracture mode as shown in Figure 4(a). The Cr fibers, assumed to be single crystal, could sustain a significant amount of plastic deformation before fracture, which resulted in a remarkable fracture toughness. However, these Cr fibers failed in a brittle mode at room temperature, but some improvement of fracture resistance caused by the alignment of fibers was observed. A careful examination of fracture surfaces found that every Cr fiber cleaved on a featureless, crystallographic plane as shown in Figure 4(b). The cleavage plane for bcc Cr crystals is known to be {100}, in contrast to {110} cleavage in B2 ordered NiAl [4]. Since Cr fibers had a cubic-cubic orientation relationship with the NiAl matrix, the cleavage planes would not coincide with each other. The toughening effect can be attributable to the geometric constraint -- the cracking direction switches whenever the crack encountered a Cr fiber. As a result, the NiAl-Cr composite developed a longer crack length and a better fracture resistance than the monolithic NiAl.

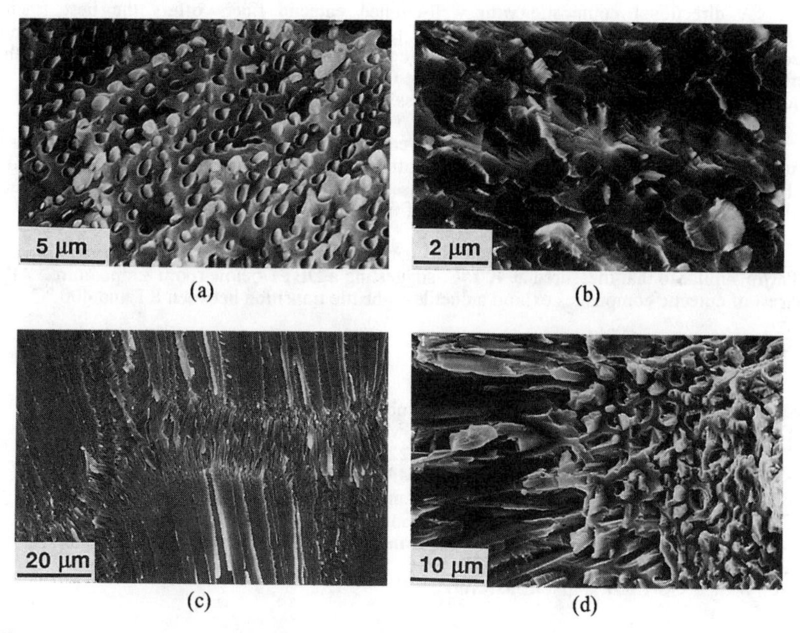

Figure 4 Fractography of DS eutectic composites: (a) CK-9, 400 °C; (b) CK-9, room temperature; (c) CK-9A, room temperature; (d) CK-12, room temperature.

The formation of cellular morphology along cell or grain boundaries can be very detrimental to the fracture properties of eutectic composites. Figure 4(c) shows the fracture surface of CK-9A prepared by a fast growth rate of 254 mm/h. Compared to CK-9 grown in 12.7 mm/hr, its fracture toughness was reduced by 60%, though the fracture modes were brittle in both cases. The good alignment of Cr fibers was one of the major causes for the composite toughening. The reduction of fracture resistance was partially caused by the microcracks initiated easily along the interfaces of those cellular structures. As seen in Figure 4(c), these microcracks extended from the cellular area into the area with aligned Cr fibers.

The high fracture toughness of the ternary eutectic alloy, CK-12, came primarily from the ductile fcc NiCrAl phase, which forms conjunct fibers with the B2 NiAl phase. If the composite had a well-aligned structure, a fully ductile fracture can be observed at room temperature. This was the case when the ternary eutectic was grown at 3.0 mm/h. However, a reduction of fracture toughness at room temperature was observed when the cellular structure was developed along cell or grain boundaries. Figure 4(d) shows both the brittle failure along the cellular interfaces and the ductile fracture of the aligned structure on the fracture surface of CK-12 grown at 12.7 mm/h. There was an indication that the desired fracture mode with fiber pull-out occurred in the ductile failure area.

SUMMARY

Four NiAlCr eutectic compositions that form pseudobinary NiAl-Cr and ternary Cr-Ni-NiAl were directionally solidified using various growth rates. Fracture behavior of these eutectic microcomposites has been studied at room temperature and 400 °C. The fracture resistance can be enhanced remarkably through several composite toughening mechanisms. DS microstructure plays a predominant role on the measured fracture toughness.

A directional composite with well-aligned eutectic fibers offers the best fracture resistance even for brittle fracture. In case of pseudobinary NiAl-Cr eutectics, the toughening mechanism is associated with different cleavage planes for eutectic phases. The fracture path is deflected repeatedly whenever it encounters the Cr fiber. On the other hand, the ductile fcc Ni-solution fibers provide a good fracture toughness for the ternary eutectic composites.

A cellular structure is developed along cell or grain boundaries when the DS processing is not appropriate. The fibers turn away from the crystal growth direction and head normal to cell boundaries. The interface of eutectic phases in the cellular structure initiates microcracks under loading, which resulted in the decrease of fracture resistance.

The well-aligned ternary eutectic has a room temperature fracture toughness above 20 $MPa\sqrt{m}$ similar to that measured at 400 °C, suggesting a DBTT below room temperature. All of the rest of eutectic composites exhibit a ductile-to-brittle transition between RT and 400 °C.

REFERENCES

1. D.A. Woodford, JOM, **42-11**, 50-55 (November 1990).
2. E.R. Thompson and F.D. Lemkey, Trans. ASM, **62**, 140-154 (1969).
3. J.L. Walter and H.E. Cline, Met. Trans., **1**, 1221-1229 (1970).
4. K.-M. Chang, R. Darolia, and H.A. Lipsitt, MRS Symp. Proc., **213**, 597-602 (1991)
5. E.P. Abrahamson II and N.J. Grant, ASM Trans., **50**, 705-721 (1958).
6. S.M. Merchant and M.R. Notis, Mater. Sci. Eng., **66**, 47-60 (1984).
7. M.R. Jackson, *Conference on in-situ Composites - II*, ed. M.R. Jackson, et. al., Xerox Individualized Publishing, MA, 67-75 (1976).
8. J.L. Walter, GE-CRD, unpublished work.

FRACTURE TOUGHNESS OF NiAl *IN-SITU* EUTECTIC COMPOSITES.

F. E. HEREDIA* and J. J. VALENCIA**
* Materials Department, University of California, Santa Barbara, CA 93106.
** Metalworking Technology, Inc., Johnstown, PA 15904.

ABSTRACT

Mechanical tests were performed on directionally solidified (DS) NiAl *in-situ* eutectic composites in order to evaluate the effect of ductile reinforcements on the fracture resistance of the B2 ordered intermetallic compound NiAl. Reinforcements consisted of i) Mo fibers, ii) Cr fibers, and iii) Cr(Mo) solid solution plates. Near stoichiometric NiAl ingots were prepared by induction melting as reference material to compare with the eutectic composites. Resistance curves were obtained for the NiAl/Mo fibrous eutectic alloy as well as for the NiAl/Cr(Mo) layered material. The initiation fracture toughness of the DS NiAl/Mo and NiAl/Cr(Mo) eutectic composites is larger than that of the stoichiometric NiAl, with the layered material producing the better properties. The mechanisms for such increase in fracture toughness are discussed.

INTRODUCTION

There has been a renewed interest during the last several years aimed to improve the ductility and fracture toughness of intermetallic alloys with potential as high temperature structural materials. One of these intermetallic compounds is the B2 NiAl. A well documented approach to improve the fracture resistance characteristics of such brittle intermetallics is by reinforcing them with ductile refractory metals [1-7]. Present understanding indicates the important effects of the flow strength and ductility of the refractory metal, its morphology and dimensions, as well as the interface debond resistance [5-10]. In the present study, two morphological classes of refractory reinforcement are investigated for the NiAl matrix, using materials produced by directional solidification [11-13]: i) NiAl/Mo and NiAl/Cr are used to provide a system with a continuous NiAl matrix, plus *aligned rods* of Mo or Cr reinforcement; ii) NiAl/Cr(Mo) is used to create a system with *alternating layers* of NiAl and Cr(Mo). The fracture resistance of these materials is measured and related to microstructure.

EXPERIMENTAL

Stoichiometric NiAl alloys of nominal compositions 45Ni-45Al-10Mo and 32.5Ni-32.5Al-35Cr (at%) were induction melted under a positive argon atmosphere in an Al_2O_3 crucible and cast into copper molds to produce bars of 12.7 mm diameter. The cast bars were remelted under argon in a high purity Al_2O_3 crucible positioned inside a Bridgeman-type directional solidification apparatus. The melts were withdrawn from the heat zone of the directional solidification device at a rate of 3 cm per hour. The directionally solidified NiAl/Cr(Mo) eutectic material of nominal composition 32.5Ni-32.5Al-33Cr-1Mo (at%) was kindly provided by GE Aircraft Engines Division, Cincinnati, Ohio. This alloy was grown by the *Edge-defined Film-Fed Growth* (EFG) process at a rate of 10 cm per hour, producing slabs 25 mm wide and 3 mm thick [14]. Binary stoichiometric NiAl alloys were also induction melted under a positive argon atmosphere and cast in Al_2O_3 crucibles. The chemical composition of the resulting alloys was characterized by Energy dispersive spectroscopy (EDS) with standards using a Tracor-Northern system attached to a JEOL-840 Scanning Electron Microscope (SEM). Table I lists the chemical compositions of the alloys used in this study (EDS analysis).

Table I
Chemical Composition of the Directionally Solidified Alloys
(at%)

MATERIAL	Ni	Al	Mo	Cr
NiAl	51.8	48.2		
NiAl/Mo	44.0	45.0	11.0	
NiAl/Cr	34.3	33.7		32.0
NiAl/Cr(Mo)	34.8	33.9	1.0	30.3

Flexure specimens with dimensions of 3 mm x 7 mm x 25 mm were cut from the as-grown material by electric discharge machining (EDM). A narrow straight 3 mm-deep notch was introduced, also by EDM. Tests were conducted in three-point flexure upon a 20 mm span in a MTS-810 servohydraulic machine, with a displacement gauge in contact with the specimen. Loads were imposed using displacement control on the actual specimen. This technique allowed pre-cracks to be introduced from the notch. Subsequent to pre-cracking, loads and displacements were recorded at constant displacement rates, and the crack length was monitored using a high resolution optical microscope All tests were conducted in air, at room temperature (25°C) and using a displacement rate of 3 x 10^{-4} mm/min. Fracture surfaces were examined in the SEM and quantitative measurements made of the plastic stretch of the refractory metal.

RESULTS AND DISCUSSION

Microstructures

Transverse and longitudinal sections of the directionally solidified alloys were examined in the SEM using backscattered electron imaging. Both the NiAl/Mo and the NiAl/Cr materials show a cellular microstructure and within the cells, the morphology is fully eutectic. Figure 1 shows low and high magnification micrographs of the microstructure of the directionally solidified alloys. For the NiAl/Mo alloy the cells consist of aligned Mo rods of radius, $R \approx 0.6$ μm and a volume fraction, $f_m \approx 0.17$. For the NiAl/Cr alloy the cells are comprised of Cr rods of radius, $R \approx 0.8$ μm and a volume fraction $f_m \approx 0.34$ (all values are average). The NiAl/Cr(Mo) provided by GE revealed a layered-type microstructure in which the NiAl layers have an average thickness $h_1 \approx 0.75$ μm, and the Cr(Mo) layers have a thickness, $h_2 \approx 0.35$ μm corresponding to a volume fraction of refractory metal $f_m \approx 0.30$. The volume fraction of the refractory fibers and plates were measured using an image analyzer. To better reveal the shape and continuity of the fibers and/or plates, the NiAl eutectic alloys were electroetched in a solution containing 3% in volume of Oxalic acid. SEM observations (see Figure 2) showed continuity of the fibers and the plates in the growth direction, although there is some branching of the Mo fibers and some faulting of the plates.

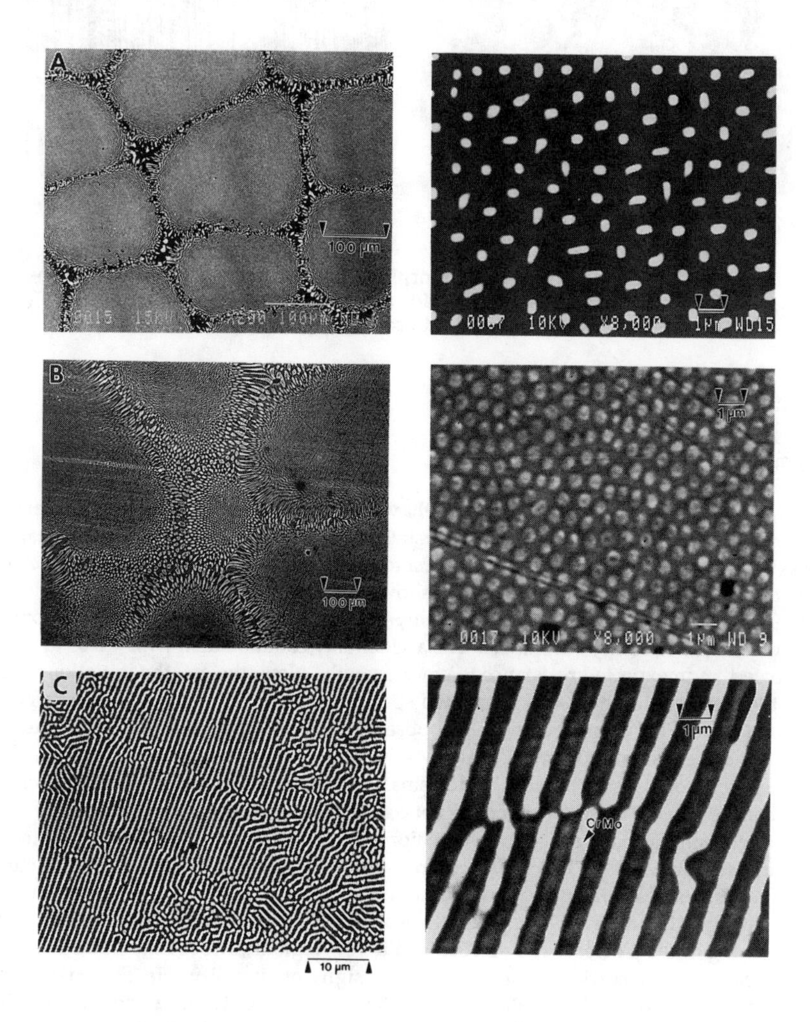

Fig. 1. Low and high magnification backscattered SEM micrographs of the transverse sections of the directionally solidified alloys: a) NiAL/Mo; b) NiAl/Cr; c) NiAl/Cr(Mo).

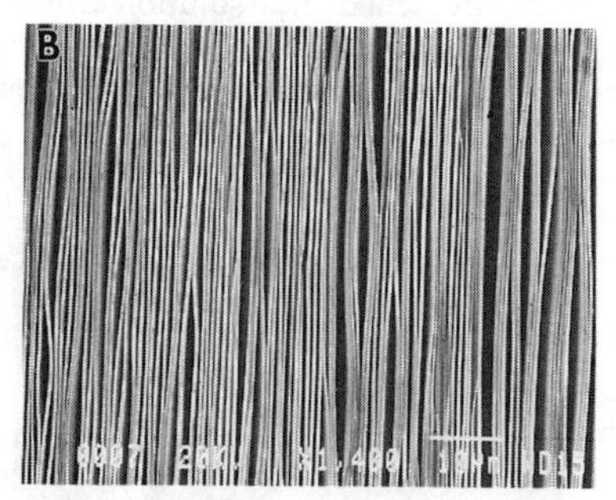

Fig. 2. SEM micrographs of two morphologically diferent alloys with deep-etched matrix showing: a) Mo rods; b) Cr(Mo) plates. Note the branching on some of the Mo rods in micrograph a).

Fracture Toughness Measurements

Different crack resistance behaviors were obtained in the materials investigated. The stoichiometric polycrystalline NiAl alloy has an initiation resistance, $K_o \approx 6\ \mathrm{MPa}\sqrt{\mathrm{m}}$ with no resistance enhancement upon crack extension. The crack path is planar and the fracture surface shows the typical intergranular appearance of the brittle nickel aluminides at room temperature. However, it is worth mentioning that a couple of binary NiAl specimens accidentally contaminated with ~ 0.05 at% Mo showed an unexpected resistance to crack growth. Nevertheless, the initiation value of the fracture toughness for these two specimens were the same as the clean NiAl alloy ($K_o \approx 6\ \mathrm{MPa}\sqrt{\mathrm{m}}$). Further SEM observations of the contaminated specimens revealed the presence of small local NiAl/Mo eutectic pools at the grain boundaries. The exact mechanisms by which such small amounts of Mo slowed down the crack growth process in these specimens in not completely understood.

In similar fashion to the stoichiometric NiAl alloy, the NiAl/Cr eutectic composite showed no crack growth resistance behavior. The initiation fracture toughness for this system was measured at $K_o \approx 7\ \mathrm{MPa}\sqrt{\mathrm{m}}$, slightly higher than for the binary NiAl alloy. The crack path was also planar and although some plastic stretching of the Cr fibers can be observed in few regions, most of the fibers failed at the wake of the crack with practically no signs of ductility. In Figure 3, SEM fractographs of the NiAl/Cr alloy show both regions, the brittle area and the one with plastic stretching of the Cr fibers. The degree of purity of the starting Cr material used for the casting (99.5% purity) could explain the rather brittle behavior since the mechanical properties of Cr are sensitive to impurities.

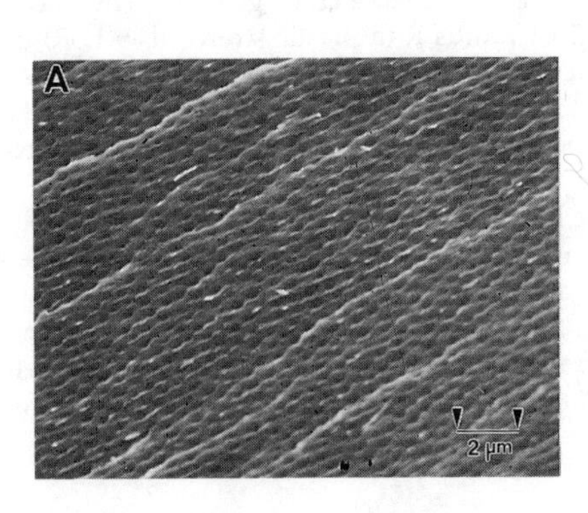

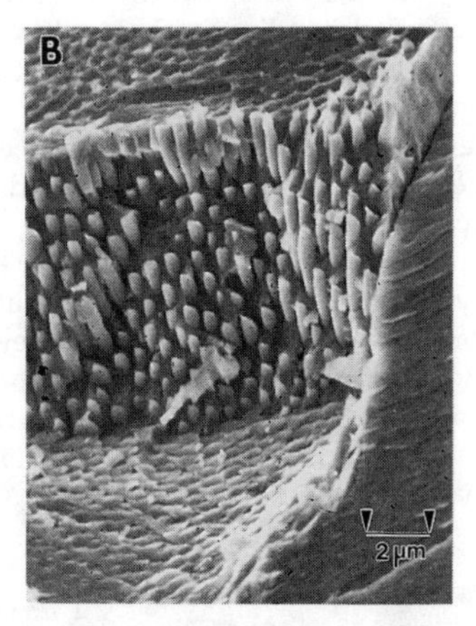

Fig. 3. SEM fractograph of the NiAl/Cr alloy: a) no plastic stretching of the fibers; b) isolated areas with some stretching/debonding of the Cr fibers.

On the other hand, crack growth resistances were measurable on the NiAl/Mo and the NiAl/Cr(Mo) eutectic systems. The NiAl/Mo material has an initiation resistance, $K_0 \approx 11\ \mathrm{MPa}\sqrt{\mathrm{m}}$. The resistance raises to a level $K_S \approx 17\ \mathrm{MPa}\sqrt{\mathrm{m}}$ after a crack extension $\Delta a \approx 500\ \mu\mathrm{m}$, point at which it grows unstably across the specimen. The NiAl/Cr(Mo) layered material has a much larger initiation resistance, $K_0 \approx 17\ \mathrm{MPa}\sqrt{\mathrm{m}}$ followed by a raising resistance that reaches a $K_S \approx 22\ \mathrm{MPa}\sqrt{\mathrm{m}}$ at a crack extension of $\Delta a \approx 500\ \mu\mathrm{m}$ (similar to the NiAl/Mo case). In this material, crack growth is stable. These results are shown in Figure 4.

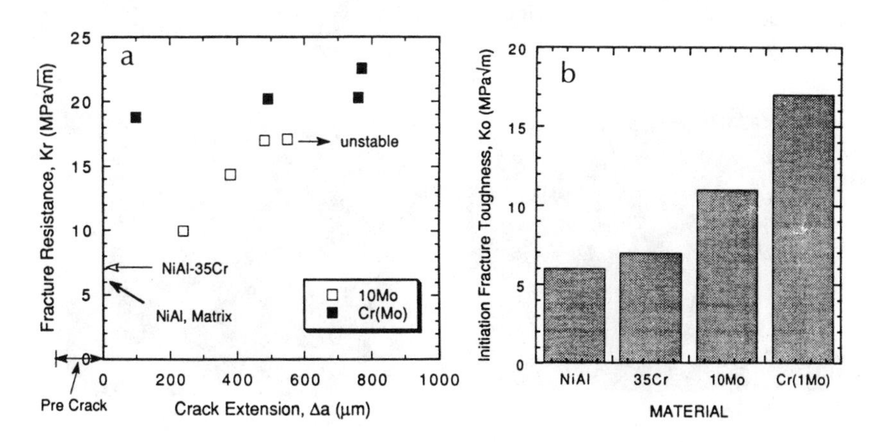

Fig. 4. a) Resistance curve results; b) Initiation fracture toughness data .

In the NiAl/Mo fibrous material, plastic stretching of the Mo fibers and interface debonding is observed in the region of stable crack growth. The debond length is usually on the order of the fiber radius R (a plastic stretch $u_c \approx 1.5R$) (see Figure 5a). At locations where the crack interacts with the Mo, steps can be seen in the NiAl matrix manifest as trails emanating from the reinforcements (Figure 5b). These trails are typical of those induced by *crack trapping* [15-17] as the crack front circumvents the reinforcements.

The fracture morphology for the NiAl/Cr(Mo) layered material is complex. A key feature is the intermittent splitting that occurs at the NiAl/refractory metal interface evident both from side views and from fracture surfaces (Fig. 6a). The split length can be relatively large, but the majority of the interface decohesions are small, typically 10 μm. The ductility of the refractory metal is evident from the plastic stretch exhibited on the fracture surface (Fig. 6b), as well as the distortion found between adjacent interface decohesions (Fig. 6a).

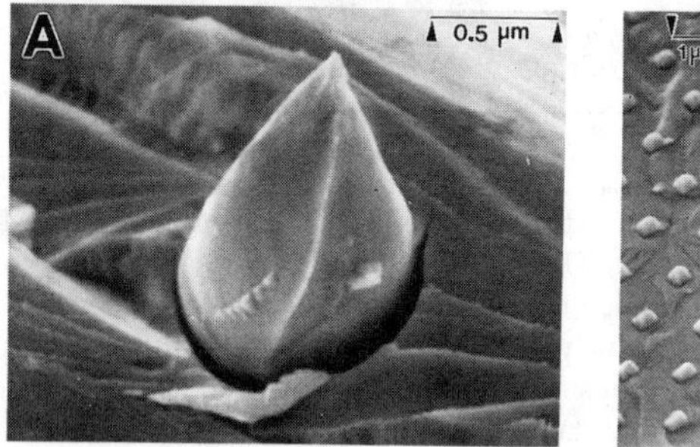

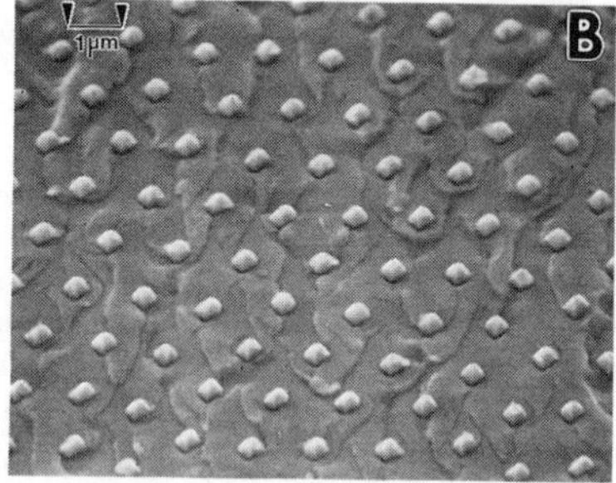

Fig. 5. SEM fractograph of the NiAl/Mo alloy: a) plastic stretching and debonding in a Mo fiber; b) crack trapping marks.

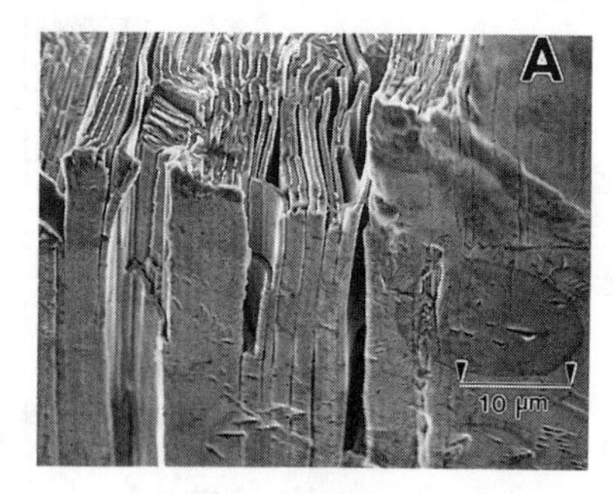

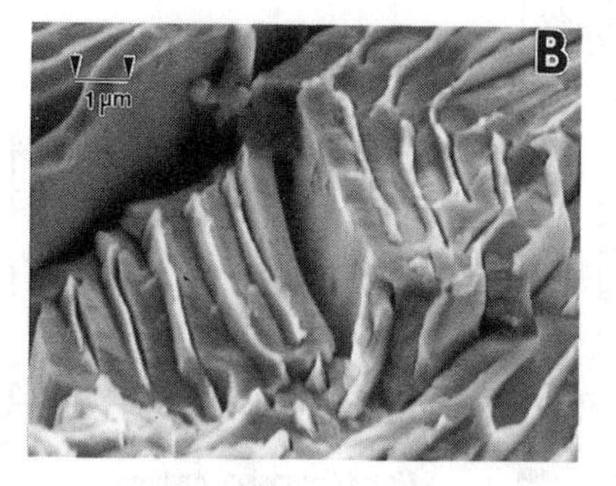

Fig. 6. SEM fractograph of the NiAl/Cr(Mo) alloy: a) side view; b) plastic stretch of the refractory metal.

Interpretation of the experimental results can be rationalized by addressing the effects that promote the elevation of the *initiation toughness* above the matrix toughness. Toughness enhancement involved *crack trapping* in the fibrous material and *crack renucleation* in the layered material where a crack that first forms in the brittle layer must renucleate in the next brittle layer [18]. For the NiAl/Mo system, preliminary *crack trapping* calculations [19], in conjunction with a NiAl matrix toughness of 5-6 MPa$\sqrt{m}$, predict an initiation toughness, $K_0 \approx 10$ MPa$\sqrt{m}$, value which is consistent with the resistance measured at crack extensions up to ~ 250 μm.

For the NiAl/Cr(Mo) system, initial crack extension involves renucleation which occurs at stresses that are sensitive to the incidence of splitting, phenomena that is seen everywhere along the crack path.

A more detailed analysis [19] indicates that ductile bridging does *not* fully account for the increase in resistance found experimentally for either material. The extra contribution to the crack growth resistance probably arises from the *non-planarity* of the fracture. This phenomenon is known to increase the ductile phase bridging contribution to the crack growth resistance, as well as introducing possible frictional contributions when the crack faces are in contact. A quantification of these effects has not been attempted.

CONCLUDING REMARKS

Continuous refractory reinforcements can be introduced into NiAl via directional solidification. Such reinforcements have been shown to provide an increase in the fracture resistance of the nickel aluminide, particularly in the case of Mo fibers and Cr(Mo) plates. The in-situ eutectic with the layered morphology, NiAl/Cr(Mo), provided the best properties of the three systems studied, as well as the most complex type of fracture. The increase in initiation toughness due to the reinforcements involves interactions of the crack front with the refractory metal. Those interactions vary according to the reinforcement and matrix morphologies, and involve crack trapping, as well as renucleation phenomena.

ACKNOWLEDGMENTS

The authors would like to thank Dr. Douglas Konitzer from GE Aircraft Engines Division for providing the NiAl/Cr(Mo) material used in this study, and Drs. Anthony G. Evans, Carlos G. Levi and Hervé Dève for helpful discussions. This research was supported by the Defense Advanced Research Project Agency (DARPA) through a University Research Initiative Grant N00014-86-K-0753 and the National Science Foundation through Grant DMR 89-15209.

REFERENCES

[1] G. R. Odette, H.E. Dève, C.K. Elliott, A. Hasegawa and G.E. Lucas, in Interfaces in Ceramic Metal Composites, eds. R.J. Arsenault, R.Y. Lin, G.P. Martins and S.G. Fishman (TMS-AIME, Warrendale, PA, 1990), p. 443.

[2] K. Vedula, V. Pathare, I. Aslanidis and R. H. Titran, in High Temperature Ordered Intermetallic Alloys, eds. C.C. Koch, C.T. Liu and N.S. Stoloff (Mat. Res. Soc., 39, Pittsburgh, 1985), pp. 411.

[3] H.E. Dève, A.G. Evans, G.R. Odette, R. Mehrabian, M.L. Emiliani and R.J. Hecht, Acta Metall. Mater., 38, 1491 (1990).
[4] P.R. Subramanian, M.G. Mendiratta, D.B. Miracle and D.M. Dimiduk, in Intermetallic Matrix Composites, eds. D.L. Anton, P.L. Martin, D.B. Miracle and R. McMeeking (Mat. Res. Soc., 194,Pittsburgh, 1990), pp. 147.
[5] H.E. Dève and M. Maloney, Acta Metall. Mater., 39, 2275 (1991).
[6] T.C. Lu, A.G. Evans, R.J. Hecht and R. Mehrabian, Acta Metall. Mater., 39, 1853 (1991).
[7] R.M. Nekkanti and D.M. Dimiduk, in Intermetallic Matrix Composites, eds. D.L. Anton, P.L. Martin, D.B. Miracle and R. McMeeking (Mat. Res. Soc., 194,Pittsburgh, 1990), pp.175.
[8] M.F. Ashby, F.J. Blunt and M. Bannister, Acta Metall. Mater., 37, 1847 (1989).
[9] M. Bannister and M.F. Ashby, Acta Metall. Mater., 39, 2575 (1991).
[10] B.D. Flinn, C. Lo, F.W. Zok and A.G. Evans, J. Am. Ceram. Soc., (1992), in press.
[11] J.L. Walter and H.E. Cline, Met. Trans., 1, 1221 (1970).
[12] H.E. Cline and J.L. Walter, Met. Trans., 1, 2907 (1970).
[13] H.E. Cline, J.L. Walter, E. Lifshin and R.R. Russell, Met. Trans., 2, 189 (1971).
[14] D. Konitzer (private communication).
[15] F.F. Lange, Phil. Mag., 22, 983 (1970).
[16] H. Gao and J.R.Rice, J. appl. Mech., 56, 828 (1989).
[17] N. Fares, J. appl. Mech., 56, 837 (1989).
[18] H.-Cao and A.G. Evans, Acta Metall. Mater., 39, 2997 (1991).
[19] M.Y. He, F.E. Heredia, D.J. Wissuchek and A.G. Evans, in preparation.

ON THE MECHANISMS OF DUCTILITY ENHANCEMENT IN $\beta+\gamma'$- $Ni_{70}Al_{30}$ AND $\beta+(\gamma+\gamma')$-$Ni_{50}Fe_{30}Al_{20}$ *in situ* COMPOSITES

A. Misra*, R.D. Noebe** and R. Gibala*
* Department of Materials Science and Engineering, The University of Michigan, Ann Arbor, MI 48109-2136
** NASA Lewis Research Center, Cleveland, OH 44135.

ABSTRACT

Ductile phase reinforcement is an attractive approach for improving room temperature ductility and toughness of intermetallics. Two alloys of nominal composition (at.%) $Ni_{70}Al_{30}$ and $Ni_{50}Fe_{30}Al_{20}$ were directionally solidified to produce quasi-lamellar microstructures. Both alloys exhibit ~10% tensile ductility at 300 K when the ductile phase is continuous, while the $Ni_{70}Al_{30}$ alloy has a tensile ductility of ~4% when the γ' phase is discontinuous. Observations of slip traces and dislocation substructures indicate that a substantial portion of the ductility enhancement is a result of slip transfer from the ductile phase to the brittle matrix. The details of slip transfer in the two model materials and the effect of the volume fraction and morphology of the ductile phase on the ductility enhancement in the composite are discussed.

INTRODUCTION

Ductile phase reinforcement has become a common method for improving the room temperature toughness and ductility of high temperature structural materials. In the case of ceramics, toughening is imparted by the interaction of the ductile phase with the propagation of matrix cracks. However, in the case of intermetallics like β-NiAl where room temperature ductility is limited by dislocation generation and mobility, ductility and toughness enhancement may result from an efficient generation of mobile dislocations at the matrix/reinforcement interface [1] or substrate/ surface film interface [2]. Two successful ductile phase reinforced model materials in the Ni-Al and Ni-Al-X systems are the $\beta+\gamma'$-$Ni_{70}Al_{30}$ and $\beta+(\gamma+\gamma')$-$Ni_{50}Fe_{30}Al_{20}$ alloys [1,3,4]. In the present investigation, we have used directionally solidified $\beta+\gamma'$- $Ni_{70}Al_{30}$ and $\beta+(\gamma+\gamma')$-$Ni_{50}Fe_{30}Al_{20}$ alloys to study the role played by the ductile reinforcing phase in the ductility enhancement. The effects of the differences in deformation behavior of the ductile γ' phase and $(\gamma+\gamma')$ phase mixture and the volume fraction and morphology of the reinforcing phase are discussed.

EXPERIMENTAL

The ingots of the $Ni_{70}Al_{30}$ alloy used in this study were directionally solidified at rates of 0.5 cm/h and 4.5 cm/h, while those of $Ni_{50}Fe_{30}Al_{20}$

alloy were solidified at 0.5 cm/h. Tensile and compression specimens were EDM cut from the ingots parallel to the growth direction, then centerless ground and finally electropolished prior to testing. Details of the experimental procedures are presented elsewhere [5]. Observations of the deformed and fractured specimens were performed on a Hitachi S520 SEM or an ElectroScan E-3 ESEM. TEM observations were performed on a JEOL 2000FX microscope.

CHARACTERISATION OF THE AS-CAST ALLOYS

Slow directional solidification of the melt resulted in quasi-lamellar microstructures shown in Fig. 1. For a withdrawal rate of 0.5 cm/h, the reinforcing phase is essentially continuous and aligned parallel to the solidification direction for both alloys (Fig.1a). At a withdrawal rate of 4.5 cm/h, the reinforcing γ phase in the $Ni_{70}Al_{30}$ alloy is discontinuous and not parallel to the solidification direction (Fig. 1b), the deviation being as large as ~45° in some regions. A part of the β matrix in the $Ni_{70}Al_{30}$ alloy is present as martensite and the surface relief associated with the martensite is shown in Fig. 1b. The β matrix is continuous in both alloys and in single crystalline form with an approximate [001] growth direction. An important difference in the microstructures of the $Ni_{70}Al_{30}$ and the $Ni_{50}Fe_{30}Al_{20}$ alloys is that in the former the volume fraction of the ductile phase is ~60% and in the latter it is ~35%.

In both alloys the matrix and the reinforcing phase exhibit the Kurdjumov-Sachs orientation relationship : $(111)_\gamma$ // $(110)_\beta$ // interface plane, $[0\bar{1}1]_\gamma$ // $[1\bar{1}1]_\beta$. The tendency of the interfaces in these directionally solidified eutectics to adopt a largely planar morphology (Fig.1a) and their preference for the $(111)_\gamma$ // $(110)_\beta$ orientation suggests that these may be the preferred (low-energy) interfaces.

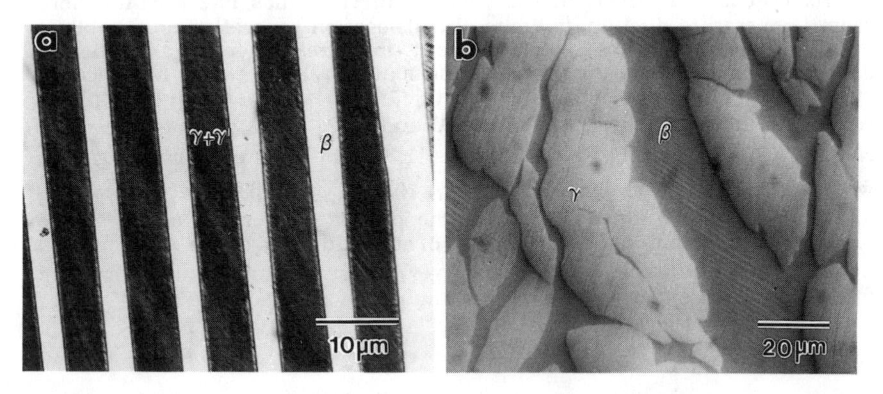

Fig. 1 Optical micrograph of the longitudinal section of (a) $Ni_{50}Fe_{30}Al_{20}$ alloy, DS rate = 0.5 cm/h and (b) $Ni_{70}Al_{30}$ alloy, DS rate = 4.5 cm/h. Note the martensite surface relief in the β phase in (b).

ROOM TEMPERATURE MECHANICAL BEHAVIOR OF THE AS-CAST ALLOYS

At room temperature, both $Ni_{70}Al_{30}$ and $Ni_{50}Fe_{30}Al_{20}$ alloys solidified at 0.5 cm/h exhibit tensile ductilities on the order of 10% compared to zero tensile ductility for [001] oriented single crystal β [1]. The volume fraction of the reinforcing phase (in the range of ~60% to ~35%) appears to have little effect on the tensile ductility. However, the $Ni_{70}Al_{30}$ alloy solidified at 4.5 cm/h has a discontinuous ductile phase and a lower strain to failure, on the order of 4%. Details of the mechanical tests results are presented elsewhere [1].

In addition to the enhanced ductility, both alloys show a significant reduction in flow stress compared to the single phase β matrix. The flow stress is not affected by the morphology of the reinforcing phase. In the binary alloy, the 0.2% offset compressive yield stress is 360 MPa whereas the yield stress is 1900 MPa for single phase β and 120 MPa for single phase γ' of same composition and orientation as the β and γ' phases in the composite [1]. Using the iso-strain rule of mixtures and taking the volume fraction of the γ' phase as 60%, the 0.2% offset yield stress is estimated to be 832 MPa. Thus, the composite exhibits ~ 60% reduction in flow stress as compared to the rule of mixtures flow stress. Reduction in flow stress of the same magnitude was also seen in oxide film coated [001] β-NiAl [2] and was attributed to change in slip vector from <111> in unconstrained β to <100> in film coated β-NiAl [6]. TEM investigations (presented later in this paper) reveal that a<100> dislocations dominate the deformation behavior of the β phase in both the $Ni_{70}Al_{30}$ and $Ni_{50}Fe_{30}Al_{20}$ alloys and hence, explain the softening observed in these composites.

DEFORMATION AND FRACTURE BEHAVIOR

For iso-strain deformation, the strain in the matrix may be accommodated by plastic deformation, by micro-cracking, or by plastic deformation followed by cracking. Surface observations of the deformed specimens reveal slip traces of the type shown in Fig. 3a and b for tensile specimens ($Ni_{50}Fe_{30}Al_{20}$ alloy) and compression specimens ($Ni_{70}Al_{30}$ alloy) respectively. The most significant feature in these micrographs is the indication of slip transfer from the ductile second phase to the brittle matrix. Thus, the matrix phase also undergoes plastic deformation and contributes to the overall ductility of the composite. In both binary and ternary alloys which have 10% tensile ductility, the side surface observations of fractured tensile specimens also show some secondary cracks in the matrix, bridged by the ductile phase. Thus, the 10% strain in the matrix is accommodated by deformation followed by secondary cracking. No clear evidence of any crack front or wake debonding is seen, indicating that the interface is very strong, as expected from low energy interfaces in directionally solidified materials. The plastic stretching of the ductile phase in the crack wake is expected to contribute significantly to the fracture toughness enhancement in the composite [7] and also provide additional tensile strain in the composite. In the case of $Ni_{70}Al_{30}$ alloy directionally solidified at 4.5 cm/h and having tensile ductility of ~4%, the side surface observations of fractured tensile specimens show secondary cracks in the matrix close to the fracture surface only. Thus, fracture initiates by matrix cracking and crack propagation involves deflection along the β/γ' interfaces. Since the γ' phase is discontinuous and not aligned with the tensile axis, no significant crack bridging is observed. Hence, discontinuous reinforcement provides lesser overall ductility than continuous reinforcement.

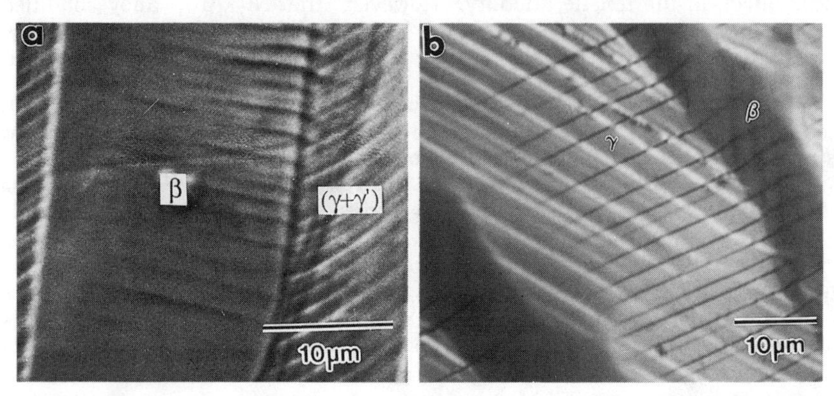

Fig. 2 Slip traces on the surface of specimens deformed to ~10% plastic strain, indicative of a slip transfer mechanism. (a) SEM micrograph, tensile specimen ($Ni_{50}Fe_{30}Al_{20}$) and (b) optical micrograph, compression specimen ($Ni_{70}Al_{30}$).

Examination of the dislocation substructures in the deformed specimens supports the slip trace data which suggest a slip transfer mechanism. In the case of the $Ni_{50}Fe_{30}Al_{20}$ alloy, the $(\gamma+\gamma')$ phase mixture exhibits extensive planar slip. Fig. 3a shows several dislocation pile-ups extending across the entire width of a γ lamellae in a compression specimen deformed 5%. The stress concentration due to the dislocation pile-ups at the interface causes dislocation nucleation in the β phase. Fig. 3 b is a higher magnification micrograph of the region marked by an arrow in Fig. 3a and shows extensive dislocation nucleation in the β phase ahead of a γ phase dislocation pile-up. The interaction of the planar γ phase dislocations with the interface has created a ledge at the interface, as suggested by Huang et al [3]. The formation of these ledges at the interface facilitates the shape change of plastically deforming γ phase constrained by the matrix.

The results of the TEM trace analysis indicate that the slip systems in the γ phase are of the type $\{111\}\langle 110\rangle$ and the slip systems in the β phase are of the type $\{110\}\langle 001\rangle$. The stereographic projection of the Kurdjumov-Sachs orientation relationship shows that for the geometry in Fig. 3 the active slip planes of the two phases, determined by TEM trace analysis, are nearly parallel. Thus, the pile-up of γ phase dislocations at the interface activates slip in the β phase on a parallel slip plane. The manifestation of this easy slip transfer is a reduction in the flow stress of the brittle matrix, allowing plastic deformation at an applied stress lower than the cleavage stress. This is consistent with research on α(soft)/β(hard) brass bicrystals and duplex crystals where a 40% reduction in the flow stress of the β phase relative to the unconstrained β was seen when the slip planes in the two phases were parallel [8].

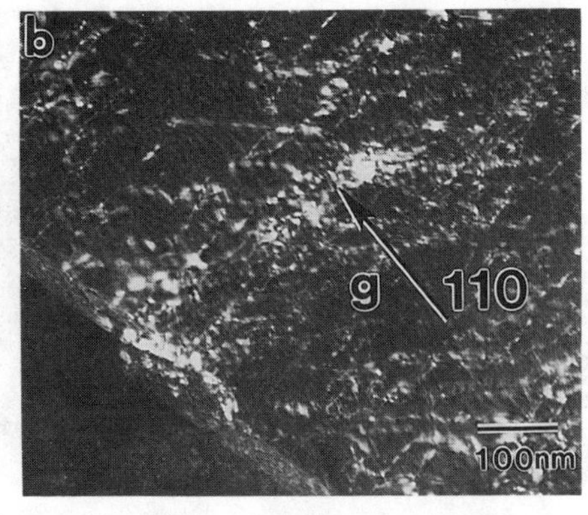

Fig. 3 (a) BF TEM micrograph showing dislocation pile-ups extending across the entire width of a (γ+γ') lamellae in the $Ni_{50}Fe_{30}Al_{20}$ alloy. **B**=[101].
(b) WB TEM micrograph showing a<100> dislocation nucleation in the β phase ahead of a γ phase dislocation pile-up. **B**=[111].

In the γ' phase of the $Ni_{70}Al_{30}$ alloy, deformation is controlled mainly by the a<101> screw superdislocations and SISF dipoles, consequently extensive planar slip is not obseved. Fig. 4a shows the typical dislocation substructure in the γ' phase near the interface in a specimen deformed 3.5% in tension. In Fig. 4a, the dislocations labelled 'a' have Burgers vector **b** = a[$0\bar{1}1$], and the dislocations labelled 'b' have **b** = a[$\bar{1}01$]. Further, in Fig. 4a the directions [$0\bar{1}1$] and [$\bar{1}01$] point towards the interface. Hence, the dislocations in the γ' phase appear to move towards the interface and finally get absorbed in it. Fig. 4b shows the β phase region across the interface from the γ' phase region shown in Fig. 4a. Here **b** = a[$0\bar{1}0$] dislocations are seen to generate at the interface and move out in the β phase. Notice that the direction [$0\bar{1}0$] in Fig. 4b points away from the interface into the β phase. The line direction of <100> dislocations in β phase was close to <111> and slip plane was {110} type. No evidence of any <111> slip was found , consistent with the observed single set of slip traces. Thus, slip by <100> dislocations in the β phase was seen in both alloys even though the loading axis was nearly parallel to <001>. Field et al [9] have estimated that a rotation of ~3.4^{o} from the <001> loading axis may give enough resolved shear stress for {010}<100> slip, with the slip on {110} planes requiring 43% higher resolved shear stress than cube slip. In the present investigation, X-ray data and determination of foil normals in the TEM specimens cut normal to the growth direction suggest that the loading axis in the β phase was within 2-3° of <001>. Therefore, the major fraction of the resolved shear stress for <100>{011} slip in the β phase is expected to come from the stress concentration at the interface by the γ phase dislocations. Future work will involve analytical treatment of the stress concentration at the interface.

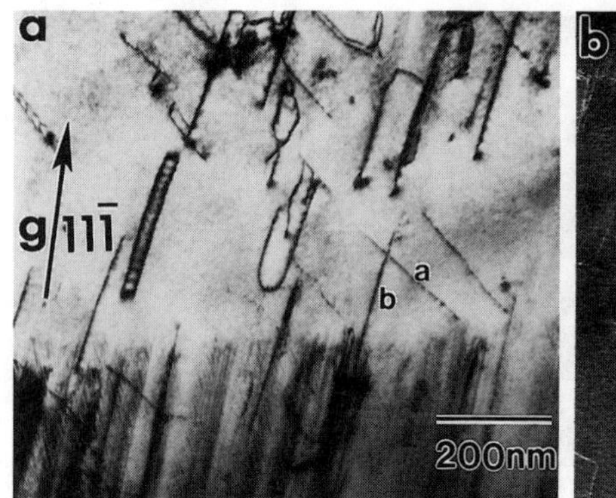

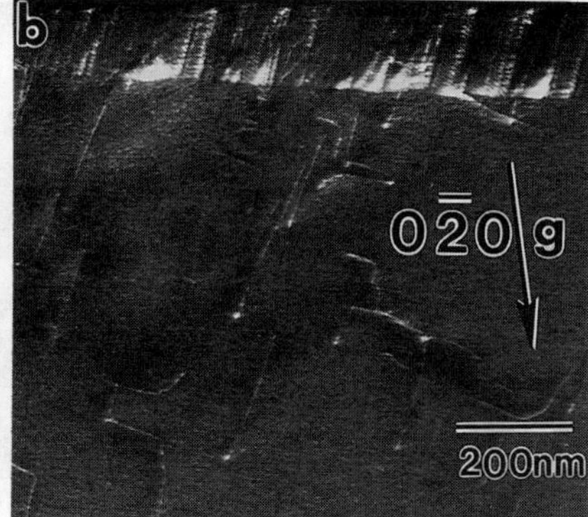

Fig. 4(a) BF TEM micrograph showing a typical dislocation substructure in γ' phase of the $Ni_{70}Al_{30}$ alloy deformed 3.5% in tension. **B**= [101]. (b) WB TEM micrograph showing the nucleation of a<010> dislocations in the β phase from the interface. **B**= [001].

SUMMARY

The efficient generation of mobile <100> dislocations by transfer of slip from the γ' and (γ+γ') ductile reinforcing phases can significantly enhance the room temperature ductility of β-NiAl. Plastic stretching in the crack wake imparts additional ductility to the composite when the ductile reinforcement is continuous.

REFERENCES

1. R.D. Noebe, A. Misra and R. Gibala, ISIJ International, 31, 1172 (1991).
2. R.D. Noebe and R. Gibala, Scripta Met., 20, 1635 (1986).
3. S.C.Huang, R.D.Field and D.D.Krueger, Met.Trans., 21, 959 (1990).
4. S.Guha, P.R.Munroe and I.Baker, Mat. Sci. and Eng., A131, 27 (1991).
5. M. Larsen, A. Misra, S. Hartfield-Wunsch, R.D. Noebe and R. Gibala, in Intermetallic Matrix Composites , edited by D.L. Anton et al (Mater. Res. Soc. Proc. 194, Pitttsburgh, PA 1990) p. 191.
6. J.T. Kim, R.D. Noebe and R. Gibala, in Intermetallic Compounds: Structure and Mechanical Properties, edited by O. Izumi (Proc. Sixth JIM International Symposium, JIMIS-6, 1991) p. 591.
7. M.F. Ashby, F. J. Blunt and M. Bannister, Acta Met., 37, 1847 (1989).
8. A.K. Hingwe and K.N. Subramanian, J. Mater. Sci., 10, 183 (1975).
9. R.D. Field, D.F. Lahrman and R. Darolia, Acta Met., 39, 2951 (1991).

This research was funded by the NSF Grant NO. DMR-9102414.

AN EVALUATION OF THE CREEP PROPERTIES OF AN Al_2O_3/Ni_3Al COMPOSITE AND THE EFFECT OF DISORDER ON MECHANICAL PROPERTIES

P.C. BRENNAN*, W.H. KAO* AND J.-M. YANG**
*Mechanics and Materials Technology Center, The Aerospace Corp.,El Segundo, CA 90245
**Department of Materials Science and Engineering, University of California, Los Angeles, CA 90024

ABSTRACT

Ordered Ni_3Al alloys and their composites are attractive materials for elevated-temperature structural applications due to their many favorable properties. The addition of alloying elements can significantly lower the Ni_3Al order-disorder transition temperature and also result in the formation of a Ni solid solution. As the percentage of Ni solid solution increases, the composite's room-temperature flexural strength increases. The effect of processing parameters on the material's microstructure is discussed. The complex matrix microstructure also has a significant effect on the composite's creep properties. Normal power-law creep was exhibited by the composite material when tested in compression.

INTRODUCTION

Future hypersonic aircraft and spacecraft will require advanced materials with temperature and strength capabilities far in excess of those possessed by the Ni-based superalloys in current use. Ordered intermetallic alloys and their composites are attractive candidates for many of these applications. The Ni_3Al intermetallic, which has a $L1_2$-type crystal structure, maintains its ordered structure to, or at least near, its peritectic melting point at 1395°C. The ordered structure means that dislocation motion is more complex, diffusion is more difficult, and, thus, creep deformation is much slower than in a similar disordered material. This is critical for materials to be used in high-temperature applications under load.

The addition of certain alloying elements can lower the order-disorder transition temperature (T_t) for Ni_3Al significantly below the melting point [1]. The alloying additions can also result in the formation of a Ni solid solution referred to as gamma. The ordered Ni_3Al intermetallic phase is often referred to as gamma prime.

Most critical to a material to be used at high temperatures is its creep properties. Creep mechanisms in Ni_3Al depend on the stress orientation and whether the creep is occurring above or below T_p, the temperature at which the yield strength peaks. When tested in tension at $T<T_p$, Ni_3Al displays primary creep followed by inverse creep [2]. Inverse creep refers to the condition where the creep rate continuously increases with strain without a prior steady-state creep regime and without a nonuniform reduction in the material's cross section. At $T>T_p$, normal creep behavior with a steady-state creep regime is typically seen.

It is the purpose of this paper to investigate the multi-phase microstructure of an Al_2O_3 particulate-reinforced IC-221 composite and the effect of microstructure on room-temperature (RT) mechanical properties, and to evaluate the composite's compression creep properties.

EXPERIMENTAL PROCEDURE

The IC-221 matrix used in this study, developed by Oak Ridge National Laboratories, has a composition of Ni, 16.1 Al, 8 Cr, 1 Zr, and 0.1 B. The IC-221 powder was mixed with 25 vol. % Al_2O_3 particulates in a Turbula Shaker Mixer.

Billets were prepared by vacuum hot pressing at 900°C and 30 MPa for 1 h. The billets were then hot extruded. Extrusion #1 was made at 1260°C with an extrusion ratio of 7.5:1. This extrusion produced large grain size samples that were used for flexural and creep testing. Extrusions #2 was made at 1100°C with an extrusion ratio of 9:1. Creep specimens were produced with a L/D ratio of 2.5:1. The standard anneal after fabrication was 1 h at 1050°C followed by 24 h at 800°C.

The composite microstructure was analyzed using SEM/EDX. The effect of extrusion temperature and annealing temperature on IC-221 was analyzed through X-ray diffraction

(XRD). Powders were also analyzed in the as-received condition and after water quenching from isochronal anneals between 800 and 1250°C.

Three-point bend tests were conducted at RT after thermal exposures at 1000°C for varying durations to investigate the composite's strength retention and also after quenching from isochronal anneals at temperatures between 800 and 1250°C. Constant-load compression creep tests were conducted at 700, 800, 900, and 1000°C in vacuum. Initial stress levels ranged between 48 and 510 MPa (7-74 Ksi).

RESULTS AND DISCUSSION

The matrix grain size of the first composite extrusion was 25 to 50 μm, and the second was approximately 5 μm. The grain size of the matrix extrusion was approximately 5 μm. The grain sizes were very stable, and no significant grain growth was seen in the samples with 25 to 50 μm diameter grain size after exposures as long as 1000 h at 1000°C.

The final portion of the standard IC-221 anneal was a 24-h hold at 800°C. As indicated by an isothermal section taken through the Ni-Cr-Al ternary phase diagram at 750°C [3], this anneal should produce approximately a 25% gamma, 75% gamma-prime microstructure. (Fig. 1). However, Liu et al. [4] reported that the IC-221 microstructure consisted of only 10 to 15% gamma. Similarly, Cahn et al. [5] reported a tendency for single-phase gamma specimens to retain their single-phase structure after high-temperature anneals even though the material compositions lay well within the two phase regions. This is possible because gamma prime is metastable over a wide compositional range (Fig. 2), and antiphase domain boundaries (APDBs) must be present for the gamma phase to nucleate heterogeneously [5]. Metastable-ordered gamma-prime that contains less than 24.5% Al (the equilibrium content at the eutectic) can remain as one phase if there are no APDBs, and, thus, no nucleation sites for the gamma phase, provided it does not exceed the metastable transition temperature exhibited in Fig. 2 [5].

Figure 3 exhibits the microstructure of a typical sample from extrusion #1. The dark phase is Al_2O_3. The light-colored phase is gamma prime. The extrusion temperature was sufficiently high that the matrix could flow easily through the extrusion die without generating many dislocations and APDBs. Thus, the matrix appears to be largely one phase, and the grains are undeformed.

Figure 4 exhibits the microstructure of a typical sample from extrusion #2. The difference in the matrix microstructure is immediately apparent. The gamma phase is randomly distributed throughout the matrix and appears as a light-colored phase. The grains are heavily deformed and much smaller than in extrusion #1. The extrusion temperature for this billet was significantly lower than for the first billet. Thus, diffusion was relatively harder, and the material did not flow as easily through the extrusion die. Significantly higher stresses and more dislocations were generated as a result. The gamma phase had ample heterogeneous nucleation sites, and a large amount of small gamma precipitates resulted.

The microstructure generated from specific extrusion parameters is greatly affected by the presence or absence of a reinforcement. The Al_2O_3 particulates, for instance, made the extrusion process more difficult and induced additional dislocations that would not have been generated if the particulates were absent. Thus, the relative amounts of gamma and gamma prime differ for reinforced and unreinforced materials extruded with the same parameters. Any comparison of the materials' mechanical properties should take this into consideration.

In light of the work by Cahn et al. [1,5] discussed above, it is interesting to note that XRD of extrusion #1 after processing only detected the gamma-prime phase, both the fundamental and superlattice (SL) lines, but not the gamma phase. In contrast, XRD of extrusion #2 detected both the gamma and gamma-prime phases immediately after processing. After thermal exposures at elevated temperatures, however, a fine distribution of the gamma phase nucleated in the extrusion #1 samples (Fig. 5). It was difficult to unambiguously identify the gamma, gamma-prime, and Al_2O_3 phases in the extruded composites when they were analyzed using XRD because of overlapping intensity peaks. IC-221 powders, however, were used to effectively describe the effect of processing and heat treatment on the matrix microstructure.

No gamma phase was detected in the as-processed powders. However, all of the powders heat treated above 800°C indicated the presence of both gamma and gamma-prime phases. The normalized intensities of the SL lines did not follow a definite trend, and the order parameter, S, could not be determined. The volume of material sampled and its relative amount of

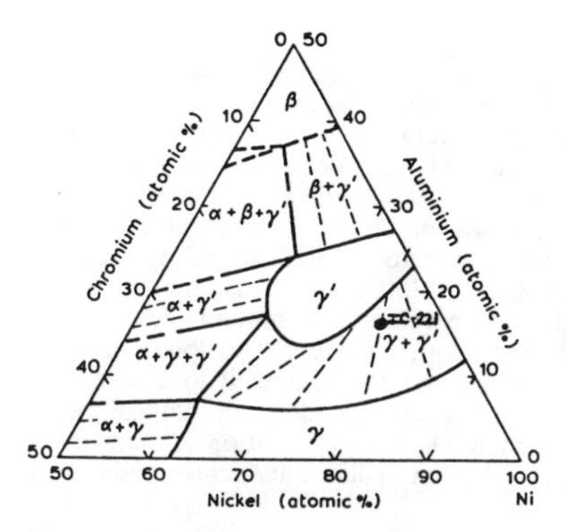

Figure 1. 750°C isothermal section of the Ni-Cr-Al ternary-phase diagram [3].

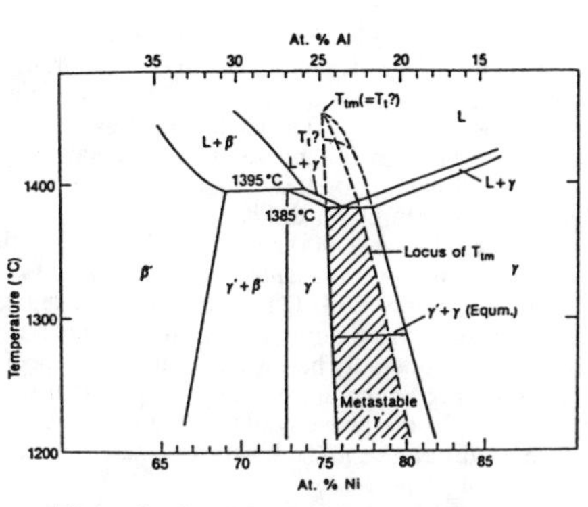

Figure 2. A portion of the Ni-Al binary-phase diagram exhibiting empirically determined gamma-prime metastable region [4].

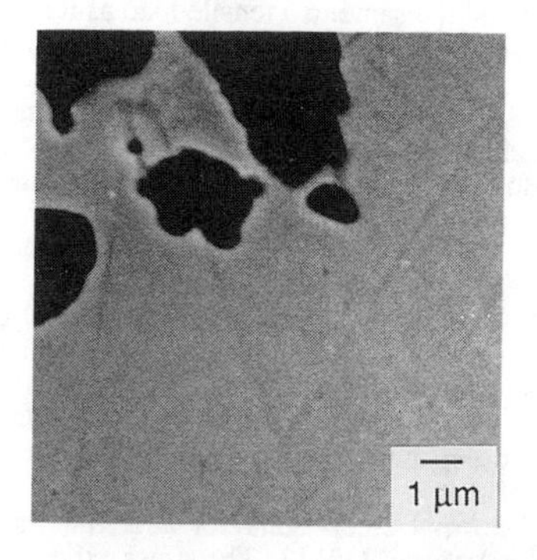

Figure 3. Electron micrograph exhibiting the composite microstructure of extrusion #1 after being extruded at 1260°C and annealed at 1050°C for 1 h and 800°C for 24 h. Dark phase is Al_2O_3. Gray phase is gamma prime.

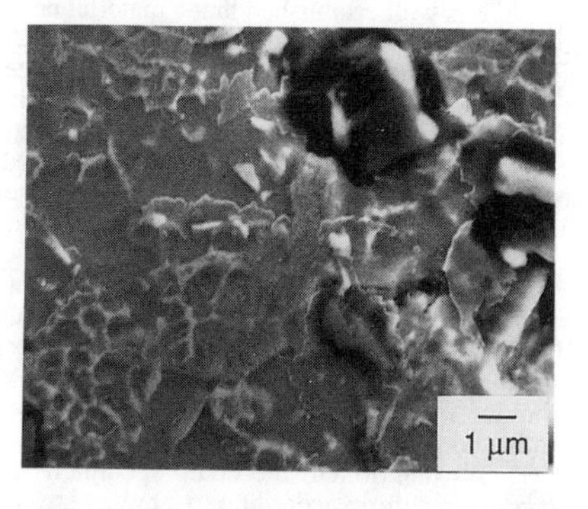

Figure 4. Electron micrograph exhibiting the composite microstructure of extrusion #2 after being extruded at 1100°C and annealed at 1050°C for 1 h and 800°C for 24 h. Black phase is Al_2O_3. Light gray phase is gamma. Dark gray phase is gamma prime.

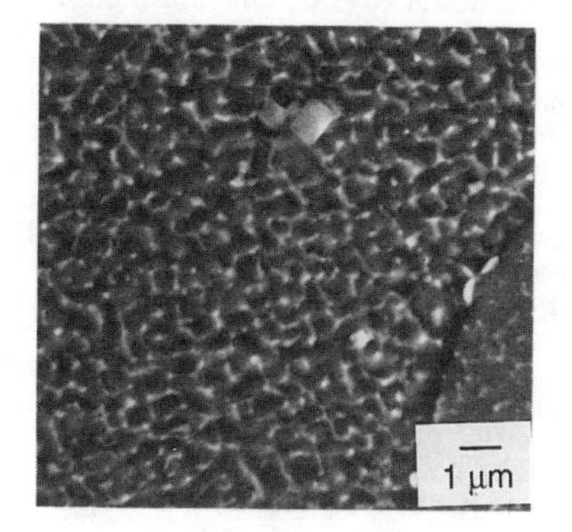

Figure 5. Secondary electron micrograph exhibiting the two-phase gamma/gamma-prime microstructure in extrusion #1 after being extruded at 1260°C, annealed at 1050°C for 1 h, 800°C for 24 h, and tested in 4-point bending at 700°C. Gamma appears as the white phase; gamma prime is the gray phase.

gamma and gamma-prime phases would also have a large effect on all attempts to determine the state of order in the gamma-prime phase alone.

The effect of gamma phase on RT strength is indicated in Fig. 6. The samples represented in Fig. 6a were quenched to RT from their respective annealing temperatures and tested in three-point bending. The annealing allowed for the equilibrium amount of gamma and gamma prime to form at each temperature. The quenching retained this structure to RT and allowed for the effect of gamma on low-temperature strength to be demonstrated. Gamma is more stable at higher temperatures for the IC-221 composition. There was no observed effect for the 800°C samples, but between 900 and 1200°C, the RT yield strength increased with annealing temperature. The samples represented in Fig. 6b were first heated to 1250°C and then cooled in 100°C decrements below 1200°C to their respective annealing temperatures. A similar increase in RT yield strength was exhibited with higher annealing temperatures; however, the lower strength limit was considerably below the limit for the specimens that were heated directly to their respective annealing temperatures. The samples that were continuously heated had decomposition of the gamma-prime phase while the samples that were cooled had decomposition of the gamma phase. A difference in the kinetics of these reactions could result in different microstructures and mechanical properties.

Creep Properties

It is well established that a material need not respond the same to a tensile load as it does to a compressive load. It is also known that Ni_3Al violates Schmid's law and exhibits a tension/compression asymmetry for the critical resolved shear stress on the [101](111) slip system [6]. Thus, inverse creep need not be expected in compression creep based solely on its existence under tensile creep conditions. Also, the general creep behavior of a reinforced material may differ from that of an unreinforced material. Indeed, the results generated in the present study failed to indicate inverse creep in the composite material when the tests were conducted in compression. However, the multiple slip systems operating in the polycrystalline samples may mask any inverse creep that is present.

Typical creep curves for three different stress levels are given in Fig. 7 for the 1000°C creep tests. The curves are typical of all the curves generated at temperatures between 700 and 1000°C. All of the curves exhibit normal creep with primary and steady-state creep regimes. The unreinforced IC-221 material (discussed elsewhere [7]) also exhibited normal power-law behavior in compression. Moreover, the composite creep properties appear to be matrix dominated [7].

As mentioned, the creep specimens exhibited normal creep behavior and, thus, are expected to deform according to a power law such as $\dot{\varepsilon} = A\sigma^n \exp(-Q/RT)$. The stress exponent, n, is shown as a function of temperature in Fig. 8. The creep-controlling mechanism transitioned from low-temperature dislocation-climb-controlled creep towards high-temperature diffusion-controlled creep. Diffusion-controlled creep mechanisms exhibit stress exponents less than 3 and operate at $T>0.8T_{m.p.}$, which for IC-221 is 1037°C. At 1000°C, the stress exponents for the three materials in Fig. 8 are all below 3. Superplasticity is possible when the stress exponent is less than 3. Superplasticity has been documented in Ni_3Al alloys at temperatures near 1050°C [8]. Thus, the creep-controlling mechanism is in a transition at 1000°C, and further decreases in n are expected with further increases in temperature or refinement of the microstructure.

The activation energy for the composite material is best described by two values. In the 700 to 900°C range, the activation energy is approximately 310 KJ (Fig. 9). This is attributed to Ni diffusion-controlled dislocation climb. The activation energy for Ni diffusion in Ni_3Al is 306 KJ [9] while the value for Al diffusion is thought to be considerably higher since Al atoms are surrounded by Ni atoms while Ni atoms are surrounded by both Al and Ni atoms in the Ni_3Al lattice. In the 900 to 1000°C range, the activation energy increases to approximately 450 KJ (Fig 9) and is attributed to Al diffusion-controlled dislocation climb.

XRD indicated that extrusion #1 consisted of only gamma prime after processing. However, during creep, dislocations and APDBs are generated that will act as the heterogeneous nucleation sites for the gamma phase. Due to the imposed stress and the high thermal energy available, the gamma phase would be able to nucleate and grow within a nonequilibrium microstructure. Differences in the relative amounts of the gamma and gamma-prime phases would allow for different deformation rates. Thus, differences in deformation rates should be expected between samples tested under the same conditions until the equilibrium gamma/gamma-prime microstructure can be established at each temperature. Furthermore, the two-phase matrix can affect the apparent activation energies that are determined.

SUMMARY

Processing parameters for the composite materials have been demonstrated to have a large effect on the matrix microstructure and the resultant materials' mechanical properties. The extent of the gamma phase is dependent on the dislocations and APDBs generated during extrusion. Low extrusion temperatures promote high dislocation densities and a fine gamma phase distribution. As the gamma content increases, the RT yield strength of the composite increases. However, the high-temperature creep strength apparently decreases as the amount of disordered gamma phase increases. The two-phase matrix microstructure complicates the creep-controlling mechanisms; however, Ni diffusion appears to control dislocation climb in the 700 to 900°C

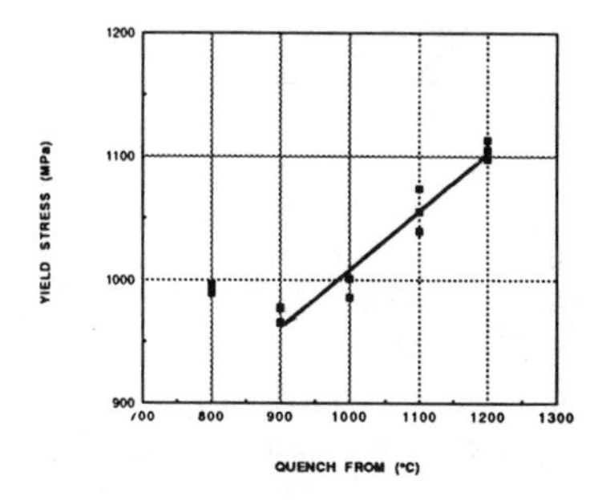

(a) Quenched to RT from annealing temperature.

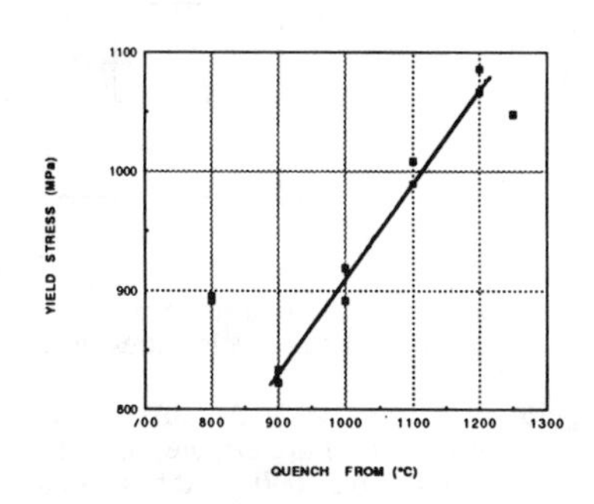

(b) Heated to 1250°C; cooled in 100°C decrements below 1200°C to annealing temperature.

Figure 6. Composite yield strength as measured in three-point bending at room temperature (RT) after isochronal thermal exposures and water quenching to RT.

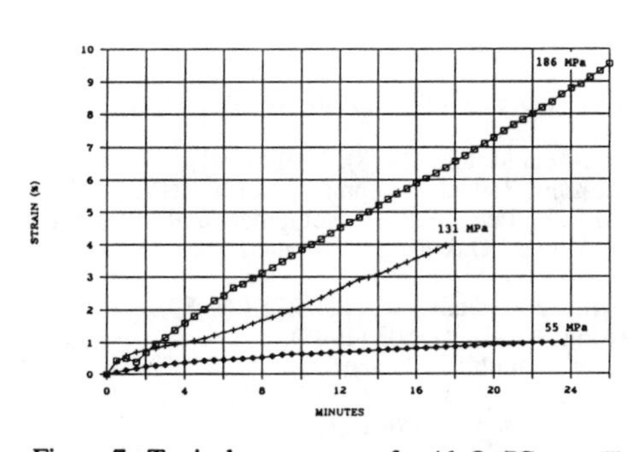

Figure 7. Typical creep curves for Al_2O_3/IC-221 composites tested in vacuum at 1000°C under constant compressive loads. Stress levels refer to initial loading.

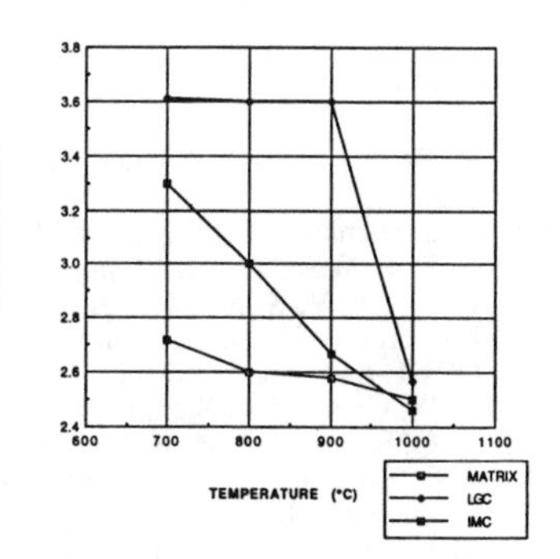

Figure 8. Stress exponent, n, as a function of temperature for IMC, matrix, and LGC materials.

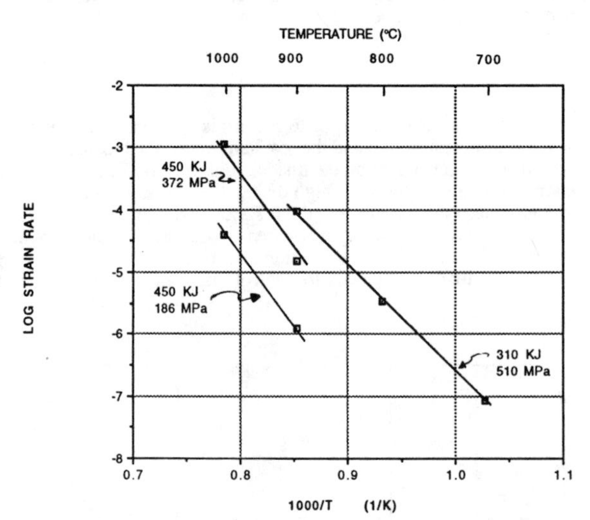

Figure 9. Log strain rate vs reciprocal absolute temperature for three stress levels. Stress levels refer to initial loading.

temperature regime while Al diffusion controls dislocation climb in the 900 to 1000°C temperature regime. Furthermore, the creep-controlling mechanism transitions to diffusion-controlled creep above 1000°C. It has also been demonstrated that Al_2O_3 particulate-reinforced IC-221 exhibits normal power-law creep when tested in compression. The composite creep properties are matrix dominated. No inverse creep behavior was noted in any of the tests for the reinforced or the unreinforced nickel aluminide.

ACKNOWLEDGMENTS

The authors would like to acknowledge the financial support of the Aerospace Sponsored Research (ASR) Program.

REFERENCES

1. R. W. Cahn, P. A. Siemers, J. A. Geiger, and P. Bardhan, Acta Metall., **35** (11), 2737 (1987).
2. K. J. Hemker, M. J. Mills, and W. D. Nix, Acta Metall., **39** (8), 1901 (1991).
3. D. R. F. West, *Ternary Equilibrium Diagrams*, 2nd ed. (Chapman and Hall, NY, 1982).
4. C. T. Liu, V. K. Sikka, J. A. Horton, and E. H. Lee, "Alloy Development and Mechanical Properties of Nickel Aluminide (Ni_3Al) Alloys," Oak Ridge National Laboratories Paper ORNL-6483, prepared for U. S. DOE, (1968).
5. R. W. Cahn, P. A. Siemers, and E. L. Hall, Acta Metall., **35**, (11), 2753 (1987).
6. Y. Q. Sun, and P. M. Hazzledine, Phil. Mag. A, **58**, (4), 603 (1988).
7. P. C. Brennan, W. H. Kao, and J.-M. Yang, (unpublished paper).
8. T. G. Nieh, M. J. Mayo, M. Kobayashi, and J. Wadsworth (eds.), "Superplasticity in Metals, Ceramics, and Intermetallics," Mater. Res. Soc., **196** 189 (1990).
9. G. F. Hancock, Phys. Stat. Solidus (a), **7** 535 (1971).

MECHANICAL PROPERTIES AND DEFORMATION MECHANISMS OF AN Al_2O_3 FIBER-REINFORCED NiAl MATRIX COMPOSITE

S. M. Jeng*, J.-M. Yang* and R. A. Amato**
* Department of Materials Science and Engineering, University of California, Los Angeles, CA 90024-1595
** GE Aircraft Engines, Cincinnati, OH 45215

ABSTRACT

The mechanical behavior of a continuous Al_2O_3 fiber-reinforced NiAl matrix composite was investigated. The interfacial mechanical properties were measured at room temperature using a fiber pushout test. Tensile test was conducted at room and elevated temperatures. Four-point bending creep test was also conducted at 700 °C. The key microstructural parameters controlling the mechanical behavior and fracture processes were identified. The effect of fiber surface coating on the interfacial properties and mechanical behavior of the composite was also studied.

INTRODUCTION

Ordered intermetallic compounds have emerged as a new class of materials for advanced structural applications. NiAl-based intermetallics, in particular, has been recognized as one of the most promising candidate materials for high temperature applications. It possess several attractive properties including low density (~ 6 g/cm^3), high melting point (1638 °C), high modulus (189 GPa) and excellent oxidation resistance up to 1300 °C. The polycrystalline NiAl exhibits a brittle-to-ductile transition at temperatures ranging from 300 to 600 °C, the exact temperature depends on the stoichiometry and grain size. However, to make the NiAl as a viable material, it is necessary to overcome some of its inherent problems. These include low ductility and fracture toughness at ambient temperatures, and inadequate strength at elevated-temperatures. Accordingly, a significant effort has centered on enhancing the mechanical properties of the NiAl through grain refinement, micro- and macro-alloying as well as incorporating second phase reinforcements[1-4].

Recent studies indicated that the interfacial requirements in the intermetallic matrix composites are more complicated than those in a typical metal or ceramic matrix composite [5]. At ambient temperatures, the intermetallics are brittle as a ceramic exhibiting a catastrophic fracture behavior. Therefore, a weak bond is required to activate the toughening mechanisms such as crack defection, fiber bridging and fiber pullout. However, at elevated-temperatures, the intermetallics behave more like a metal exhibiting plastic yielding behavior. A strong bond is needed to effectively transfer the load from the weak, ductile matrix to the strong, stiff reinforcing fibers. The optimum interfacial properties for a NiAl matrix composite with a combination of room-temperature toughness and high-temperature strength have not been determined yet.

This work is conducted to study the mechanical behavior of an Al_2O_3 fiber-reinforced NiAl matrix composite at room- and elevated- temperatures. The single-crystal aluminum oxide fiber has been shown as the most promising reinforcements for several intermetallic matrix composites. Both coated and uncoated fibers will be used to assess the effect of interfacial properties on the mechanical properties and fracture behavior of the resulting composites.

EXPERIMENTAL

The single crystal Al_2O_3 fiber produced by Saphikon were used as reinforcing materials. The fiber has a diameter of 150 μm and has a tensile strength of 3.15 GPa at room temperatures. To assess the effect of coating on the behavior of the composite, a thin layer of SiC was coated on the fiber using chemical vapor deposition. The matrix alloy consisted of Ni-32 at% Al-20 at% Fe which is a single phase (β) alloy. The 6-ply unidirectional composites with both coated and uncoated fibers were consolidated using hot isostatic pressing. A monolithic NiAl matrix alloy was also consolidated under the same processing conditions.

The interfacial shear strength was measured using an indentation (pushout) test [6]. A scanning electron microscope was used to examine the morphology of debonded interface and to locate the origin of debonding. Tensile tests were conducted at room temperature, 760 °C (1400 °F) and 870 °C (1600 °F). Specimens having dimensions of 50 mm in length with a diameter of 12.5 mm were prepared with fiber parallel to the length direction. The gauge section had a dimension of 25.4 x 6.25 x3.15 mm. The strain-to-failure was monitored using an extensometer. The tensile test was carried out on an Instron testing machine operated at a crosshead speed of 0.5 mm/min. Four-point bending creep tests were performed on monolithic NiAl and Al_2O_3/NiAl composite with dimensions 5 x 5 x 50 mm. The deflection at the center of the beam was measured with a linear variable-deflection transducer.

Fractographic analysis was performed on all test specimens using a scanning electron microscope. The fractured specimen parallel to the tensile loading direction was further polished to reveal the damage occurred near the fracture surface. The mechanisms of damage initiation and propagation were determined.

RESULTS & DISCUSSION

Microstructure and Interfacial Mechanical Properties

Figures 1(a) and (b) show the microstructure of the as-fabricated Al_2O_3/NiAl composite along transverse and longitudinal directions. The composites were fully dense and the Al_2O_3 fibers were randomly distributed among the NiAl matrix. The fiber volume was estimated to be approximately 3.5 - 4.5 %. Fiber misalignment and fiber fracture during composite consolidation were also evident. Figures 2(a) and (b) are the microstructure of the interfacial region in the uncoated and SiC-coated Al_2O_3/NiAl composites, respectively. No evidence of the reaction between the fiber and matrix was observed in the uncoated composites. However, some voids were found at the interfacial region in the uncoated composite. In the SiC-coated composite, the SiC coating layer is absent after consolidation. Instead, a second phase (dark phase in Fig.2(b)) randomly distributed along the interfacial region was observed. By EDS analysis, this dark phase contains Ni, Al, Fe and Si. The x-ray intensity of Al is much lower than that in the matrix alloy. This is believed to be the reaction product between the SiC coating layer and matrix. A detailed microstructural analysis by TEM is being conducted to further characterize the reaction products.

The interfacial shear strength of the uncoated and SiC-coated Al_2O_3/NiAl composites are listed in Table I. The results indicated that the uncoated Al_2O_3/NiAl composite has a higher interfacial shear strength than that of the SiC-coated Al_2O_3/NiAl composite. Figure 3(a) and 3(b) are the interfacial debonding location of the uncoated and SiC coated Al_2O_3/NiAl composites, respectively. In the uncoated Al_2O_3/NiAl composite, the fiber surface is quite clean with only a small amount of matrix adhered. Meanwhile, some scratches resulting from the relative sliding between the fiber and matrix was observed. However, in the SiC-coated Al_2O_3/NiAl composite, the fiber surfaces are rough with some reaction products of the coating layer and matrix adhered.

Table I. Tensile properties and interfacial shear strength of the NiAl alloy and its composites

	Interfacial shear strength, τ_i (MPa)	Modulus (GPa)			UTS (MPa)		
		RT	760 °C	870 °C	RT	760 °C	870 °C
Matrix		136	88	68	422	160	85
Al_2O_3/NiAl	136 ± 8	152	102	77	333	181	96
SiC-coated Al_2O_3/NiAl	107± 10	168	NA	112	285	NA	85

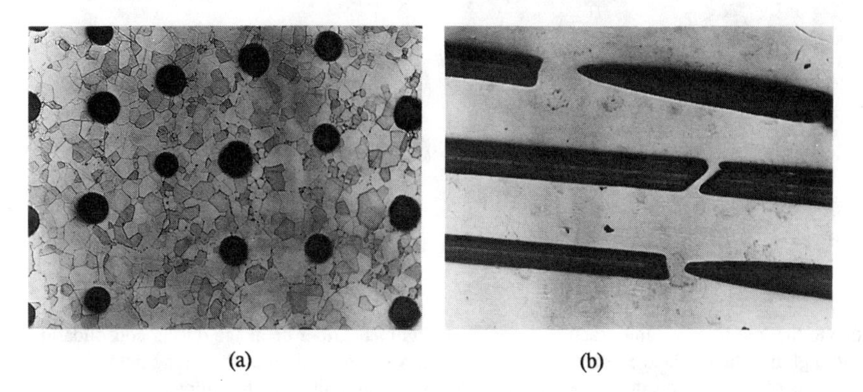

(a) (b)

Fig. 1 Microstructure of the Al_2O_3/NiAl composite along (a) transverse, and (b) longitudinal direction.

Tensile Behavior

The tensile stress-strain curves of the monolithic NiAl, uncoated and SiC-coated Al_2O_3/NiAl composites tested at 25, and 870 °C are shown in Fig. 4. The elastic modulus and strength obtained from these curves are listed in Table I. At room temperature, the monolithic NiAl alloy and the Al_2O_3/NiAl composite exhibited a linear elastic stress-strain behavior till fracture at room temperature. However, when tested at elevated-temperature, the NiAl and composites exhibited an extensive plastic deformation before fracture. The stain-to-failure of the NiAl alloy increased from $< 1\%$ to~ 20% at 870 °C. The room-temperature tensile strength of the monolithic alloy is higher than that of both composites. This may be

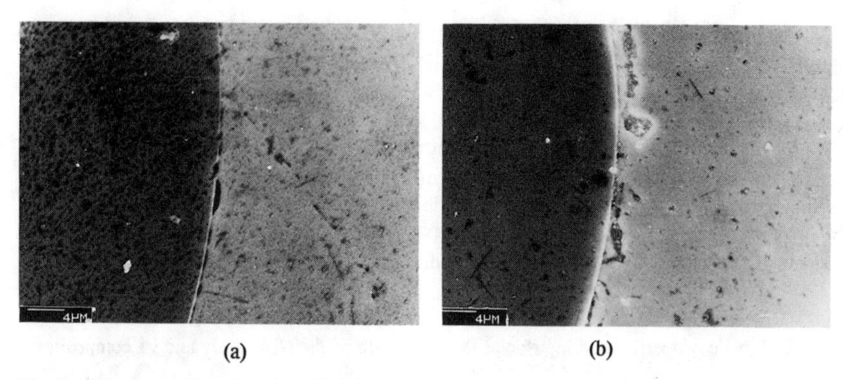

Fig.2 Microstructure of the interfacial region in the (a) uncoated, (b) SiC-coated Al_2O_3/NiAl composites.

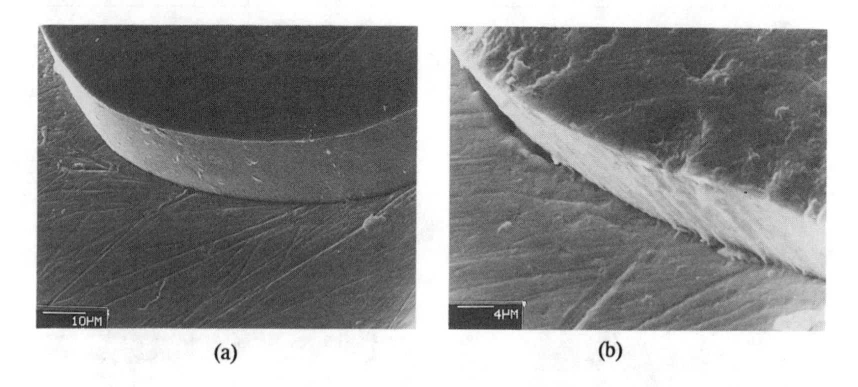

Fig. 3. Interfacial debonding location of the (a) uncoated, (b) SiC-coated Al_2O_3/NiAl composites.

due to the low fiber volume fraction, poor fiber alignment, fiber breakage during consolidation and high interfacial shear strength. However, at elevated temperature, the strength and modulus of both composites are higher than that of the monolithic matrix alloy.

The fracture surface of the monolithic NiAl tested at different temperatures are shown in Figures 5(a) to 5(c), respectively. It clearly shows that the fracture modes of NiAl alloy have transformed from transgranular cleavage at room temperature to intergranular at intermediate temperature and ductile dimple failure at high temperature. This is due to the increasing of the dislocation slip system and reduction of the degree of ordering at high temperature of the NiAl crystal [7]. Figures 6(a) to 6(c) are tensile fracture surface of the uncoated Al_2O_3/NiAl composite at three different testing temperatures. It clearly shows that the fiber pull-out length of the composites increases as the testing temperature increases. Meanwhile, transgranular cleavage and intergranular failure were observed at the matrix near the interfacial region for the composites tested at 760 and 870 °C, respectively (Fig. 7(a) and

7(b)). These failure modes were quite different from those of the monolithic matrix tested at these temperatures. Furthermore, examination of the microstructure beneath the fracture surface showed that multiple fiber fracture was observed only in the composite tested at high temperature. Meanwhile, matrix cracking and relative displacement between the fiber and matrix are also found near the broken fiber end (Fig. 8(a) and 8(b)). The failure mechanisms of the SiC-coated Al_2O_3/NiAl composite at each temperature are similar to those of uncoated composite. This suggests that the reduction of the interfacial shear strength through SiC coating did not affect the failure modes of the composite at both room and high temperatures.

From the above microstructural observation, the damage mechanisms of the Al_2O_3/NiAl composite tested at different temperatures can be summarized as follows. Upon subjected to tensile loading, due to the brittle characteristic of the NiAl matrix (fracture strain $< 1\%$), the crack will be initiated at the matrix. With a strong interfacial shear strength of the composite at room temperature, the crack can easily propagate across the fibers. As a result, the composite shows catastrophic failure behavior with limited amount of fiber pull-out. However, as the testing temperature increases, the matrix ductility increases and the fiber strength decreases. Meanwhile, the residual clamping stress resulting from the CTE mismatch would decrease, leading to the decrease of the interfacial shear strength. As a result, the fiber would fail at early stage of tensile loading. In case of the composite tested at 760 °C, the interfacial shear strength is still strong enough to prevent the interfacial debonding. The cracks could easily extend into matrix as the applied loading increases. However, in case of the composite tested at 870 °C, the interfacial shear strength is further decreased which allows debonding and fiber pull out from the matrix to occur as the loading increases. For the above two conditions, final failure of the composite is controlled by the void coalescence mechanism in the matrix.

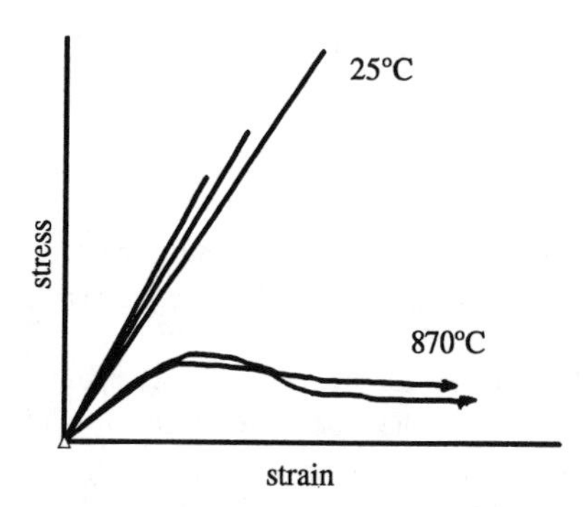

Fig. 4. Tensile stress-strain curves of the monolithic NiAl and composites at 25 and 870 °C.

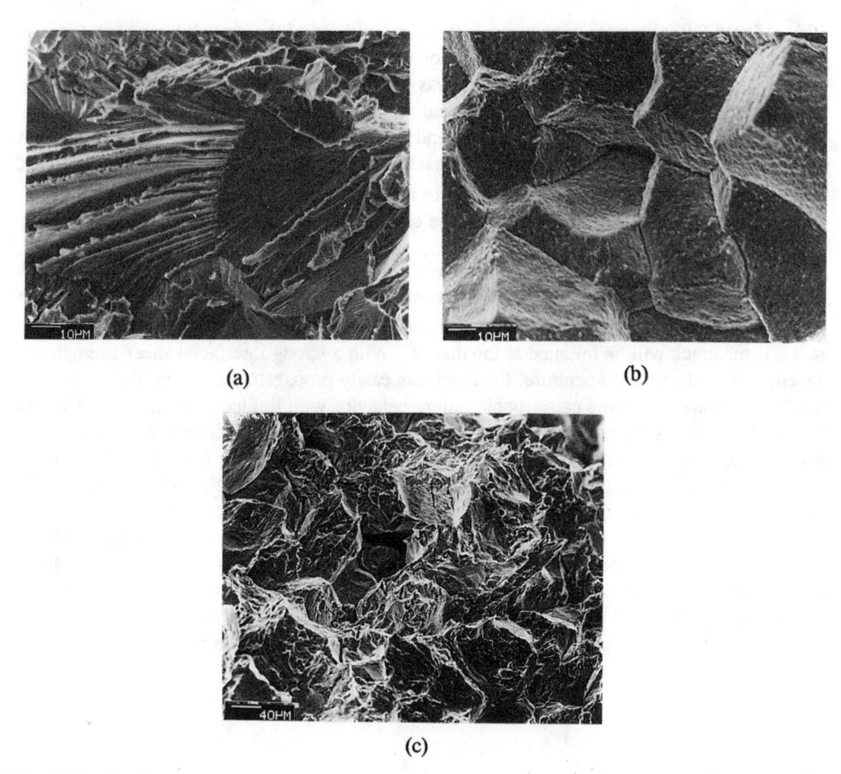

(a) (b)

(c)

Fig. 5 Tensile fracture surface of the monolithic NiAl tested at (a) 25, (b) 760 and (c) 870 °C.

Creep Behavior

Fig. 9 is the creep curves of the monolithic matrix and Al_2O_3/NiAl composite tested at 700 °C. It clearly shows that both materials exhibited primary and steady-state creep. The steady-state creep rates at 100 and 136 MPa are 2.64×10^{-7}, 4.1×10^{-7} s^{-1} for the monolithic matrix, and 1.86×10^{-7}, 3.66×10^{-7} s^{-1} for the Al_2O_3/NiAl composite, respectively. It is evident that the creep resistance of the NiAl alloy can be improved through the incorporation of Al_2O_3 fibers. A close examination of the crept composite showed that a significant multiple fiber breakage occurred during creep deformation. This phenomena is similar to that of the SCS-6/Si_3N_4 composite under creep deformation [8].

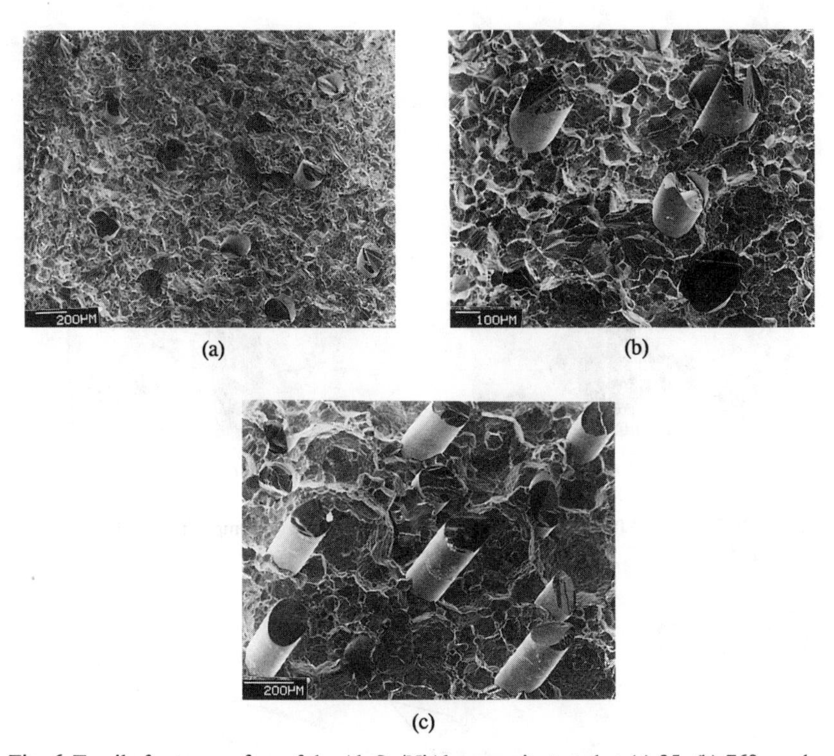

(a) (b)

(c)

Fig. 6 Tensile fracture surface of the Al_2O_3/NiAl composite tested at (a) 25, (b) 760 , and (c) 870 °C.

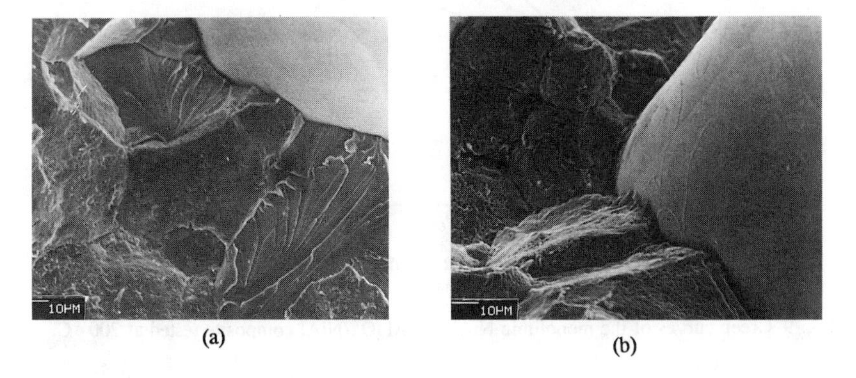

(a) (b)

Fig. 7 Failure morphology near interfacial region of the Al_2O_3/NiAl composite tested at (a) 760 and (b) 870 °C.

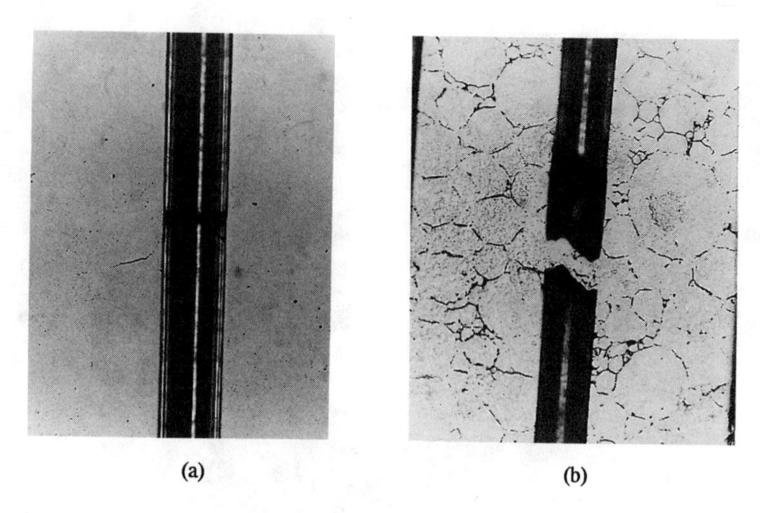

Fig. 8 Microstructure near the broken fiber end in Al_2O_3/NiAl composite tested at (a) 760 and (b) 870 °C.

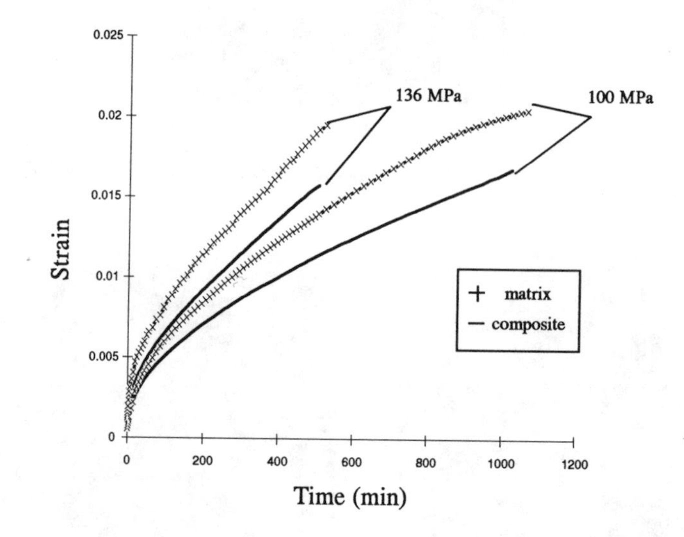

Fig. 9 Creep curves of the monolithic NiAl and Al_2O_3/NiAl composite tested at 700 °C.

REFERENCES

1. R. D. Field, D. F. Lahrman and R. Darolia, Acta Met., to be published.
2. E. M. Schulson and D. R. Barker, Scripta Met., 17, pp. 519-304, 1983.
3. E. P. George and C. T. Liu, J. Mater. Res., 5, pp.754-762, 1990.
4. S. C. Jha, R. Ray and D. J. Gaydosh, Script Met., 23, pp. 805-810, 1989.
5. J.-M. Yang and S. M. Jeng, J. Mat. Res., 6[3], pp.505, 1991.
6. C. J. Yang, S. M. Jeng and J.-M. Yang, Scripta Met. 24, 3, p.468, 1990.
7. R. Darolia, J. of Metals, pp. 44-49, March 1991.
8. J.-M. Yang, R. B. Thayer, S. T. Chen and W. Lin, in Proc. of ICCM VIII, p. 23-C1, 1991.

PART III

$MoSi_2$ Composites

A PERSPECTIVE ON $MoSi_2$ BASED COMPOSITES

J.J. PETROVIC* AND A.K. VASUDEVAN**
*Materials Science and Technology Division, Group MST-4, Los Alamos National Laboratory, Los Alamos, NM 87545
**Office of Naval Research, Code 1222, 800 North Quincy Street, Arlington, VA 22217-5000

ABSTRACT

$MoSi_2$ based composites represent an important new class of "high temperature structural silicides", with significant potential for elevated temperature structural applications in the range of 1200-1600 ^{o}C in oxidizing and aggressive environments. The properties of $MoSi_2$ which make it an attractive matrix for high temperature composites are described and the developmental history of these materials traced. Latest results on elevated temperature creep resistance, low temperature fracture toughness, and composite oxidation behavior are summarized. Important avenues for future $MoSi_2$ based composite development are suggested.

INTRODUCTION

$MoSi_2$ based composites are attracting increasing attention as high temperature structural materials [1]. The purpose of the present discourse is to provide a perspective on these silicide-based materials, describing their characteristics, highlighting key recent research results, and suggesting important avenues for further development.

The intermetallic compound $MoSi_2$ was first investigated by Hoenigschmid in 1907 [2]. Due to its excellent high temperature oxidation resistance and brittle characteristics at low temperatures, it was initially used as an oxidation resistant coating material for metals. Later, based on its electrical conductivity, it was employed as a heating element material for electrical resistance furnaces [3]. This heating element application continues to the present.

In the early 1950's, Maxwell first suggested the use of $MoSi_2$ as an elevated temperature structural material [4], and determined some of its high temperature mechanical properties. However, due to the low temperature brittle nature of $MoSi_2$, Maxwell's work was not continued, since the materials community at that time did not possess the tools to deal with brittle structural materials.

In the 1970's, Fitzer began examining $MoSi_2$ composites reinforced with Al_2O_3, SiC, and Nb wire as a means to improve the mechanical properties of $MoSi_2$ [5]. Encouraging results obtained by Fitzer lead to a review article by Schlichting [6], suggesting the use of $MoSi_2$ as a matrix material for high temperature structural composites.

Two key articles were published in 1985. The first was an article by Fitzer and Remmele [7], describing in detail their investigations of Nb wire-$MoSi_2$ matrix composites. In particular, they showed that Nb wire reinforcements significantly improved the room temperature mechanical properties of such composites. The second article was by Gac and Petrovic [8], in which they established the feasibility of SiC whisker-$MoSi_2$

matrix composites, demonstrating improvements in room temperature strength and fracture toughness.

In 1988, Carter et.al. [9] established that submicron SiC whisker-$MoSi_2$ matrix composites exhibited mechanical properties within the range of high temperature engineering applications. This lead to an acceleration of research interest in these materials by industry, academia, and government laboratories. In November 1991, the First High Temperature Structural Silicides Workshop was held at NIST in Gaithersburg, MD under the sponsorship of the Office of Naval Research. Many of the recent research results discussed here were presented at that Workshop.

PROPERTIES OF $MoSi_2$

Of the many known silicide compounds, $MoSi_2$ appears the most promising for elevated temperature structural applications, due to the following combination of properties. $MoSi_2$ possesses a high melting point, 2030 ^oC, and excellent high temperature oxidation resistance. Although brittle at low temperatures, it exhibits a brittle-to-ductile transition at approximately 1000 ^oC (compressive deformation of single crystals), with deformation by dislocation plasticity above this temperature. $MoSi_2$ is thermodynamically stable with potential ceramic reinforcements such as SiC, ZrO_2, Si_3N_4, Al_2O_3, TiB_2, and TiC [10], and may also be alloyed with other high melting point silicides such as WSi_2 [11]. $MoSi_2$ is an abundant, low cost material which is non-toxic and environmentally benign. Additionally, due to its low electrical resistivity (10 micron-ohm cm at room temperature), it can be electro-discharge machined, which is a significant benefit for the low cost fabrication of components.

The crystal structure of $MoSi_2$ is tetragonal, with c = 0.785 nm and a = 0.32 nm. Dislocations in $MoSi_2$ have been observed to have <100>, <110>, and 1/2<111> burgers vectors [12]. Studies of the elevated temperature compressive deformation of single crystals have indicated an onset of ductility at 1000 ^oC and yield stress levels of 300-800 MPa at 1100 ^oC and 50-270 MPa at 1500 ^oC, depending on crystallographic orientation [13]. Bend tests on low oxygen content, large grained, polycrystalline $MoSi_2$ indicate a brittle-to-ductile transition in flexure in the vicinity of 1350 ^oC, with a flexural yield stress at 1400 ^oC of 210 MPa [14]. The room temperature fracture toughness of polycrystalline $MoSi_2$ is 3 MPa $m^{1/2}$, and the fracture mode is 75% transgranular and 25% intergranular at room temperature [15]. Its hardness is 9 GPa. The thermal expansion coefficient of $MoSi_2$ is 7-10 x 10^{-6} C^{-1} in the range of 20-1400 ^oC, and is a reasonably close match to that of Al_2O_3. Its thermal conductivity is 65 W/mK at room temperature and decreases to 30 W/mK at 1400 ^oC. These conductivity values are intermediate between Si_3N_4 and SiC.

The elevated temperature oxidation resistance of $MoSi_2$ is similar to that of SiC, since it forms a very protective, adherent, and coherent SiO_2 layer. Recent work [16] has shown that the maximum oxidation rate occurs at 500 ^oC, an intermediate temperature range where oxidation pesting can occur under certain conditions. However, pest behavior is not observed in stress-free single crystals of $MoSi_2$, or in dense polycrystalline materials. Pest behavior can also be minimized by the addition of $MoGe_2$ [7], which alters the viscosity and thermal expansion coefficient of the oxidation layer.

$MoSi_2$ may be alloyed with other high melting point silicides as a means of improving mechanical properties [11]. It has been shown that solid solution alloying with WSi_2 improves high temperature mechanical properties [17]. A recent interesting observation is the ubiquity of the C40 hexagonal phase in alloys of $MoSi_2$ with other disilicides such as $TiSi_2$ [18]. However, it should be noted that all other disilicides and trisilicides have an oxidation resistance inferior to that of $MoSi_2$.

$MoSi_2$ COMPOSITES

$MoSi_2$ possesses excellent elevated temperature oxidation resistance. However, for $MoSi_2$ to be used as an oxidation-resistant elevated temperature structural material, both its high and low temperature mechanical properties must be significantly improved. This dictates improvements in high temperature strength and creep resistance, and low temperature fracture toughness.

The composite approach with $MoSi_2$ as the matrix can produce such mechanical property improvements. $MoSi_2$ is stable with a large number of carbide, nitride, oxide, and boride ceramic reinforcements, such as SiC, TiC, Si_3N_4, ZrO_2, Al_2O_3, Y_2O_3, TiB_2, and ZrB_2 [1,10]. It is reactive with refractory metals such as Nb, Ta, Mo, and W. $MoSi_2$ also reacts with carbon.

Significant issues for $MoSi_2$ composites include reactivity of reinforcement and matrix, thermal expansion coefficient mismatch, low temperature fracture toughness, elevated temperature creep resistance, and both intermediate and high temperature oxidation resistance. To date, the most extensively studied $MoSi_2$ based composites have been SiC-$MoSi_2$, ZrO_2-$MoSi_2$, C-$MoSi_2$, Al_2O_3-$MoSi_2$, TiB_2-$MoSi_2$, and refractory metal-$MoSi_2$. Key research findings on the properties of selected $MoSi_2$ composites will now be discussed.

COMPOSITE ELEVATED TEMPERATURE CREEP RESISTANCE

Adequate creep resistance is a central issue for high temperature structural materials. The creep behavior of SiC-$MoSi_2$ composites has been examined the most extensively to date [19,20,21]. Observed creep rates are shown in Table I.

A number of aspects may be noted in Table I. Absolute creep rate values reported by the different investigators are not totally consistent with one another. This may be related to differences in processing of the various materials examined. Reinforcement with SiC whiskers significantly reduces the creep rate by more than an order of magnitude, while incorporating WSi_2 in solid solution with $MoSi_2$ also leads to lower creep rates. Low creep rates are observed in SiC reinforced composites with a $MoSi_2$-WSi_2 alloy solid solution matrix, suggesting additive effects of reinforcement and solid solution on creep rate. Creep rates in tension are higher than in compression. The creep rate of a <210> oriented $MoSi_2$ single crystal is similar to that of polycrystalline $MoSi_2$.

The lowest creep rates observed at 1200 ^oC and 50 MPa for current $MoSi_2$ based composites are of the order of 10^{-8} s^{-1}. Such creep rate levels would lead to 1% creep strain in approximately 300 hours at this temperature and stress. By way of comparison, the creep rate under these conditions for a MAR-M-509 superalloy is 3 x 10^{-5} s^{-1}, three orders of magnitude higher than the $MoSi_2$ based composites.

Table I. Creep Rates for $MoSi_2$ Materials at 1200 ^oC and 50 MPa

Material	Test Type	Creep Rate (s^{-1})	Ref.
$MoSi_2$ HP	Compression	1.5 x 10^{-6}	19
$MoSi_2$ HP	Compression	9 x 10^{-6}	20
$MoSi_2$ HIP	Compression	4 x 10^{-7}	20
50/50 $MoSi_2$-WSi_2 HP	Compression	1.5 x 10^{-6}	19
50/50 $MoSi_2$-WSi_2 HP	Compression	1.6 x 10^{-7}	20
20% SiC(w)-$MoSi_2$ HP	Compression	1.6 x 10^{-8}	19
20% SiC(w)-$MoSi_2$ HP	Compression	1.6 x 10^{-7}	20
20% SiC(w)-$MoSi_2$ HIP	Compression	1.8 x 10^{-8}	20
20% SiC(p)-$MoSi_2$/WSi_2 HIP	Compression	6 x 10^{-8}	20
20% SiC(w)-$MoSi_2$/WSi_2 HP	Compression	5 x 10^{-9}	21
20% SiC(w)-$MoSi_2$/WSi_2 HP	Tension	2.5 x 10^{-8}	21
$MoSi_2$ single crystal <210>	Compression	1 x 10^{-7}	20

HP = hot pressed HIP = hot isostatically pressed

Observed creep parameters for $MoSi_2$ materials are summarized in Table II. The data in Table II provide some initial insight into creep mechanisms in $MoSi_2$ materials. Creep stress exponents are in the approximate range of 2-3. A creep stress exponent of 1 is indicative of viscous flow processes. Exponents in the range of 1-3 suggest a combination of viscous flow and power law dislocation creep. Stress exponents of 3-4 indicate dislocation recovery controlled by self diffusion, while exponents greater than 4 are taken to indicate the glide/climb of dislocations.

The stress exponent results for polycrystalline $MoSi_2$ materials suggest that dislocation processes play a major role in creep deformation. It is also possible that grain boundary sliding with cavitation due to the presence of viscous silica phases at the grain boundaries is an additional creep mechanism in current materials. This may account for the observed creep rate differences in tension and compression. Self diffusion activation energies in $MoSi_2$ are not well known. The activation energy for Si diffusion in $MoSi_2$ is indicated to be approximately 250 kJ/mole [22,23]. Based on creep data, Sadananda et.al. [19] have inferred an activation energy of 350-540 kJ/mole for Mo diffusion in $MoSi_2$, suggesting that Mo diffusion may be the rate controlling process for dislocation creep mechanisms.

Table II. Creep Parameters for $MoSi_2$ Materials

Material	Test Type	Temp. (ºC)	Stress Exponent	Activation Energy (kJ/mole)	Ref.
$MoSi_2$	Compression	1100-1300	1.75	380	19
$MoSi_2$	Compression	1200	3	306	20
50/50 $MoSi_2$-WSi_2	Compression	1100-1300	2.27	540	19
50/50 $MoSi_2$-WSi_2	Compression	1200	3	306	20
20% SiC(w)-$MoSi_2$	Compression	1100-1450	2.63	460	19
20% SiC(w)-$MoSi_2$	Compression	1200	3	306	20
20% SiC(w)-$MoSi_2$/WSi_2	Compression	1150-1225	2.3	312	21
20% SiC(p)-$MoSi_2$/WSi_2	Compression	1200	3	306	20
20% SiC(w)-$MoSi_2$/WSi_2	Tension	1100-1200	3.2	557	21
$MoSi_2$ single crystal	Compression	1200	3	251	20

COMPOSITE FRACTURE TOUGHNESS

Since $MoSi_2$ is a brittle material below its brittle-to-ductile transition temperature, composite approaches to improve low temperature fracture toughness generally follow those employed for structural ceramic materials. Evans [24] has recently summarized such approaches, indicating that composite toughening mechanisms decrease in effect in the following order: continuous fibers, metal dispersed particles, transformation toughening, whiskers/platelets/particles, microcracking.

Table III. Room Temperature Fracture Toughness of $MoSi_2$ Based Composites

Type of Reinforcement	Highest Fracture Toughness ($MPa\ m^{1/2}$)	Ref.
Refractory metal (Nb,W,Mo) wires	Greater than 15	7, 25
20 vol.% Ta particles	10	26
20 vol.% ZrO_2 particles	7.8	27
20 vol.% SiC whiskers	4.4	28
20 vol.% SiC particles	4.0	29
Polycrystalline $MoSi_2$	3	15

Current low temperature fracture toughness results for $MoSi_2$ based materials are summarized in Table III. It is evident from Table III that $MoSi_2$ composites can possess significantly higher room temperature fracture toughness than polycrystalline $MoSi_2$.

Refractory metal wires and particles have exhibited the highest toughness levels to date. However, the oxidation resistance of such composites can be poor. In addition, $MoSi_2$ is not thermodynamically stable with the refractory metals, and thus coatings are required on refractory metal reinforcements in order to minimize reaction effects. Concerning such effects, Maloney and Hecht [25] have observed that oxide "plugs" form over the refractory metal fibers of W-3%Re fiber reinforced $MoSi_2$ composites when exposed to air at 1400 ^{o}C, which protect the fibers from further oxidation. They have employed Al_2O_3 coatings on the fibers to minimize reaction effects with $MoSi_2$.

ZrO_2 transformation toughening effects can produce substantial toughening in $MoSi_2$ based materials. Highest toughness levels to date have been obtained with unstabilized ZrO_2, and appear to be associated in part with microcrack toughening mechanisms. A very intriguing additional aspect occurring in unstabilized ZrO_2-$MoSi_2$ composites is the "pumping" of dislocations into the $MoSi_2$ matrix as a result of the volume change associated with the spontaneous ZrO_2 tetragonal-to-monoclinic phase transformation [27]. Upon cooling from the fabrication temperature (1700 ^{o}C), this transformation initiates in the vicinity of 1175 ^{o}C. The unstabilized zirconia transformation temperature is above the brittle-to-ductile transition of $MoSi_2$, and so dislocation "pumping" occurs as a result of the spontaneous ZrO_2 transformation strains. R-curve behavior and synergistic toughening effects with combined ZrO_2-SiC reinforcements have also been observed [27]. Presence of ZrO_2 does not significantly degrade the elevated temperature oxidation resistance of ZrO_2-$MoSi_2$ composites [30].

Only moderate room temperature toughening effects are derived from submicron SiC whiskers and particles. Toughening levels for these reinforcements are similar to those observed in ceramic matrix composites [31], and are associated with mechanisms such as crack deflection and crack bridging.

A very important recent observation is that grain boundary silica phases, resulting from oxygen on the surfaces of commercial $MoSi_2$ powders, have a detrimental effect on the elevated temperature fracture toughness of polycrystalline $MoSi_2$ materials [32,33]. This occurs because presence of the silica phase promotes grain boundary sliding deformation mechanisms. When the grain boundary silica is removed by reaction with carbon additions, fracture toughness increases with increasing temperature, due to the operation of dislocation plasticity mechanisms. The fracture toughness of polycrystalline $MoSi_2$ containing 2 wt.% carbon has been reported to be 11.5 MPa $m^{1/2}$ at 1400 ^{o}C, with a "graceful failure" stress-strain response due to plasticity effects [33].

COMPOSITE OXIDATION BEHAVIOR

It is important that $MoSi_2$ composite development avenues lead to materials with acceptable oxidation behavior at both elevated and intermediate temperatures. The oxidation of $MoSi_2$ based composites has recently been investigated by Cook et.al. [34], for composites synthesized by the XD^{TM} process with various types of reinforcements. Cyclic oxidation results from this study are summarized in Table IV.

Table IV. 72 Hour Cyclic Oxidation of $MoSi_2$ Composites at 1200 °C and 1500 °C

Material	1200 °C Weight Gain (mg/cm^2)	1500 °C Weight Gain (mg/cm^2)
$MoSi_2$	0.1	0.4
30% SiC-$MoSi_2$	0.01	0.5
30% TiB_2-$MoSi_2$	0.6	2.8
30% HfB_2-$MoSi_2$	0.7	2.0
30% ZrB_2-$MoSi_2$	2.7	7.3

These cyclic oxidation results indicate that SiC-$MoSi_2$ composites possess the best elevated temperature oxidation resistance in comparison to pure $MoSi_2$, and that TiB_2-$MoSi_2$ and HfB_2-$MoSi_2$ also possess reasonable oxidation resistance.

Cook et.al. [34] also performed intermediate temperature oxidation studies at 500 °C, for both static and cyclic conditions. With the exception of the SiC reinforcement, none of the materials in Table IV exhibited oxidation pest behavior. Pest behavior was not observed either statically or after 1200 °C cycling for the SiC-$MoSi_2$ composites, but was observed after 1500 °C cycling. The occurrence of pest behavior was attributed to the large particle size of the SiC reinforcement, since particles on the order of 50 microns were present in the composite matrix. It is likely that these large particles in combination with the thermal expansion mismatch between SiC and $MoSi_2$ lead to thermal stress-induced cracking and associated pesting.

These intermediate temperature oxidation results make it clear that thermal expansion coefficient mismatch and reinforcement size effects must be kept in view in $MoSi_2$ based composite development. For SiC reinforcements, this dictates that the size of the SiC phase should be kept as small as possible, so as to eliminate thermally induced cracking. It may also indicate the use of pest inhibiting additives such as $MoGe_2$ [7] in developmental composite approaches.

IMPORTANT AVENUES FOR FUTURE $MoSi_2$ BASED COMPOSITE DEVELOPMENT

Important aspects of future $MoSi_2$ based composite development can be roughly divided into the three general areas of materials, processing, and properties. Key research avenues for each area are indicated in Table V. In pursuing the various aspects of $MoSi_2$ based composite development indicated in Table V, it is important to keep in view the fact that $MoSi_2$ is a borderline intermetallic compound, due to its mixed metallic/covalent/ionic atomic bonding state. This material possesses ceramic-like brittleness at room temperature, and metal-like plasticity at elevated temperatures. In addition, $MoSi_2$ composites require significant volume fractions of ceramic reinforcements, and both ceramic and metal processing techniques are important for their development. All the above factors dictate that the optimization of $MoSi_2$ based composites will require the combined skills of both ceramists and metallurgists.

ACKNOWLEDGEMENTS

Support for the preparation of this perspective on $MoSi_2$ based composites was provided by the Office of Naval Research and the DOE Advanced Industrial Concepts Materials Program.

Table V. Key Research Areas for $MoSi_2$ Based Composite Development

Materials

- O Elimination of $MoSi_2$ grain boundary silica phases
    - o Higher quality starting powders (submicron, high purity, low oxygen content)
    - o Removal with carbon additions
- O $MoSi_2$ alloying with other high melting point silicides
    - o Binary and ternary silicide phase diagrams
- O Investigations of $MoSi_2$ single crystals
    - o Basic information on mechanical behavior, diffusion, oxidation

Processing

- O Development of important $MoSi_2$ composite processing technologies
    - o Discontinuous fiber/particle, continuous fiber, layered composites
    - o Pressureless sintering
    - o HIP, Sinter/HIP
    - o Plasma spraying
    - o Hot working
    - o Osprey process
    - o In-situ composites
    - o Mechanical alloying
    - o Require fine, uniform dispersions of second phase constituents

Properties

- O Improve $MoSi_2$ composite mechanical properties while retaining elevated and intermediate temperature oxidation resistance
    - o Improve elevated temperature strength/creep resistance
    - o Increase low temperature fracture toughness
    - o Lower brittle-to-ductile transition temperature
    - o Improve thermal shock/thermal fatigue resistance
    - o Optimize mechanical fatigue behavior

REFERENCES

1. A.K. Vasudevan and J.J. Petrovic, "A Comparative Overview of Molybdenum Disilicide Composites", in High Temperature Structural Silicides, Eds. A.K. Vasudevan and J.J. Petrovic (Elsevier Science Publishers, Amsterdam, 1992).

2. O. Hoenigschmid, Monatsh. Chem., 28, 1017 (1907).

3. Kanthal, Swedish Patent No. 155836, 1953.

4. W.A. Maxwell, "Some Stress-Rupture & Creep Properties of Molybdenum Disilicide in the Range of 1600-2000 F", NACA Research Memorandum NACA RM E52DO9, June 1952.

5. E. Fitzer, O. Rubisch, J. Schlichting, and I. Sewdas, Special Ceramics, Vol. 6, 1973.

6. J. Schlichting, High Temp.-High Press., 10, 241 (1978).

7. E. Fitzer and W. Remmele, Proceedings 5th International Conference on Composite Materials, ICCM-V, Eds. W.C. Harrigan Jr., J. Strife, and A.K. Dhingra (AIME Publications, Warrendale, PA, 1985), p. 515.

8. F.D. Gac and J.J. Petrovic, J. Amer. Ceram. Soc., 68, C200 (1985).

9. D.H. Carter, M.S. Thesis, MIT, 1988; D.H. Carter, W.S. Gibbs, and J.J. Petrovic, Proceedings 3rd International Symposium on Ceramic Materials & Components for Engines, (The American Ceramic Society, Inc., Westerville, OH, 1989), p.977.

10. P.J. Meschter and D.S. Schwartz, Journal of Metals, November 1989, 52.

11. J.J. Petrovic, R.E. Honnell, and A.K. Vasudevan, Mat. Res. Soc. Symp. Proc., 194, 123 (1990).

12. O. Unal, J.J. Petrovic, D.H. Carter, and T.E. Mitchell, J. Amer. Ceram. Soc., 73, 1752 (1990).

13. Y. Umakoshi, T. Sakagami, T. Hirano, and T. Yamane, Acta Metall. Mater., 38, 909 (1990).

14. R.M. Aikin, Jr., Scripta Met., 26, 1025 (1992).

15. R.J. Wade and J.J. Petrovic, "Fracture Modes in $MoSi_2$", J. Amer. Ceram. Soc., in press.

16. D.A. Berztiss, R.R. Cerchiara, E.A. Gulbransen, F.S. Pettit, and G.H. Meier, "Oxidation of $MoSi_2$ and Comparison to Other Silicide Materials", in High Temperature Structural Silicides, Eds. A.K. Vasudevan and J.J. Petrovic (Elsevier Science Publishers, Amsterdam, 1992).

17. J.J. Petrovic and R.E. Honnell, Ceram. Eng. Sci. Proc., 11, 734 (1990).

18. W.J. Boettinger, J.H. Perepezko, and P.S. Frankwicz, "Application of Ternary Phase Diagrams to the Development of $MoSi_2$-Based Materials", in High Temperature Structural Silicides, Eds. A.K. Vasudevan and J.J. Petrovic (Elsevier Science Publishers, Amsterdam, 1992).

19. K. Sadananda, C.R. Feng, H. Jones, and J. Petrovic, "Creep of Molybdenum Disilicide Composites", in High Temperature Structural Silicides, Eds. A.K. Vasudevan and J.J. Petrovic (Elsevier Science Publishers, Amsterdam, 1992).

20. S. Bose, "Engineering Aspects of Creep Deformation of Molybdenum Disilicide", in High Temperature Structural Silicides, Eds. A.K. Vasudevan and J.J. Petrovic (Elsevier Science Publishers, Amsterdam, 1992).

21. S.M. Wiederhorn, R.J. Gettings, D.E. Roberts, C. Ostertag, and J.J. Petrovic, "Tensile Creep of Silicide Composites", in High Temperature Structural Silicides, Eds. A.K. Vasudevan and J.J. Petrovic (Elsevier Science Publishers, Amsterdam, 1992).

22. R.W. Bartlett, P.R. Gage, and P.A. Larssen, Trans. AIME, 230, 1528 (1964).

23. P. Kofstad, High Temperature Oxidation of Metals, (John Wiley & Sons, New York, 1966).

24. A.G. Evans, J. Amer. Ceram. Soc., 73, 187 (1990).

25. M.J. Maloney and R.J. Hecht, "Development of Continuous Fiber Reinforced Molybdenum Disilicide Base Composites", in High Temperature Structural Silicides, Eds. A.K. Vasudevan and J.J. Petrovic (Elsevier Science Publishers, Amsterdam, 1992).

26. R.G. Castro, R.W. Smith, A.D. Rollett, and P.W. Stanek, "Ductile Phase Toughening of Molybdenum Disilicide by Low Pressure Plasma Spraying", in High Temperature Structural Silicides, Eds. A.K. Vasudevan and J.J. Petrovic (Elsevier Science Publishers, Amsterdam, 1992).

27. J.J. Petrovic, A.K. Bhattacharya, R.E. Honnell, T.E. Mitchell, R.K. Wade, and K.J. McClellan, "ZrO_2 and ZrO_2/SiC Particle Reinforced-$MoSi_2$ Matrix Composites", in High Temperature Structural Silicides, Eds. A.K. Vasudevan and J.J. Petrovic (Elsevier Science Publishers, Amsterdam, 1992).

28. J.J. Petrovic and R.E. Honnell, in Ceramic Transactions, Volume 19, (The American Ceramic Society Inc., Westerville, Ohio, 1991), p. 817.

29. A.K. Bhattacharya and J.J. Petrovic, J. Amer. Ceram. Soc., 74, 2700 (1991).

30. W.L. Worrell, University of Pennsylvania, private communication.

31. P.F. Becher, C-H Hsueh, P. Angelini, and T.N. Tiegs, J. Amer. Ceram. Soc., 71, 1050 (1988).

32. S. Maloy, A.H. Heuer, J. Lewandowski, and J. Petrovic, J. Amer. Ceram. Soc., 74, 2704 (1991).

33. S.A. Maloy, J.J. Lewandowski, A.H. Heuer, and J.J. Petrovic, "Effects of Carbon Additions on High Temperature Mechanical Properties of Molybdenum Disilicide", in High Temperature Structural Silicides, Eds. A.K. Vasudevan and J.J. Petrovic (Elsevier Science Publishers, Amsterdam, 1992).

34. J. Cook, A. Khan, E. Lee, and R. Mahapatra, "Oxidation of $MoSi_2$ Based Composites", in High Temperature Structural Silicides, Eds. A.K. Vasudevan and J.J. Petrovic (Elsevier Science Publishers, Amsterdam, 1992).

THE STRUCTURE AND PROPERTIES OF $MoSi_2$-Mo_5Si_3 COMPOSITES

H. KUNG, D.P. MASON, A. BASU, H. CHANG, D.C. VAN AKEN, A.K. GHOSH AND R. GIBALA
Department of Materials Science and Engineering
The University of Michigan, Ann Arbor, MI 48109

ABSTRACT

The addition of Mo_5Si_3 as a reinforcing second phase in a $MoSi_2$ matrix has been investigated for possible high temperature strengthening effects. $MoSi_2$ with up to 45 vol % Mo_5Si_3 was fabricated using powder metallurgy (PM) and arc-casting (AC) techniques. Effects of processing routes, which result in different microstructures, on their mechanical properties are given. PM composites, which have an equiaxed microstructure, exhibit a limited increase in hardness. Higher hardnesses are observed in script-structured AC eutectics and Er-modified-eutectics throughout the temperatures studied (25-1300°C). Crack propagation paths induced by indentation show long transphase cracks in the AC materials vs short intergranular and interphase cracks in the PM composites at high temperatures.

Transmission electron microscopy discloses that the interface in the AC composites has a low-index orientation relationship between the two phases and shows regularly faceted interfacial structures, while planar interfaces are found in the PM composites. These observations suggest the interface is stronger and lower in energy in the AC composites, which is consistent with the higher hardness values and long transphase cracks observed.

Dislocation analysis shows the presence of ordinary dislocations (<100>, <110> and 1/2<111>) in $MoSi_2$ in the as-fabricated composites. These types of dislocation are also responsible for the high temperature plastic deformation in compression in both the monolithic $MoSi_2$ and the composites. <331> types of dislocation are only found in $MoSi_2$ either near the interface of the AC composites or in materials deformed below 1000°C.

INTRODUCTION

The intermetallic compound $MoSi_2$ is a potential matrix material for high temperature structural composites due to its high melting point and good oxidation resistance at elevated temperatures [1]. One major drawback for its structural application is inadequate high temperature (≥1200°C) strength at significant applied stresses. The need for composite additions for strengthening at high temperature has become the focus of extensive investigations in recent years [2]. Mo_5Si_3 is one of several elastically hard, brittle phases which can potentially strengthen $MoSi_2$ at high temperatures.

Materials with different microstructures, resulting from various processing routes, are expected to exhibit different mechanical behaviors. Ashby et al. [3] have shown that the toughness of the ceramic-metal composite can be influenced by the strength of the interface through changes of the local deformation behavior in the metal. A fundamental understanding of the structure-mechanical properties relationships is critical to the development of materials for high temperature structural applications. In the present investigation, we report some initial results directed towards understanding the roles of microstructure, interfacial structure and defect structure in influencing the mechanical properties of $MoSi_2$-Mo_5Si_3 composites.

EXPERIMENTAL PROCEDURE

PM processed $MoSi_2$ and composites containing 15 and 30 vol % of Mo_5Si_3 were made by slurry milling (in acetone) mixtures of the $MoSi_2$ (-325 mesh, from CERAC, Inc.) and Mo (-325 mesh, Johnson & Matthey) for ~24 h. The mixtures were then hot pressed at 1700°C at 25 MPa for 2 h. This treatment was sufficient to form the 15 and 30 vol % of Mo_5Si_3 expected from the phase diagram. Fig. 1(a) shows the typical equiaxed microstructure of the polycrystalline PM composites. Typical grain sizes in the $MoSi_2$ and Mo_5Si_3 are 25-30 μm and 10-15 μm, respectively. $MoSi_2$-Mo_5Si_3 single crystal composites of the eutectic composition (45 vol % of Mo_5Si_3) were prepared by arc-melting either a mixture of elemental powders or a mixture of $MoSi_2$ and Mo. The as-fabricated eutectic shows the script-lamellar microstructure (Fig. 1(b)) which was formed upon solidification from the melt. Erbium (1 weight percent) was added to some AC eutectics as a potential microstructural refiner and possible oxygen getter.

Microhardness indentation was used to evaluate the temperature dependence of the strength of the composites. Microhardness was determined at temperatures 25-1300°C using a Nikon QM hot hardness tester. A diamond Vickers indenter with a load of either 0.5 or 1 kg was used in all experiments and an average of 8 measurements were made at each temperature.

The microstructure of the composites were characterized by scanning electron microscopy (SEM) and transmission electron microscopy (TEM) techniques [4,5]. SEM was used primarily for determining the grain size, grain shape and second phase distribution. Crack propagation paths induced by indentation were also examined using SEM. Conventional TEM (on a JEOL 2000FX) was employed to characterize the defect structure in $MoSi_2$ in the composites and the orientation relationships between the two phases. High resolution TEM (HRTEM) observations of the atomic structure of the interface and atomic-level defects were made on a JEOL 4000EX microscope.

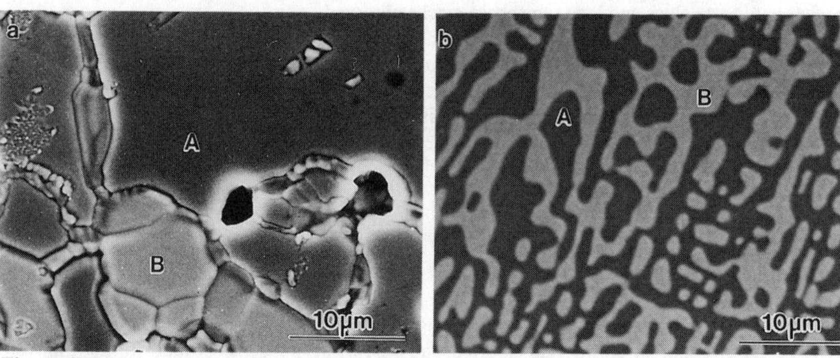

Fig. 1 SEM micrographs of the $MoSi_2$-Mo_5Si_3 composites. (a) PM composite with 15 vol % Mo_5Si_3. (b) Er-free AC eutectic composite. (Phase identification: A-$MoSi_2$ and B-Mo_5Si_3)

RESULTS AND DISCUSSION

Microhardness

The temperature dependence of the hardness of the PM and AC composites and the monolithic $MoSi_2$ is shown in Fig. 2. Fig. 2(a) shows that the addition of 15 and 30 vol % of Mo_5Si_3 in the PM composites increases the hardness slightly between 25 and 1000°C. The higher the Mo_5Si_3 content the more prominent the strengthening is, especially in the intermediate temperature range (400-800°C). No increase in hardness due to the addition can be seen above 1000°C, which represents the lower end of the reported ductile to brittle transition temperature (DBTT) range of $MoSi_2$ [6]. It is likely that the strength of Mo_5Si_3 also decreases above 1000°C.

However, since above the DBTT $MoSi_2$ deforms by dislocation motion, the composite deformation is probably dominated by the plastic yielding of $MoSi_2$.

Fig. 2(b) shows that the AC eutectics have higher hardnesses than the unreinforced $MoSi_2$ at all temperatures from 25 to 1300°C. The modification by erbium, which was shown previously to refine the script microstructure [5], is found to increase the hardness substantially. The higher hardness observed in the AC composites as compared to the PM materials is probably due in part to the distribution and volume fraction of the second phase. It can be seen from Fig. 1 that the AC composite has a more uniformly distributed Mo_5Si_3 phase and also a larger volume fraction (45 vs 15 and 30 vol %). It has been suggested that during indentation, dislocation glide plasticity is a principal deformation mechanism due to the high hydrostatic stress involved [7]. The strength increase in the composites can thus be explained at least in part by the constraint placed by the second phase on the matrix. The more constraint experienced by the matrix the higher the achievable strength. However, the volume fraction may be of less significance than the distribution in influencing the strength. This is supported by the small increase in hardness observed upon doubling the second phase volume fraction in Fig. 2(a). Interfacial structure may also play a large role in determining the differences in hardness shown in Fig. 2 because the strength of the interface has direct influence on the deformation mechanisms of the composites [3]. In the limit of a weak interface, which is characterized to be high in energy, debonding occurs during deformation or fracture. Unstable crack growth and/or premature microcracking may be produced at the interface, which can greatly reduce the composite strength. From the creep strengthening point of view, a strong interface in the composite is desired.

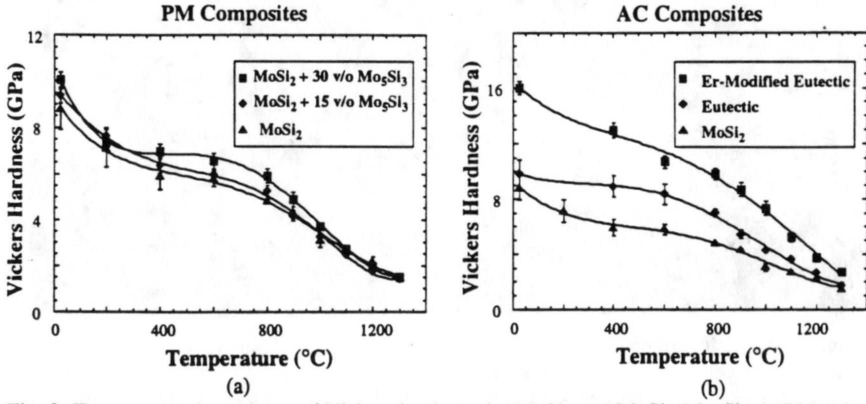

Fig. 2 Temperature dependence of Vickers hardness for $MoSi_2$ and $MoSi_2$-Mo_5Si_3. (a)PM, 15 and 30 vol %, composites and (b) AC, eutectic and Er-modified eutectic composites.

Crack Propagation Path

The strength of the interface was monitored qualitatively by examining the indentation-induced crack propagation routes such as shown in Fig. 3. Fig. 3(a) shows a room temperature (RT) indentation and the crack paths in the PM material. Both intergranular and transgranular cracks are observed, as indicated by Ig and Tg respectively. Fig. 3(b) shows a RT indentation for the AC composites, where long (15-30 μm) radial cracks with transphase (Tp) character are observed. Fig. 3(c) and (d) give the crack paths observed in-situ at 1300°C for the PM and AC composites, respectively. Short (10-20 μm) intergranular (Ig) and interphase (Ip) cracks in the PM composites and long (20-30 μm) transphase (Tp) ones in the AC materials were observed. The observation of long transphase cracks at all temperatures corresponds well with the higher hardness values. The intergranular and interphase cracking found in the PM materials suggests that these interfaces are intrinsically weaker than those of the AC composites. Transgranular cracks which were sometimes seen when indenting the PM composites below the DBTT are probably related to the lack of plasticity of the matrix. The intergranular/interphase vs transphase fracture of the two sets of composites suggests the interface strength and structure are different. Hence, HRTEM analysis of the interfacial structure was undertaken.

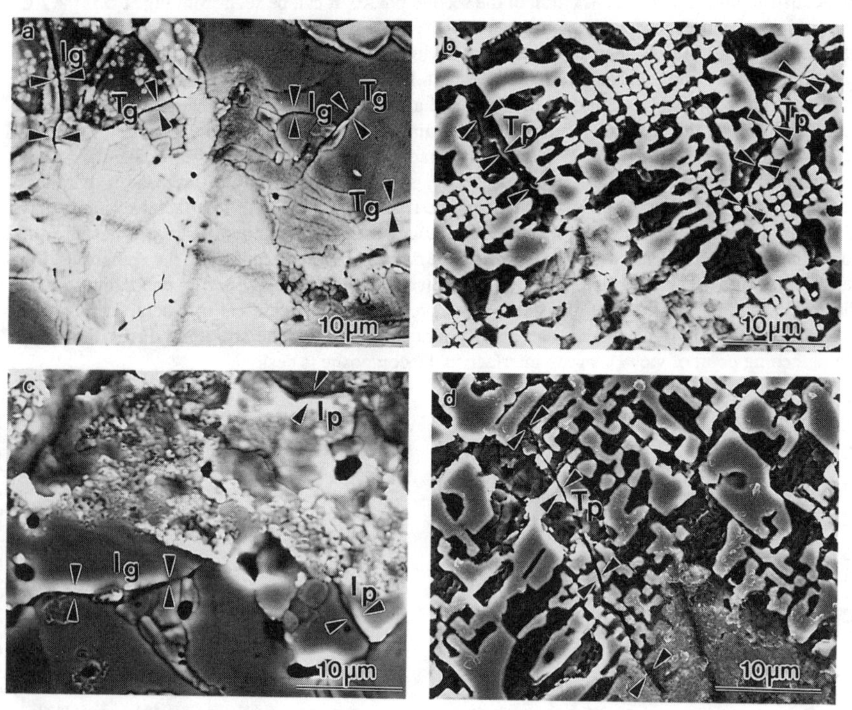

Fig. 3 SEM micrographs of 1 kg Vickers pyramid indentation and crack propagation paths (indicated by arrows) in PM (15 vol %) composites at (a) 25°C and (c) 1300°C, and in AC (Er-modified eutectic) composites at (b) 25°C and (d) 1300°C. (Crack type identification: Ig-intergranular, Ip-interphase, Tg-transgranular and Tp-transphase)

Interfacial Structure

Fig. 4(a) shows the lattice image of an interface in the PM composites. The lattice fringes from the {110} planes of Mo_5Si_3 and {002} planes of $MoSi_2$ extend from the bulk to the interface. The interface is planar and does not contain a thin glassy phase as seen in $MoSi_2$-Si_3N_4 composites [8]. Pockets of amorphous silica are sometimes found at the interfaces and grain boundaries of $MoSi_2$, but are well-dispersed and too small in content to affect the hardness. Moreover, silica particles were rarely seen on the crack paths. Fig. 4(b) shows the faceted interfacial structures of the AC composites which were made from elemental powders. The interface is faceted on a fine scale, with a long facet of ~4 nm separated by steps each with a height of ~0.6 nm. A detailed characterization of the interfacial structure was reported previously [4]. Specifically, the orientation relationship $(1\bar{1}0)[001]_{1\text{-}2} \parallel (3\bar{3}0)[110]_{5\text{-}3}$ has been observed between the two phases and the Burgers vectors of interface dislocations are <100> and 1/2<111> types. The low index orientation relationship and the faceted interfacial structure suggest that the interface in the AC composites is stronger and lower in energy than that of the PM material, which is consistent with the higher hardness values of the AC material.

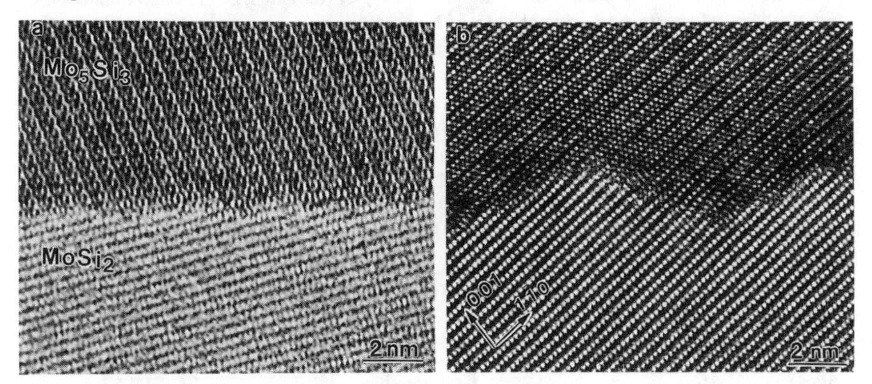

Fig. 4 HRTEM images of $MoSi_2$-Mo_5Si_3 interfaces in (a) a PM (15 vol % Mo_5Si_3) composite showing a planar interface and in (b) an Er-free AC eutectic composite showing regular faceting.

Defect Structures

Dislocation analysis of both the as-fabricated and deformed composites and the baseline matrix material were performed using standard g•b analysis. Table I summarizes the results. Basically, <100> types of dislocation are present in $MoSi_2$ in the as-fabricated composites. Ordinary dislocations of Burgers vectors of <100>, <110> and 1/2<111> are seen in both the monolithic $MoSi_2$ and in $MoSi_2$ of the composites (PM and Er-free AC) deformed in compression at 1300°C. These results are in agreement with the high temperature deformation studies of Unal et al. [9] and Umakoshi et al. [10]. <331> types of dislocation are observed in three specific situations. One is in the monolithic $MoSi_2$ after testing in compression at 900°C, where short segments of 1/2<331> dislocations were observed. Similar results have been shown in single crystals of $MoSi_2$ deformed at 900°C [10]. In the second case, 1/2<331> dislocations were seen near interfaces in the AC composites. The third situation involves observations of <331> shear faults, possibly consisting of six 1/6<331> superpartials coupled by antiphase boundaries, extending from the interface into a $MoSi_2$ grain [4]. It is interesting that the <331> faults are found either in the constrained matrix, e.g. near the interface, or in specimens deformed at or below the DBTT, i.e. before any significant plastic flow occurred.

SUMMARY AND CONCLUSIONS

Mo_5Si_3-reinforced $MoSi_2$ has been investigated for possible high temperature strengthening effects. Composites prepared by PM techniques show a small increase in hardness below 1000°C and no significant increase at higher temperatures. Higher hardnesses were observed in the AC composites at all temperatures relative to the monolithic $MoSi_2$ and the PM materials. Higher hardnesses can be achieved through refined AC microstructure as a result of Er additions to the eutectics. The strength increase can be related to the more uniformly distributed second phase and probably the stronger interface in the AC composites. The latter is supported by 1) transphase crack paths observed in the AC composites compared to intergranular and interphase cracking modes in the PM materials and 2) by observations of a low index orientation relationship and faceted interfacial structures observed in the AC composites.

Dislocation analysis discloses that <100> types of dislocation are present in the as-fabricated composites. Ordinary dislocations (<100>, <110> and 1/2<111>) dominate the high temperature deformation in compression of both $MoSi_2$ and the composites. <331> types of dislocation were only seen near the interface in the AC materials and in monolithic $MoSi_2$ deformed below its DBTT.

Table I Dislocation Structure of $MoSi_2$ and $MoSi_2$-Mo_5Si_3 Composites

Materials	Conditions	Burgers Vector
$MoSi_2$	Compression tested at 1300°C	<100>
		<110>
		1/2<111>
	tested at 900°C	1/2<331>
PM Composites (15 and 30 vol %)	As-Fabricated	<100>
	Compression tested at 1300°C	<100>
		<110>
		1/2<111>
AC eutectic composites	As-Fabricated	<100>
		1/2<331>
		<331>
	Compression tested at 1300°C	<100>
		<110>
		1/2<111>

REFERENCES

1. S.M. Tuominen, J. Less-Common Met. 81, 249 (1981).
2. R.M. Aikin, Jr., Ceram. Eng. Sci. Proc. 12, 1643 (1991).
3. M.F. Ashby, F.J. Blunt and M. Bannister, Acta metall. 37, 1847 (1989).
4. H. Kung, H. Chang and R. Gibala in Interfacial Structure and Properties of Interfaces in Materials, edited by W.A.T. Clark, U. Dahman, and C.L. Briant, Mater. Res. Soc. Proc. 238, Pittsburgh, PA 1992, p. 599.
5. R. Gibala, A.K. Ghosh, D.C. Van Aken, D.J. Srolovitz, A. Basu, H. Chang, D.P. Mason and W. Yang, Proceedings, First High Temperature Structural Silicides Workshop, A.K. Vasudevan and J.J. Petrovic, eds., Mater. Sci. and Eng. A155 (1992).
6. R.M. Aikin, Jr., Scripta metall. mater. 26, 1025 (1992).
7. W.B. Li, J.L. Henshall, R.M. Hooper and K.E. Easterling, Acta metall. mater. 39, 3099 (1991).
8. A.K. Ghosh, A. Basu and H. Kung, this meeting.
9. O. Unal, J.J. Petrovic, D.H. Carter and T.E. Mitchell, J. Amer. Ceram. Soc. 73, 1753 (1990).
10. Y. Umakoshi, T. Sakagami, T. Hirano and T. Yamane, Acta metall. mater. 38, 909 (1990).

ACKNOWLEDGEMENTS This research is supported by the Air Force Office of Scientific Research (AFOSR) under the AFOSR URI Program, Grant No. DoD-G-AFOSR-90-0141, Dr. A.H. Rosenstein, Program Manager.

EFFECT OF DUCTILE PHASE REINFORCEMENT MORPHOLOGY ON TOUGHENING OF $MoSi_2$

D.E. Alman and N.S. Stoloff
Materials Engineering Department, Rensselaer Polytechnic Institute, Troy NY 12180

ABSTRACT

Niobium was added to $MoSi_2$ in the form of particles, random short fibers and continuous aligned fibers. It was found that the morphology of Nb played a role in the toughening that occurred (as measured by the area under load displacement curves from room temperature three point bend tests and the examination of fracture surfaces). The Nb particles did not toughen $MoSi_2$. The random short fibers appeared to toughen $MoSi_2$ via crack deflection along the fiber matrix interface. Aligned fibers imparted the greatest toughness improvements, as toughening resulted from fiber deformation. However, larger diameter fibers displayed a greater ability to toughen $MoSi_2$ than smaller diameter fibers. This was attributed to the constraint resulting from the interfacial layer between the $MoSi_2$ matrix and the Nb fiber. Maximum toughness occurs when the fiber is able to separate from the matrix and freely deform.

INTRODUCTION

The attractive combination of low density (6.25 g/cm^3), high melting temperature (2030°C) and excellent resistance to oxidation (up to 1700°C) makes the compound $MoSi_2$ very attractive as a potential elevated temperature structural material. Recent research employing $MoSi_2$ as a matrix material for composites has resulted in improving the inadequate creep resistance and poor low temperature fracture toughness of monolithic $MoSi_2$. Petrovic and co-workers [1] have reinforced $MoSi_2$ with SiC whiskers. The SiC whiskers markedly improved the mechanical behavior of $MoSi_2$ at elevated temperatures. More importantly, the room temperature fracture resistance of $MoSi_2$ has been improved by the addition of Nb fibers, first reported by Fitzer and Remmele [2] in 1985. Room temperature fracture toughness for $MoSi_2$ with Nb fibers has been reported to be 12 MPa•m$^{1/2}$, compared to 3.3 MPa•m$^{1/2}$ for monolithic $MoSi_2$ [3]. These successes have generated enormous research activity on $MoSi_2$ and other silicides over the past few years, as evidenced by a recent workshop devoted solely to structural silicides [4].

Other researchers have published results on $MoSi_2$/Nb foil composites tested in tension in situ in the scanning electron microscope [3,5]. In these studies, the Nb foils would deform during testing. However, when powder-consolidated composites with Nb filaments [3] or particles [5] were tested in bending at room temperature, the Nb reinforcement showed limited ductility. The explanation for the behavior of Nb was a combination of both embrittlement of the Nb by oxygen and the stress state around the Nb reinforcement [5].

The goal of the present effort was to systematically study the role of Nb reinforcement morphology on the mechanical behavior of $MoSi_2$/Nb composites. Thus, $MoSi_2$ was fabricated reinforced with Nb particles, random Nb fibers, and aligned Nb fibers with two different diameters.

EXPERIMENTAL PROCEDURE

The characteristics of the starting powders, particles and fibers are listed in

Table I. The powders and particles were sized using a laser light scattering technique (microtrac) and shapes determined by viewing the powders or particles with a scanning electron microscope. Short fibers were produced by manually chopping the fibers to roughly 10 mm in length.

Particulate and random short fiber composites were produced by incorporating the appropriate amount, twenty volume percent, of Nb with $MoSi_2$ powder with the aid of a turbula mixer (mixing was performed for 60 minutes). Continuously aligned composites were produced by a combination hand layup-infiltration technique described in detail in a previous publication [6]. Briefly, a Cold Isostatic Press (CIP) mold bag was filled to half its capacity with $MoSi_2$ powder. The fibers were layed up within the CIP mold, after which, the CIP mold was filled to capacity with $MoSi_2$ powder infiltrating the fibers. The amounts of powder and fiber were determimed prior to layup and infiltration so to produce a composite with 20 v% fibers.

The mixtures were then CIPed at a pressure of 241 MPa. Cylindrical CIP bars were produced, approximately 25 mm in diameter and 65 mm long. CIP mold bags were lined with Nb foil. The Nb foil had two purposes, first to prevent fragmentation of the CIP bar, since $MoSi_2$ powder tend not to press well. Second to prevent a reaction with the Hot Isostatic Press (HIP) cans and the $MoSi_2$ powder during consolidation. Specimens were vacuum degassed at 600°C for 10 hours and then vacuum encapsulated in Ti HIP cans. Consolidation occurred by HIP at parameters 1350°C-172 MPa.

Smooth bar bend specimens (cylindrical, 6.35 mm in dia. by 38 mm long) were produced by electrodischarge machining (EDM). All bend specimens were containerless re-HIPed after EDM at 1350°C-172 MPa for 3 hours after which they were mechanically polished though 0.3μm Al_2O_3 powder. Room temperature three point bend tests were performed at a crosshead velocity of 0.127 mm/sec. The distance between the lower supports of the bend jig was 25.4 mm. Three tests were performed for each composite condition.

TABLE I
MATERIALS

POWDER	SOURCE/DESIGNATION	SHAPE	SIZE
$MoSi_2$	H.C. Stark Grade C Germany	Angular	2μm
Nb	Teledyne -80 Mesh	Angular	108μm
FIBER	**SOURCE**	**DIAMETER**	
Nb	Teledyne	800 μm	
Nb	NRC	400 μm	

RESULTS AND DISCUSSION

Typical microstructures of the consolidated composites are shown in Fig. 1. Representative results of the bend tests are shown in Fig. 2. Note, that no toughening was observed from the Nb particles. The maximum toughening (as measured by area under the load displacement curves) occurred with the continuously aligned, large diameter Nb fibers. Examination of fracture surfaces reveals that the Nb particles fractured by cleavage and did not deform in the $MoSi_2$ during testing. The graceful failure exhibited by the random short fibrous composite resulted from extensive crack deflection by decohesion at the Nb/$MoSi_2$ interface (Fig. 3a) for fibers that were off axis and for fibers which perchance happened to be on-axis (aligned perpendicular to the motion of bending) by fiber fracture. For the small diameter continuous aligned specimen the toughening mechanism appears to be fiber failure, with extra energy required to either fracture or deform the fiber (Fig 3b). However, note from Fig. 3c that both fiber failure and considerable separation between the fiber and the matrix occurred with the large diameter fibers.

Microprobe traces across the interface between $MoSi_2$ and Nb indicate that the size and chemical nature (mixed silicides) of the reaction layer were identical in both the large and small diameter fiber composites. Oxygen traces across the interface reveal similar oxygen contents in both fibers and even a lower oxygen content in the Nb particles, implying that the Nb is not being embrittled by oxygen during consolidation. Therefore, the differences in the observed mechanical behavior of the composites cannot be attributed to differences in the matrix/reinforcement interfaces that develop.

The results indicate that maximum toughening will occur when fiber deformation is accompanied by fiber matrix separation (debonding), as occurred with the large diameter continuous aligned Nb fibers. Similar observations have

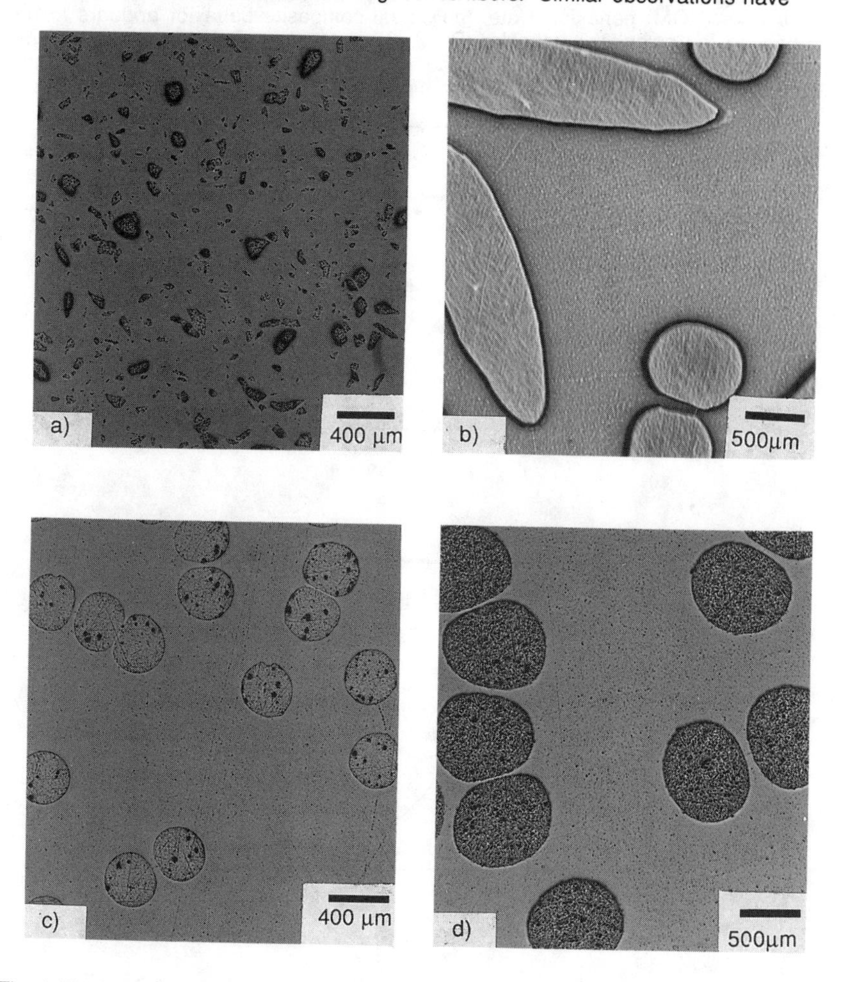

Fig. 1 Typical microstructures of $MoSi_2$/20v% Nb a) Nb particles b) random Nb fibers c) continuous aligned 400µm dia. Nb fibers d) continuous aligned 800µm dia. Nb fibers.

been noted for other brittle matrices reinforced with refractory fibers (Si_3N_4/Ta [7] and Al_3Nb/Nb[8]). For fiber matrix separation to occur a critical fiber diameter must be exceeded so that the fibers may deform. This critical diameter results from a balance between the interfacial strength, which impedes fiber-matrix debonding and prevents the fiber from deforming, and the load carrying capacity of the fiber. The smaller diameter fiber did not exceed this diameter and thus failed without debonding. Notice from the fracture surface for the small diameter aligned fibrous composites (Fig. 3b), that some fibers appeared to have pulled out from the matrix. These fibers tended to be in areas of high fiber agglomeration. As the fibers agglomerate together, they act as a fiber of effectively larger diameter, and thus are able to debond.

Fig. 4 shows a comparison of actual composite behavior to the calculated rule of mixtures (ROM) behavior. Note, the actual composite behavior appears to be superior to the calculated behavior. ROM calculations assume that once a crack propagates through the matrix all load is transfered to the fiber; there is no contribution of force necessary to debond the fiber from the matrix. The force

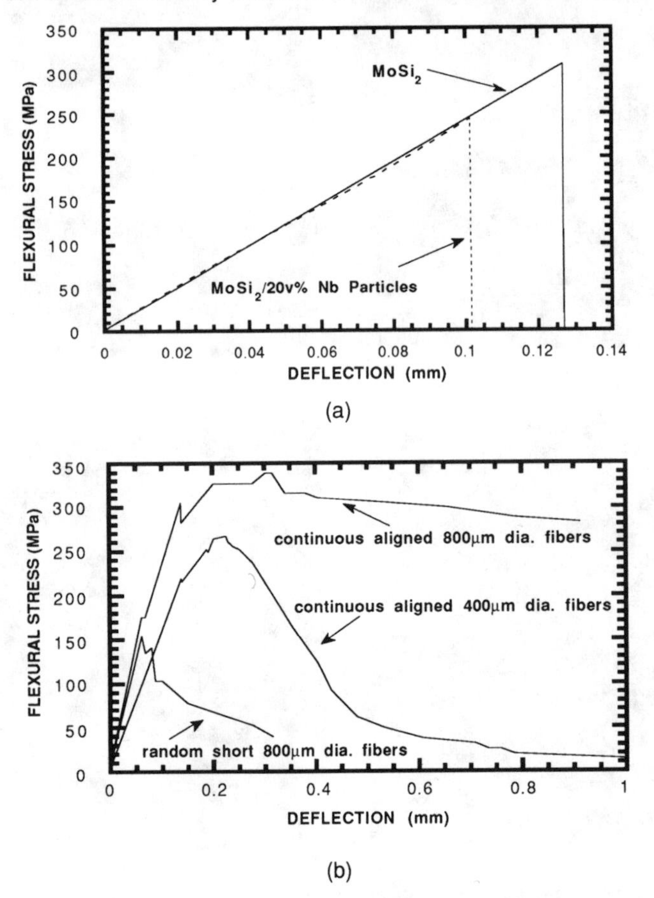

Fig. 2 Results of room temperature bend tests on $MoSi_2$/20v% Nb composites a) monolithic $MoSi_2$ and $MoSi_2$/Nb particles b) $MoSi_2$ /Nb fibers.

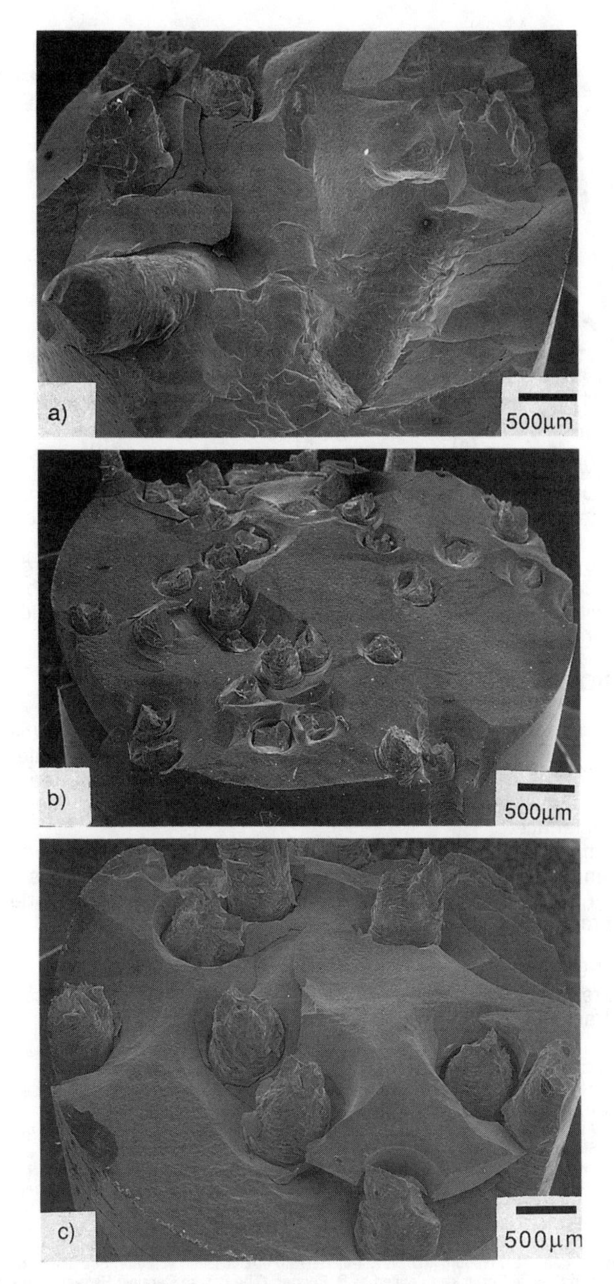

Fig. 3 Fracture surfaces of $MoSi_2$/20v% Nb tested in bending at room temperature a) random Nb fibers b) Nb continuous aligned 400μm dia. fibers c) Nb continuous aligned 800μm dia. fibers.

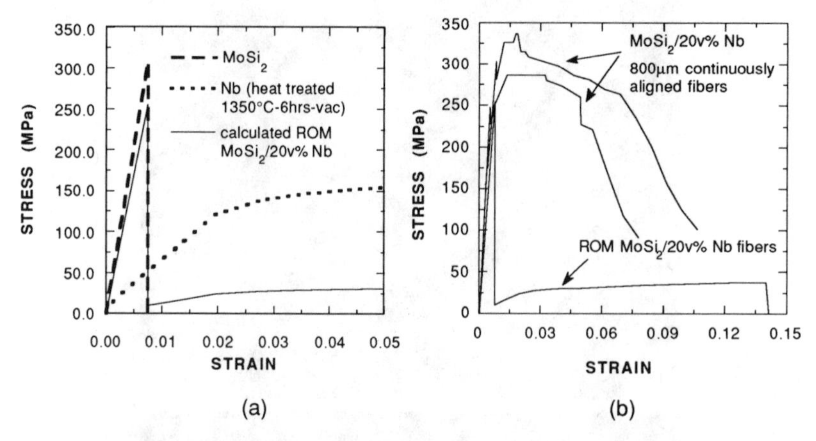

Fig. 4 Comparision of calculated rule of mixtures (ROM) to actual composite behavior. (a) ROM calculation (b) comparison to composite behavior

necessary to pull the fiber from the matrix results in an added energy required to fracture the composite. This suggests that any coating applied to Nb fibers to prevent a reaction with the matrix should be designed to provide some chemical bonding to both the fiber and the matrix.

CONCLUSIONS

The morphology of the Nb reinforcement plays a large role in the toughening mechanisms that occured in $MoSi_2$/Nb composites. The large diameter aligned Nb fibrous composites had the highest toughness (as measured by area under a load displacement curve) due to a combination of fiber deformation and separation at the fiber-matrix interface.

Maximum toughness will occur in brittle matrices reinforced with ductile phases when the reinforcement phase is able to separate from the matrix and freely deform. The behavior of these types of composites is very much dependent on interfacial properties.

ACKNOWLEDGEMENTS

This research was supported by a DARPA/ONR University Research Initiative under Contract N00014-86-K-0770.

REFERENCES

1. D.H. Carter, W.S. Gibbs and J.J. Petrovic, in Proc. 3rd. Intl. Symp. on Ceramic Matls & Components for Engines (the American Ceramic Society, Westerville, OH, 1989) p. 977.
2. E. Fitzer and W. Remmele, in Proc. 5th Intl. Conf. on Composite Materials, ICCM-V, edited by W.C. Harrigan, J.Strife and A.K. Dhingra (TMS, Warrendale, PA, 1985) p. 515.
3. L. Xiao, Y.S. Kim, R. Abbaschian and R.J. Hecht, Mater. Sci. and Engr. A144, 277 (1991).
4. High Temperature Structural Silicides WorkShop, Gaithersburg MD, Nov. 4-6, 1991 (proceedings to be published in Mater. Sci. and Engr., 1992).
5. T.C. Lu, A.G. Evans, R.J. Hecht and R. Mehrabian, Acta Metall. 39, 1853 (1991).
6. D.E. Alman. K.G. Shaw, N.S. Stoloff and K. Rajan, presented at the High Temperature Structural Silicides WorkShop, Gaithersburg MD, Nov. 4-6, 1991 (to be published in Mater. Sci. and Engr., 1992).
7. J.J. Brennan, in Special Ceramics 6, edited by P. Popper (The Brittish Ceramic Research Association, Stoke-On-Trent, England, 1975) p. 123.
8. L. Lu, Y.S. Kim, A.B. Gokhale and R. Abbaschian, in Intermetallic Matrix Composites, edited by D.L. Anton, P.L. Martin, D.B. Miracle and R. McMeeking (Mater. Res. Soc. Proc. 194, Pittsburgh, PA, 1990) p. 79.

PLASTICITY ENHANCEMENT OF $MoSi_2$ AT ELEVATED TEMPERATURES THROUGH THE ADDITION OF TiC

H. CHANG, H. KUNG AND R. GIBALA
Department of Materials Science and Engineering
University of Michigan, Ann Arbor, MI 48109

ABSTRACT

Monolithic $MoSi_2$ and $MoSi_2$-TiC particulate composites with 10 vol % and 15 vol % TiC were tested in compression between 950°C and 1200°C. The $MoSi_2$-TiC composites can be deformed plastically at lower temperatures than $MoSi_2$ can before brittle fracture occurs. The composites exhibit much lower strain hardening rates and attain zero strain hardening rate at much lower strains than monolithic $MoSi_2$. The differences between the composites and monolithic $MoSi_2$ in plasticity and in strain hardening behavior is attributed to efficient dislocation generation into the matrix from sources at the $MoSi_2$-TiC interfaces.

INTRODUCTION

$MoSi_2$ exhibits limited plasticity at temperatures from 900°C to 1300°C such that various authors have ascribed the brittle to ductile transition temperature (BDTT) to be in this temperature range [1]. Between about 900°C and 1300°C, $MoSi_2$ exhibits limited deformation capability. Above about 1300°C, substantial plastic flow in compression has been observed [2]. In this paper, we investigate how the plasticity of $MoSi_2$ can be enhanced in this transition region of limited plasticity by the addition of a more deformable phase.

TiC is a candidate phase for enhancing the plasticity of $MoSi_2$ in this temperature range. It deforms to large strains in compression around 900°C and above [3,4]. This plasticity is attributed to the change from slip on NaCl slip systems to slip on fcc slip systems at its BDTT of 600°C to 800°C. Previous experimental work [5,6] and thermodynamic calculations [5] have shown that TiC and $MoSi_2$ are chemically stable. No high temperature interfacial reaction products have been observed between these two compounds. TiC has a lower density than $MoSi_2$ (4.9 g cm^{-3} vs 6.3 g cm^{-3}) and a higher melting point (about 3140°C vs 2030°C). The coefficient of thermal expansion of TiC at 1000°C is 7.7 x 10^{-6}°C^{-1} and of $MoSi_2$ at 1000°C is 8.5 x 10^{-6}°C^{-1} [7]. In addition, TiC exists in a wide range of substoichiometries from the maximum stoichiometry $TiC_{0.97}$. Substoichiometric TiC_x decreases in strength with decreasing carbon content x [4]. This variability in strength may be a useful parameter in optimizing any plasticity enhancement effect of the carbide.

EXPERIMENTAL PROCEDURE

Monolithic $MoSi_2$ and $MoSi_2$-TiC composites with 10 vol % TiC and 15 vol % TiC have been compression tested between 950°C and 1200°C. These three materials were prepared by hot pressing commercially available powders at 1700°C for one hour at 29.4 MPa. The $MoSi_2$ is 5 to 6 micron size powder from CERAC, Inc.; the TiC is 2.5 to 4 micron size powder from Johnson Matthey, Inc.

with a nominal stoichiometry of $TiC_{0.95}$. The monolithic $MoSi_2$ and 10 vol % TiC composite were in addition hot isostatically pressed (HIPed) at 1800°C for 1.5 hours at 200 MPa.

Constant crosshead speed tests were performed at an initial strain rate of 10^{-4} s^{-1}. For certain temperatures and materials, the tests were repeated once or twice. The 6 mm x 3 mm x 3 mm specimens were tested in an Instron 4507 with a furnace attachment in an argon atmosphere. Transmission electron microscopy was performed on a JEOL 2000FX microscope operating at 200 kV. Foils were prepared by hand grinding, dimpling, and ion milling.

The densities of hot pressed $MoSi_2$ and $MoSi_2$-10 vol % TiC are approximately 95%. Hot pressed $MoSi_2$-15 vol % TiC has a lower density of 90% to 91%. The densities of HIPed $MoSi_2$ and $MoSi_2$-10 vol % TiC are both approximately 98%.

The average grain size of both the hot pressed and the HIPed $MoSi_2$ is 28 μm. Average grain sizes of the matrix in the hot pressed 10 vol % and 15 vol % TiC composites are 21 μm and 15 μm, respectively, while in the HIPed 10 vol % composite the average grain size is 25 μm.

Carbide size distributions in both composites range from 5 μm to 45 μm, with the average size in the HIPed 10 vol % composite at 9 μm, the hot pressed 10 vol % composite at 10 μm, and the hot pressed 15 vol % composite at 11 μm. All materials were ball milled in an attempt to uniformly distribute the carbide. This uniform distribution was achieved, but particle agglomeration resulted in the size distribution. Both dry blending of the powders and blending in a solvent slurry were attempted to try to obtain a uniform size distribution, but the results from the two methods are not substantially different from each other. Figure 1 shows the microstructure of HIPed $MoSi_2$-10 vol % TiC. Interfacial reaction products were not observed.

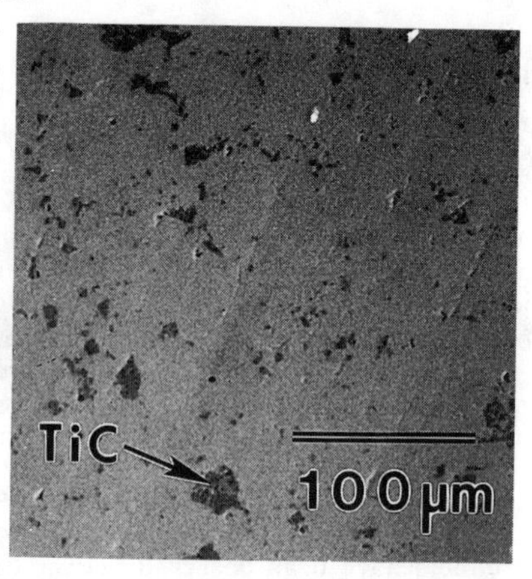

Figure 1. Microstructure of HIPed $MoSi_2$-10 vol% TiC. The TiC is the dark phase.

RESULTS AND DISCUSSION

Selected stress-strain curves for the three materials are shown in Figure 2. In the materials which have been only hot pressed, the flow stresses of the composites are lower than those of monolithic $MoSi_2$ (Figure 2a). The addition of 10 vol % TiC strengthens $MoSi_2$ over the temperature range of 1050°C to 1200°C in the HIPed materials (Figure 2b). Because the hot pressed materials have more porosity than the HIPed materials, the strengthening in the HIPed composite is more likely representative of the true effect of TiC addition. These results show that the strengths of these materials are sensitive to the defect populations present. The strengthening of the composite is consistent with previous studies on the compression of monolithic TiC, which show strengths of

$TiC_{0.97}$ [3] and $TiC_{0.93}$ [4] from 1000 °C to 1200 °C comparable to or above that of the currently known strengths of $MoSi_2$.

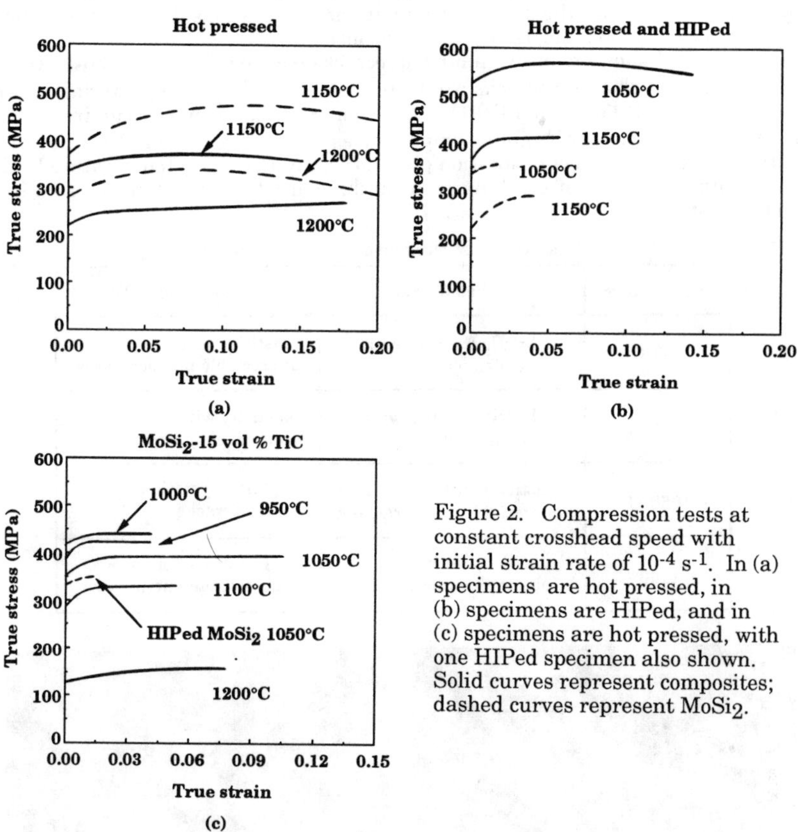

Figure 2. Compression tests at constant crosshead speed with initial strain rate of 10^{-4} s^{-1}. In (a) specimens are hot pressed, in (b) specimens are HIPed, and in (c) specimens are hot pressed, with one HIPed specimen also shown. Solid curves represent composites; dashed curves represent $MoSi_2$.

Table I summarizes qualitatively the temperature dependence of the plasticity and the presence of microcracking in $MoSi_2$ and $MoSi_2$-15 vol % TiC. $MoSi_2$ exhibits no plasticity in compression at 1000°C or at lower temperatures. At 1050°C, $MoSi_2$ exhibits brittle fracture either with no plasticity, or with limited plasticity and appreciable microcracking. At 1100°C, $MoSi_2$ exhibits appreciable plasticity with no microcracking observable at a magnification of 400X. Figure 3 shows the dislocation substructure of $MoSi_2$ deformed at 1100°C to 16% plastic strain. The high dislocation density demonstrates the dominant contribution of dislocation plasticity in $MoSi_2$ deformed at this temperature.

The $MoSi_2$-15 vol % TiC composite examined in this investigation exhibits some plasticity at 1000°C prior to brittle fracture, unlike the behavior of $MoSi_2$. The composite exhibits plasticity even at 950°C. Microcracking makes the major contribution to the deformation of the composite at 950°C and 1000°C, but an increased dislocation plasticity leads to more deformation before brittle fracture at these temperatures in comparison to monolithic $MoSi_2$. TiC particles act as dislocation sources with the result that the relative contribution of cracking to deformation is reduced by the contribution from dislocations generated at the TiC interface. Monolithic $MoSi_2$ is a dislocation deficient material at 1050°C and below, so that without the presence of additional dislocations during deformation microcracking dictates the deformation response.

Figure 4 shows transgranular microcracks (see arrows) in $MoSi_2$-15 vol % TiC at 1050°C. The microcracks are transgranular in both this material and in $MoSi_2$ tested at 950°C to 1050°C. Some microcracking may occur in all the materials even at the higher temperatures of 1100°C to 1200°C, but the increasing contribution of dislocation plasticity with increasing temperature leads to a diminishing role of cracking as a mechanism of deformation for small to moderate strains.

Table I. Observed plasticity and occurrence of microcracking.

Temperature	HIPed $MoSi_2$	$MoSi_2$-15 vol % TiC
950°C	brittle fracture with no plasticity	plasticity with appreciable microcracking
1000°C	brittle fracture with no plasticity	plasticity with appreciable microcracking
1050°C	plasticity with appreciable microcracking	plasticity with few microcracks
≥ 1100°C	plasticity with no observable microcracks	plasticity with no observable microcracks

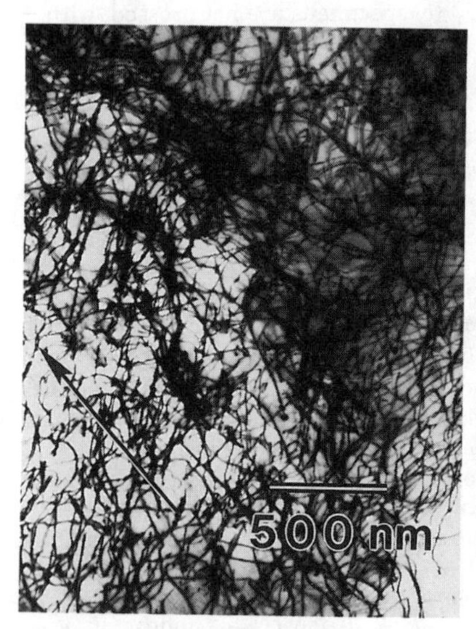

Figure 3. Dislocation substructure in $MoSi_2$ deformed 16% at 1100°C. B = [331] g = $\bar{1}10$.

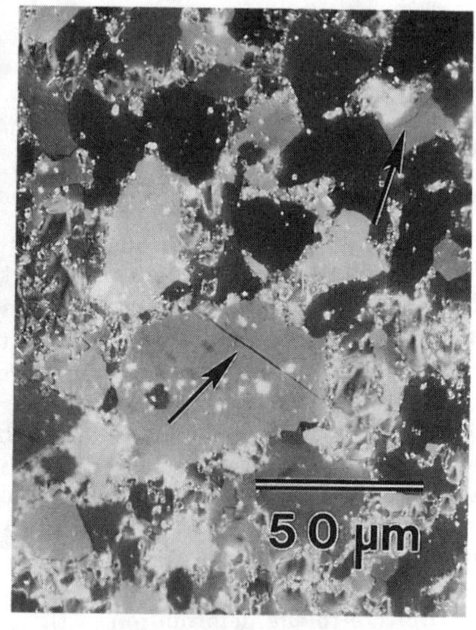

Figure 4. Microcracking in $MoSi_2$-15 vol % TiC deformed 11% at 1050°C. Compression direction is vertical.

The steady state flow stresses (i.e., the stresses at which strain hardening becomes zero) occur at substantially different strains in $MoSi_2$ and in the composites. Both composites reach steady state flow stresses much more rapidly than $MoSi_2$ does (Figure 2). Steady state flow commences at 2% to 3% strain and can continue to much larger strains in the composites. Apparently efficient dislocation multiplication processes begin at small strains in the composites and are maintained to large deformations.

In contrast, $MoSi_2$ exhibits strain hardening over larger initial deformations and attains steady state flow only after 6% to 14% strain. The dislocation multiplication processes present in the composites either do not operate in $MoSi_2$ until greater amounts of deformation have occurred, or a different multiplication process occurs in the absence of TiC particles. The dislocation substructure of $MoSi_2$ shown in Figure 3 consists of a high density of extremely tangled dislocations, which is consistent with the level of strain hardening in this material.

TEM observations of the undeformed composite show a very low density of dislocations. Figure 5 shows the dislocation substructure in the matrix around a TiC particle in $MoSi_2$-10 vol % TiC deformed 5% at 1150°C. Dislocations are also observed within the TiC and will be reported in a later publication. The matrix dislocations in the vicinity of the interface with TiC have been characterized, using **g•b** and trace analyses, to have <100> {011} and <100> {013} slip systems. These slip directions correspond to the shortest lattice vectors in $MoSi_2$, and these slip planes are among the high atomic density planes. Additional slip systems in $MoSi_2$ have been reported [2,9] but have not been observed in the present study. The dislocations are all mixed in nature, with either large edge-oriented or screw-oriented components. However, a statistical analysis of the occurrence of various types of dislocations has not yet been performed.

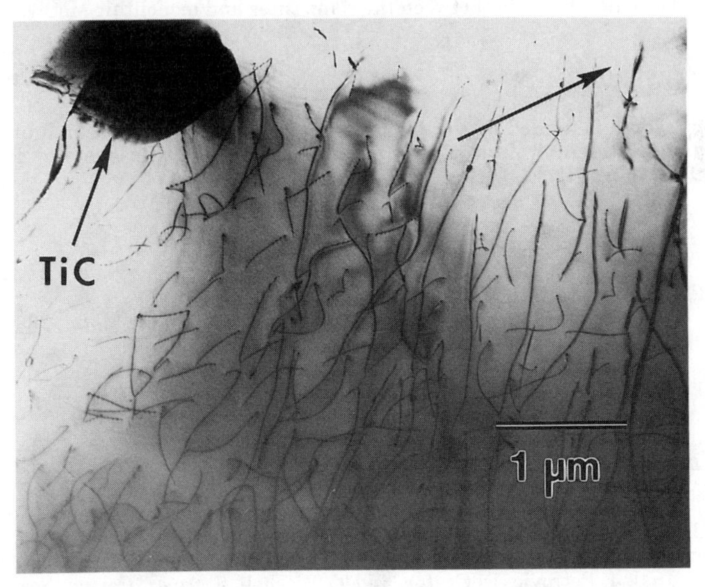

Figure 5. Dislocation substructure in $MoSi_2$-10 vol % TiC deformed 5% at 1150°C. B = [331] g = $\bar{1}10$.

Figure 6 shows a decremental step strain rate test on HIPed $MoSi_2$-10 vol % TiC at 1150°C. The stress exponent is approximately 3 at the strain rate of 10^{-4} s^{-1} used in the compression tests of this study. Previous step strain rate tests in compression on $MoSi_2$ have shown, for this strain rate, approximate stress exponents of 5 at 1200°C and 8 at 1100°C [8]. These results suggest an operating mechanism of dislocation creep, with both glide and climb processes occurring, for these materials at these temperatures and strain rates.

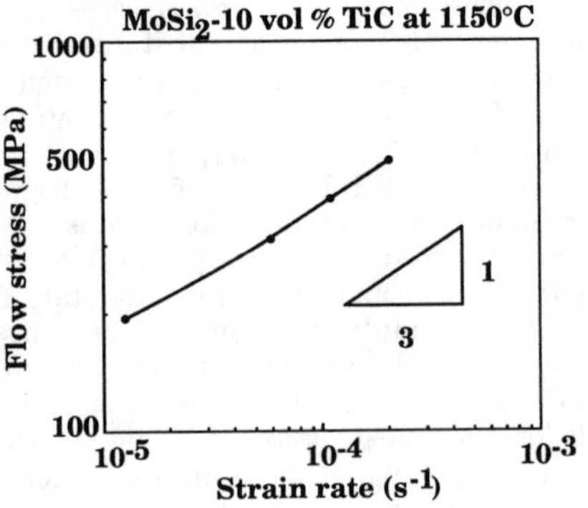

Figure 6. Decremental step strain rate test on $MoSi_2$-10 vol% TiC at 1150°C.

SUMMARY

The addition of 10 vol % TiC strengthens $MoSi_2$ in compression between 1050°C and 1200°C. The strengths of the materials are sensitive to the flaw populations present. $MoSi_2$-TiC composites can be deformed plastically at lower temperatures than $MoSi_2$ can before brittle fracture occurs. The composites reach steady state flow behavior, characterized by a zero strain hardening rate, at much lower strains than in $MoSi_2$. Dislocations generated in the matrix from sources at the TiC interface are likely responsible for the difference in plasticity and in steady state flow behavior between the composites and monolithic $MoSi_2$.

ACKNOWLEDGEMENT

This research is funded by the Air Force Office of Scientific Research under the University Research Initiative Program, Dr. Alan H. Rosenstein, Program Director. Grant No. DOD-G-AFOSR-90-0141.

REFERENCES

1. R. Aikin, Jr., Scripta Metall. et Mater., 26 (7), 1025 (1992).
2. Y. Umakoshi, T. Sakagami, T. Hirano, and T. Yamane, Acta Metall., 38 (6), 909 (1990).
3. G. Hollox and R. Smallman, J. Appl. Phys., 37 (2), 818 (1966).
4. D. Miracle and H. Lipsitt, J. Am. Ceram. Soc., 66 (8), 592 (1983).
5. P. Meschter and D. Schwartz, J. of Metals, 41 (11), 52 (1989).
6. J. Yang and S. Jeng, J. Mater. Res., 6 (3), 505 (1991).
7. Smithells Metals Reference Book, 6th ed. (Butterworths Publishers, London, 1983), p. 27-10.
8. A. Basu and A. Ghosh, in Proceedings, International Symposium on Advanced Metal Matrix Composites for Elevated Temperatures, edited by M. Gungor, E. Lavernia, and S. Fishman (ASM International, Materials Park, OH, 1991), p. 1.
9. O. Unal, J. Petrovic, D. Carter, and T. Mitchell, J. Amer. Ceram. Soc., 73 (6), 1752 (1990).

THE EFFECT OF SECOND PHASE PROPERTIES ON THE COMPRESSION CREEP BEHAVIOR OF $MoSi_2$ COMPOSITES

A.K. GHOSH, A. BASU and H. KUNG
Department of Materials Science & Engineering, University of Michigan, Ann Arbor, MI 48109-2136

ABSTRACT

In an effort to enhance the toughness and creep strength of $MoSi_2$, the role of various metallic and ceramic reinforcements is being examined. In this work, the effects of an oxide, a carbide and a nitride reinforcement on the compression creep behavior of $MoSi_2$ are explored. Variations in the deformability of reinforcements and their relative strength and flaw population appear to influence the creep strength of the composites. Refinements in grain size also improve crack tolerance of the composite during deformation at 1200°C.

INTRODUCTION

The silicides of refractory metals such as Mo, Nb and W have great potential as matrix materials for composites with service capabilities at temperatures greater than 1000°C. Of these materials, molybdenum disilicide exhibits excellent high temperature oxidation resistance, approaching that of SiC, because of the formation of a protective silica film.[1,2] While a high melting point of 2030°C makes it an attractive choice as a high-temperature material, $MoSi_2$ is extremely brittle at low temperatures. At ~1000°C $MoSi_2$ begins to yield and deforms by dislocation motion, similar to the behavior of metals.[3] At an even higher temperature (~1200°C) the yield strength of $MoSi_2$ drops further and its creep resistance is not extremely high.[4] Consequently, attempts are being made to improve the mechanical properties of $MoSi_2$ by synthesizing composites with hard ceramic phases as well as ductile metallic phases.[5-8] It is believed that by variations in the strength, deformability, and the coefficient of the thermal expansion (CTE) of particulate reinforcements, it would be possible to alter composite strength and interfacial stress. Consequently, the toughness of the composite may also be influenced. Additionally, reinforcements, when distributed in a fine scale through the microstructure, can change the grain size of the matrix material. The objective of this study was to examine the effect of reinforcement properties and the microstructural scale in $MoSi_2$ composites reinforced by oxide, nitride and carbide reinforcements, and their effects on creep behavior. This report describes progress to date, while a comprehensive view will be presented later.

Material Selection

SiC and Si_3N_4 particulate reinforcements were selected because of their high strength (or hardness) and compatibility with silicides. Since $MoSi_2$ is known to contain small amounts of SiO_2 impurities, a CaO reinforcement was selected partly because of its ability to clean the matrix of this impurity by compound formation with SiO_2, and secondly, because of its easier deformability. Table I lists the melting points, room temperature hardness and CTE of the matrix and the various reinforcement phases. In addition to the strength issue, the difference in CTE was believed to induce tensile vs. compressive residual hoop stresses around the particles of SiC (and Si_3N_4) and CaO, respectively. This was thought to be important for toughening of the composites.

Table I. Physical Properties and Sources of $MoSi_2$ and Various Ceramic Reinforcements

Material (Avg. Particle size)	Melting Point, °C	Hardness at room temp. GPa	Coefficient of thermal expansion x 10^{-6} °C^{-1}	Source of Material
$MoSi_2$ (4.9 μm)	2030	12.4	8.6	Cerac Inc. Milwaukee, WI
CaO (<3 μm)	2572	5.5	13.7	Fischer Sci. Fair lawn, NJ
$CaO.SiO_2$	1540		11.2	
Si_3N_4 (<2 μm)	1900 (decomposes)	15.7	3.7	Cerac Inc. Milwaukee, WI
SiC (<0.7 μm)	2500 (decomposes)	24.5	4.7	H.C. Stark Germany

EXPERIMENTAL WORK

Material Synthesis

$MoSi_2$ powder (-325 mesh) from Cerac was used for consolidating monolithic $MoSi_2$. Particulate reinforced composites were prepared by mixing $MoSi_2$ powder with powders of CaO, SiC, and Si_3N_4 of nominally the same powder size. For the composites, a slurry milling approach was used to develop a homogeneous intimate mixture of fine scale constituents. In the case of CaO reinforcement, dry milling was used which led to coarser CaO particles and larger interparticle spacing. The powder mixtures were hot pressed in a graphite die at 1700°C for 2 hours under 28 MPa pressure. Hot pressed billets were cut and a portion HIP'ed at 1700°C for 90 min under 200 MPa pressure.

Characterization of Microstructure

The optical micrographs of $MoSi_2$ and three particulate composites are shown in Fig. 1. Cross-polarized light clearly shows the grain boundaries, and the microstructure is found to become more refined as the particulate volume fraction is increased. The monolithic material shows small spherical SiO_2 particles at grain boundaries and a small amount of porosity, the total volume of both was approximately 2%. CaO was found to clean up the grain boundaries considerably; small spherical features were removed and grain coarsening was observed. CaO also produced a second phase which was more rounded in comparison to the jagged particles of Si_3N_4 and SiC. Table II lists the size of microstructural features determined via quantitative microscopy technique. Fine SiC dispersions produced the finest grain size of 3 μm, while the CaO addition provided a $MoSi_2$ grain size of 30 μm.

Fig. 2 shows transmission electron microscopy images of clean grain boundary structures in CaO and SiC containing composites. In general, more dislocations were found in both $MoSi_2$ and CaO phases, Figs. 2(a) and (b), than in the SiC and Si_3N_4 containing composites. It is clear that the CTE differences lead to greater interfacial stresses in the CaO bearing composite. In some areas the interface was found to have separated at room temperature. The CaO phase was found to possess a lamellar two phase structure with one of the phases being Ca_2SiO_5, which formed possibly by reaction with SiO_2 particles in the grain boundaries of the matrix. SiC interface with $MoSi_2$ seems to contain no other phase, as indicated by the high resolution microscopy photograph in Fig. 2(d). This interface is believed to be strong, however, SiC particles were polycrystalline, possibly with weak grain boundaries due to insufficient consolidation temperature.

On the other hand, the interface structure of $MoSi_2/Si_3N_4$, using high resolution transmission microscopy, indicates a continuous straight layer of amorphous material at the interface. The composition of this layer has not been identified, however, this is likely to be a silica or silicate layer.

Table II. Microstructural Features of Monolithic $MoSi_2$ and its Composites

Material	Avg. Matrix Grain Intercept, μm	Avg. Reinforcement particle intercept, μm	Average Interparticular Spacing, μm
Base Material $MoSi_2$	26	-	-
$MoSi_2$ + 15% CaO Dry Milled	30.	17	30.5
$MoSi_2$ + 15% Si_3N_4 Wet Milled	13.	5.6	10.3
$MoSi_2$ + 20% SiC Wet Milled	3	3.2	3.3

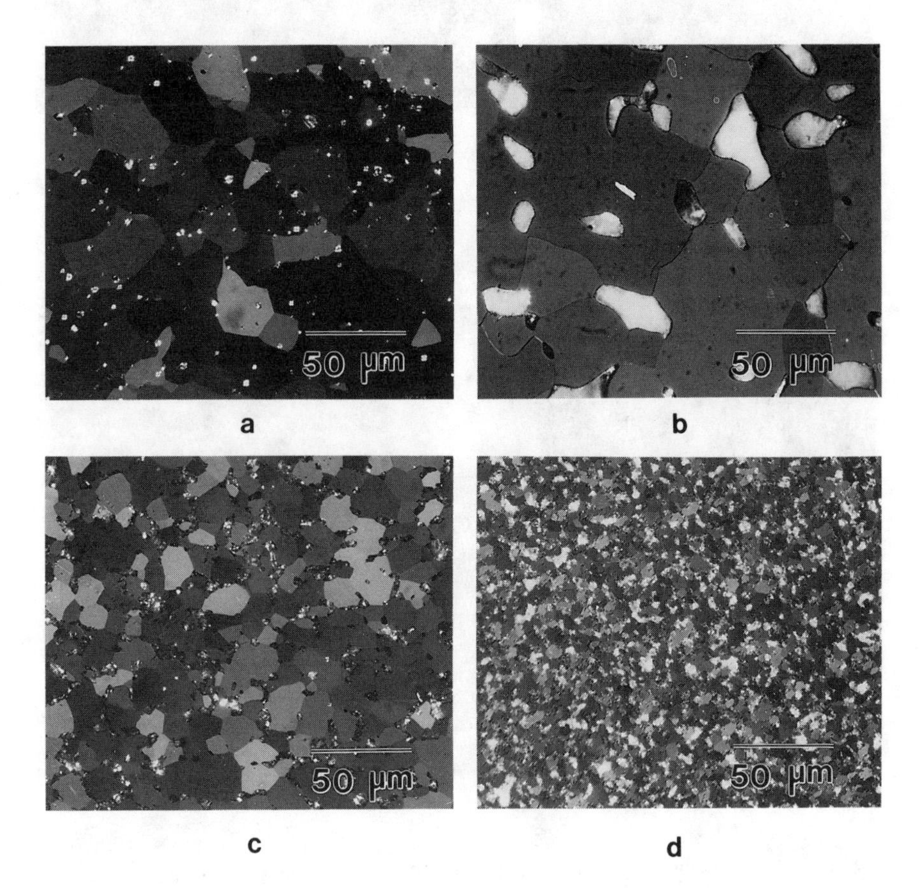

Fig. 1. Optical micrographs taken with cross-polarized light of (a) Monolithic $MoSi_2$, (b) $MoSi_2$/15%CaO, (c) $MoSi_2$/15% Si_3N_4 and (d) $MoSi_2$/20% SiC.

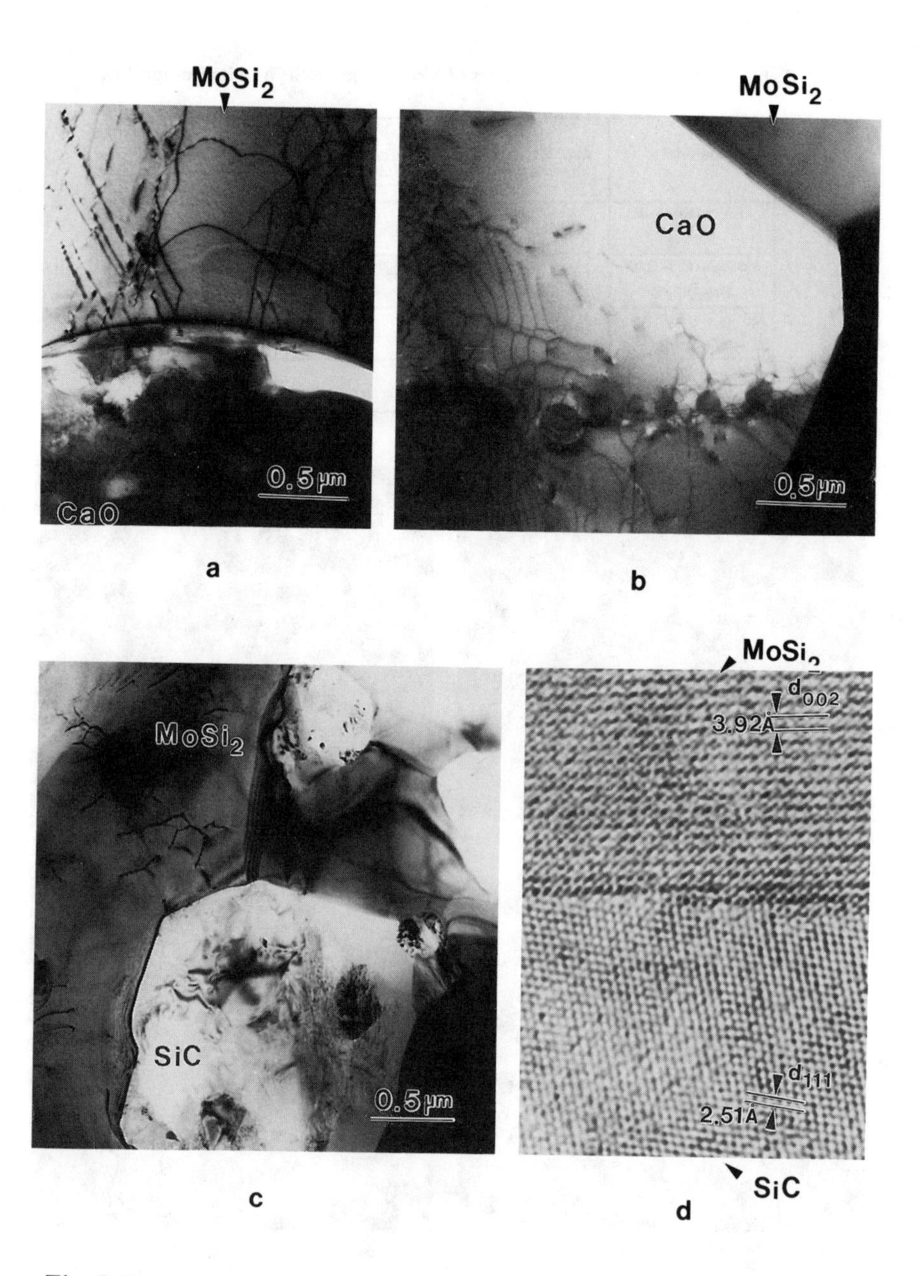

Fig. 2. Transmission electron micrographs of $MoSi_2$/CaO composite. (a) and (b) show dislocations in both phases and a region of interface debond. Transmission electron micrograph of $MoSi_2$/SiC composite (c) and high resolution photograph (d) of lattice structure near the $MoSi_2$/SiC interface showing excellent bonding.

Compression Tests

Rectangular samples, 3 mm x 3 mm x 6 mm, were cut from the billets and two types of compression tests were conducted in temperature ranges of 1200-1400°C by pressing the samples between SiC rigid platens in an argon environment. The tests were conducted in an Instron machine fitted with a Centorr furnace. The first set of tests were step strain rate tests in which the samples were first strained at a strain rate of $10^{-4}s^{-1}$ to a steady level of stress, followed by decremental step strain rate tests. Steady stress values were achieved at each strain rate level which were subsequently used to plot log (strain rate) vs. log (stress) plots. Constant stress creep tests were also performed by subjecting separate samples to controlled load levels. A constant stress on the creeping specimen was maintained by increasing this load with continued creep strain under the assumption of constant specimen volume. Crosshead displacement corrected for machine compliance was used to determine sample strain. Minimum creep rates were determined from the strain-time plots, and these were plotted as a function of stress. Typically, creep tests on composites led to an earlier onset of tertiary creep than that for monolithic $MoSi_2$, which was later found to be related to weakening due to microcracking.

Hot Hardness Tests

A limited amount of hot hardness tests have been conducted to date on the materials selected for study using a Nikon hot hardness test apparatus. The indenter was made of diamond and the test environment was vacuum. The indentation size was measured under the microscope to determine their hardness, which was later expressed in GPa as a measure of resistance to deformation.

RESULTS AND DISCUSSION

Figure 3 shows strain rate vs. stress plots for the various materials tested at 1200°C. (The results from the creep tests were in close agreement with step strain rate tests and therefore, only an average data range is shown.) Monolithic $MoSi_2$ and CaO containing composite, both of relatively coarse grain size (26-30 μm), exhibit the highest creep strength. The SiC containing composite, having the finest grain size (~3 μm), shows the lowest creep strength. The $MoSi_2/Si_3N_4$ composite with an intermediate grain size (12 μm) has intermediate strength level. The grain size dependence of creep strength follows a $\sigma \sim d^{\frac{1}{2}}$ relationship, which is not consistent with diffusional creep. All materials exhibited a stress exponent, n (as in $\dot{\varepsilon} \alpha \sigma^n$), of approximately 3, indicating climb-controlled dislocation creep.

Figure 4 shows intergranular microcracks developing in the monolithic $MoSi_2$ during compressive deformation. Thus, the grain boundary sliding process is believed to be operative but it is not accommodated by diffusional processes. Some transgranular cracking has also been seen, however, intergranular cracking is predominant and its extent increases with increasing strain. In the $MoSi_2/CaO$ composite the grains of both phases were found to have deformed to a great extent, and interfacial cracks, as shown in Fig. 2 at room temperature, did not seem to have any weakening effect.

In addition to climb-controlled creep, it is believed that microcracking provides an additional strain accommodating mechanism in the $MoSi_2$ composites. Figure 5 shows that SiC particles in $MoSi_2/SiC$ composite crack along their grain boundaries and eventually produce cracks at the interface with $MoSi_2$. In finer grain material, such cracking can be distributed in a fine scale throughout the material so that a significant amount of deformation is possible before cracks become visible. Similar intergranular cracking or flaw-driven cracking has been observed within the Si_3N_4 phase in the composite which contains this reinforcement. However, because the microstructural scale is coarser, the accommodating strains are smaller in the $MoSi_2/Si_3N_4$ composite. This increases the creep strength of the composite but failure becomes more catastrophic in this material. Thus, microcracking as an additional mechanism for accommodating deformation in conjunction with climb-controlled creep can yield the same stress exponent of 3, but lower creep strength results from the microcracking effect. Consequently, measurements of activation energy is meaningless for such materials.

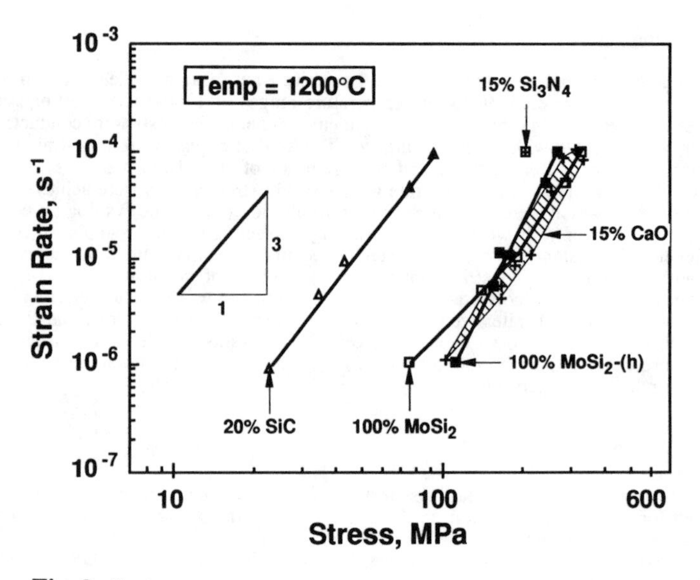

Fig. 3. Strain rate vs. stress plots for $MoSi_2$ and its composites at 1200°C.

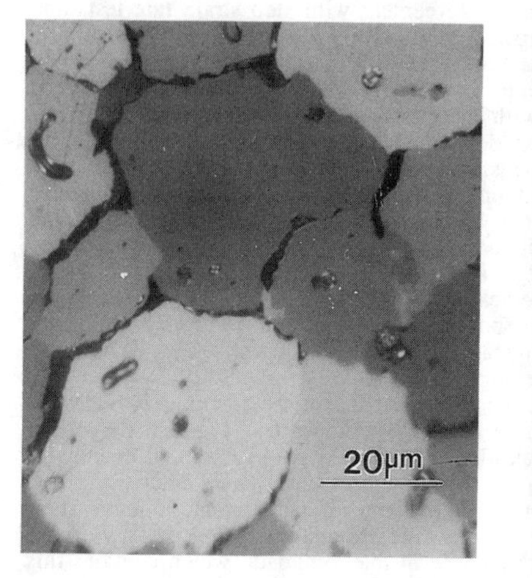

Fig. 4. Intergranular microcracks in monolithic $MoSi_2$ after compression creep at 1200°C.

Fig. 5. Microcracking of SiC particles in the $MoSi_2$/SiC composite.

Figure 6 shows hot hardness results indicating that the $MoSi_2/SiC$ composite has greater resisitance to deformation than the monolithic $MoSi_2$, except at temperatures as high as 1400°C. This apparent contradictory result appears to be due to a difference in stress state between pure compression and hardness tests. The state of hydrostatic compression underneath the indenter can suppress microcracks and allow the higher strength of SiC to become effective in composite strengthening.

Figure 7 is a comparison of present creep data with those of Sadananda et al.[9] and Wiederhorn et al.[10] Our study found that the monolithic $MoSi_2$ and $MoSi_2/CaO$ composites have creep strength close to those of $MoSi_2/SiC_w$ and (Mo, W)Si_2/SiC_w in their

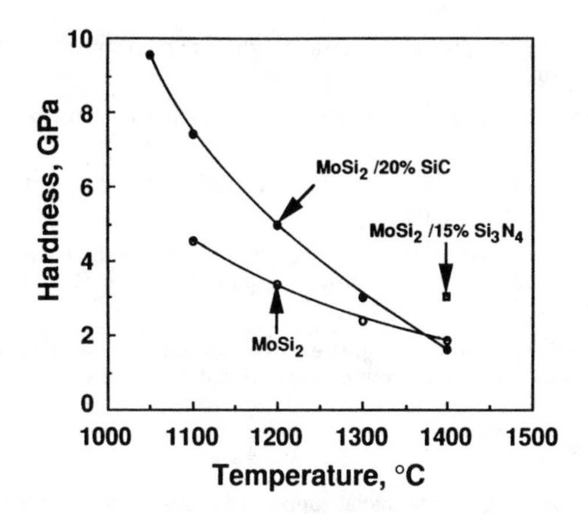

Fig. 6. Hot hardness data for $MoSi_2$ and $MoSi_2/SiC$ composite

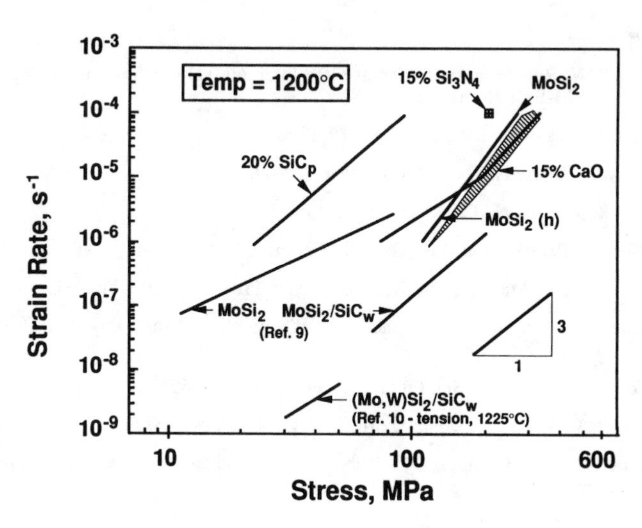

Fig. 7. Comparison of creep data for $MoSi_2$ and its composites with those of other investigators

work. However, the monolithic $MoSi_2$ from Sadananda is weaker than that obtained in the present study, which may be due to a higher amount of SiO_2 in their material. In any event, the strengthening effect from the SiC whiskers as reported in their work has not been duplicated here for particulate reinforcements because of the weak particle problem cited above. It is suggested that the key to creep strengthening these materials might lie in improving the distribution of hard particles (e.g., SiC), and avoiding polycrystalline particles, and producing clean grain boundaries (e.g., CaO). An approach for creep rupture resistance or toughness is to refine the grain size or use a deformable reinforcement phase which is reasonably strong (e.g., CaO or Ca-silicate).

CONCLUSIONS

- Composites of $MoSi_2$ with oxide, nitride and carbide reinforcements and a range of grain sizes have been synthesized by the P/M process.
- The stress exponents for creep of $MoSi_2$ and its composites have been found to vary between 2 and 4.
- Coarse grain structure and clean grain boundaries produced by a CaO addition can improve creep resistance of $MoSi_2$.
- The brittleness associated with ceramic reinforcements, such as SiC and Si_3N_4, can weaken $MoSi_2$ matrix composites unless particle distribution is improved and polycrystalline particles are avoided.
- Hot hardness of $MoSi_2$/SiC is greater than monolithic $MoSi_2$ due to reduced microcracking under hydrostatic compression present under the indenter.

ACKNOWLEDGEMENTS

The authors acknowledge the financial support of this work through the U.S. Air Force URI Grant No. DOD-G-AFOSR-90-0141 (Dr. A.H. Rosenstein, Grant Monitor).

REFERENCES

1. E. Fitzer and W. Remmele, presented at Fifth International Conference on Composite Materials. (ICCM-5, 1985) p. 515.
2. Berkowitz-Mattuck et al., Met. Trans. 1, 479, 1970.
3. W.A. Maxwell, Report No. NACA RM E9G01, 1949.
4. R.A. Long, Report No. NACA RM E50F22, 1950.
5. W.S. Gibbs, J.J. Petrovic and R.E. Honnell, (Ceram. Eng. Sci. Proc. 8, 1987) p. 645.
6. D.H. Carter and P.L. Martin, (Mat. Res. Soc. Symp. Proc., 194, 1990) p. 131.
7. D.H. Carter, J.J. Petrovic, R.E. Honnell and W.S. Gibbs, (Ceram. Eng. Sci. Proc., 10, 1989) p. 1121.
8. R.M. Aikin, (J. Ceram. Eng. Sci. Proc., 12, 1991) pp. 1643-1655.
9. K. Sadananda, C.R. Feng, H. Jones, J. Mat. Sci. and Eng., in press (1992).
10. S.M. Wiederhorn, R.J. Gettings, D.E. Roberts, C. Ostertag and J.J. Petrovic, J. Mat. Sci. and Eng., in press (1992).

CREEP BEHAVIOR OF SiC-REINFORCED XD[TM] $MoSi_2$ COMPOSITE

M. SUZUKI[1], S. R. NUTT[1] AND R. M. AIKIN, Jr.[2]
[1]Brown University, Division of Engineering, Providence, RI 02912
[2]Case Western Reserve University, Department of Materials Science and Engineering, Cleveland, OH 44108

ABSTRACT

The compressive creep behavior of $MoSi_2$ reinforced with 30v/o SiC fabricated by *in situ* XD[TM] process was investigated at 1050°C-1300°C in anaerobic and aerobic test ambients. Creep experiments performed with the composite in dry nitrogen and in air showed power-law type constitutive behavior and a stress exponent of ~3.5. Creep deformation occurred by dislocation glide accompanied by cavitation, and the apparent activation energy for creep at 1100°C-1300°C was bi-valued with a threshold temperature of ~1170°C. Microstrucural observations by TEM indicated that the rate-controlling process changed from dislocation glide to dislocation climb at higher temperature, corresponding to the change in activation energy. Creep damage occurred by cavitation at SiC-matrix interfaces and at grain boundaries within polycrystalline SiC particles. This process was apparently facilitated by the accumulation of glassy phase at these sites during creep. Creep experiments in air showed there was no appreciable atmospheric effect on the response of the composite, while an increased strain rate was observed in the base alloy due to an increase in glass phase resulting from thermal oxidation.

INTRODUCTION

$MoSi_2$ is a candidate material for high-temperature structural applications in oxidizing ambients (>1200°C). The compound has a high melting point of 2030°C and exhibits excellent oxidation resistance, which arises from the formation of a protective SiO_2 film. Several attempts have been made to fabricate $MoSi_2$-based composites with the goal of improving strength and creep resistance at high temperatures, and enhancing fracture toughness and ductility at room temperature. However, there have been relatively few studies of the relation between creep properties and microstructure of these composites [1, 2].

Recently, a novel process for *in situ* fabrication of composites was developed at Martin Marietta Laboratories. The process, designated XD[TM], can be used to fabricate SiC-reinforced $MoSi_2$ composites with substantially less glass phase than silicides produced by conventional powder-processing. This SiC-reinforced $MoSi_2$ composite showed improved compressive strength (2x) compared to the $MoSi_2$ base alloy [3].

Atmospheric effects on the high-temperature creep behavior are often dramatic for carbide-reinforced composites. Creep rates in aerobic and anaerobic test ambients can differ by factors of 5-10 because of thermal oxidation and the associated reaction products [4]. However, the creep behavior of $MoSi_2$ may show only minor atmospheric effects because of the excellent oxidation resistance described above.

In this report, we present results showing atmospheric effects on the compressive creep behavior of SiC-reinforced XD[TM] $MoSi_2$ and the base alloy. Microstructural observations of pre-test and crept materials are used to determine the mechanism of creep deformation and damage.

EXPERIMENTAL PROCEDURE

The two materials included in this study were a composite of $MoSi_2$ reinforced with 30 v/o SiC fabricated by the *in situ* process (XD[TM]) and a base alloy of polycrystalline $MoSi_2$. The $MoSi_2$ composite was produced at Martin Marietta Laboratories, and although the XD[TM] process is considered proprietary, the basis is as follows. First, a $MoSi_2$ composite sponge was produced by XD[TM] synthesis, and the sponge was then crushed to produce a -50 mesh $MoSi_2$ composite. The bulk of the $MoSi_2$ composite was produced by hot-pressing the -50 mesh $MoSi_2$ composite powder at 1800°C and 20 MPa for 4 hours in a graphite die, followed by hot isostatic pressing at

1600°C and 310 MPa for 2 hours. The base alloy was also produced at Martin Marietta Laboratories by hot-pressing a -325 mesh $MoSi_2$ powder (Consolidated Astronautics, a division of CERAC) at 1800° C and 20 MPa for 4 hours in a graphite die, followed by hot isostatic pressing at 1800° C and 207 MPa for 4 hours. Table 1 shows the chemical composition of the composite and the base alloy [3]. Note the higher content of oxygen, nitrogen and carbon in the base alloy.

Cylindrical specimens 3 mm in diameter and 6 mm in length were cut by EDM and end-polished for creep experiments. Compressive creep experiments were conducted at 1050°C-1300°C under applied stresses of 35-300 MPa. The experiments were conducted under constant load in test atmospheres of continuously circulated dry nitrogen or continuously circulated air. The displacement of the upper SiC platen was monitored with a linear variable-differential transducer, and a plot of strain rate versus time was used to determine that a constant creep rate (or minimum strain rate) was reached and maintained. The stress (or temperature) was then changed incrementally, and the procedure was repeated. The test temperature was maintained to within ±1°C of the preset value, and specimens were held at the test temperature for at least 30 min before the load was applied. Upon completion of a test, specimens were cooled rapidly under constant stress to retain the deformed microstructure. The strain in each specimen was kept under 3% to minimize non-uniform deformation (barreling).

Thin foil specimens were prepared for TEM observation by sectioning a disk from the center of each cylindrical specimen The discs were ground and polished and finally thinned by Ar ion-milling. The thin foils were examined using a Philips EM420T operated at 120 kV and equipped with a double-tilt specimen holder and an energy dispersive X-ray (EDS) spectrometer. Additional experiments were carried out using a JEOL 2010 operated at 200 kV. The 2010 was equipped with a parallel detection electron energy loss spectrometer (PEELS, made by GATAN). The $MoSi_2$ composite and the base alloy were examined in pre-test and crept conditions to determine changes associated with creep.

RESULTS AND DISCUSSION

Microstructure

A variety of precipitate phases were detected within the composite and the base alloy. An optical micrograph of the $MoSi_2$ composite is shown in Fig. 1(a). The specimen was etched by a modified Murakami's reagent (15g $K_3Fe(CN_6)$, 2g NaOH, 100ml water). In Fig. 1(a), the ß-SiC reinforcement (gray contrast) appears as equiaxed particles ranging in size from 1-15 µm, and the distribution is slightly non-uniform. An additional phase (1-10 µm) adjacent to the SiC particles was revealed by etching, and this was identified in TEM foils as CMo_5Si_3. The morphology and proximity of this phase, which was always adjacent to the SiC, suggested that CMo_5Si_3 resulted from a change of local composition associated with the formation of the SiC. Similar morphologies were observed in C-modified $MoSi_2$ in which SiC phase formed *in situ* during the fabrication process [5, 6]. No glassy phase was observed at triple grain junctions or at grain boundary-interface (GBI) junctions in the composite. ("Grain boundary-interface junctions" refers to triple junctions formed by $MoSi_2$ matrix grains and matrix-particle interfaces). Fig.1(b) is an image of a triple grain junction formed by $MoSi_2$, CMo_5Si_3 and SiC, showing an absence of glassy phase within the resolution of conventional TEM used in this study. The average grain size of the composite was ~10 µm, almost half the grain size in the base alloy (20 µm). The grain refinement was attributed to the presence of SiC particles, which pinned migrating grain boundaries during hot-pressing. In addition, the hot pressing of the composite was conducted at a lower temperature.

Fig. 2(a) shows an optical micrograph of the base alloy in which second-phase particles are unexpectedly present. The majority of the the second-phase particles exhibit the same gray contrast of the SiC particles in the composite, although some particles exhibit a darker contrast. TEM observations confirmed that the light-gray particles were ß-SiC, although small amounts of glassy phase were sometimes observed within particles, as shown in Fig. 2(b) (The dark phase adjacent to SiC is CMo_5Si_3.). The darker particles in Fig.2(a) were identified as SiO_2 glass, as shown in Fig. 2(c). Unlike P/M silicides produced by hot-pressing, the glass phase was not present at triple grain junctions. Additional small precipitates (30-80 nm) were present in some matrix grains in both materials, as shown in Fig. 3(a). EDS spectra (Fig. 3(b)), PEELS spectra, and microdiffraction patterns were used to identify the phase as η-Al_2O_3, a spinel-type structure

Table 1 Results of wet chemical analysis [3]

Material	Mo/Si Ratio		Interstitial Elements			Reinforcement (vol%)
	Mo (at%)	Si (at%)	O (wt%)	N (wt%)	C (wt%)	
Composite	33.4	66.6	0.13	0.018	n.a.	29.1
Base alloy	32.2	67.8	0.65	0.100	0.376	-

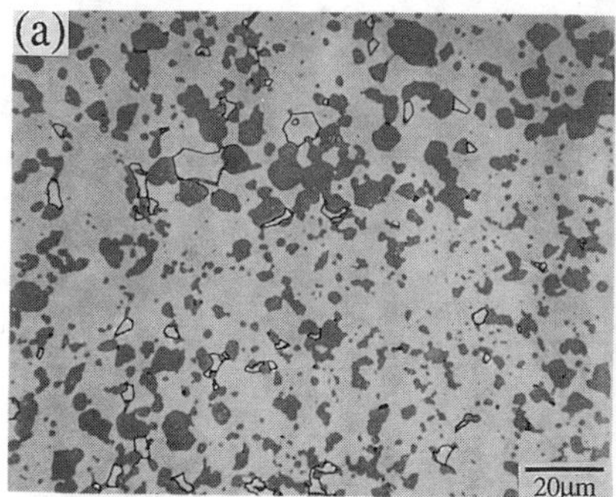

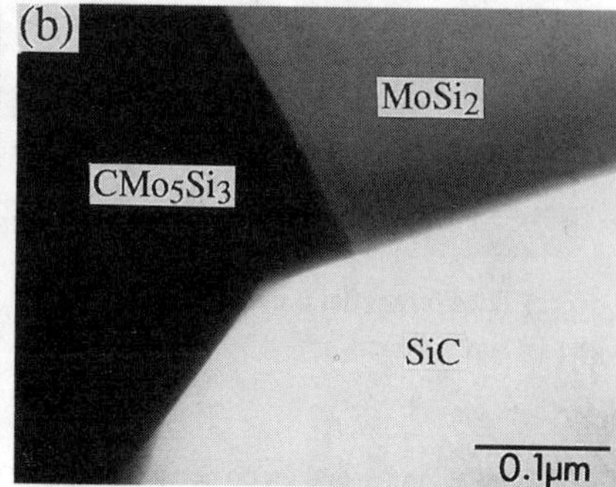

Fig. 1. Microstructure of the $MoSi_2$ composite: (a) optical micrograph, (b) TEM image of a triple grain junction consisting of $MoSi_2$, CMo_5Si_3, and SiC showing an absence of glassy phase.

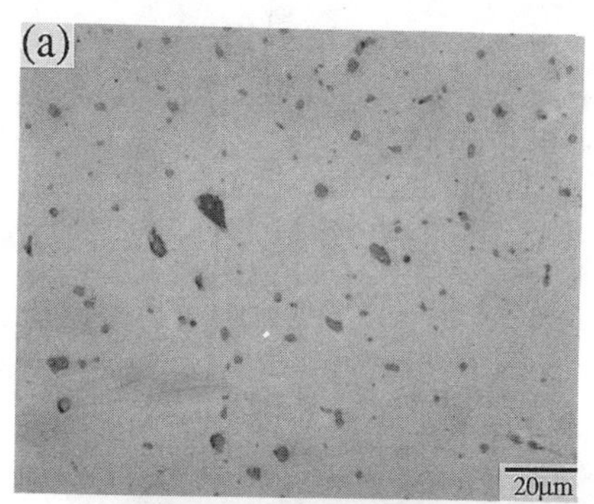

Fig.2 Microstructure of the the base alloy: (a) optical micrograph, (b) TEM bright field image of ß-SiC particle containing glassy phase and CMo_5Si_3, corresponding to the gray phase in (a), (c) SiO_2 glass particle, corresponding to the darker phase in (a).

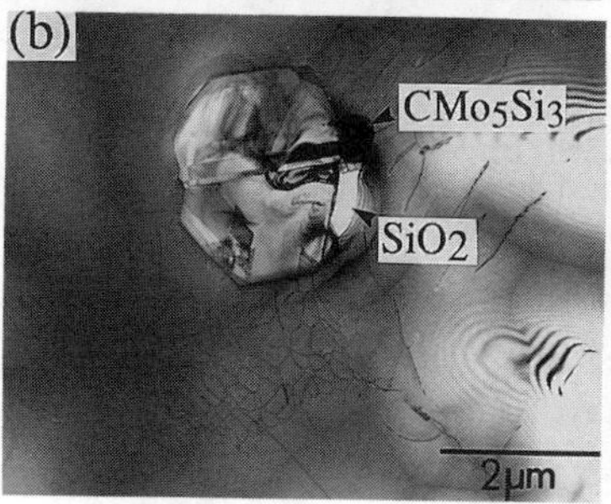

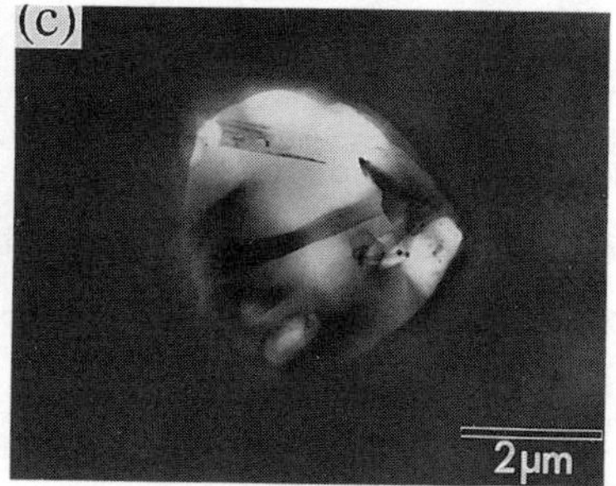

with a=0.794 nm. The observation of oxide phase in the base alloy is consistent with the high oxygen content mentioned previously.

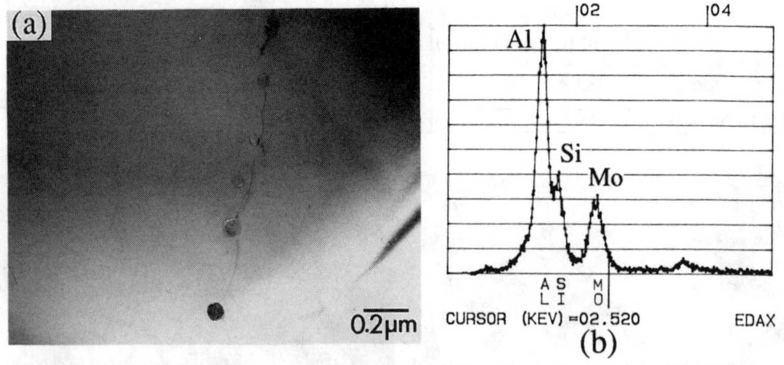

Fig. 3(a) TEM bright field image of η-Al_2O_3 precipitate found in both materials, (b) EDS spectra of η-Al_2O_3 precipitate.

Creep behavior

Typical isothermal creep behavior for the composite is shown in Fig. 4, a plot of strain rate versus time for stress levels that were increased incrementally. From Fig. 4, a steady-state strain rate is achieved at 60 and 100 MPa, and a minimum creep rate is achieved at 150 MPa. These creep rates are plotted as a function of applied stress in Fig. 5(a), which shows data obtained from experiments in dry nitrogen and in air for various applied stress levels, 35-300 MPa. The data conform to a power-law constitutive relation for creep at 1050°C-1300°C. The stress exponent n was approximately 3.5, similar to results reported elsewhere. K. Sadananda *et al.* measured a stress exponent of 3 for a SiC whisker-reinforced $MoSi_2$ composite at 1200°C-1400°C and monolithic $MoSi_2$ at 1100°C-1300°C [1], while Umakoshi *et al.* reported a stress exponent of 3 for a $MoSi_2$ single crystal at 1200°C-1400°C [2]. A stress exponent of 3 often implies a dislocation creep mechanism [7], although more rigorous documentation is required for the case of multi-phase materials.

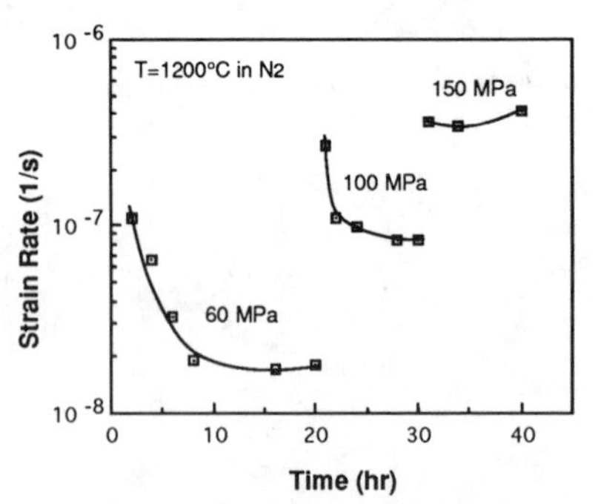

Fig. 4. Strain rate-time curve for creep experiment at constant temperature (1200°C) with incrementally increased stress (60, 100, and 150 MPa).

The apparent activation energies for creep were obtained from the Arrehenius plot shown Fig. 5(b). At high temperature, the apparent activation energies were 494 kJ/mole in dry nitrogen, and 436 kJ/mole in air, similar to the value reported by K. Sadananda *et al.*, who measured an activation energy of 590 kJ/mole for creep of SiC whisker composite at 1200°C-1450°C [1]. However, below ~1170°C the apparent activation energies were 283 kJ/mole in dry nitrogen, and 246 kJ/mole in air, a finding not reported previously. The marked change in activation energy at ~1170°C implies a change of the rate-controlling process for creep deformation at this temperature. The processes associated with the measured values of activation energies can only be determined by careful examination of the crept microstructure, as described below.

TEM observations of composite specimens deformed at 1050°C-1300°C revealed increased activity of matrix dislocations, and it was concluded that creep deformation was occurring by dislocation glide [7]. At 1100°C, matrix dislocations were distributed uniformly and there was a slight tendency to form networks, as shown in Fig. 6(a). However, at 1300°C, there was a strong tendency to form networks and subgrain boundaries, as shown in Fig. 6(b). Similar observations were reported by Umakoshi *et al.* for uniaxial compression of $MoSi_2$ single crystals at 1300°C [8]. Because formation of subgrain boundaries involves climb, our observations were interpreted to indicate that the rate-controlling process for creep deformation changed from dislocation glide (below ~1170°C) to dislocation climb above ~1170°C, corresponding to the observed change in activation energy.

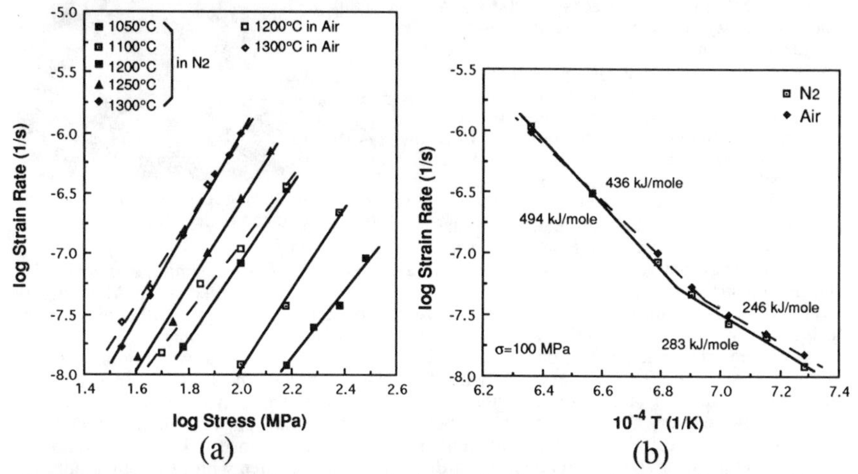

Fig. 5(a) Steady-state strain rates versus applied stress for the composite in dry nitrogen and in air, (b) Arrehenius plot of steady-state strain rates versus inverse temperature for the composite in dry nitrogen and in air.

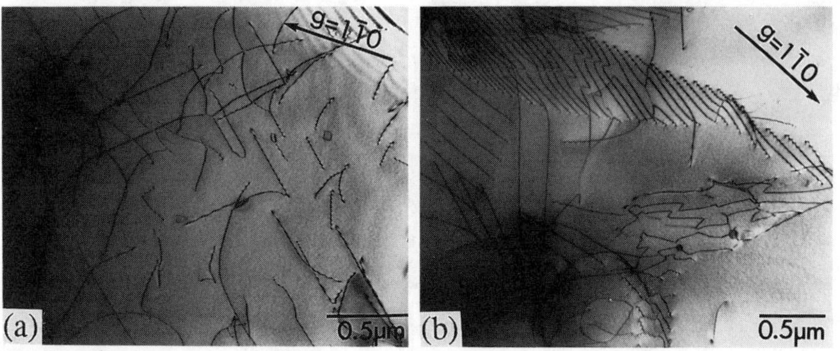

Fig. 6. Dislocation structures in the composite after creep in dry nitrogen: (a) at 1100°C, and (b) at 1300°C.

Several types of second-phase particles were effective in impeding the glide of dislocations during creep. The observed density of dislocations was generally higher in the vicinity of SiC particles and η-Al_2O_3 precipitates, indicating effective obstruction of glide dislocations. Observations of creep-deformed specimens also revealed dislocation loops, as shown in Fig. 7. The loop configurations apparently resulted from Orowan bowing around coherent precipitates, the structure of which has not been determined yet. Bowed segments of dislocations linked the precipitates, which were often encircled by one or more loops. Controlling the distribution of fine-scale precipitates (η-Al_2O_3 and coherent precipitates), either by modifying alloy compositions and/or processing parameters, could lead to improved creep resistance and high-temperature strength.

Creep damage at 1050°C-1300°C occurred by cavitation at SiC-matrix interfaces and at grain boundaries within polycrystalline SiC particles. Fig. 8(a) shows the cavity formed at SiC-matrix interfaces at 1200°C. A glassy phase was typically observed around the cavities, and the process of cavitation was apparently facilitated by the accumulation of the glass phase during creep. Fig. 8(b) shows a glass accumulation at a GBI junction that developed during creep deformation. PEELS (Fig. 8(c)) showed that the composition of the glass pocket consisted of silicon, oxygen, and a small amount of molybdenum. As stated previously, no glass phase was detected in the microstucture prior to creep. However, a very thin glassy layer (native silica) is often present at SiC-matrix interfaces in similar composites [4], and the redistribution of such films might lead to cavitation at the observed sites. Alternatively, oxygen dissolved in the matrix could conceivably lead to internal oxidation. Further experiments (using HREM) are presently underway to determine the origin of the glassy phase.

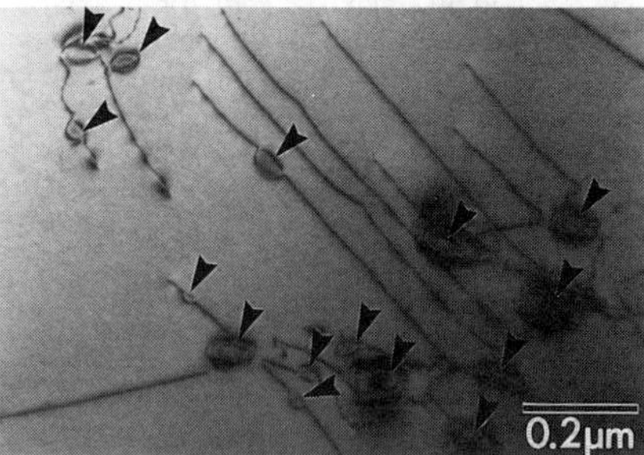

Fig. 7. Loop configurations found in the composite after creep in dry nitrogen at 1200°C.

The environmental effects on creep rates of the composite were relatively minor, as shown in Fig. 5(a) and (b). There was essentially no atmospheric effect on the stress dependence of the strain rate and the activation energies, although the activation energies in air tended to be slightly lower than those in dry nitrogen. The strain rates in dry nitrogen and air were virtually identical, a result that may be attributable to the oxidation resistance of the composite. TEM observations of specimens crept in air showed no noticeable difference when compared with TEM observations of specimens crept in dry nitrogen. The similarity is consistent with the measured creep response.

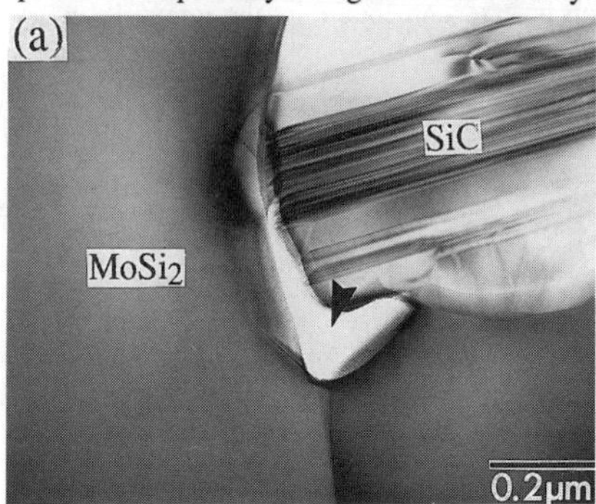

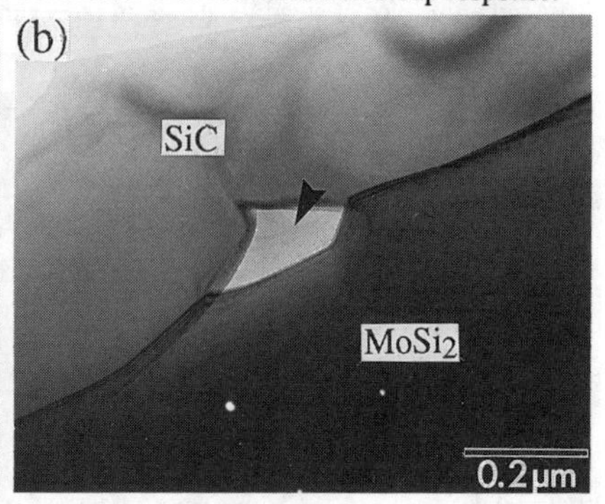

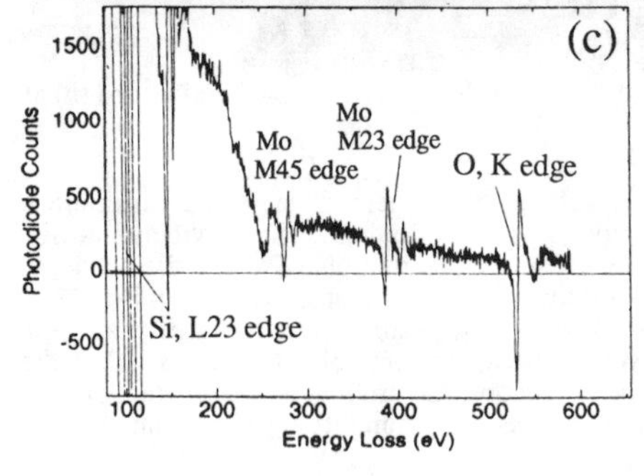

Fig. 8. Creep damage: (a) cavitation at SiC-matrix interfaces after creep at 1200°C, (b) glass pocket at SiC-matrix interfaces, and (c) PEELS spectrum acquired from the glass pocket showing Si, and a small amount of Mo.

Fig.9 shows a comparison of atmospheric effects on creep of the composite and the base alloy at 1200°C. The difference in creep rates in different test ambients is greater for the base alloy than for the composite, and the base alloy shows a higher strain rate in air than in dry nitrogen. Optical micrographs of the specimen crept in dry nitrogen (Fig. 10(a)) and in air (Fig. 10(b)) showed an increase in the number of second-phase particles after creep in air. Many second-phase particles are present in both micrographs, particularly Fig. 10(b). In this figure, many of the particles exhibit dark contrast, and these are glassy phase particles as discussed in the previous section. The increase in the number density of these particles indicates that oxidation occurred during creep *even in dry nitrogen* . However, the oxidation was far more extensive and rapid for creep experiments conducted in air. TEM observations of the specimens supported this contention and showed an increase in the number of glassy phase particles after creep. This was particularly true for specimens crept in air, where the glass phase formed during creep was apparently different from the glass phase formed during fabrication process, as shown in Fig. 10(c). The oxidation behavior of the composite is not well-understood, but the accelerated creep of the base alloy in air is undoubtedly associated with the viscous flow of glass phase generated during creep.

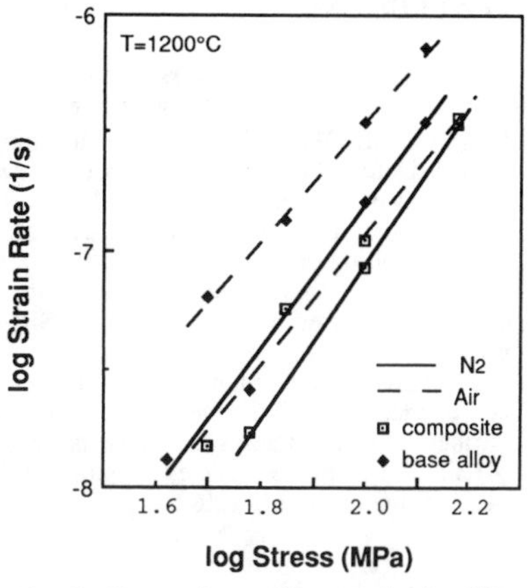

Fig. 9. Comparison of atmospheric effects on creep of the composite and the base alloy.

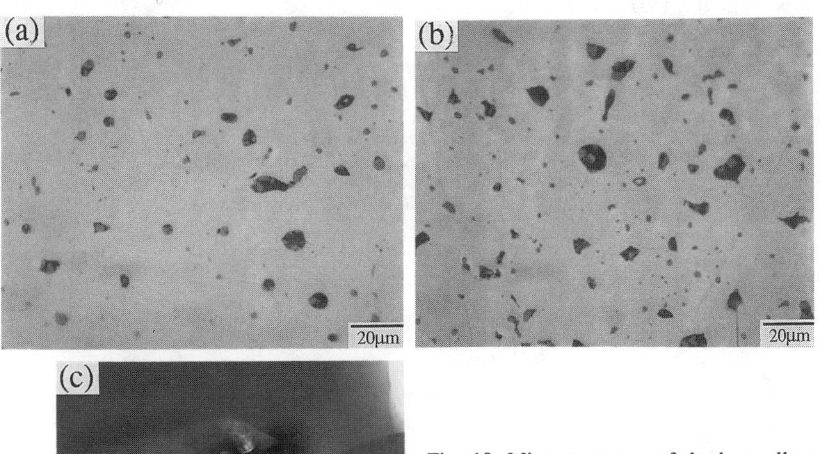

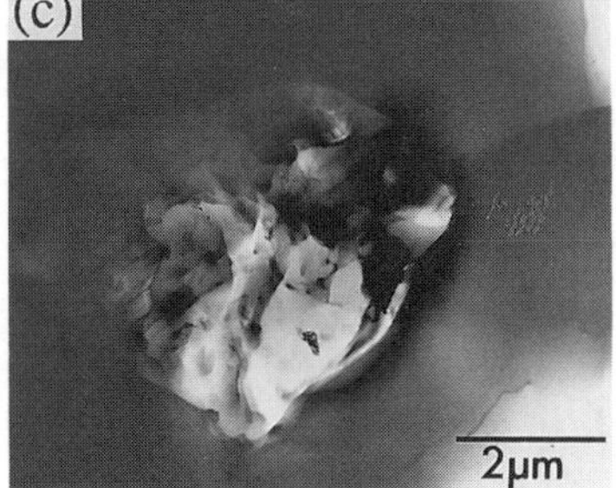

Fig. 10. Microstructures of the base alloy after creep: (a) optical micrograph after creep in dry nitrogen, (b) optical micrograph after creep in air, and (c) TEM bright field image of glassy phase found after creep in air.

CONCLUSIONS

The results of compressive creep experiments on SiC-reinforced XD^{TM} $MoSi_2$ and the base alloy at 1050°C-1300°C in dry nitrogen and in air can be summarized as follows.
(1) In the XD^{TM} processed composite, equiaxed particles of ß-SiC were present and ranged in size from 1-15 μm. Glass phase was not observed in the as-fabricated composite, although some glass particles were present in the base alloy.
(2) The creep deformation mechanism of the composite involved dislocation glide in which the rate-controlling process changed from dislocation glide to dislocation climb at higher temperature, corresponding to a change in activation energy at a threshold temperature (~1170°C). Creep damage occurred by cavitation at SiC-matrix interfaces and at grain boundaries within polycrystalline SiC particles.
(3) The creep test ambients of dry nitrogen and air had a negligible effect on creep behavior of the composite, although the base alloy showed slightly accelerated creep deformation in air. This was attributed to an increase glass-phase particles resulting from accelerated oxidation in air.
(4) Small precipitate particles developed (fortuitously) in the microstucture and appeared to inhibit glide of dislocations during creep.

ACKNOWLEDGMENTS

The research was supported by DARPA through NADC contract No. N62269-89-C-0233 (RA) and the Office of Naval Research contract No.N00014-91-J-1480 (SRM and MS).

REFERENCE

1. K. Sadananda, H. Jones, J. Feng, J. J. Petrovic, and A. K. Vasudevan, *Ceram. eng. Sci. Proc. 12[9-10],* 1671-1678 (1991).
2. Y. Umakoshi, J. Sennami, T. Yamane, *Proceedings of fall meeting of The Japan Institute of Metals,* 279 (1990).
3. R. M. Aikin, Jr., *Ceram. Eng. Sci. Proc. 12[9-10],* 1643-1655 (1991).
4. P. Lipetzky, S. R. Nutt, D. A. Koester, and R. F. Davis, *J. Am. Ceram. Soc., 74[6],* 1240-1247 (1991).
5. S. Jayashankar and M. J. Kaufman, *Scr. Metall.,26,* 1245-1250 (1992).
6. S. Maloy, A. H. Heuer, J. J. Lewandowski, *J. Am. Ceram. Soc., 74[10],* 2704-2706 (1991).
7. J. Cadek, Creep in Metallic Materials, Elsevier, 1988.
8. Y. Umakoshi, T. Sakagami, T. Hirano and T. Yamane, *Acta metall., 38,* 909-915 (1990).

SYNTHESIS AND PROPERTIES OF Mo/$MoSi_x$ MICROLAMINATES USING ION BEAM ASSISTED DEPOSITION

A. MASHAYEKHI,* L. PARFITT,* C. KALNAS,* J. W. JONES,* G. S. WAS * and D. W. HOFFMAN **
* The University of Michigan, Ann Arbor, MI 48109
** Research Staff, Ford Motor Company, Dearborn, MI 48124

ABSTRACT

Films of Mo, $MoSi_x$ and Mo/$MoSi_x$ (1.22<x<1.35) multilayers were formed by physical vapor deposition (PVD) and ion beam assisted deposition (IBAD) onto (100) Si, glass and graphite substrates. Ion to atom arrival rate (R) ratios for IBAD varied from 0.01 to 0.1 and film thicknesses varied from 200 to 1100 nm. The Si/Mo ratio decreased with increasing R ratio. The oxygen content of Mo films was greater than silicide films, but both decreased substantially with increasing R ratio. Ar incorporation increased with increasing R ratio to a maximum of 1 at% in Mo and 5 at% in $MoSi_{1.22}$. Mo films exhibit a strong (110) fiber texture at low R ratios. At the highest R ratio, a tilting of the (110) fiber texture by 15° occurs, along with the development of a distinct azimuthal texture indicative of planar channeling of the ion beam along (110) planes. The microstructure of the multilayer consists of small Mo grains and an amorphous silicide. Average film stress in Mo films increases from tension to a maximum value of 0.63 GPa and becomes compressive with increasing normalized energy. The stress in the $MoSi_x$ films decreases with increasing normalized energy and saturates at a compressive stress of -0.24 GPa at 25 eV/atom. Indentation fracture experiments using a Vickers indenter with a 300 g load show a fracture behavior that is consistent with a residual stress effect for the IBAD monolithic $MoSi_x$ and microlaminate, but which is influenced by additional factors in the PVD microlaminate.

INTRODUCTION

In recent years considerable interest in aluminides and silicides as structural materials suitable for service in temperature ranges in excess of 1000°C. $MoSi_2$, in particular, and $MoSi_2$-based composites have received considerable attention. Problems with low temperature toughness and elevated temperature creep are currently under study. The incorporation of a ductile phase into the $MoSi_2$ system offers the possibility of substantially improving toughness by crack bridging, interface delamination and plastic deformation of the ductile phase. For example the addition of Nb to $MoSi_2$, have been shown to significantly increase toughness [1]. Microlaminates offer the potential to tailor composition, microstructure and interface properties to enhance strength, toughness and oxidation resistance and are, therefore, interesting candidates for study in these systems and substantial increases in strength and toughness have been reported in microlaminate systems [2-5].

Vapor deposition has traditionally been used to produced microlaminates. More recently, it has been recognized that significant improvement in properties of films can be achieved by ion beam assisted deposition (IBAD) [6]. By proper control of ion/atom arrival rate ratio, ion energy and type, control of a wide range of film morphologies can be achieved. IBAD has been shown to increase film density, promote equiaxed grain formation, promote film/substrate adherence, control film texture and provide control over film growth stresses [6].

The objective of the current research is to synthesize Mo/$MoSi_2$ microlaminates by IBAD with the intent of developing an understanding of microstructural and geometric variables on strength and fracture resistance. The initial studies reported here describe the synthesis and characterization of monolithic Mo, $MoSi_x$ films and Mo/$MoSi_x$ microlaminates produced by ion beam assisted deposition.

EXPERIMENT

Ion beam assisted deposition was conducted in a chamber containing two 6 kw electron beam guns and a 3 cm Kaufman ion gun with a base pressure of 2 x 10^{-9} torr. Mo was deposited at a rate of 0.25 nm/s and $MoSi_x$ was codeposited at rates of 0.25 nm/s for Mo and 0.65 nm/s for

Si, normal to the substrate surface. Ion bombardment was performed using 250 to 516 eV Ar^+ ions at current densities in the range 2 to 75 $\mu A/cm^2$, giving a range of R ratios from 0.01 to 0.1 at an angle of 45° to the substrate normal. The chamber pressure during deposition was 1-3 x 10^{-5} torr. Deposition rates from each source were measured independently with quartz crystal monitors and the beam current was monitored with a Faraday cup. Depositions ranged in thickness from 200 to 1000 nm for Mo, $MoSi_x$ films and 1100 nm for Mo/$MoSi_x$ multilayers. The silicide composition was intended to be $MoSi_2$.

The amount of material deposited, and the composition or stoichiometry of the films were determined by Rutherford backscattering spectrometry at an angle of 165°. Spectra were fit with the aid of the program, RUMP [7]. Crystallinity and texture were determined by x-ray diffraction (XRD) using Cu K_α radiation in a Rigaku diffractometer operated at 40 kV and 100 mA. Theta-two theta scans of rotating samples were conducted for both Mo and $MoSi_x$ films of thickness 1000 nm. Pole figures were generated for the (110) pole. Grain size and morphology were determined from transmission electron microscopy (TEM) of planar and cross-section samples. Planar samples were prepared by grinding the substrate side of the silicon substrate and ion milling. Cross-section specimens were prepared by grinding, polishing and ion-milling a glued "sandwich" of two films on their substrates. TEM analysis was conducted on a JEOL 2000 FX TEMSCAN at 200 kV.

Residual stress was determined by optical interferometry on 200 nm films deposited onto glass cover slips (150 μm thick). The shape of the cover slip was determined before and after deposition with reference to an optical flat by observation in monochromatic light. The resulting pattern of Newton's rings was photographed and the fringe spacing measured in orthogonal directions. Curvatures were determined by least squares fitting of parabolas to the recorded fringe spacings. The resulting quadratic coefficients, k_x and k_y are linearly related to the average stress by the formula:

$$\sigma = \frac{Eh^2}{3(1-\nu)t}(\Delta k_x + \Delta k_y) \quad (1)$$

where E is Young's modulus of the cover slip, ν is Poisson's ratio, h is the coverslip thickness, $t(\ll h)$ is the thickness of the deposited layer and the Δks indicate the changes in the quadratic coefficients induced by the deposition. All calculations of stress were made using the nominal film thicknesses.

Indentation was conducted with a Vickers diamond pyramid indenter using 100, 300 and 500 gram loads on depositions made on (100) silicon wafers. The diagonals of the pyramid indenter were aligned along {110} and crack lengths were measured along each of the diagonals for each indent. A minimum of 7 indents were made on each sample.

RESULTS AND DISCUSSION

A description of the films used in this study is given in table I. Monolithic films were deposited with nominal thicknesses of 200 and 1000 nm and multilayers were made with a thickness of 1100 nm. All determinations of residual stress and fracture toughness are based on the nominal film thicknesses.

Characterization of Films

In monolithic Mo films, there is a decrease in the oxygen content with increasing R ratio. The PVD films (R=0) showed large amounts of oxygen incorporation (17 at%) which drops to 5 at% at R=0.1 This is in qualitative and quantitative agreement with the results of Cuomo et al. [8] who found that O incorporation into Nb was at a maximum at R=0 (33 at%) and dropped to about 10 at% at R~0.09. The base pressure used in Cuomo's experiments was about an order of magnitude higher, accounting for the larger magnitudes of incorporated oxygen. Ar entrapment remained low, amounting to only 1 at% at the highest R ratio, also consistent with observations on Nb [8]. The Si/Mo ratio of the silicide ranged between 1.22 and 1.35, considerably below the target value of 2.0. This may be due to the shape of the Mo plume during deposition. Due to uneven melting of the charge material, it is possible that the deposition rate of Mo was greater at the substrate than at the thickness monitor, resulting in a Mo enriched deposition. The Si/Mo ratio decreased with increasing R ratio indicating preferential sputtering of Si, in agreement with

Table I. Deposition parameters and composition analysis of 200 nm Mo and $MoSi_x$ films.

Film	R ratio (ion/atom)	Energy (eV/ion)	E_n (RxE) (eV/atom)	Si/Mo (<±0.01)	Ar (at%) (±0.01)	O (at%) (±0.1)
Mo	0	0	0	-	0	17.0
	0.01	251	2.5	-	0.69	12.2
	0.025	340	8.5	-	0.0	14.9
	0.05	376	18.8	-	0.57	9.0
	0.075	409	30.7	-	1.0	6.6
	0.10	461	46.1	-	1.0	5.1
$MoSi_x$	0	0	0	1.35	0	3.8
	0.01	345	3.5	1.36	0.75	4.6
	0.025	415	10.4	1.37	1.43	3.8
	0.05	471	23.6	1.34	2.58	2.9
	0.075	488	36.6	1.27	3.49	2.4
	0.10	516	51.6	1.21	4.94	2.4

results of Liau et al. [9] for PtSi. The oxygen concentration remained low in $MoSi_x$, and tended to decrease with increasing R ratio. The Ar concentration was significantly higher than for Mo metal with a peak of about 5 at% at R=0.1. These results are higher by about a factor of 2 than those of Yee et al. [10] for amorphous WSi_2. However, they are consistent with observations of greater incorporation of Ar in amorphous materials due to the high concentration of voids which are larger than the interstices in close-packed crystalline materials [10].

The pole figure for the 1 μm, PVD film, Fig. 1a, shows a high intensity area in the center, representative of a strong (110) fiber texture perpendicular to the film surface and consistent with the growth texture of bcc films [11]. The pole figure for the 1 μm, R=0.1 Mo film, Fig. 1b, exhibits a strong (110) fiber texture at an angle of 15° to the substrate, as well as a distinct azimuthal texture represented by the areas of high intensity 60 radial degrees from the fiber axis. The evolution of the azimuthal texture is indicative of planar channeling of the ion beam along (110) planes in the film, resulting in a preferred orientation of the grains along the ion beam direction [11,12]. This channeling behavior is indicated by the high intensity areas located 90° to the ion beam direction. The tilt in the fiber axis is not well understood but correlates with the

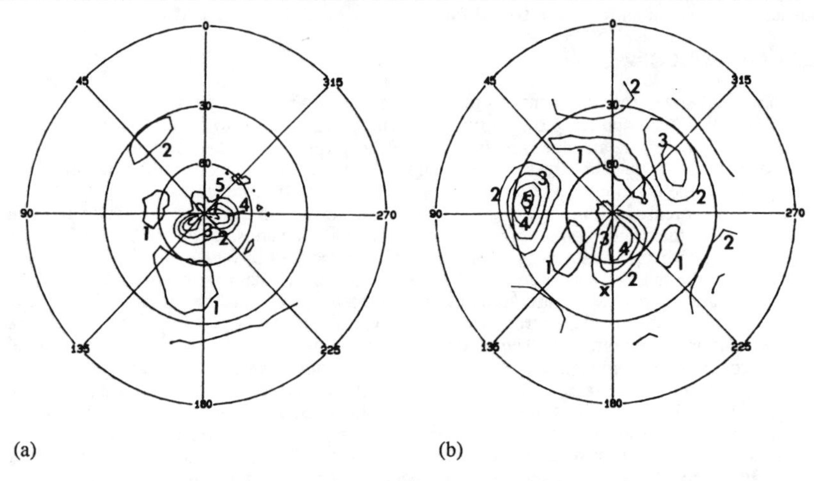

Figure 1. Pole figures of 1.1 μm thick Mo films deposited by a) PVD (R=0), showing the strong (110) fiber texture, and b) using 500 eV Ar^+ ions at 45° to the substrate normal (x marks beam direction) showing a tilt of the (110) fiber texture and an azimuthal texture as well.

beam direction. XRD of the $MoSi_x$ samples indicated that they were predominantly amorphous. Several low intensity peaks were found in the scan suggesting a crystalline phase, but the peaks do not correspond to Si, Mo or their compounds or oxides.

Bright field transmission electron micrographs from cross-sections of Mo/$MoSi_x$ multilayers deposited by PVD and IBAD are shown in Fig. 2. The microstructure shows a fine grain structure for the Mo phase (<10 nm) and a featureless but "streaked" structure for the silicide phase. The streaking is most likely due to density variations caused either by cracking or the formation of voids. There is little difference between the PVD and IBAD microstructures except that in the IBAD microlaminate, the streaking is less pronounced. These observations are consistent with those of Harper et al. for Nb deposition [12].

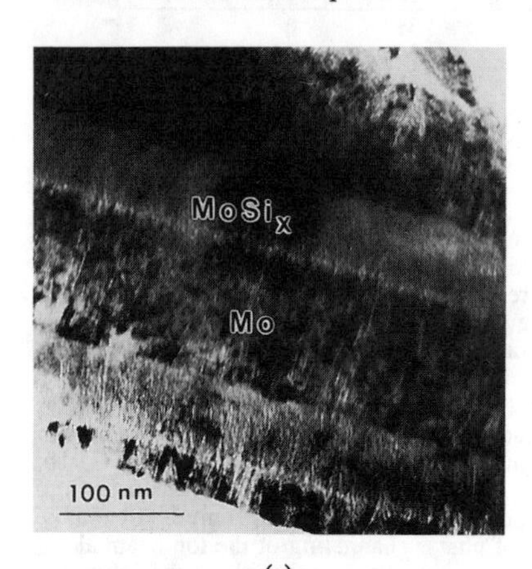

(a)

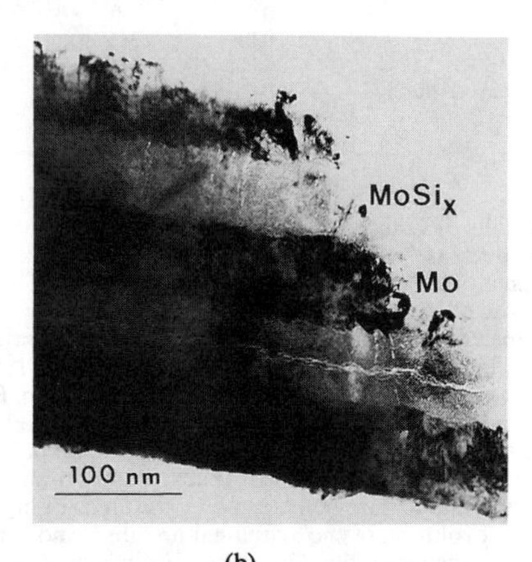

(b)

Figure 2. Bright field transmission electron micrograph of a cross-section of a) a 1.1 μm thick PVD (R=0) Mo/$MoSi_x$ multilayer film, and b) a 1.1 μm thick Mo/$MoSi_x$ multilayer film deposited using Ar^+ at an R ratio of 0.07 for the Mo layers and 0.01 for the $MoSi_x$ layers.

Mechanical Properties

Residual stress measurements were conducted on Mo and $MoSi_x$ films over an R ratio range of 0 to 0.1, corresponding to a normalized energy range of 0 to 52 eV/atom. Figure 3 shows the variation of the average film stress in GPa as a function of the normalized energy E_n (energy/deposited atom). The stress in the metal film without simultaneous ion bombardment is tensile with a value of about 0.33 GPa. At low E_n, the stress first increases to a value of 0.63 GPa, and then decreases with increasing E_n, driving the stress into a compressive state at about 25 eV/atom. The maximum compressive stress is -0.9 GPa. The stress of the PVD film (R=0) is in agreement in both sign and magnitude with Nb films deposited by Yee et al. [10] in a similar manner. Since the relative magnitudes of the coefficients of thermal expansion for Mo and the cover slip dictate a compressive stress, the measured tensile stress must be primarily intrinsic in nature. The initial rise in tensile stress, the stress reversal and the large compressive stress are also consistent with the literature. The rise in tensile stress is believed to be due to the removal of open porosity in the film resulting in a porous, but compact network in which the short range attractive interatomic forces across small defects can act most effectively [13]. These results are also consistent with observations of a critical film thickness by PVD, at which the tensile stress increases abruptly due to interconnection of the growing nuclei [14]. Further increases in E_n continue to densify the film forcing the stress toward zero. The compressive stress at higher values of E_n is probably due to point defect production by irradiation damage. Although oxygen incorporation and Ar entrapment have been suggested as possible causes for residual stress changes [15], neither is able to account for the observed behavior of the Mo films.

The silicide films behave very differently from the metal films in that the stress decreases at the very lowest value of E_n, crosses into the compressive range earlier, and saturates by 25 eV/atom. The immediate decrease in stress and the lower crossover value of E_n (3.5 eV/atom) may be related to the lower value of the activation energy for self-diffusion in an amorphous structure as compared to a crystalline lattice. Yee et al. [10] found that the R ratio required to cause a transition from tensile to compressive stress was as much as a factor of 3 smaller for amorphous WSi_2 than for crystalline Nb. The saturation in the value of the compressive stress with increasing R ratio is expected for an amorphous structure where the mechanism is due to densification of the microstructure and not the introduction of lattice defects.

In an initial attempt to characterize the fracture properties of the various films studied here, indentation fracture techniques have been used with the change in length of radial cracks generated by Vickers indentation serving as a relative indicator of the fracture resistance of the film/Si substrate system. Residual stresses in the films and intrinsic fracture resistance of the films, along with the intrinsic fracture resistance of the substrate should determine the equilibrium crack length resulting from indentation. From the analysis of Lawn and Fuller [16], residual stress is expected to exert a significant influence on the equilibrium radial crack length developed by Vickers indentation. They show that for an average residual stress, σ in a surface layer of thickness, d, the radial crack length, C, is given by:

$$2\Psi\sigma\sqrt{d} / K_c = 1 - (C_o / C)^{3/2} \tag{2}$$

where K_c is the substrate fracture toughness, Ψ is a dimensionless constant and C_o is the radial crack length at zero residual stress. Thus, a compressive surface stress will result in a reduction in radial crack length while a tensile residual stress will cause an increase in crack length compared to the unstressed condition. Although analyses which accurately predict the magnitude of the effect of changes in toughness of near-surface layers on equilibrium crack length are not available, it is reasonable to assume that surface layers with fracture resistances greater than the substrate will result in a decrease in equilibrium crack length.

In Fig. 4 the relative crack lengths C/C_o are presented as a function of normalized energy for different film types. In this graph $C/C_o<1$ implies an increase in fracture resistance of the film/substrate combination compared to that of the substrate. Although it is not possible to obtain a complete understanding of fracture behavior from these preliminary tests, several interesting observations can be made. For the monolithic silicide films, the fracture resistance increases with increasing normalized energy. Although residual stresses for these thick films were not measured, Fig. 3 shows that thinner films become progressively more compressive at higher values of E_n. Assuming that this holds for the thicker films, then the residual stress is indeed affecting the fracture behavior of the film in the expected manner. The results are also consistent with companion experiments in our lab on Al_2O_3 films produced and tested in the same instruments [17]. The results for the microlaminate films are also presented and show the opposite trend. Residual stress measurements of these films show that the PVD film and the R=0.1 film are in tension (~0.4 GPa) while the intermediate film is at a nearly zero stress state. The crack length results are consistent for all but the PVD film indicating that factors other than

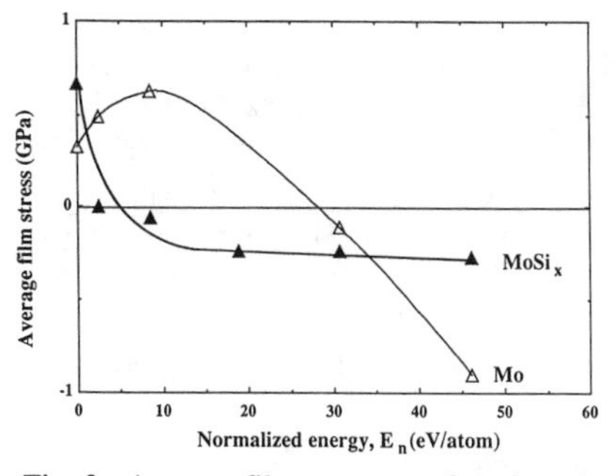

Fig. 3. Average film stress as a function of the normalized energy for Mo and $MoSi_x$ films of thickness 200 nm.

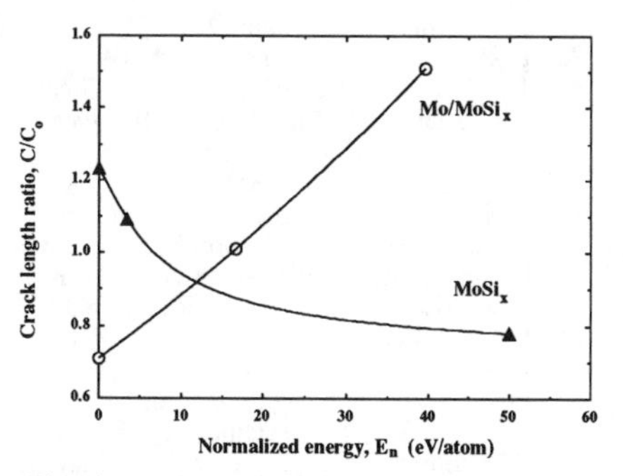

Fig. 4. Variation in crack length ratios for monolithic $MoSi_x$ and $Mo/MoSi_x$ microlaminates with E_n.

residual stress are affecting the fracture behavior of microlaminate films, such as the nature of the substrate-film interface. Additional studies using additional mechanical test techniques, including three and four-point bending and tension testing of free standing films are underway to more completely characterize film fracture characteristics.

CONCLUSIONS

- Increasing R ratio leads to an increase in Ar entrapment and a reduction in oxygen incorporation in Mo and $MoSi_x$ films. In $MoSi_x$ films, the stoichiometry becomes increasingly Mo rich with increasing R ratio due to preferential sputtering of Si.
- The (110) fiber texture characteristic of deposited Mo is intensified with an ion bombardment and an azimuthal texture develops, indicative of planar channeling of the ion beam along (110) planes in the film, resulting in a preferred orientation of the grains along the ion beam direction.
- Ion bombardment during deposition results in little visible change in the Mo layers but a decrease in contrast variations in the silicide layers, presumably due to densification.
- Residual film stress changes from tension to compression with increasing normalized energy for both Mo and $MoSi_x$. This is probably due to ion beam induced densification of the film, or or possibly entrapped Ar gas.
- Radial crack length as measured by indentation fracture experiments using a Vickers indenter indicate that the cracking behavior of the $MoSi_x$ film is strongly controlled by the residual stress, but that other factors are influencing the fracture behavior of the microlaminate structure.

ACKNOWLEDGEMENTS

The authors acknowledge the Michigan Ion Beam Laboratory for Surface Modification and Analysis and the Electron Microbeam Analysis Laboratory. This research was funded by the Air Force Office of Scientific Research under the University Research Initiative Program, contract #DOD-G-AFOSR-90-0141, Dr. Alan H. Rosenstein, Program Director.

REFERENCES

1. T. C. Lu, A. G. Evans, R. J. Hecht and R. Mehrabian, Acta metall. 39, 1853 (1991).
2. A.T. Alpas, J. D. Embury, D. A. Hardwick and R. W. Springer, J. Mater. Sci. 25, 1603 (1990).
3. D. Tench and J. White, Met. Trans. A 15A, 2039 (1984).
4. R. S. Bhattacharya, A. K. Rai and M. G. Mendiratta, in Intermetallic Matrix Composites, edited by D. L. Anton, P. L. Martin, D. B. Miracle and R. McMeeking (Mat. Res. Soc. Proc., Pittsburgh, PA 1990) pp. 71-78.
5. D. A. Hardwick and R. C. Cordi, in Intermetallic Matrix Composites, edited by D. L. Anton, P. L. Martin, D. B. Miracle and R. McMeeking (Mat. Res. Soc. Proc., Pittsburgh, PA 1990) pp. 65-70.
6. F. A. Smidt, Inter. Mater. Rev. 35, 61 (1990).
7. L. R. Dolittle, Nucl. Instr. Meth. B9, 334 (1985).
8. J. J. Cuomo, J. M. E. Harper, C. R. Guarnieri, D. S. Yee, L. J. Attanasio, J. Angilelio and C.
9. Z. L. Liau, J. W. Mayer, W. L. Brown and J. M. Poate, J. Appl. Phys. 49, 5295 (1978).
10. D. S. Yee, J. Floro, D. J. Mikalsen, J. J. Cuomo, K. Y. Ahn, and D. A. Smith, J. Vac. Sci. Techn. A. 3, 2121 (1985).
11. L. See, J. M. E. Harper, J. J. Cuomo and D. A. Smith, J. Vac. Sci. Technol. A 4, 443 (1986).
12. J. M. E. Harper, D. A. Smith, L. S. Yu and J. J. Cuomo, in Beam-Solid Interactions and Phase Transformations, edited by H. Kurz, G. L. Olsen and J. M. Poate, (Mater. Res. Soc. Proc. 51, Pittsburgh, PA 1986) pp. 343-348.
13. K-H. Müller, J. Appl. Phys. 62, 1796 (1987).
14. R. W. Hoffman, in Physics of Thin Films, edited by G. Hass and R. E. Thun, (Academic Press, New York, 1966), p. 230.
15. T. Wu, J. Vac. Sci. Technol. 20, 349 (1982).
16. B. R. Lawn and E. R. Fuller, Jr., J. Mat. Sci. 19, 4061 (1984).
17. C. E. Kalnas, L. J. Parfitt, A. Mashayekhi, J. W. Jones, G. S. Was and D. W. Hoffman, in Intermetallic Matrix Composites II, edited by D. Miracle, J. Graves and Don Anton, (Mater. Res. Soc. Proc., Pittsburgh, PA 1992) in press.

IN SITU SYNTHESIS OF A $MoSi_2/SiC$ COMPOSITE USING SOLID STATE DISPLACEMENT REACTIONS

C. H. HENAGER, JR.*, J. L. BRIMHALL*, J. S. VETRANO*, and J. P. HIRTH**
*Pacific Northwest Laboratory, Battelle Blvd., Richland, WA 99352
**Washington State University, Dept. of Mechanical and Materials Engr., Pullman, WA 99164

ABSTRACT

A high-strength *in situ* composite of $MoSi_2/SiC$ was synthesized using a solid state displacement reaction between Mo_2C and Si by blending Mo_2C and Si powders and vacuum hot-pressing the powders at 1350°C for 2 h followed by 1 h at 1700°C. The resulting microstructure consisted of 1-µm diameter β-SiC particles (30 vol%) uniformly dispersed in a fine grained $MoSi_2$ matrix. Transmission electron microscopy was used to study the fine-scale morphology and phase distribution of the composite. Evidence for a small amount of grain boundary glass phase was observed using diffuse dark field imaging. The β-SiC particles were distributed mainly on grain boundaries and triple points within the $MoSi_2$ matrix. These findings were used to rationalize the observed mechanical property behavior.

INTRODUCTION

Solid state displacement reactions can be used to produce intermetallic/ceramic matrix composites *in situ* [1-3]. Interwoven and dispersed microstructures, important for composites, can be produced using displacement reactions [1-4]. The reinforcement phases are produced *in situ*, an increasingly important idea in composite synthesis because of the potential for low cost processing and improved mechanical properties. Issues of microstructural control and cleanliness can potentially be addressed using displacement reactions in ways that are not possible with other synthesis techniques. A similar approach, usually termed reaction sintering, has been used as a method of synthesizing ceramic composites, but with limited success in producing high-quality structural materials [4]. An example of a ceramic composite produced by reaction sintering is t-ZrO_2 + mullite from $ZrSiO_4$ + alumina [5].

EXPERIMENTAL PROCEDURES

Powders of Mo_2C (d <45 µm, 99+% purity) and Si (d <45 µm, 99.99% purity) were blended in a 5:1 Si:Mo_2C ratio using a vibratory ball mixer and hot-pressed at 27.5 MPa using graphite dies under a vacuum of about 10^{-2} Pa. The hot-pressing temperature cycle was 1350°C for 2 h followed by 1 h at 1700°C. A hot-press die diameter of 2.2 cm was used with 10 to 15 g of blended powders to produce initial materials. A 7.62 cm diameter die was used with 100 g of powders to produce materials for mechanical property tests.

Fracture strengths were determined from 4-point bend tests in air as a function of temperature from ambient up to 1200°C using fully articulated SiC fixtures with extensometry at the specimen tensile-face mid plane. Bend specimens measuring 4 mm x 4 mm x 50 mm were made from a 7.62-cm (3.0-inch) diameter hot-pressed disk that had a density of 5.34 g/cm^3. In addition, chevron-notched specimens were machined from the bend bars and tested to obtain fracture toughness values as a function of temperature. Four-point bend and chevron-notch testing were performed at strain rate of 1.3 x 10^{-5} s^{-1}.

Transmission electron microscopy (TEM) was performed on thin-foil specimens cut from the large hot-pressed disk using ultrasonic cutting, mechanical dimpling, and ion milling. These specimens were examined in a JEOL 1200 analytical TEM. Energy dispersive X-ray (EDX) spectra and electron diffraction patterns were used for phase identification. Diffuse dark field techniques were used to examine grain boundary regions for amorphous (glass) phases.

RESULTS

Composites fabricated by hot-pressing the blended powders indicated that the following displacement reaction occurred:

$$Mo_2C + 5Si \rightarrow 2MoSi_2 + SiC \quad (1)$$

The SiC particles are nearly equiaxed and are uniformly distributed (Figure 1). Larger regions, up to 20 μm in diameter, of what is likely Mo_5Si_3 are dispersed throughout the material. X-ray diffraction (XRD) revealed strong $MoSi_2$ peaks, β-SiC as a second phase, and some faint peaks that could not be indexed but likely belong to the Mo_5Si_3, which was identified in the scanning electron microscope using EDX (Figure 1). The β-SiC particles are smaller than 1 μm in diameter. The composite appeared to be near full-density from observing polished surfaces in the SEM. The density was determined to be 5.53 g/cm^3. An estimated theoretical density for a $MoSi_2$/SiC (30 vol% SiC) composite is 5.35 g/cm^3.

The TEM investigation revealed that the β-SiC particles were mainly situated on $MoSi_2$ grain boundaries and at triple points (Figure 2a). The particles were heavily faulted (Figure 2b) and were observed to contain a dispersed crystalline phase that could not be identified (Figure 3). The $MoSi_2$ matrix material appeared very clean, with a low dislocation density and many 120-degree angle grain boundary intersections. Electron diffraction substantiated the XRD results, but no Mo_5Si_3 phase has been observed in the TEM.

Diffuse dark field imaging was performed on several $MoSi_2$/$MoSi_2$ grain boundaries to determine if any amorphous, or glassy, phases were present. A thin film of amorphous material was observed on selected boundaries (Figure 4), but was not present at all the boundaries examined. The amorphous phase is most likely SiO_2, which has been observed in other $MoSi_2$/SiC composites produced by hot-pressing [6]. The speckled appearance of the thin foil surface under diffuse dark field conditions in Figure 4 is typical of ion-milled materials and is caused by amorphization of the $MoSi_2$ during the ion milling process.

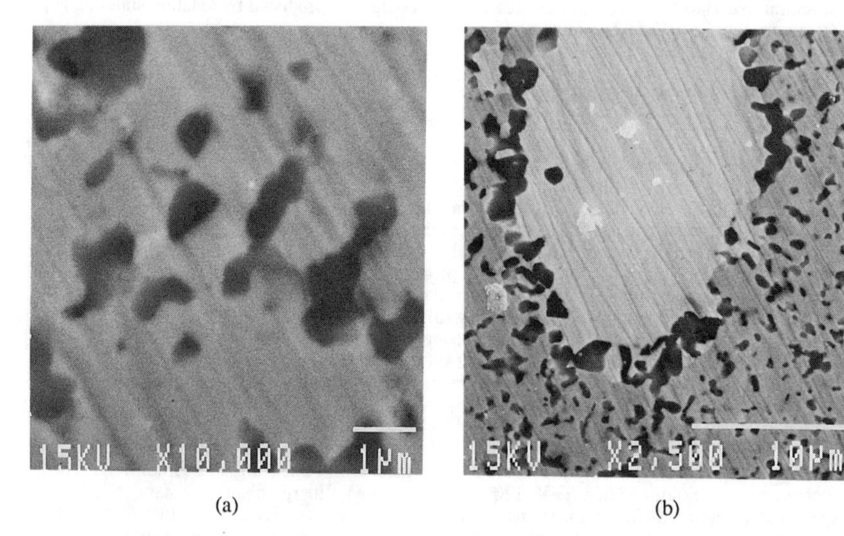

Figure 1. SEM micrographs of $MoSi_2$/SiC composite produced by hot-pressing blended powders at 1350°C for 2 h followed by 1 hour at 1700°C. a) β-SiC morphology, and b) Mo_5Si_3 inclusion and β-SiC dispersion in $MoSi_2$ matrix.

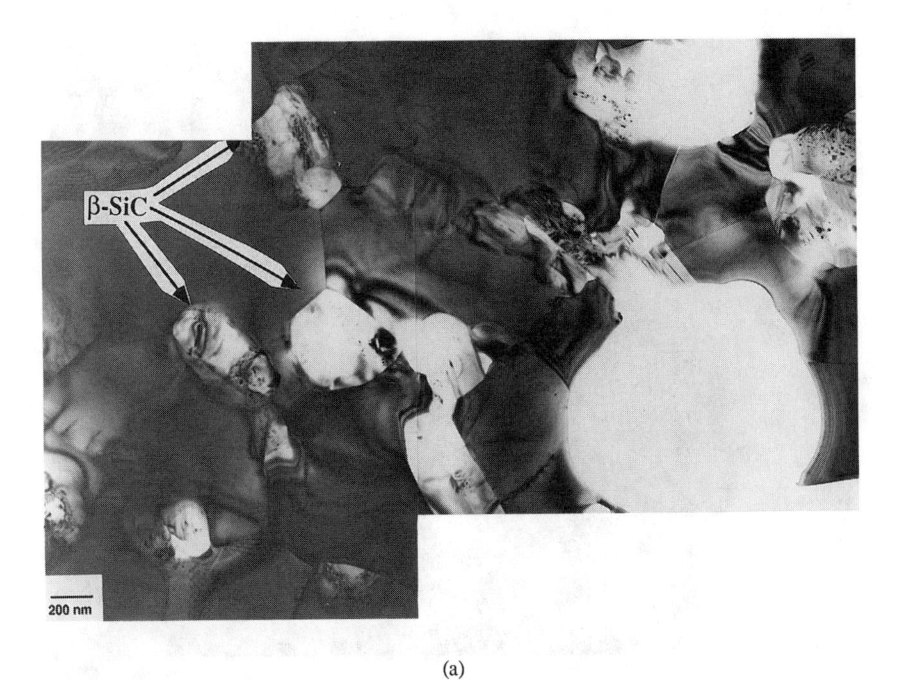

(a)

(b)

Figure 2. TEM micrographs of $MoSi_2$/SiC composite produced by hot-pressing blended powders at 1350°C for 2 h followed by 1 hour at 1700°C. a) Showing β-SiC on $MoSi_2$ grain boundaries, and b) Showing heavily faulted β-SiC particles.

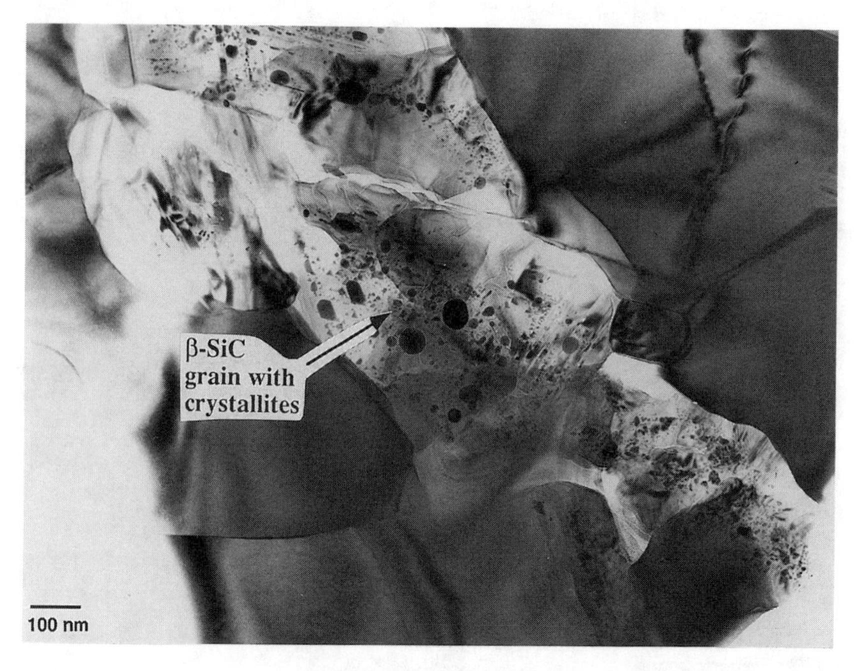

Figure 3. TEM micrograph of unidentified crystalline phase observed in the β-SiC particles. The small crystallites appears to be aligned along the stacking faults of the β-SiC phase.

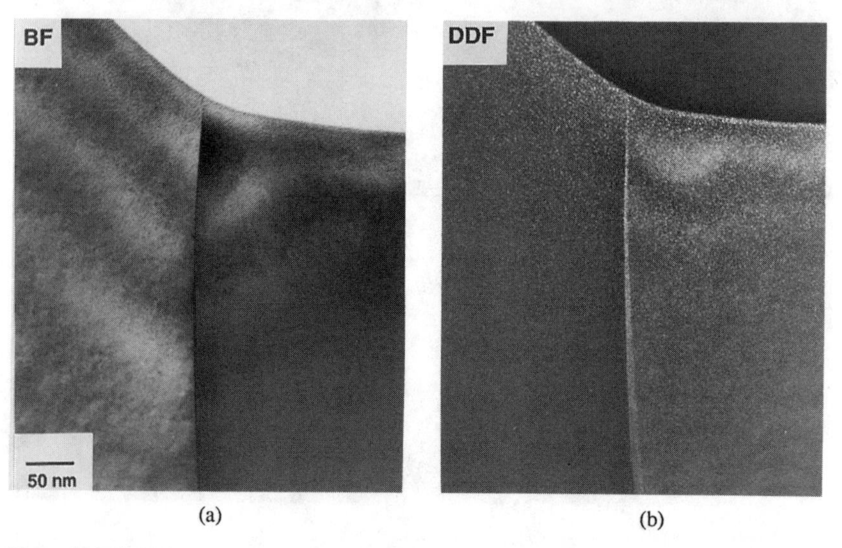

Figure 4. TEM micrographs of $MoSi_2/MoSi_2$ grain boundary in hot-pressed composite. a) Bright field micrograph of grain boundary, and b) diffuse dark field micrograph showing thin layer of amorphous phase (light region at boundary).

Strengths were determined from 4-point bend tests in air as a function of temperature from ambient up to 1200°C. Brittle fracture was observed at ambient temperature, and a bend strength of 475 MPa was obtained. A modulus of 296 GPa was measured at ambient temperature. At 1000°C, a fracture strength of 550 MPa was measured and yielding was observed in the load-deflection curve. At 1050°C and above, the material yielded quite readily and exhibited extensive plastic deformation. Fracture toughness was measured at ambient, 1000°C, and 1050°C using chevron-notched bend bars. A toughness value of 6.7 MPa√m was measured at room temperature. The toughness increased to 10.5 MPa√m at 1050°C. Bend strength and toughness as functions of temperature are shown in Figure 5, where 0.2% offset strengths are reported for test temperatures above ambient.

A hardness of 14.2 ± 0.1 GPa (1440 ± 12 HV_{1000}) was measured, and an indentation fracture toughness of 8.7 ± 0.1 MPa√m was calculated for this composite using the observed median cracking under Knoop indents made at a load of 15 kg. The cracking pattern was irregular, however, involving comminution to a rubble-like state at the tip so the fundamental significance of the toughness value is questionable. Crack interactions with the β-SiC particles, such as crack deflection and crack-wake bridging, were observed and provided direct evidence of the increased toughness due to crack-wake effects involving the β-SiC particles.

DISCUSSION

The formation of SiC particles in $MoSi_2$ is desirable since both creep strength and fracture toughness can be improved by this particulate-reinforced microstructure. Prior work [1-3] involving diffusion couples of Mo_2C and Si revealed that the SiC morphology formed at 1350 C by interdiffusion was ribbon-like and aligned in the direction of the diffusing species. The alignment of the SiC platelets suggested a cooperative precipitation reaction was occurring during interdiffusion of Si and Mo_2C to form $MoSi_2$. It was suggested that a reaction path between Mo_2C and Si would probably include the Mo_5Si_3C phase, which would provide a source of

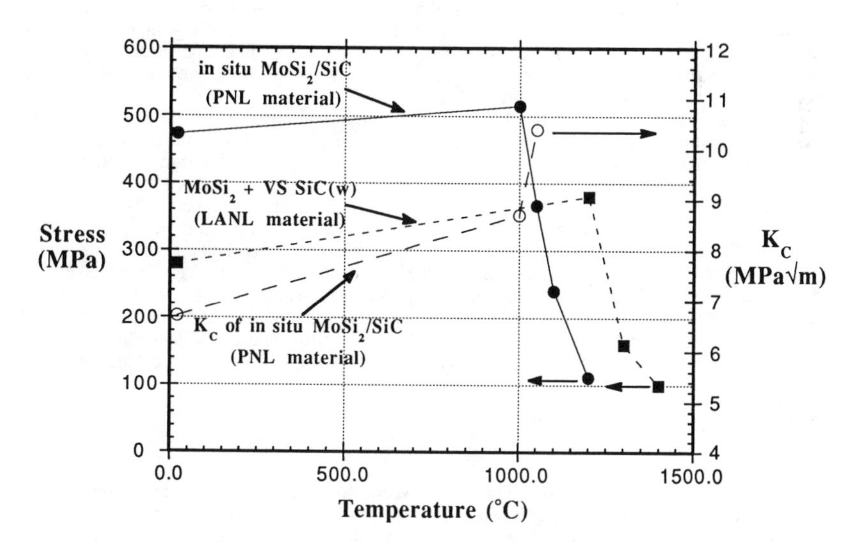

Figure 5. Bend strength and fracture toughness as functions of temperature for the *in situ* $MoSi_2$/SiC composite material. Also shown are data on SiC whisker-reinforced $MoSi_2$ [6]. Reported strengths are 0.2% offset yield strengths except at ambient temperature.

carbon from the reaction $Mo_5Si_3C \rightarrow Mo_5Si_3 + C$. Further interdiffusion of Si was suggested to result in the formation of SiC and $MoSi_2$, with retained Mo_5Si_3 as necessary to maintain stoichiometry.

The reaction between the blended powders during hot-pressing is expected to follow the same reaction path, resulting in randomly oriented SiC platelets in the hot-pressed body. The equi-axed particles are formed by coarsening of the SiC platelets at 1700°C. Observations of hot-pressed bodies at 1350°C without the 1700°C densification step revealed that the reaction was completed in 2 h at 1350°C and that the particles were still elongated at that stage of the process. While the volume fraction can be estimated from the balanced reaction (Eq. 1), the actual SiC morphology is not predictable with the available models and data. We expect the platelet size to depend on the reaction temperature as for other discontinuous precipitation reactions [7].

The TEM observations show that the β-SiC particles are located mainly on grain boundaries (Figure 2a), although occasionally an isolated β-SiC particle is observed within a $MoSi_2$ grain. This suggests that the fine-grained $MoSi_2$ material is effectively pinned at the boundaries by the β-SiC particles and that $MoSi_2$ grain growth is sluggish. The β-SiC particles are polycrystalline and heavily faulted (Figure 2b) and do not show any particular orientation relation with the $MoSi_2$ grains. This may indicate that the preferential alignment observed in the diffusion couples is due to the growth mechanisms occurring at the reaction front and not due to any crystallographic orientation relation.

The fracture toughness value of 6.7 MPa√m at room temperature is consistent with the SiC whisker-reinforced $MoSi_2$ and represents a 26% increase in toughness compared to unreinforced $MoSi_2$ [6]. The bend strengths reported here show a strength increase at 1000°C compared to room temperature strengths, which indicates that strengths are flaw dominated at low temperatures and that fracture strengths increase with increasing fracture toughness (Figure 5).

$MoSi_2$, though brittle at low temperatures, becomes ductile above a ductile-brittle transition temperature (DBTT) and deforms readily by dislocation slip accompanied by recovery associated with creep and/or cross slip. The strength decrease observed at temperatures above 1000°C indicates that the DBTT for this material is about 1000°C, which is somewhat lower than expected for this material. At temperatures above 1300°C, $MoSi_2$ single crystals deform on {110} planes with <100>- and <1$\bar{1}$0>-type dislocations observed [8]. Above the DBTT for this material, therefore, strengths are determined by dislocation flow and, thus, are reported as offset yield strengths. The *in situ* composite material shows not only a lower DBTT but also a lower temperature to achieve a given strength in the ductile region at elevated temperatures compared to the whisker-reinforced material (Figure 5). This suggests the presence of a grain boundary glass phase. The TEM observation of a small quantity of amorphous grain boundary phase (Figure 4) does tend to support this conjecture, but the amount of amorphous phase is quite small and does not readily explain a 300°C shift in the DBTT.

Another reason for a lower DBTT is the fine grain size observed for the *in situ* composite material. Fracture surface micrographs reveal a grain size of 1 to 2 μm for the *in situ* composite compared to about 5 to 10 μm for the whisker-reinforced composite [6]. Grain boundary sliding may be playing a role in the decreased elevated temperature strength of the *in situ* composite material since fine grain sizes increase grain boundary sliding deformation contributions.

A possible reason for the loss of strength above 1000°C compared to the SiC-whisker-reinforced material is that the *in situ* β-SiC particles are less effective than the SiC whiskers in impeding dislocation motion. The small whiskers (0.1 μm x 10 μm) used for the whisker-reinforced materials would have a smaller spacing in the slip plane than the roughly equiaxed particles in the *in situ* composites so that the former should give greater Orowan strengthening in resisting dislocation bowout. Also, the equiaxed particles provide less resistance than whiskers to recovery (or creep) by climb or interface relaxation because of their lower particle aspect ratio.

SUMMARY AND CONCLUSIONS

The solid state displacement reaction between Mo_2C and Si was used to synthesize a $MoSi_2$/SiC composite *in situ*. Composites made by vacuum hot-pressing Si and Mo_2C powders consisted of 1-μm diameter β-SiC particles uniformly dispersed in a $MoSi_2$ matrix. TEM examinations revealed a fine-grained $MoSi_2$-matrix containing a dispersed β-SiC phase located primarily on $MoSi_2$ grain boundaries. Evidence of a grain boundary amorphous phase was seen on some $MoSi_2$/$MoSi_2$ boundaries. Four-point bend and chevron-notched specimens gave a bend strength of 475 MPa and a fracture toughness of 6.7 MPa√m at room temperature. The bend strength increased to 515 MPa at 1000°C and then decreased to 112 MPa at 1200°C, while the fracture toughness increased to 10.5 MPa$\sqrt{m}$ at 1050°C.

Displacement reaction processing is a promising technique for producing *in situ* composites. Proper optimization of this processing technology can produce microstructures that are effective at dislocation pinning, retarding recovery at elevated temperatures, and increasing low-temperature toughness, such as is observed for whisker-reinforced composites. The advantage of the displacement reaction processing is lower cost (no expensive or hazardous whiskers to handle) and, possibly, near-net-shape fabrication if liquid phase consolidation can be implemented.

ACKNOWLEDGMENTS

This work was supported by internal U. S. Department of Energy (DOE) funds and by the U. S. Air Force at Wright Laboratory, Wright-Patterson AFB under U. S. DOE Contract DE-AC06-76RLO 1830 with Pacific Northwest Laboratory, which is operated for DOE by Battelle Memorial Institute.

REFERENCES

1. C. H. Henager, Jr., J. L. Brimhall, and J. P. Hirth, Scripta Metall. et Mater., 26 (1992) 585-589.
2. C. H. Henager, Jr., J. L. Brimhall, and J. P. Hirth, Mater. Sci. and Engr, A155 (1992) in press.
3. C. H. Henager, Jr., J. L. Brimhall, and J. P. Hirth, Mater. Res. Soc. Symp. Proc. (1992) in press.
4. J. S. Haggerty and Y. -M. Chiang, Ceram. Eng. Sci. Proc., 11[7-8], 757 (1990).
5. N. Claussen and J. Jahn, J. Amer. Ceram. Soc., 63, 228 (1980).
6. D. H. Carter, W. S. Gibbs, and J. J. Petrovic, Ceramic Materials and Components for Engines, V. J. Tennery, ed., p. 977, The American Ceramic Society, Westerville, OH (1989).
7. R. D. Doherty, Physical Metallurgy, 3rd edition, R. W. Cahn and P. Haasen, eds., p. 933, North-Holland, New York (1983).
8. Y. Umakoshi, T. Sakagami, T Hirano, and T. Yamane, Acta Metall. et Mater., 38, 909 (1990).

ELECTRON MICROSCOPY STUDIES OF Mo_5Si_3-$MoSi_2$ EUTECTIC COMPOSITES MODIFIED BY ERBIUM ADDITIONS

D. P. Mason and D. C. Van Aken, Department of Materials Science and Engineering, The University of Michigan, 2300 Hayward St., Ann Arbor, MI 48109-2125
J. F. Mansfield, Electron Microbeam Analysis Laboratory, The University of Michigan, 2455 Hayward St., Ann Arbor, MI 48109-2143

ABSTRACT

Erbium additions to Mo_5Si_3-$MoSi_2$ eutectic composites have been investigated in an effort to reduce oxygen impurity levels found in the $MoSi_2$ starting powders. Additions of 0.35 atomic percent Er were found to refine the script lamellar microstructure and increase the hardness of the composite as compared with the unmodified material. This increase in hardness was maintained over the temperature range 25°C to 1300°C. Oxygen rich Er particles have been observed in the Er-modified material suggesting that Er has acted as a gettering agent for oxygen. An intermetallic phase $Er_2Mo_3Si_4$ was also observed in the Er-treated material. The character of this phase was determined by convergent beam electron diffraction (CBED), wavelength dispersive spectroscopy (WDS) and X-ray dispersive spectroscopy (XEDS). The particles of $Er_2Mo_3Si_4$ were found to be heavily faulted which caused a reduction in the experimental diffraction symmetries. Thus, ambiguities arose during CBED analysis. Therefore, a standard sample of $Er_2Mo_3Si_4$ was prepared and used to aid in the identification of the faulted particles.

INTRODUCTION

High oxygen contents have been reported in commercial $MoSi_2$ powders and is present in the form of SiO_2[1]. This has been suggested as one of the reasons for the poor mechanical properties in $MoSi_2$-based materials[1,2]. Thus, SiO_2 may be the cause of the low strength values reported for Mo_5Si_3-$MoSi_2$ eutectic composites[2]. Therefore, alloying with the addition of Er was investigated in an effort to reduce the oxygen concentration in these materials. It was believed that the Er would act as a gettering agent and remove the oxygen from the melt by forming Er_2O_3. Furthermore, the presence of rare-earth oxides may contribute to the creep strength of the eutectic microstructure. Thus, arc-melted samples of the eutectic composition were modified by the addition of 0.35 at% Er. Two microstructural changes resulted from the addition of Er. First, the scale of the script microstructure was reduced; and second, the occurrence of pro-eutectic $MoSi_2$ was greatly diminished. (Fig. 1a,b) The results of the solidification study are reported elsewhere[3].

The addition of Er also had a beneficial effect on the mechanical properties. Hot hardness tests performed on Er-treated and untreated samples showed that the treated material had a higher hardness at all test temperatures (Fig. 2a). This increase in hardness may be due to an increasing constraint placed upon the $MoSi_2$ by the Mo_5Si_3 as the microstructure becomes more refined. However, in monolithic $MoSi_2$, Maloy et. al. have attributed the increase in hardness to the elimination of SiO_2 by adding 2 wt% C [1]. The role that material constraint and SiO_2 removal play in increasing the hardness of the composite are still under investigation.

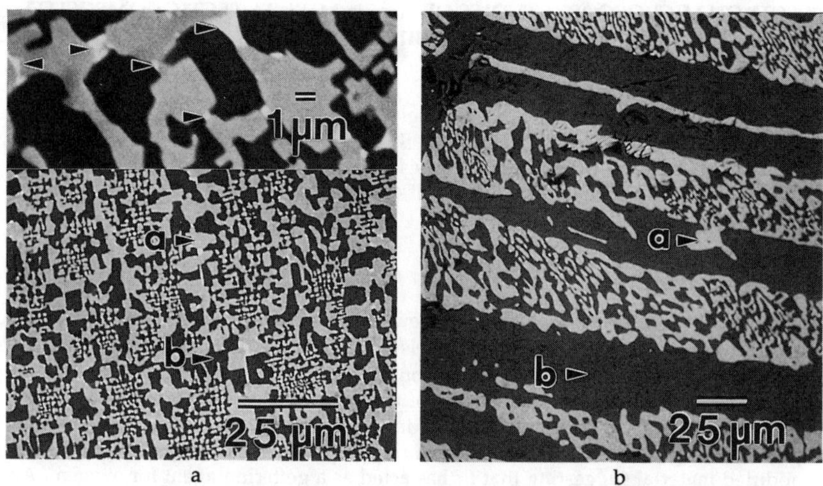

a b

Fig.1 Back-scattered electron (BSE) micrograph of a) Mo_5Si_3-$MoSi_2$ + 0.35 at.% Er. Er-rich particles appear as bright phases within the Mo_5Si_3-$MoSi_2$ eutectic and are indicated by arrows in the inset. b) unmodified Mo_5Si_3-$MoSi_2$ eutectic. (phase identification a-Mo_5Si_3 and b-$MoSi_2$)

It was the purpose of this investigation to determine what microstructural changes the addition of Er produced in an effort to better understand the observed changes in the hot hardness tests.

EXPERIMENTAL

Alloys of the eutectic composition (44.5 vol% Mo_5Si_3 and 55.5 vol% $MoSi_2$) were prepared by arc-melting elemental Mo and $MoSi_2$ powders in an argon gas atmosphere. Samples of the eutectic with additions of 0.35 to 1.75 at% Er and samples of the intermetallic compound $Er_2Mo_3Si_4$ were also prepared via arc-melting. Scanning electron microscopy (SEM) revealed that the desired script lamellar microstructure occured only in the 0.35 at% Er samples (Fig 1a). The microstructure became progressively more blocky in nature as the Er content was increased above 0.35 at% [3].

Electron microscopy studies were performed using the facilities of the Electron Microbeam Analysis Laboratory at the University of Michigan. A CAMECA model MBX electron microprobe analyzer, equipped with a three crystal spectrometer, was used for chemical analysis of the $Er_2Mo_3Si_4$ samples. Chemical analysis was performed using elemental standards of Er, Mo and Si. Thin foils of the Er-modified eutectic and the $Er_2Mo_3Si_4$ material were prepared by mechanical dimpling and room temperature ion-milling. Analytical electron microscope (AEM) studies were performed in a JEOL 2000FX. The AEM was equipped with a Tracor-Northern TN-5500 analysis system which was used to perform X-ray energy dispersive spectroscopy (XEDS) studies. A liquid nitrogen cold stage was used for CBED experiments.

Hardness as a function of temperature was determined using a Nikon QM Hot Hardness Tester under a vacuum of $2x10^{-5}$ torr. Five to ten indentations were made

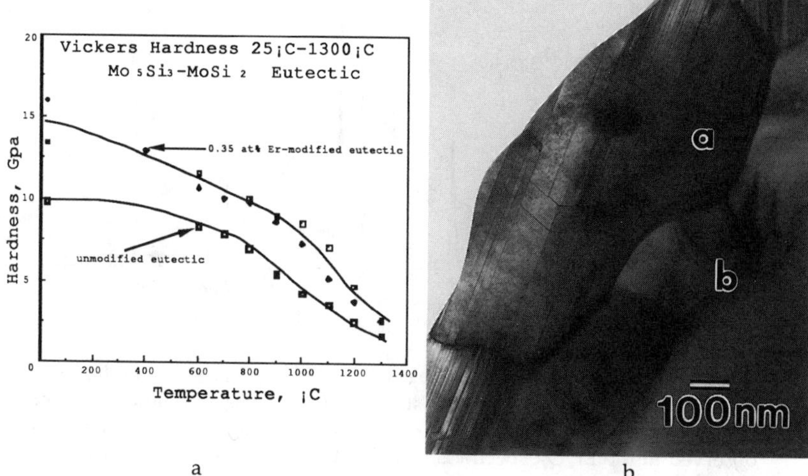

a b

Fig.2 a) Hot hardness tests of the Er-modified and untreated eutectic alloys. Results show an increased hardness for the Er-modified alloys at all test temperatures. b) Bright field micrograph showing the typical heavily faulted microstructure of the $Er_2Mo_3Si_4$ particles. (phase identification a-$Er_2Mo_3Si_4$ and b-$MoSi_2$)

at each test temperature in the range of 25°C to 1300°C using a 1000g load and an indentation time of five seconds.

RESULTS and DISCUSSION

Backscattered electron images (BSE) revealed small particles, which were Er-rich, in the modified material (Fig 1a). Microprobe analysis of the Er-rich particles revealed that ten percent contained oxygen, which may indicate that the Er was forming erbium oxide, while the remaining particles were an erbium, molybdenum, and silicon intermetallic. Two AEM techniques, XEDS and CBED, were employed to determine the composition and structure of these particles. XEDS analysis was also able to qualitatively identify an erbium-oxygen phase and an erbium-molybdenum-silicide intermetallic (Fig 3a,b). The oxygen-rich Er particles were approximately 150-750nm in size. These particles are most likely Er_2O_3 since there is only one stable oxide of Er reported in the open literature[4]. The $Er_2Mo_3Si_4$ particles ranged in size from 0.5 µm to 3 µm and primarily resided in the Mo_5Si_3 phase of the eutectic. They also exhibited a heavily faulted morphology (Fig 2b). Lattice parameters determined from selected area diffraction patterns (SADPs) were consistent with those reported for $Er_2Mo_3Si_4$[5]. However, analysis of CBED patterns of the faulted particles did not yield sufficient symmetry information to unambiguously identify the point or space group of the phase (Fig 4a,b). The extensive faulting occluded the true pattern symmetries.

Often CBED analysis is performed by comparing patterns from an unknown phase with those recorded from a standard. In this way any correlation between the

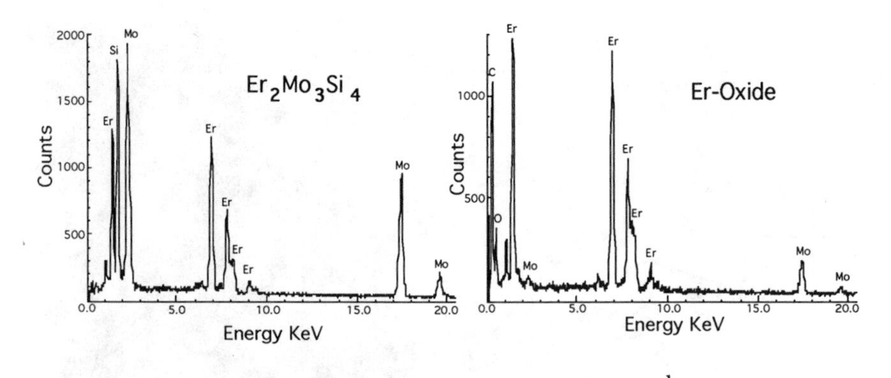

Fig.3 XEDS spectra from a) $Er_2Mo_3Si_4$ particle and b) Er_2O_3 particle in the eutectic. The peak ratios from the standard $Er_2Mo_3Si_4$ sample were almost identical to that of the faulted phase shown here. The large carbon peak observed in the Er_2O_3 spectra is from contamination during acquisition.

standard and faulted CBED patterns can be observed. Kaufman et. al. have demonstrated the utility of this type of analysis on heavily faulted Ni_3Mo precipitates in a Ni-Al-Mo alloy[6]. Thus, a standard sample of $Er_2Mo_3Si_4$ was prepared from Er, Mo and $MoSi_2$ powders by arc-casting in an argon atmosphere. Wavelength dispersive spectroscopy (WDS) analysis of the standard material was consistent with the phase $Er_2Mo_3Si_4$, and although XEDS analysis of the faulted material was not quantifiable (due to a lack of Er cross-section data) the peak ratios from the standard were almost identical to that of the faulted phase. CBED symmetry information was attainable from the standard material, as the microstructure was generally free of faults. CBED patterns of the [100], [001] and [010] zones are shown in figure 5. The whole pattern and bright field symmetries in the [100] and [001] patterns (Fig 5a and 5c) are both m, where the mirror is perpendicular to the 010 reciprocal vector. The whole pattern symmetry in the [010] zone is that of a single two fold axis. With reference to the tables in Buxton et. al. it may be deduced that the possible diffraction point groups are either 2_rmm_r or m[7]. It is important to note here that no higher symmetry elements were observed in the patterns taken from the standard material. The unit cell was determined to be primitive by projecting the first order laue zone reflections onto the zero order layer.

The space group was determined by a detailed examination of the dynamic absences present in the [100] and [001] CBED patterns (Fig 5a,c). In the case when a line of dynamic absences is parallel to a BF mirror line then a glide plane parallel to the incident beam may be deduced and when the line of dynamic absences is orthogonal to the BF mirror this indicates a 2_1 screw axis (or its equivalent) perpendicular to the mirror line [8]. There are two orthogonal dynamic absence lines in the 010 and 100 directions of the the [001] CBED pattern. This indicated a glide plane parallel to the [001] and a screw axis in the [010] direction. The [100] CBED pattern revealed only one line of dynamic absences in the 010 direction orthogonal to the mirror plane. This indicated the existence of a 2_1 screw axis in the [010]

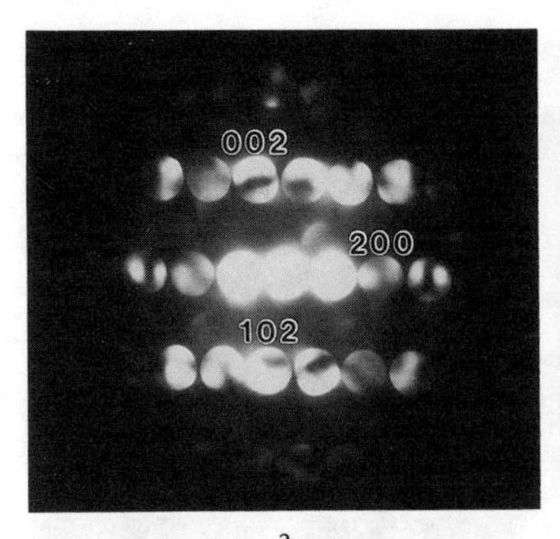

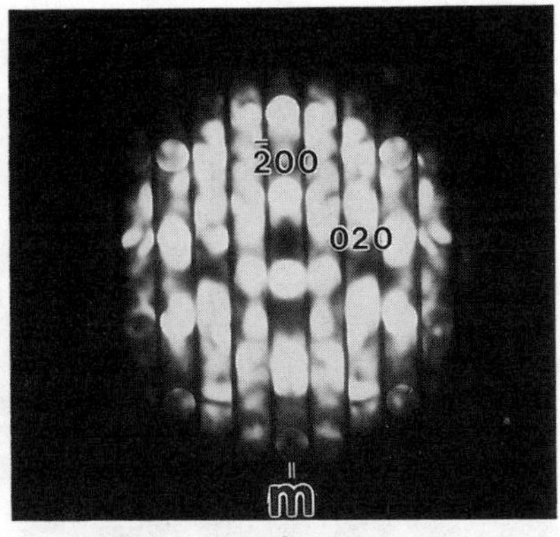

a b

Fig.4 CBED patterns taken from the faulted particles in the eutectic material. The zone axes were later determined to be (a) [010] and (b) [001] of $Er_2Mo_3Si_4$.

direction. There is only one possible monoclinic space group with this configuration of screw axis and glide plane, so these observations were sufficient to identify the space group as $P2_1/c$ which is consistent with previous studies of $Er_2Mo_3Si_4$[5]. The [010] CBED pattern obtained from the faulted particles (Fig. 4a) compares well with the respective CBED results from the [010] pattern of the standard sample (Fig 5b). Much of the symmetry information was occluded in the CBED pattern shown in figure 4b, however, the pattern was determined to be that of the [001]. This was deduced by a detailed examination of SADPs and the zone's relative position to other known zone axes in the reciprocal lattice. Therefore, it was evident that the heavily faulted particles found in the Er-modified eutectic composite were $Er_2Mo_3Si_4$.

Oxygen-rich Er particles observed in the modified material may indicate that the eutectic has been purified by the addition of Er. These particles may also contribute directly to the increase in hardness of the material by eliminating the glassy SiO_2 phase. However, this increase in hardness may also be a result of the refined eutectic microstructure of the Er-treated samples. The role that the $Er_2Mo_3Si_4$ phase plays in the mechanical properties has yet to be determined. High resolution electron microscopy (HREM) is being used to determine the nature of the faulting.

CONCLUSIONS

Two Er-rich phases have been observed in the Mo_5Si_3-$MoSi_2$ + 0.35 at.% Er eutectic composites, an oxygen rich phase and the $Er_2Mo_3Si_4$ phase. The presence of the oxygen-rich Er particles may indicate that the erbium is acting as a purifying agent for the silicide phases. The $Er_2Mo_3Si_4$ phase was characterized by CBED, WDS and XEDS methods.

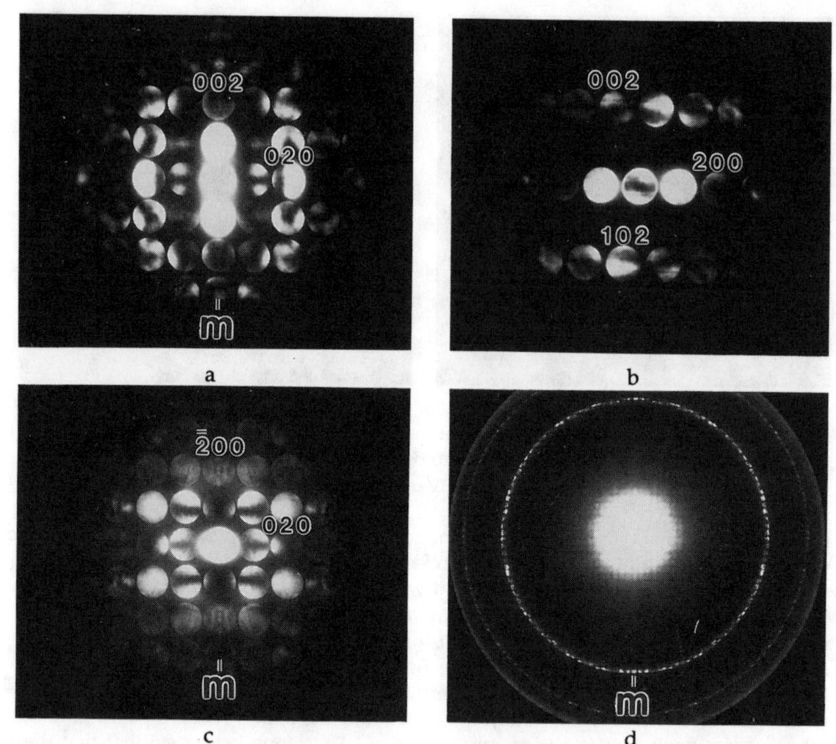

Fig.5 CBED patterns obtained from the standard $Er_2Mo_3Si_4$ sample. Zero order laue zones (ZOLZ) are shown in (a) [100], (b) [010] and (c) [001] where the m's refer to mirror lines and d) the first order laue zones of the [001] (FOLZ)

REFERENCES

1. S. Maloy, A. H. Heuer, J. Lewandowski, and J. Petrovic, J. Am. Ceram. Soc.,74 (10), pp. 2704-706 (1991)
2. R. Gibala, A. K. Ghosh, D. C. Van Aken, D. J. Srolovitz A. Basu, H. Chang, D. P. Mason, and W. Yang, to appear in Mater. Sci. and Eng., (1992)
3. D. P. Mason, D. C. Van Aken and J. Mansfield, to be published in J. Mat. Research
4. Pearson's Handbook of Crystallographic Data for Intermetallic Phases, P. Villars and L. D. Calvert (eds), 2nd ed, vol. 3, ASM, Materials Park, OH (1991) pp. 3183
5. O. I. Bodak, Yu. K. Gorelenko, V. I. Yarovets and R. v. Skolozdra: Izvestiya AkademII Nauk SSSR, Neorgan Icheskie Materialy, 20(5), pp.741-43 (1984)
6. M. J. Kaufman, J. A. Eades, M. H. Loretto and H. L. Fraser: Metallurgical Transaction A, 14A, pp.1561-71 (1983)
7. B. F. Buxton, J. A. Eades, J. W. Steeds and G. M. Rackham: Phil. Trans. R. Soc. London, A281, pp. 171 (1976)
8. J. W. Steeds, Quantitative Electron Microscopy, J. N. Chapman and A. J. Craven (eds.), Scottish Universities Summer School of Physics, Glasgow (1983) pp. 49

IN-SITU GROWTH OF SiC IN $MoSi_2$ BY MELT PROCESSING

DANIEL J. TILLY, J.P.A. LöFVANDER and C.G. LEVI
Engineering Materials Department, University of California,
Santa Barbara CA 93106

ABSTRACT

The growth of SiC during solidification of Mo-Si-C melts was investigated to explore the potential for developing *in-situ* refractory reinforcements for $MoSi_2$ matrices. Volume fractions of up to 20% were readily incorporated by arc melting. Primary β-SiC grows as equiaxed particles, plates and hopper crystals, whereas secondary SiC grows with ribbon, thin flake or whisker morphologies. The dominant facets of both primary and secondary SiC are of the {111} and {002} type, which are also characteristic of the equilibrium crystal shape. There is a clear orientation relationship between the phases wherein the close packed planes and directions of the $C11_b$ and B3 structures are parallel.

INTRODUCTION

Molybdenum disilicide ($MoSi_2$) is attractive for high temperature structural applications owing to its high melting point (T_M = 2300 K) and good oxidation resistance associated with the formation of a protective silica scale [1,2]. The ordered tetragonal $C11_b$ structure offers the potential of high strength and stiffness at elevated temperatures, but its intrinsic brittleness hinders the application of monolithic $MoSi_2$ as a structural material. Compositing schemes based on ceramic fibers or ductile refractory metal reinforcements are under investigation to improve both the toughness and creep strength of $MoSi_2$ [3-6]. Thermal expansion mismatch between $MoSi_2$ and the reinforcements may cause matrix cracking during processing or thermal cycling [7]. Introduction of SiC into $MoSi_2$ has been successfully used to lower the thermal expansion coefficient of the matrix and suppress cracking in Mo-wire reinforced composites [6]. SiC whiskers and particulates have also been shown to improve the creep resistance of $MoSi_2$ [2,8]. Thus, SiC additions to $MoSi_2$ are of significant interest, but often lead to less than optimal microstructures if implemented by powder blending approaches. This paper reports on a study aimed at growing SiC-reinforcements *in-situ* by solidification processing of Mo-Si-C melts of suitable compositions.

EXPERIMENTAL TECHNIQUES

The alloys investigated in this study were produced with high purity elemental components; 99.97% Mo, 99.995% Si and 99.9% C (all in weight percent). Alloys were prepared by arc-melting under an atmosphere of purified argon containing less then 0.1 ppb oxygen. The buttons, weighing approximately 15 g each, were flipped and remelted several times to ensure macroscopic chemical homogeneity. The materials were characterized by optical, scanning (SEM) and transmission (TEM) electron microscopy. Specimens were prepared for optical and SEM microscopy by standard grinding and polishing techniques; a solution consisting of 2 % hydrofluoric acid, 25 % nitric acid (by volume) and lactic acid making up the balance was used for deep-etching to reveal the SiC phase morphologies. SEM examination was performed in a JEOL SM840A microscope equipped with a Tracor Northern TN5500 Energy Dispersive X-ray Spectroscopy (EDS) system. Samples suitable for TEM were produced by cutting with a low-speed diamond wafering blade, polishing to ~80 μm and subsequently dimple ground to a final thickness of < 10 nm. The disks were thoroughly washed in acetone and methanol before ion milling. Subsequent TEM analysis was performed in a JEOL 2000FX microscope operated at 200 kV equipped with a Link Analytical eXL EDS system.

RESULTS AND DISCUSSION

A tentative liquidus projection for the Mo-Si-C system is given in Figure 1. This diagram modifies the original version of Nowotny et al. [9] by incorporating recent revisions of the binary systems and experimental work in the vicinity of the $MoSi_2$-SiC eutectic, as discussed in greater detail elsewhere [10]. This liquidus projection has three main regions representing the solidification of primary graphite, SiC and $MoSi_2$. The graphite and SiC liquidi are separated by the monovariant line L+SiC+C emerging from the peritectic in the Si-C binary. A two-fold saturation line $L \rightarrow MoSi_2+SiC$ originates at the quasi-binary eutectic Mo-66Si-2C (in at.%). Thus, SiC is not only thermochemically stable in $MoSi_2$ at elevated temperatures, but may also be grown from the melt as a primary and/or secondary phase by suitable control of the alloy chemistry. According to Figure 1, the maximum volume fraction of primary SiC which can be produced through melt processing is ~35%.

Alloys with the compositions noted in Figure 1 were prepared to explore the microstructural development and morphologies of the SiC phases produced. (Both alloys were slightly richer in Si than the quasi-binary compositions to ensure that their solidification paths were on the same side of the liquidus surface.)

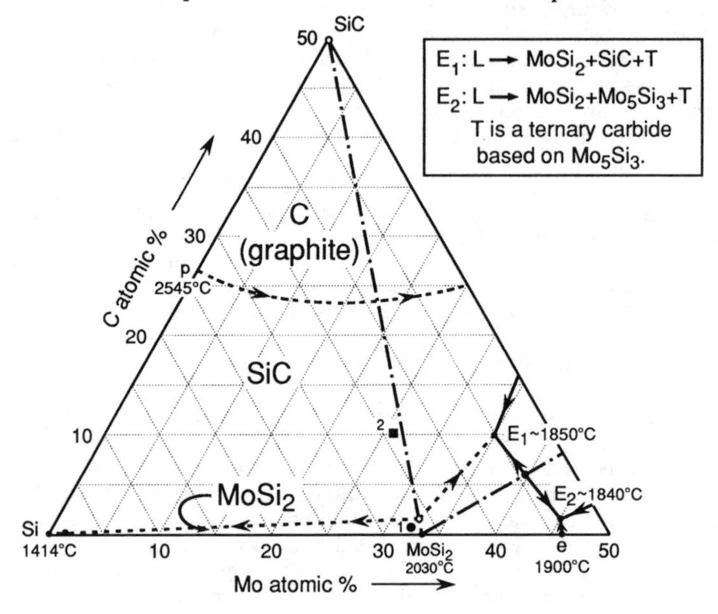

Figure 1. Tentative partial liquidus projection for the Si-rich corner of the Mo-Si-C system, from reference [10].

The hypoeutectic alloy (1) is Mo-67Si-1C and should produce primary $MoSi_2$, whereas the hypereutectic alloy (2) is Mo-63.5Si-10C and should form primary SiC. Since both phases are essentially stoichiometric compounds, the liquid composition should trace a straight path on the liquidus surface directly away from the composition of the primary phase as temperature decreases and solidification progresses. Once the liquid composition reaches the line of two-fold saturation, both phases should grow concurrently provided there is no significant kinetic hindrance for nucleation of SiC on $MoSi_2$ and vice-versa. If neither phase can accommodate the excess Si in solid solution, the liquid composition will become progressively enriched in Si as solidification proceeds along the monovariant line toward the Si-rich corner. Eventually, a small amount of elemental Si should form from the last pockets of liquid.

The hypoeutectic alloy (1) exhibited only secondary SiC phases, mostly in the form of (sub-micron) thin ribbons or flakes with very high aspect ratios, as illustrated in the SEM micrograph of a deep etched specimen in Figure 2(a). (Whiskers were also observed but are not shown in this figure). The array of parallel ribbons suggests that the two $MoSi_2$ and SiC phases can grow coupled along the line of two-fold saturation; the structure appears as an irregular eutectic in metallographic sections. TEM analysis of secondary phases consistently revealed the sphalerite (B3) structure of β-SiC (cubic, space group $F\bar{4}3m$). The cross section of the ribbon in Figure 2(b) is oriented along an $\langle 011\rangle_{\beta-SiC}$ zone axis (which is also the axis of the ribbon), whereupon the dominant facets are found to be of the {111} type and the edges of the {002} and {111} type.

An orientation relationship was observed wherein the close packed planes and directions of both phases are parallel, i.e.

$$(11\bar{1})_{SiC} \parallel (110)_{MoSi_2} \text{ and } [011]_{SiC} \parallel [001]_{MoSi_2}$$

This is consistent with the suggested coupled growth of the eutectic-like structure along the line of two-fold saturation.

The hypereutectic alloy (2) produced a substantial amount (~20%) of primary SiC, growing primarily in equiaxed and platelike morphologies as revealed in the SEM views of deep etched specimens, Figure 3(a). Hopper crystal variations of these morphologies were also observed, as illustrated in Figures 3(b-d).

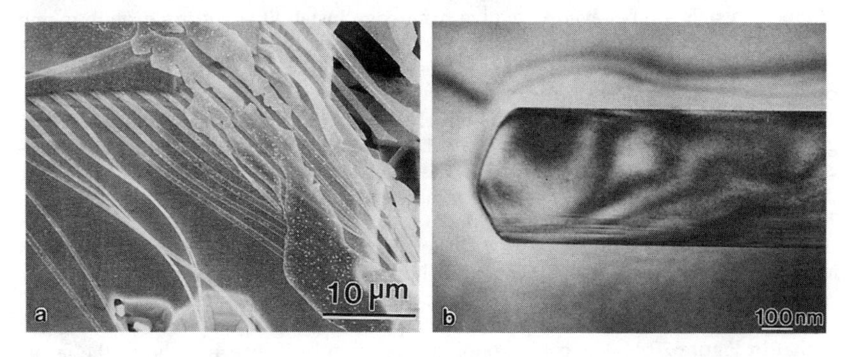

Figure 2. Secondary SiC in hypoeutectic alloy 1: (a) SEM view of secondary SiC ribbons and flakes with $MoSi_2$ matrix partially etched away. (b) TEM view of SiC ribbon along its axis, corresponding to the $[011]_{\beta\text{-}SiC}$ zone.

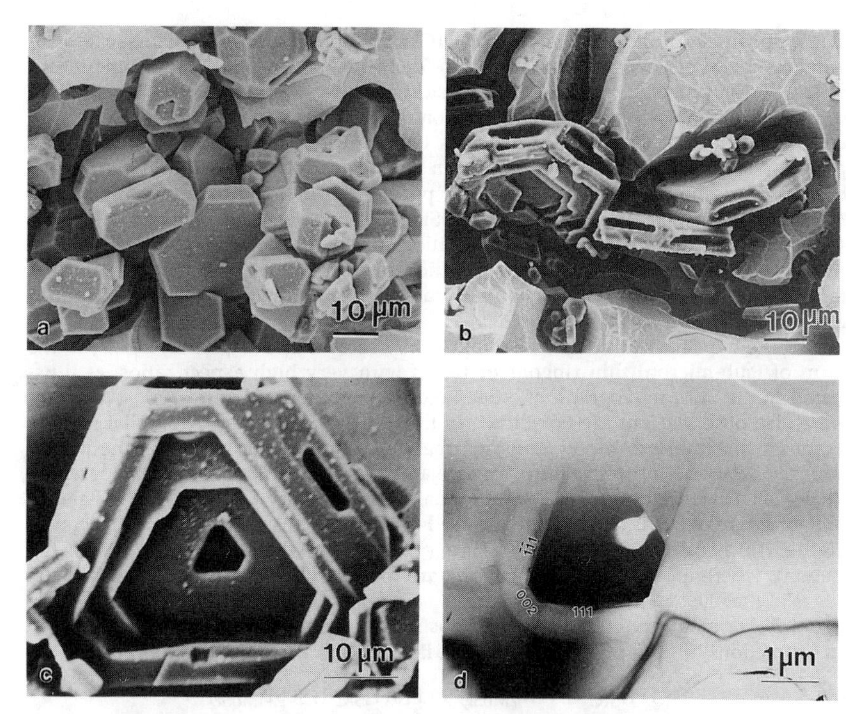

Figure 3. Primary SiC morphologies in hypereutectic alloy 2: (a) Equiaxed particles and platelets revealed by deep etching $MoSi_2$ matrix, (b) cluster of hopper crystals in the same specimen, and (c) closer view of a hopper crystal showing preferential growth at the edges of a {111} facet. (d) TEM micrograph of a hopper crystal normal to a [110] zone showing $MoSi_2$ matrix solidified within a triangular cavity similar to that in (c). Note the shrinkage porosity and the bounding {111} and {002} facets.

Hopper crystals are believed to evolve from edge instabilities during growth owing to the evolution of different supersaturations between the center and the edges of a crystal facet [11]. In the present case, these variations arise from the local buildup of Mo and concomitant depletion of Si and C in front of the SiC/melt interface. The SEM micrograph in Figure 3(c) clearly shows the preferential nucleation and growth of ledges in the periphery of a {111} crystal facet, consistent with the classical hopper growth mechanism. Note the progressive expansion of the triangular cavity in the primary SiC as growth evolves out of the plane of the micrograph. The cavity is later filled with solid $MoSi_2$ when the solidification path reaches the two-fold saturation line in Figure 1. The TEM micrograph in Figure 3(d) reveals $MoSi_2$ present in the center of a large SiC hopper crystal in a separate specimen.

TEM analysis of the primary particles confirm that they also have the β-SiC structure. Figure 4(a) shows a primary SiC platelet along a ⟨011⟩ zone axis, similar to that in Figure 2(b). The corresponding projection in Figure 4(b) shows the facets to be of the {$11\bar{1}$} and {002} type, which are thus inferred to be the planes of slowest growth. (Note the same facets on the ledges of the hopper crystal, Figure 3d.) In contrast, growth along the [011] axis seems to be relatively easy, as suggested by the

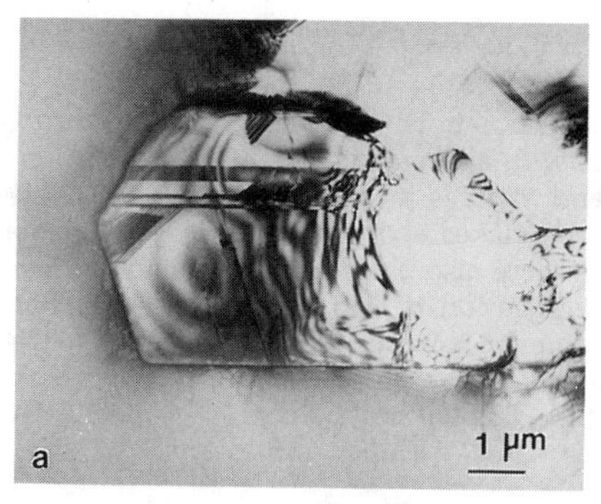

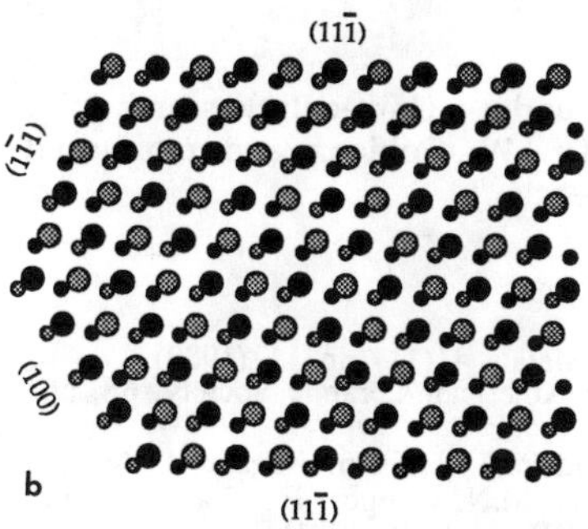

Figure 4. TEM of primary β-SiC platelet (a) showing $\langle 011 \rangle_{\beta\text{-SiC}}$ zone axis with {111} and {002} facets and twins parallel to the {111} planes. (b) Structure projection corresponding to the zone axis in (a); the larger circles are Si atoms and the smaller C atoms; the two shades represent the 0 and 1/4 levels along [011].

Figure 5. SEM view of deep-etched sample showing possible re-entrant twins at the edge of a primary SiC plate.

evolution of long ribbons in this direction. Analysis of Figure 4(b) reveals that growth along the slower $\langle 11\bar{1} \rangle$ and $\langle 002 \rangle$ directions involves the stacking of alternating layers consisting of all C or all Si atoms, whereas growth along the faster $\langle 011 \rangle$ direction involves layers containing C and Si in the proper stoichiometry. Similar relationships between stacking pattern and ease of propagation have been observed in the growth of TiB and TiB_2 from Ti alloy melts [12-14]. While these observations provide some insight on the role of crystallography on the growth behavior, the understanding of the relevant mechanisms is far from complete.

The dominant facets observed in these SiC crystals are consistent with their equilibrium shape predicted from interfacial energy considerations [15]. It is not clear, however, why plate crystals also evolve in addition to the equiaxed particles expected for the cubic β-SiC phase [15]. One possible explanation is suggested by Figure 5, which shows features reminiscent of re-entrant twins with the relevant twin planes parallel to the long {111} facets of the plate. Re-entrant twins are known to enable the platelike growth of Si [16] and could play a similar role in β-SiC. Twins are indeed present in many particles, as confirmed by high resolution TEM of the striations in Figure 4(a) reported elsewhere [10]. Understanding the evolution of these twins and their role in the growth process may open an avenue for morphological control in $MoSi_2$/SiC microstructures.

The orientation relationship noted between the $MoSi_2$ matrix and the secondary SiC is also consistently observed with the primary SiC phases. This suggests that $MoSi_2$ can readily nucleate and grow epitaxially on the primary SiC. The implications of the orientation relationship and the details of the interfacial structure have been elaborated in a separate publication [10].

CONCLUDING REMARKS

It has been shown that SiC additions can be developed *in-situ* during solidification processing of Mo-Si-C melts with appropriate chemistries. β–SiC can grow as primary or secondary phases and evolves in a variety of morphologies including equiaxed particles, plates, ribbons, flakes and whiskers. Hopper crystal forms of the plates and equiaxed particles were also observed, presumably arising from instabilities associated with the segregation of Mo during growth. Both primary and secondary particles show dominant facets of the $\{111\}_{\beta\text{-SiC}}$ and $\{002\}_{\beta\text{-SiC}}$ type, and exhibit a clear and consistent orientation relationship with the matrix. Preliminary evidence suggests that a re-entrant twin growth mechanism may be responsible for the development of platelike shapes in primary SiC.

ACKNOWLEDGMENTS

This work was sponsored by DARPA under a University Research Initiative grant N00014-86-K-0753 supervised by Dr. W. Coblenz and monitored by Dr. S.G. Fishman of the Office of Naval Research.

REFERENCES

[1] P.J. Meschter and D.S. Schwartz: Journal of Metals, **41**(11), pp. 11, (1989).
[2] F.D. Gac and J.J. Petrovic: Journal of the American Ceramic Society, **68**, pp. 200, (1985).
[3] J.P.A. Löfvander, J.Y. Yang, C.G. Levi and R. Mehrabian: Advanced Metal Matrix Composites for Elevated Temperatures, eds. M.N. Gungor, E.J. Lavernia and S.G. Fishman, ASM International, Materials Park, OH, pp. 1-10, (1991).
[4] T.C. Lu, A.G. Evans, R.J. Hecht and R. Mehrabian: Acta Metall. & Mater., **39**(8) pp. 1853-1862, (1991).
[5] J. Besson, M. De Graef, J.P.A. Löfvander and S.M. Spearing: accepted for publication in Journal of Materials Science, (1992).
[6] M. Maloney and R.J. Hecht: to be published in the Proceedings of the First High Temperature Structural Silicides Workshop, eds. A.K. Vasudevan and J.J. Petrovic, (1992).
[7] T.C. Lu, J. Yang, Z. Suo, A.G. Evans, R. Hecht and R. Mehrabian: Acta Metall. & Mater. **39**(8), pp. 1883-1890, (1991).
[8] R.M. Aikin, Jr. and L. Christodoulou: to be published in the Proceedings of the First High Temperature Structural Silicides Workshop, eds. A.K. Vasudevan and J.J. Petrovic, (1992).
[9] H. Nowotny, E. Parthé, R. Kieffer and F. Benesousky: Monatsh. Chem., **85,** pp. 255-272, (1954).
[10] D.J. Tilly, J.P.A. Löfvander and C.G. Levi: submitted to Materials Science and Engineering.
[11] I. Minkoff and B. Lux: J. Crystal Growth, **22**, pp. 163, (1974).
[12] M.E. Hyman, C. McCullough, J.J. Valencia, C.G. Levi and R. Mehrabian: Metallurgical Transactions A, **20A**, pp. 1847-1859, (1989).
[13] M.E. Hyman, C. McCullough, C.G. Levi, R. Mehrabian: Metallurgical Transactions A, **22A**(7), pp. 1647-1662, (1991).
[14] M. De Graef, J.P.A. Löfvander and C.G. Levi: Acta Metall. & Mater., **39**(10), pp. 2381-2391, (1991).
[15] G. A. Wolff: Intermetallic Compounds, ed. J. H. Westbrook, John Wiley & Sons New York, NY, pp. 91-92, (1967).
[16] G.F. Bolling and W.A. Tiller: Metallurgy of Elemental and Compound Semiconductors, Interscience, NY, (1961).

PART IV

Interface Properties, Processing, and Advanced Intermetallic Composites

MECHANICS OF INTERFACIAL FAILURE DURING THIN-SLICE FIBER PUSHOUT TESTS

D. A. Koss, M .N. Kallas, and J. R. Hellmann
Center for Advanced Materials, The Pennsylvania State University, University Park, PA 16802

ABSTRACT

Interfacial failure along the fiber-matrix interface during fiber push-out tests is examined for conditions where (1) a "thin-slice" specimen geometry is used and (2) matrix plasticity is necessary for large scale fiber displacement. The influence of both test geometry and the combination of thermally and mechanically induced stresses on the potential failure process is discussed. Experimental results are presented which illustrate a range of interfacial failure processes as a consequence of different combinations of specimen geometries, thermally induced residual stresses, and mechanically applied stresses. It is concluded that mode I crack opening and growth induced by specimen bending can be a major contributor to the interfacial failure process. Accurate quantitative measurements of interfacial shear properties must therefore rely on test configurations in which specimen bending is eliminated.

INTRODUCTION

The recognition that the interfacial shear strength can control the strength and crack growth behavior of fiber-reinforced composites has led to many studies directed at the fiber-matrix interfacial shear behavior [see, for example, ref's 1 and 2]. In metal-matrix composites (MMC) as well as intermetallic matrix composites (IMC), the shear strength is usually a result of chemical bonding across the "interface" as well as the presence of large thermally induced residual stresses arising from the thermal expansion mismatch between the fiber and matrix. The combination of these two factors usually results in a high observed interfacial shear strength, especially at low temperatures where mechanical interlocking between the matrix and fiber is accentuated due to thermally induced clamping effects[3-7]· Furthermore, in some IMC systems, it is also likely that plastic deformation of the matrix must occur before large-scale (> 1 to 2 μms) fiber displacement can take place. Despite their tendency to be brittle in tension, many intermetallics may plastically deform under local stress states consisting of the large compressive stresses and high equivalent shear stresses, which we will show are typical under the indenter load in an IMC.

The purpose of this paper is to review interfacial failure under specific test conditions where "thin-slice" fiber pushout specimens are used and matrix plasticity is necessary for large-scale fiber pushout. The use of thin-slice specimens is common in IMC and MMC studies where high interfacial shear strengths limit specimen thickness if large-scale fiber displacements (> 1 μm) are achieved prior to indenter failure[3, 8-14]. Unfortunately, for experimental convenience, such tests are usually performed by supporting the thin-slice specimen over holes or slots, which are usually a factor of two or more larger than the fiber diameter. This results in a combination of factors which at best complicate the interpretation of observed results and at worst create failure sequences that are unique to the test itself and which are not indicative of fiber pullout behavior. These factors include the following:

(1) The distributions and relative magnitudes of stress components of the normal and shear stresses along the interface depend on (a) the specimen geometry[15] and (b) the degree of thermoelastic clamping.

(2) There is a strong likelihood that mixed-mode I and II crack growth will occur induced by bending stresses characteristic of the thin-slice test geometry. The problem is that these cracks propagate along the interface at the specimen back-face opposite the indenter and can initiate the eventual failure process.

(3) Small scale fiber displacements (usually < 1 μm) should occur prior to maximum load due to mode II crack growth along the fiber-matrix interface near the topface adjacent to the indenter. However, in thin-slice pushout "back-face" cracking may precede and pre-empt the onset of the topface shear cracking.

(4) Matrix plasticity may also contribute to stable fiber displacement under rising loads at large (> 1 μm) displacements.

The issue of small scale fiber displacements due to crack growth along fiber matrix interfaces (factor 3 above) has been addressed in detail analytically for fiber push-out conditions in the absence of specimen bending and for brittle matrix composites; (see for example ref's 16-19). In this paper, we provide a brief review of the mechanics of thin-slice pushout behavior with references to the role of plasticity and especially to bending-induced crack growth (factors 1-3 above) in controlling the thin-slice fiber pushout responses in MMC's and IMC's. Specific examples which illustrate differences in experimentally observed responses will be given.

RESULTS AND DISCUSSION

The Interfacial Stress State: Computational Procedure

The stress state within the matrix adjacent to the fiber was obtained using an axisymmetric finite element model (FEM) as shown in Figure 1. We have previously performed the analysis for a range of specimen thicknesses, assuming perfect bonding at the interface and accounting for matrix yielding but ignoring thermally induced residual stresses[14]. In those MMC's or IMC's which do not exhibit a chemical bond between the fiber and matrix, at low temperatures thermoelastic clamping will maintain registry between the matrix and fiber until the applied load initiates debonding by shear. Thus analyses were also performed to simulate the combination of thermally induced stresses as well as the mechanically-induced stresses which are exerted on the inner part of the fiber by the flat indenter; some of those results will be presented in section 3 below. Several push-out specimen geometries, within the limits shown in Figure 1, were examined to evaluate the effect of specimen configuration on experimentally observed interfacial failure modes. In all cases, the specimen was supported by a thick, rigid plate with a hole whose diameter was varied; the fiber was always concentrically located with respect to the hole.

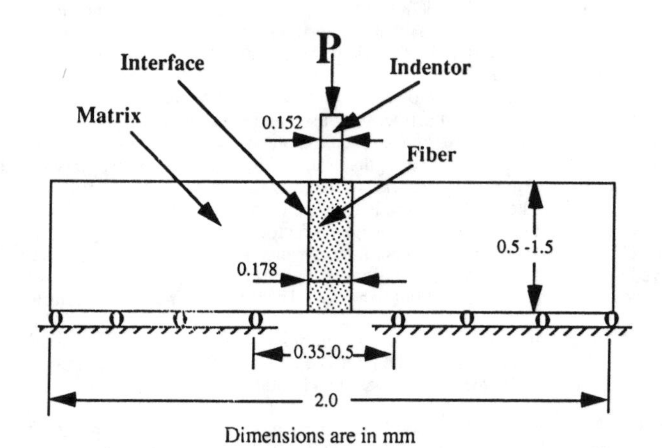

Figure 1 A schematic of a "thin-slice" fiber push-out test .

Analyses of a fiber supported over a slot whose width is equal to the hole diameter was also performed[20]. Those results indicated stress states similar to those when the fiber was supported over a hole of similar size. Furthermore, the interfacial shear stresses were relatively insensitive to whether the bottom face of the specimen was fixed to the support plate or allowed to slide, as shown in Figure 1[19,21].

"Thin-Slice" Interfacial Stress States: Case I - Negligible Thermally Induced Stresses

The combination of a well-bonded fiber experiencing minimal thermally induced stress describes the preferred condition for most creep-resistant IMC's loaded at high temperatures. For the case where the thermally induced stresses are negligible compared to the mechanical stresses, we have used FEM to analyze 152 μm diameter sapphire filaments (E = 379 GPa and υ = 0.22) embedded in a niobium matrix (E = 95 GPa and υ = 0.29 with a yield stress adjacent to the fiber of about 380 MPa)[9]. The niobium-sapphire is a convenient model system because the nearly matched thermal expansion coefficients between niobium and sapphire result in negligibly small residual thermoelastic stresses. Furthermore, it is possible to process a niobium/sapphire composite such that a very good chemical bond exists at the fiber-matrix interface[22-24].

The computational analysis was based on a thin-slice specimen geometry with a circular support hole varied between 0.35 and 1.00 mm in diameter while specimen thickness ranged from 0.2 to 1.5 mm[15]. In order to compare the stress distributions of different test geometries, the interfacial stress components (normalized with respect to indenter pressure) were plotted as a function of the position (normalized with respect to specimen thickness) along the interface. The indenter pressure (p) is defined here as the external load (P) on the indenter divided by the indenter cross-sectional area (πr_i^2).

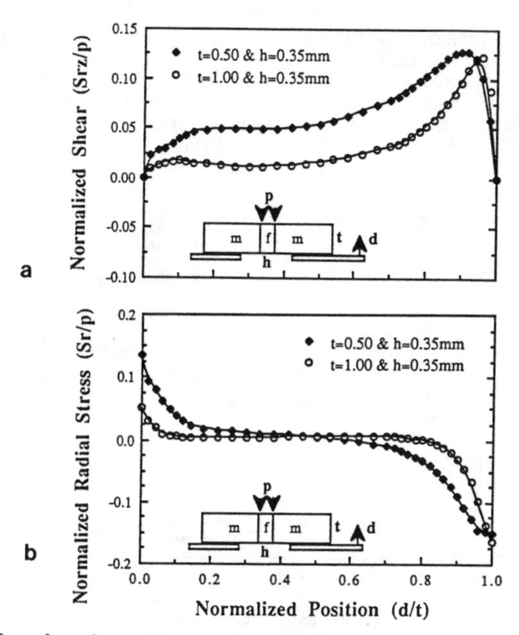

Figure 2 Effect of specimen thickness on (a) the shear stress distribution and (b) the radial stress distribution at the F/M interface of a niobium-sapphire composite.

Figure 2 shows the effect of specimen thickness on the interfacial shear and radial stress distributions for two specimen geometries. In all cases, the cylindrical coordinate system used to denote the stress components corresponds to the cylindrical geometry of the fiber. As noted in Figure 2a and predicted previously for "thick-slice" fiber push-out configurations [17-19, 25-28], the shear stress S_{rz} at the interface is not uniform along the fiber length. For the thin-slice specimens in Figure 2, S_{rz} has a maximum value of approximately 0.12p at a location approximately 0.5R (where R is the fiber diameter) below the top surface of the specimen. As is usually assumed due to the shear stress distribution, failure of the fiber-matrix interface can initiate by a shear mechanism at the point of maximum shear stress near the top of the specimen and propagate toward the bottom of the specimen. This is the generally accepted interfacial failure sequence for push-out or push-through tests, especially for "thick-slice" CMC specimens where the high stiffness matrix and generally weak interface permit testing of thick specimens (t > 1 mm) [25, 29, 30].

However, for thin-slice specimens the conventional interpretation of the above failure sequence may not be correct. Figure 2b shows that a significant radial stress component (S_r) exists at the bottom of the niobium-sapphire interface for the 0.5 mm thick specimen. A result of axisymmetrical bending, the radial stress is tensile and its magnitude approaches 0.14p at the bottom face of a 0.5 mm-thick specimen. Obviously S_r will increase for thinner specimens. We note that in the testing of IMC's with high interfacial bond strengths, specimen thicknesses of 0.2 - 0.4 mm are common. These are usually supported over holes or slots > 0.3 mm wide. Thus, if a significant indenter pressure is applied, as is likely for a well-bonded interface, large normal stresses due to bending will arise at the backface of the specimen. Even though most MMC's and IMC's are characterized by large thermally induced compressive stresses at low temperatures, these "back-face" tensile stresses due to bending may exceed the thermally-induced compressive stresses during loading.

As an example of a likely interfacial failure sequence in a thin specimen (specimen thickness/hole size < 2), we consider the possibility that the presence of large tensile stresses results in mode I crack initiation at the specimen bottom. If a crack initiates at the bottom interface, then its propagation as a mixed mode I/II crack will redistribute the interfacial stress components. Specifically, the combination of a mode II crack-tip stress field, when combined with a decreased bonded ligament length along the interface between the crack tip and the topface, will cause an increase in the S_{rz} shear stress component above that shown in Figure 2a. This increase in S_{rz} can in turn result in a "thin-slice" failure involving a sequence of interfacial crack growth up from the backface, initiating interface "shear", which is driven primarily by a large shear stress component S_{rz} along the interface down from the top face near the indenter. Thus, specimen geometry may create a stress state which can influence or even dictate the interfacial failure process.

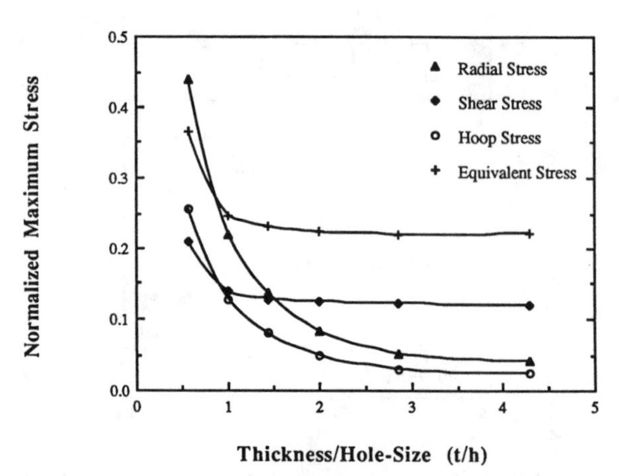

Figure 3 Summary of specimen geometry effect on the maximum interfacial stress components. Graph shows the normalized (with respect to indenter pressure p) maximum stress components vs thickness/hole size ratio (t/h). After ref 15.

The sensitivity of stress state to "thin-slice" specimen geometry is further demonstrated in Figure 3[15]. Shown in Figure 3 are the maximum stresses along the interface as a function of the specimen thickness-(t) to-hole(h) span ratio (t/h) for the range of t- and h-values in Figure 1. For thick specimens with t/h > 2, Figure 3 shows that (a) shear stresses dominate and (b) the maximum stress values are relatively independent of specimen geometry. Thus for these "thick-slice" configurations, previous analyses of interfacial failure during fiber pushout should be appropriate[17-19], at least in the absence of matrix plasticity.

Unfortunately as noted above, large-scale pushout of well-bonded ≅ 150 μm fibers in MMC's and IMC's usually require thin-slice specimens with thicknesses < 0.4 mm if large-scale

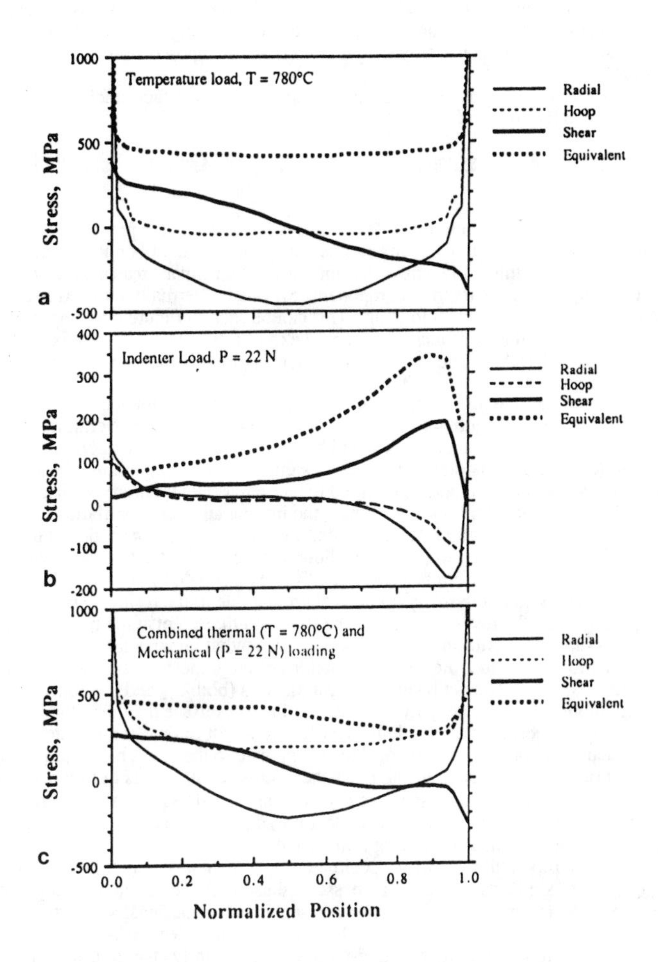

Figure 4 Normalized stress profiles in the sapphire-NiAl system due to (a) thermoelastic clamping from 780°C to room temperature (assuming matrix yielding at 380 MPa), (b) an indenter load of 22N applied to a specimen 0.5 mm thick, and (c) combined thermal and mechanical stress distribution for ΔT=780°C and P=22N for a sapphire-reinforced NiAl specimen 0.5 mm thick.

fiber displacements are to be induced prior to indenter or fiber failure at the high compressive stresses (> 3 GPa). Under these conditions, $t/h < 1.3$ if the support hole or slot is twice the fiber diameter, as is experimentally convenient. Figure 3 shows that for $t/h<1.5$, there is (a) a rapid escalation of the normal stresses both in the hoop and radial directions with decreasing specimen thicknesses and (b) the relative magnitudes of the maximum values of the shear stresses and normal stresses depend strongly on specimen configuration[15]. Thus, for these thin-slice geometries, there is not only the problem that the imposed stress state will differ from test to test,
but there is also the likelihood that the test itself may trigger back-face cracks and create a failure sequence unlike that in fiber pullout.

"Thin-Slice" Interfacial Stress States: Case II - Mechanical as well as Thermally Induced Stresses

An additional complication with "thin-slice" testing is that the presence of thermally induced stresses due to a thermal expansion mismatch between the fiber and matrix can have a very large effect on the interfacial stress distributions. The primary reason is the opposite nature of the signs of the shear stress components S_{rz} due to thermally induced loading as compared to that from mechanical loading. This can be seen in Figure 4 for sapphire-reinforced NiAl where the coefficients of thermal expansion $(CTE)_{NiAl} \cong 14.5x10^{-6}/°C > (CTE)_{sapphire} \simeq 8.7x10^{-6}/°C$. The key comparison in Figure 4 is the sign and magnitude of the shear stress in Figure 4a versus that in Figure 4b.

Given both thermal and mechanical loading at low temperatures in IMC's, the superposition of the stresses in Figure 4a with those in Figure 4b dictate the overall interfacial stress state. This implies that interfacial stresses will be sensitive to (a) test temperature (b) indenter pressure, and (c) interfacial bond strength. For example, at low temperatures in a typical IMC where a moderate indenter pressure is required to cause interfacial failure (such as for the case of high bond strengths in a thin-slice specimen, the interfacial stress distribution at fiber pushout shown in Figure 4c may result at the onset of either mode I or mode II failure at the specimen backface. Compared to Figure 2, where thermoelastic stresses were negligible, it is obvious that there are large differences between Figures 2 and 4c in the shear stress distributions. As expected, thermally induced clamping due to the residual stresses creates large compressive radial stress over most of the fiber length, except at the bottom interface where bending induces tensile stresses. However, somewhat unexpected is the form of the shear stress distributions under the combined actions of large thermoelastic and mechanical loading, as in Figure 4c. The net result are shear stresses (both S_{rz} and the equivalent shear stress) distributions which are actually smaller near the top surface than near the bottom interface. Thus, it is possible that mode II shear failure might initiate at the bottom face rather than the topface as is normally assumed. This is opposite to the result presented in Figure 2a, and is a result of the difference in the signs of shear stress components in Figures 4a and 4b.

While Figures 2 and 4c present two specific cases of stress states resulting from combined mechanical and thermally induced loading, it is obvious that a wide range of stress states can arise along the interface during thin slice push-out testing of IMC's. These will vary with (a) test temperature, (b) from specimen to specimen as indenter pressure varies at pushout, even if (c) similar specimen thicknesses are used. It is essential to recognize the importance of indenter pressure, which in turn is sensitive to interfacial shear strength during pushout, in affecting the local stress state. Thus during a thin-slice push-out test of a typical IMC, the stress distributions along the fiber-matrix are constantly changing with increasing load. If specimen bending is minimized in a thick specimen ($t/h>2$), interfacial failure will eventually occur as expected near the indenter, driven by the large mechanically-induced shear stress which exceeds the S_{rz} component from the thermal stresses. However, if the specimen is thin and significant specimen bending is present ($t/h \lesssim 1.5$), interface failure can be triggered by the introduction of a mixed mode crack propagation up from the backface of the specimen. Thus a different interfacial failure process may occur from specimen to specimen, dictated by specimen geometry and not by the intrinsic behavior of the composite.

Fiber Pullout and Thin-Slice Pushout: A Comment

Comparisons of stress state along the interfaces of thick-slice fiber pushout and fiber pullout specimens have been examined in detail for brittle matrix composites[17-19,28,31]. There are obvious differences in stress states that arise from to the Poisson expansion/contraction issue of the fiber[31], which is probably a small factor in IMC behavior. For pullout vs pushout of fibers in IMC's, we suggest that larger issues are (a) the fact that the thermally induced shear stresses (S_{rz}) near the surface assist fiber pullout, but resist fiber pushout as described in the above section, and (b) that any specimen bending during a pushout test may create large tensile stresses located at the opposite end of the fiber from the maximum shear stresses. This latter effect differs from pullout where large shear stresses as well as large thermally induced tensile stresses are both located along the interface near where the fiber exits the matrix. Thus in the pullout case, tensile plus shear stresses can combine at the same location to trigger failure, while in the other case (pushout) they act at separate locations. Different interfacial failure processes may occur as a result.

EXPERIMENTAL BEHAVIOR: Case I - Negligible Thermal Stresses: Sapphire-Reinforced Niobium.

Sapphire-reinforced niobium is a system in which interfacial shear occurs under conditions often desired in an IMC at high temperatures: a strong chemical bond between the fiber and matrix[9], but only very small residual stresses due to thermally induced clamping. The latter is a result of the good match between the coefficients of thermal expansion between niobium and sapphire, while the former depends on processing conditions[22-24].

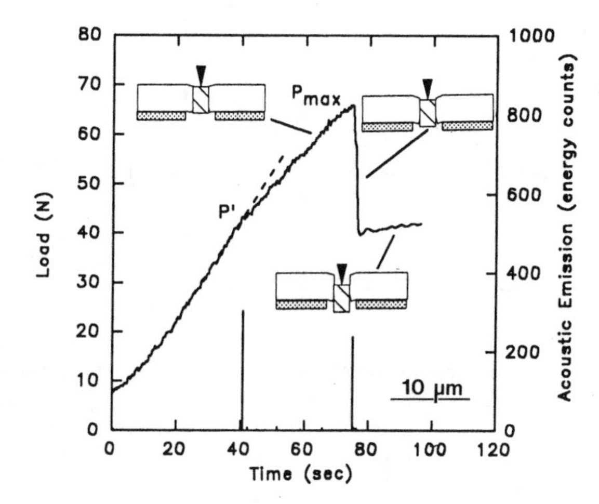

Figure 5 Load-time response for fiber pushout in the Nb-sapphire system for a thickness of 0.5 mm. The physical stages of fiber pushout are depicted in the insert. Ref. 9

Thin-slice push-out tests have been performed on sapphire-niobium resulting in the response shown in Figure 5[9]. Thin-specimen behavior is characterized initially by linearly elastic behavior, terminated by deviation from linear elastic behavior at P' at which time a strong acoustic signal is usually detected. Examinations of the specimens immediately after the load P' is attained indicate that a crack has initiated at/or near the fiber matrix interface at the backface of the specimen. The crack usually does not encompass the entire perimeter of the fiber. Calculated at P', the average interfacial shear stress+ from seventeen tests of specimens with thicknesses ranging from 0.2 mm to 0.7 mm ($0.7 \leq t/h < 2$) is 145 MPa. However there is considerable scatter in the shear stress values with a range of 105 to 175

+ The "average" interfacial shear stress is $P'/2\pi rt$ where r and t are the fiber radius and specimen thickness, respectively.

MPa being observed. We suggest that the large data scatter is due in part to the range of thickness tested and the resulting variation in stresses, as in Fig. 3.

The nonlinear load-time response, which initiates at P', continues under increasing loads to P_{max} as shown in Fig. 5. During this period, both matrix deformation near the indenter and increasing crack opening displacement at the backface are observed. At P_{max} or shortly thereafter, there is an abrupt load drop, accompanied by large-scale shear displacement of the fiber at/or very near the fiber-matrix interface. We conclude that large-scale shear is triggered by increased shear stress created by the back-face crack growth with a mode I component.

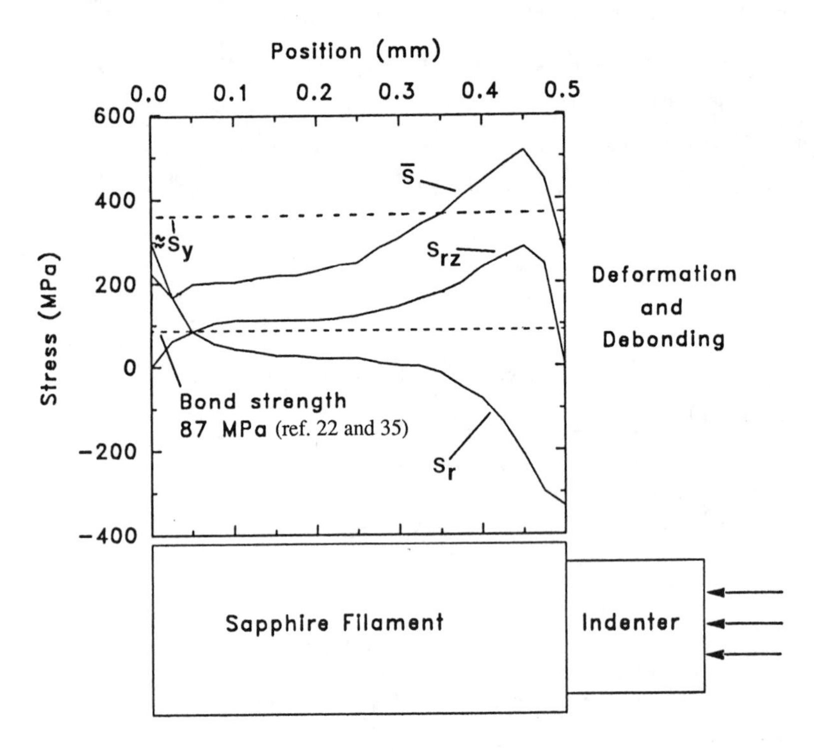

Figure 6 The stress state associated with a thin-slice fiber pushout test (sapphire-reinforced Nb, 0.5 mm thick) at a load P' of 31N. After ref 9.

The above sequence can be understood in terms of the stress state along the fiber-matrix interface, which is depicted in Fig. 6 at the load P' in Fig. 5. Several important features are apparent. First, due to biaxial bending, the magnitude of σ_{rr} exceeds the anticipated bond strength of Nb at the surface opposite the indenter. We believe that it is this stress component that initiates the interfacial crack which corresponds with the acoustic signal at P' in Fig. 5 and which is observed on the backface using SEM[9].

In addition to the radial stress component which creates a fiber-matrix debond at the bottom interface, the equivalent shear stress near the indenter increases with increasing indenter pressure as $P'<P<P_{max}$. This causes plasticity near the indenter (see $\bar{\sigma}$ curve in Fig. 2) which subsequently spreads over the length of the specimen, causing stable fiber displacement at P>P' as depicted in the inset to Fig. 6. During this stage, fiber displacement occurs under a stable manner as crack growth up from the bottom face and diffuse matrix plastically concentrated near the top face occur concurrently. Finally at $P = P_{max}$, either a shear instability or crack growth within the niobium under fully plastic conditions cause the abrupt large scale displacement of the fiber from the matrix.

In summary, under the above conditions, interfacial shear in thin-slice push-out involves a combination of "back-face" crack growth and "top-face" matrix deformation near the indenter. The result is a sustained period of fiber displacement of typically 5-20μm. It is important to recognize that the above is a direct consequence of the stress state described in Figures 2 and 6. These conditions could well describe an IMC tested at high temperatures.

We also suggest that under similar material conditions (no thermal stresses, ductile matrix bonded to the fiber), the pullout of fibers near their ends can involve a similar combination of crack growth and matrix deformation. The result should be a very high energy required to pull the fiber ends and improved fracture toughness.

EXPERIMENTAL BEHAVIOR: Case II - Large Thermal Stresses: Sapphire - Reinforced NiAl

In view of the large differences in the interfacial stress states when thermally induced residual stresses are present, compare Figures 2 and 4c, it seems likely that differences in interfacial shear behavior might be observed. An example is shown in Figure 7 in which the pushout of a sapphire fiber embedded in a NiAl matrix is depicted. In this system, there is no chemical bond after consolidation, but fiber asperities and diametrical variations dictate mechanical interlocking at room temperature.

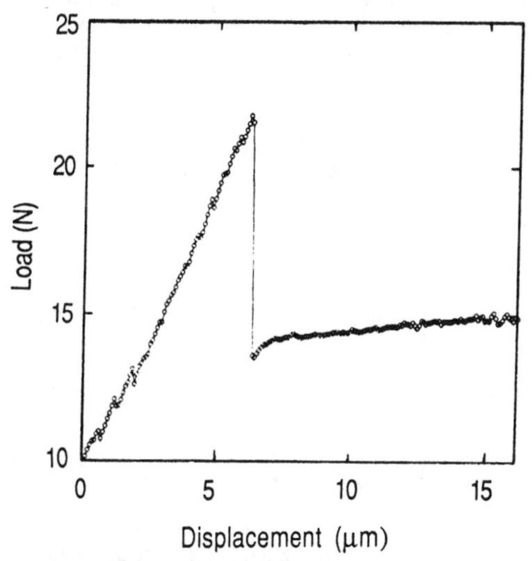

Figure 7 The load-displacement response of a sapphire-reinforced NiAl thin-slice test specimen 0.5 mm at room temperature.

We wish to make three points concerning the data in Figure 7 and the accompanying stress state in Figure 4c. First, unlike the sapphire-niobium system, there is no detectable fiber displacement occurring prior to the load drop and large scale fiber pushout. Careful examination of the specimen compliance using a technique similar to that of Parthasarathy, Jero, and Kerans[34] shows no measurable change in compliance which would be associated with fiber displacement $\geq$ 0.3 μm. Similarly, there is no acoustic signal prior to the load drop. Thus, interfacial shear appears to occur in an abrupt, catostrophic manner. As a second point, we propose that in view of the stress state in Figure 4c, it is in fact likely that abrupt fiber displacement <u>should</u> occur. Previous observations of the sapphire/NiAl system show that fiber displacement requires matrix plasticity in the form of grooves resulting from localized plastic deformation of the matrix due to asperities on the fiber surface[3]. Furthermore, at P=P', the computations in Figure 4c show that there is a ligament of matrix adjacent to the fiber near the indenter which remains elastic (i.e., $\bar{\sigma} < \sigma_y$ = 380 MPa) just prior to debond. Thus the fiber is held in place by an "elastic link". The fiber will release in an abrupt manner only if backface crack growth increases the shear stress in the "link" region to trigger plasticity. Both Figure 4 and 6 show a large radial stress component, due to specimen bending, at the specimen backface. We believe this causes crack initiation and growth along the interface, increasing the shear stress within the matrix along the uncracked ligament. Fiber pushout occurs abruptly when the shear stress elevates $\bar{\sigma}$ to a point where $\bar{\sigma} = \sigma_y$ of the matrix. Thus, backface interfacial crack growth, caused by specimen bending, triggers interfacial shear in this specific example.

SUMMARY

In view of the fact that the stress distributions such as that in Figures 2 and 4c are sensitive to specimen thickness as well as indenter pressure (assuming perfect alignment of the fiber with the support hole or trough), a series of tests of non-identical specimens will be accompanied by a range of stress states. Thus, it is not surprising that these data often are characterized by large experimental scatter of the values for average interfacial shear strengths. Furthermore, if we recognize that interfacial failure is likely to be triggered by backface crack growth in thin-slice pushout tests, then eliminating specimen bending and backface cracking should increase the loads necessary for interfacial shear. In other words, we suggest that most thin-slice pushout tests of IMC's result in average shear strength values which are lower than their true values. The current test methodology, which is relatively easy to perform, may indeed provide a useful qualitative comparison of interfacial "shear strength" values as well as an indication of the failure processes, particularly if plasticity is involved. However, accurate quantitative descriptions of the interfacial shear behavior of IMC's requires test configurations which eliminate specimen bending.

ACKNOWLEDGMENTS

We wish to acknowledge the cooperation of Clark Moose in providing the NiAl/sapphire results. Discussions with Richard Petrich are very much appreciated. This research was supported by NASA through Grant No. NAGW-1381.

REFERENCES

1. Interfacial Phenomena in Composite Material '91, edited by I. Verpoest and F. Jones (Butterworth Heinemann, Oxford, 1991).

2. Intermetallic Matrix Composites II, edited by D. B. Miracle, J. A. Graves, and D. L. Anton (Mater. Res. Soc., Pittsburgh, PA, 273, 1992).

3. C. A. Moose, D. A. Koss, and J. R. Hellmann in Intermetallic Matrix Composites, edited by D. L. Anton, P. L. Martin, D. B. Miracle, and R. McKeeking, (Mater. Res. Soc. Proc 194, Pittsburgh, PA, 1990) p. 293.

4. W. C. Carter, E. P. Butler, and E. R. Fuller, Scripta Metall. Mater. 25, 579 (1991).

5. T. J. Mackin, P. D. Warren and A. G. Evans, Acta Mettal. Mater. 40, 1251 (1992).

6. P. D. Jero and R. J. Kerans, Scripta Metall. Mater., in press.

7. P. D. Jero, R. J. Kerans, and T. A. Parthasarathy, J. Am. Ceram. Soc., in press.

8. J. I. Eldridge, this conference proceedings.

9. D. A. Koss, R. R. Petrich, J. R. Hellmann, and M. N. Kallas in Interfacial Phenomena in Composite Materials, edited by I. Verpoest and F. Jones (Butterworth Heinemann, Oxford, 1992), p. 155.

10. J. I. Eldridge and P. K. Brindley, J. Mat. Sc. Letters 8, 1451 (1989).

11. R. D. Noebe, R. R. Bowman, and J. I. Eldridge in Intermetallic Matrix Composites, edited by D. L. Anton, P. L. Martin, D. B. Miracle, and R. McMeeking (Mater. Res. Soc. 194, Pittsburgh, PA 1991, p. 323.

12. M. C. Watson and T. W. Clyne, Acta Metall. Mater. 40, 141 (1992).

13. P. D. Warren, T. J. Mackin, and A. G. Evans, Acta Metall. Mater. 40, 1243 (1992).

14. P. Kantzos, J. Eldridge, D. A. Koss, and L. J. Ghosn, this conference proceedings.

15. M. N. Kallas, D. A. Koss, H. T. Hahn, and J. R. Hellmann, J. Mater. Sci. (in print).

16. J. W. Hutchinson and H. M. Jensen, Mech. Mater. 9, 139 (1990).

17. R. J. Kerans and T. A. Parthasarathy, J. Am. Ceram. Soc. 74, 1585 (1991).

18. D. B. Marshall, Acta Metall. Mater. 40, 427 (1991).

19. C. Liang and J. W. Hutchinson, to be published.

20. L. J. Ghosn, NASA-Lewis Research Center, unpublished research.

21. M. N. Kallas, Penn State University, unpublished research.

22. S. Morizumi, M. Kikuchi, and T. Nishino, J. Mat. Sci. 16, 2137 (1981).

23. G. Elssner and H. Krohn, Z. Metallkde. 70, 81 (1979)

24. N. P. O'Dowd, M. G. Stout, and C. F. Shih, Phil. Mag. (in print).

25. D. B. Marshall and W. C. Oliver. J. Am. Ceram. Soc. 70, 542 (1987).

26. D. H. Grande, J. F. Mandell, and K. C. C. Hong, J. Mater. Sci. 23, 311 (1988).

27. J. D. Bright, D. K. Shetty, C. W. Griffior, and S. Y. Limaye, J. Am. Ceram. Soc. 72, 1891 (1989).

28. C. H. Hsueh, Acta Metall. Mater. 38, 403 (1990).

29. D. B. Marshall and W. C. Oliver, J. Mater. Sci. Eng. A126, 95 (1990).

30. T. P. Weihs and W. D. Nix, J. Am. Ceram. Soc. 74, 524 (1991).

31. C. H. Hsueh, M. K. Ferber, and P. F. Becher J. Mater. Res. 4, 1529 (1989).

32. Introduction to Ceramics, 2nd Edition, edited by W. D. Kingrey, H. K. Bowen, and D. R. Uhlmann (John Wiley & Sons, New York, 1976) p. 595.

33. J. R. Stephens in High Temperature Ordered Intermetallic Alloys, edited by C. C. Koch, C. T. Liu, and N. S. Stoloff (Mater. Res. Soc. 39, Pittsburgh, PA 1985), p. 381.

34. T. A. Parthasarathy, P. O. Jero, and R. J. Kerans, Scripta Metall et Mater. 25, 2457 (1991).

35. M. G. Stout, M. L. Lovato, A. Geltmacher, T. G. Zocco, and T. R. Jervis, Los Alamos National Laboratory, unpublished research.

STATISTICAL ANALYSIS OF THE PROPERTIES OF ADVANCED FIBERS DESIGNED FOR TITANIUM ALLOY AND INTERMETALLIC REINFORCEMENT

JOHN R. PORTER
1049 Camino Dos Rios, Rockwell International Science Center, Thousand Oaks, CA 91360

ABSTRACT

The mechanical properties of new ceramic reinforcing fibers need to be well characterized before their incorporation into composite materials. Critical fiber properties include strength and Weibull modulus, both off the spool and after matrix extraction, bundle strength, modulus and creep resistance. Important composite properties include thermochemical stability, interface debond energy and interfacial sliding resistance. Tailoring these interfacial properties invariably involves the use of a fiber coating that can, in turn, influence fiber properties. Methods of measuring strength related properties are addressed and the results of a computer simulation to assess the quality of measured data using statistical methods are presented. The simulation was developed to determine the errors associated with a strength/Weibull modulus determination based on a limited number of samples. Finally, an assessment of the effect of mixing of high and low quality fiber on bundle strength and composite properties is made.

INTRODUCTION

The available monofilament fibers for reinforcing titanium alloys and intermetallics are mostly chemically vapor deposited (CVD) silicon carbide. Current aerospace programs have identified the need for new ceramic fibers as enabling materials for their success. Consequently, a number of programs have recently addressed the development of new fibers.† The new fibers are needed for several reasons including: having fibers with a CTE greater than that of SiC, to match more closely that of titanium alloys; having fibers that have higher strength and creep resistance; and having fibers that are chemically stable in new candidate matrices. These new programs necessitate the need for characterizing fibers and, in particular, assessing the quality of the property measurements that are made both during fiber development and for a property data base of a developed fiber.

The strength of a ceramic fiber is statistical and is determined by the largest flaw (weakest link) in the flaw population of the tested length of fiber. This can have important implications for toughening a brittle matrix material such as an intermetallic. For toughening by continuous fiber reinforcement, fibers need to exhibit pull-out during matrix crack propagation, as shown schematically in Fig. 1. However, the stress distribution along a bridging fiber, as shown in the figure, has a maximum in the crack plane and for fiber failure to occur away from the crack plane, a prerequisite for pull-out, then a weak link must exist some distance away from the crack plane. [1]

† National Aerospace Plane (NASP), High Speed Civil Transport Enabling Materials Program (HSCT/EPM), Integrated High Performance Turbine Engine Technology (IHPTET) program.

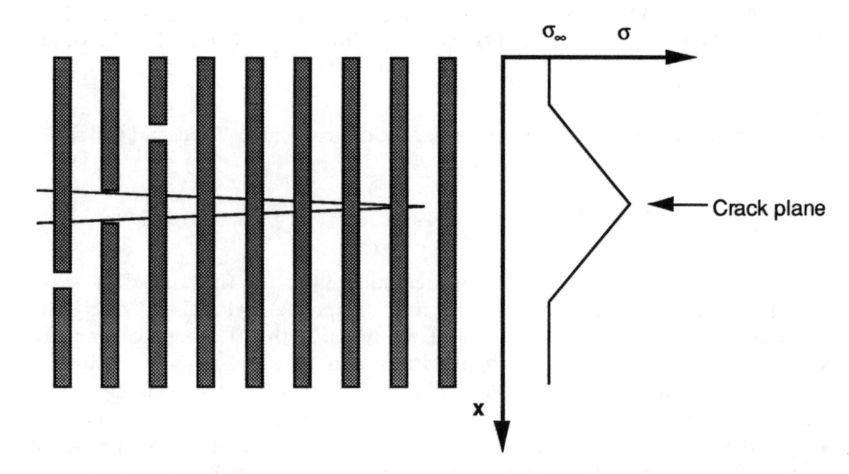

Fig. 1. Schematic of fibers bridging a matrix crack and the associated stress along an intact bridging fiber.

Consequently, a statistical approach to fiber strength reporting is necessary and one accepted approach is to use the Weibull statistics method.[2] In the Weibull formulation, the cumulative probability that fibers will fail at a given applied stress, $F(\sigma)$, is given by:

$$F(\sigma) = 1 - \exp\left\{-V\left(\frac{\sigma-\sigma_u}{\sigma_0}\right)^m\right\}$$

where V is the volume of tested material (proportional to gauge length), σ is the measured strength, σ_u is a threshold strength (set to zero in the two parameter Weibull analysis typically adopted for fibers), σ_0 is a normalizing constant and m is the Weibull modulus. In a Weibull analysis, $\ln(\ln(P_s))$, where $P_s = 1 - F(\sigma)$, is plotted against $\ln(\sigma)$ so that the Weibull modulus, m, is given by the slope. For an accurate determination of Weibull modulus and average strength, a "large" number of tests is required and an appropriate determination of the statistical estimator (the value of P_s needs to be made. Estimations of errors associated with Weibull determinations as a function of the estimator used and the number of tests exist in the literature, most recently with respect to fiber testing in a paper by van der Zwaag.[3]

The appropriate number of mechanical tests for *one fiber* depends on the statistical variability in the measured strengths. This work describes a computer simulation in which an artificial, large data set of strength values is created with a normalized 50% survival strength and fixed Weibull modulus. From this, subsets are selected and the subset's 50% survival strength and Weibull modulus are measured, thereby simulating an experiment in which only a small number of tests of fiber strength are made. This statistical analysis, discussed in detail below, has been used to determine the number of tests required for a given statistical accuracy in the data.

A similar computer model has been used to simulate *fiber bundle properties*. The bundle strength of a set of fibers is the peak load that can be carried as fibers in a bundle are uniformly strained. As the weaker fibers in a bundle fail, their load carrying ability is eliminated. The number of surviving fibers at the ultimate tensile strength of the bundle of N_0 fibers depends on the Weibull modulus, m, and has been shown by Kelly to be $N_0/\exp(1/m)$.[1] The computer simulation allows the mixing of fibers with different properties and the calculation of a resultant bundle strength and fraction of surviving fibers.

Finally, these results are used in an assessment of the validity of bundle properties as a guide to composite properties.

EXPERIMENTAL

Fiber mechanical properties were measured using a tensile testing machine in which cross-head displacement is controlled by a linear stepper motor and load and displacement are monitored by a miniature load cell and displacement transducer, respectively. The machine is computer controlled and can be run under load or displacement control.‡ Strength measurements typically are made using fibers glued to paper tabs, usually with epoxy, to minimize grip failures. The strength and Weibull modulus of a given fiber spool are then determined from a minimum of twenty tests. The diameters of tested fibers were measured with a microscope. Error estimations indicate that for some fibers, diameter determinations and the variability of fiber diameter along a fiber are the most error prone measurements in the strength determinations.

Experimentally, the grip face material and the grip pressure had to be established for each fiber investigated: since fibers often shatter at failure, establishing that failure occurred in the test zone was often not possible, precluding the possibility of discarding data where grip failure was evident. Extracting single fibers from tows also posed experimental challenges. One counterintuitive observation was that rough handling during tow separation proof tested extracted fibers such that measured fiber strengths were erroneously high.¥ Tow testing introduced new experimental variables. Unless all fibers in a tow can be equally loaded, the results of a tow or strand test become difficult to interpret. However, the direct relationship between fiber strength, Weibull modulus and bundle strength can be used to verify the quality of such experimental data.

Since fiber strength is determined by a statistical flaw population, strengths are gauge length dependent. Therefore, a standardized test gauge length had to be established, and in this work, a one inch gauge length was adopted. Grip face material, fiber handling and fiber storage conditions all affect the measured strengths of ceramic fibers. Since these factors affect fibers differently, a test methodology was designed for each fiber tested.

THE COMPUTER SIMULATION

The computer simulation was developed to determine the errors associated with a strength/Weibull modulus determination based on a limited number of samples using several different statistical estimators for the calculation of probability

‡ Now available commercially from Micropull Science.

¥ C. Lundgren of Dupont first drew our attention to this potential artifact.

of failure, $F(\sigma)$. From the analysis described below, using the estimator $F(\sigma) = (n - 1/2) / N$, where n is the ranking of the data from weakest to strongest and N is the number of tests, the standard deviation in the strength and Weibull modulus measurements were found to be related to the 50% survival strength, σ, Weibull modulus, m, by the two simple relationships:

$$\frac{StDev(\sigma)}{\sigma} = \frac{1.16}{(m\sqrt{N})} \text{ and } \frac{StDev(m)}{m} = \frac{1.04}{\sqrt{N}}.$$

This analysis was then used to determine the number of tests required for a given statistical accuracy in the data.

Details of the computer simulation follow. Various statistical estimators have been proposed for $F(\sigma)$ in Weibull modulus determinations, including the mean rank, $n / (N + 1)$; the median rank, $(n - 3/8) / (N + 1/4)$ and $(n - 1/2) / N$. Since $F(\sigma) = (n - 1/2) / N$ is the preferred estimator of many workers for reasons discussed below, it was used in these experiments.

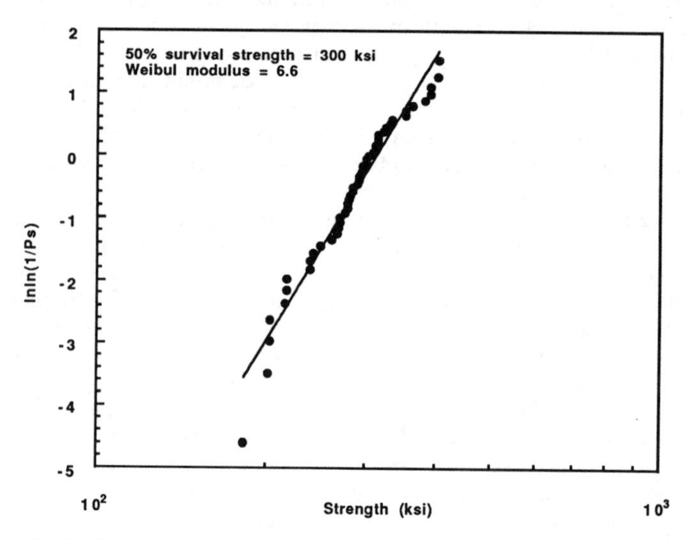

Figure 2. Weibull plot for experimentally determined strengths of PRD166 fiber. (50 tests)

As an example of a typical Weibull plot, Fig. 2 shows $\ln(\ln(1/P_S))$, where the probability of survival, $P_S = 1 - F(\sigma)$, is plotted against the measured strength data (on a log scale) and the Weibull modulus is determined by linear regression. There is a choice in how the strength of the set of fibers is calculated; the reported strength can be a simple average, an average determined from an analysis of the Weibull fit, or a 50% survival strength, determined from the fit on the Weibull plot evaluated at $P_S = 0.5$. Typically, the 50% survival strength and the numerical average are well within statistical variation of measured strength: the 50% survival strength has been used here for simplicity.

The purpose of the computer simulation was primarily to calculate the number of tests necessary for a given statistical accuracy. First, a master data set of

1000 strength values with a predetermined Weibull modulus was generated in the computer. From this master set, sub-sets of different size were randomly selected and the 50% survival strength and Weibull modulus were calculated for each sub-set. This was repeated 100 times for each sub-set size and the statistical variability of the measured 50% survival strength and Weibull modulus were then evaluated. Figure 3 shows the scatter observed for sub-sets of size N = 20 generated from master sets with Weibull moduli of 5, 10 and 20 and with nominal strengths, σ, of 100 arbitrary units. From Fig. 3, clearly significant errors in measuring Weibull modulus will result even from data sets from 20 tests and the error in the measured strength decreases as the Weibull modulus increases.

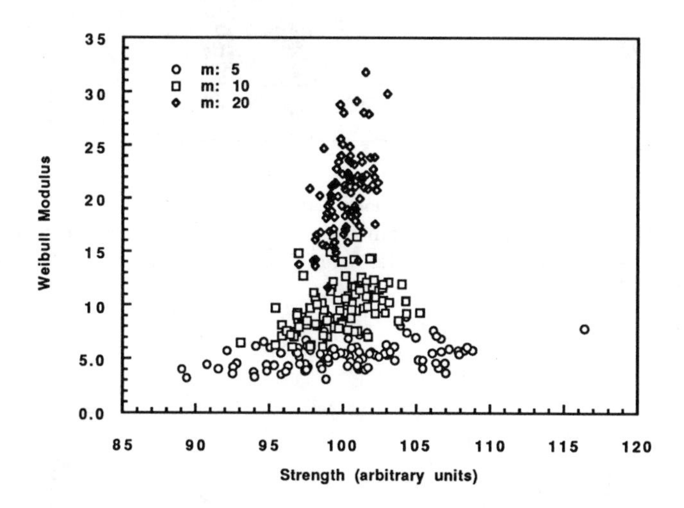

Figure 3. Calculated values of strength and Weibull modulus for sub-sets of 20 values extracted from master sets of 1000 with fixed Weibull moduli of 5, 10 and 20.

Figure 4 shows the results from a similar exercise, but in this case the Weibull modulus was fixed at 10 and the set sizes, N, were 10, 20 and 50. Clearly, the scatter was less as the set size increased.

The standard deviation of the scatter in the measured values of strength and Weibull modulus were evaluated and Fig. 5 is a plot of these standard deviations as a function of Weibull modulus and sub-set size. A simple curve fitting exercise was used to generate a relationship for the standard deviation as a function of set size and Weibull modulus. The standard deviation in the strength as a function of Weibull modulus (m) was of the form A/m, where A is a constant. The standard deviation in the Weibull modulus as f(m) was of the form $B \times m$, where B is a constant. By plotting A and B as f(N) (Fig. 6) and fitting a $C/\sqrt{N}$ function, where C

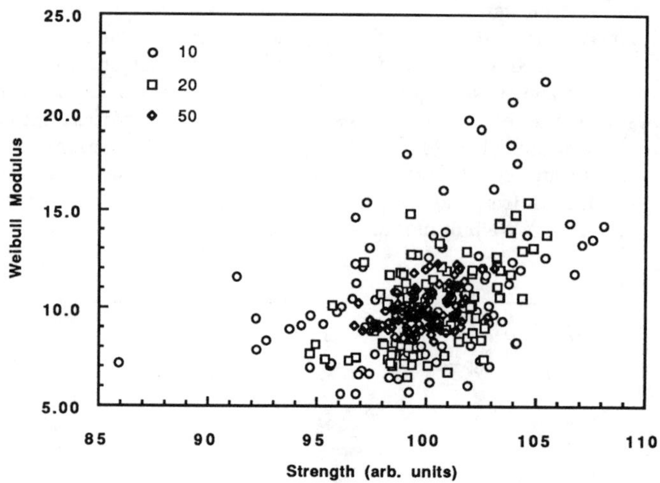

Figure 4. Measured values of strength and Weibull modulus for sub-sets of 10, 20 and 50 values extracted from a master set of 1000 values that had a fixed Weibull modulus of 10.

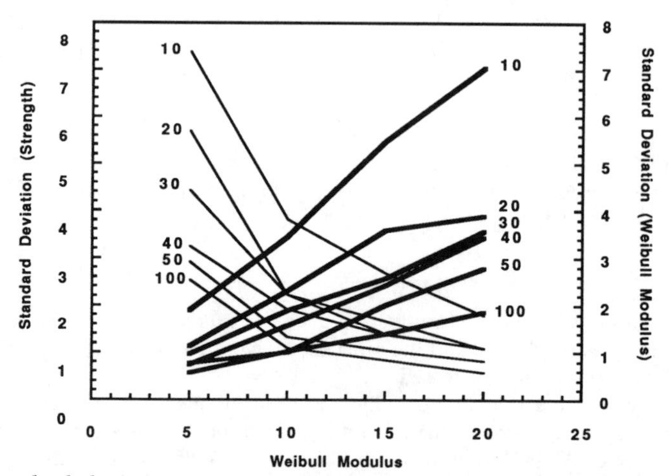

Figure 5. Standard deviation in measured values of 50% survival strength (thin lines) and Weibull modulus (thick lines) for different set sizes (indicated on lines).

is a constant, allowed the relationships: $\frac{StDev(\sigma)}{\sigma} = \frac{1.16}{(m\sqrt{N})}$ and $\frac{StDev(m)}{m} = \frac{1.04}{\sqrt{N}}$ to be determined.

This analysis has been repeated using the mean rank statistical estimator $F(\sigma) = n/(N+1)$. Figure 7 shows the average evaluated Weibull moduli for different set sizes for true Weibull moduli of 5, 10 and 15, using both statistical estimators. Neither estimator accurately evaluates the Weibull modulus for small set sizes, although the estimator (n-0.5)/N converges on the correct value with a smaller set size and for this reason is preferred.

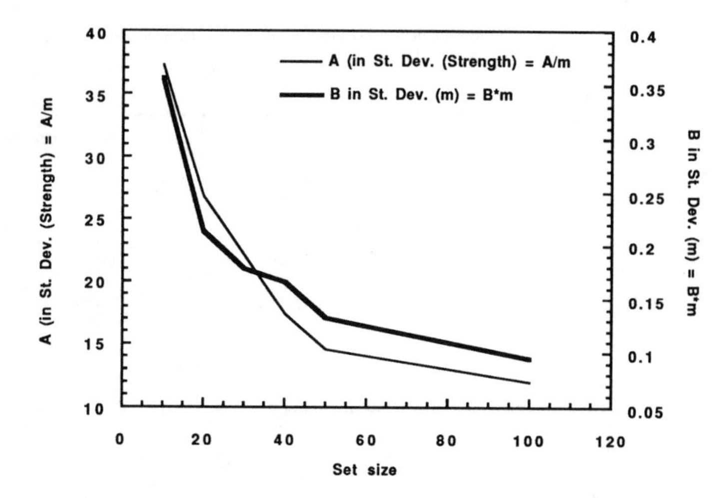

Figure 6. Calculated values of the A and B parameters described in the text as f(N), used in the determination of standard deviations of strength and Weibull modulus.

FIBER BUNDLE PROPERTIES

Fiber characterization is necessary to predict the behavior of composite materials. The two issues, therefore, that need to be addressed are, fiber degradation during composite processing and fiber bundle properties. Figure 8 shows a comparison of SiC fiber properties before composite fabrication and after extraction by dissolution from an orthorhombic Ti alloy matrix. One interpretation of Fig. 8 is that approximately one-third of the fibers were damaged during compositing (or matrix extraction), thereby introducing a new flaw population that resulted in the observed change in slope in the Weibull plot. The effect of having two fiber flaw populations in a fiber bundle was therefore addressed in a second computer simulation.

To investigate the effect of having two different flaw populations in a single fiber bundle, two groups of fibers, with identical properties except for their strength

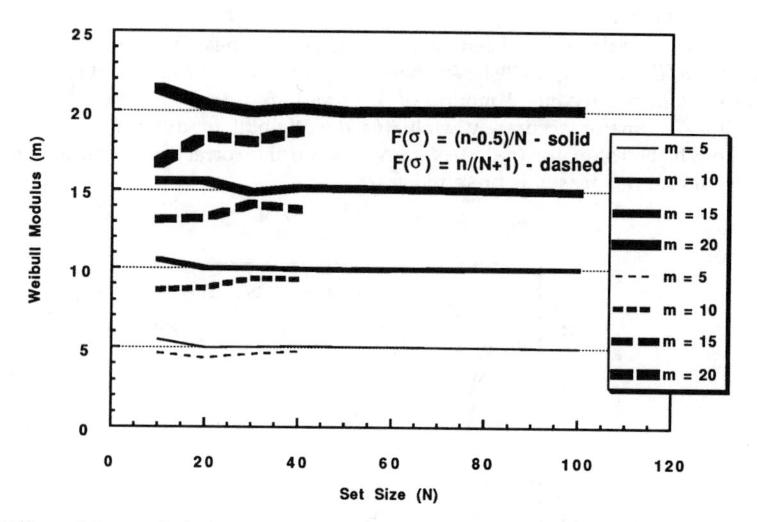

Fig. 7 Effect of the statistical estimator, F(σ), on the Weibull modulus as f(N).

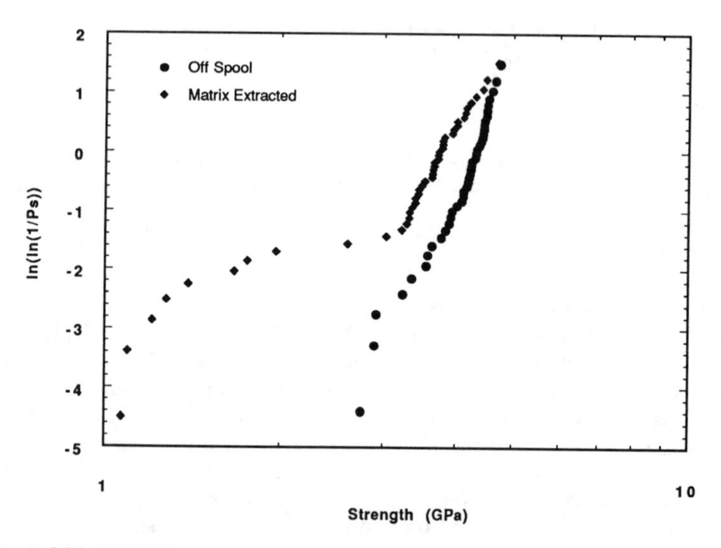

Fig. 8 SCS-6 SiC fiber strength off spool and extracted from orthorhombic Ti alloy.

distributions, were strained in parallel in the second simulation. Figure 9 is the computed Weibull plot of such a model bundle for which 80% of the fibers had a 50% survival strength of 100 arbitrary units with a Weibull modulus of 10 and the remaining 20% of fibers had a strength of 70 with a Weibull modulus of 5. The calculated bundle stress/strain behavior is shown in Fig. 10, which shows the properties of two fiber bundles, one with a fiber strength of 100 and a Weibull modulus of 10 together with a second mixed bundle with the 20% weaker fibers added. Superimposed on Fig. 10 are plots of the number of surviving fibers. As predicted by the theory described by Kelly, the bundle strength of a single population of the better fibers is 74.6 units with a fraction of surviving fibers of 90%, whereas for the poorer fibers, the bundle strength is 44.7 units with the fraction of surviving fibers of 82%.[2] However, the bundle strength of the *mixed* population is 63.7 units, significantly lower than the rule of mixtures calculated strength of 68.6 units for these two bundles acting independently.

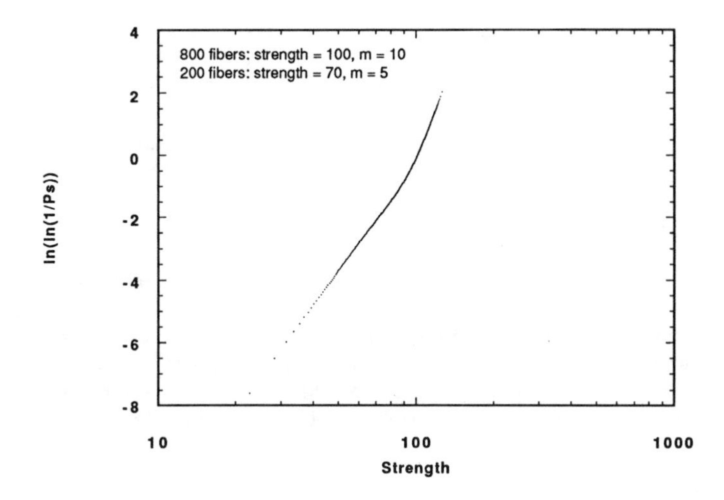

Fig 9. Weibull plot for simulated mixed population of fibers.

This simulation, in which different qualities of fiber are mixed, models an experiment described by Draper et al.[4] In that work, the intentional mixing of poor quality fibers with good quality fibers led to composite properties essentially the same as when all poor quality fibers were used. That the ultimate tensile stress in these bundle strength experiments occurs when only a small fraction of fibers have failed causes the small fraction of weak fibers to influence bundle strength more than predicted by rule of mixture calculations.

This situation is possibly different to that described in Fig. 8, which appears to show that about one third of fibers have been damaged when one inch gauge length samples are tested. If so, then the new flaw population appears to consist of one large flaw every three inches or so along the fiber, a situation that may not be deleterious to composite properties, and for which Weibull statistics may not be appropriate, since the entire new flaw population is being tested rather than just the weak links or largest flaws of a larger population. On the other hand, the situation described in Fig. 9 models a proportion of fibers having a completely different population of flaws, a situation that clearly influences the bundle properties.

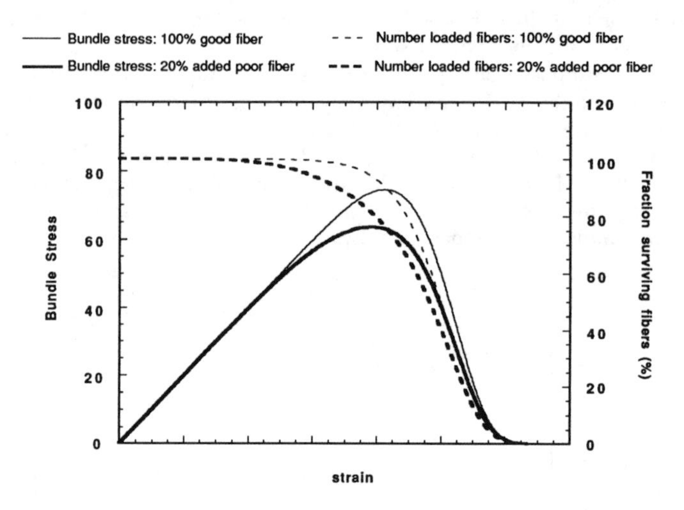

Fig. 10. Bundle stress of mixed population of fibers

One final point, that is not addressed in detail here, is the relevance of bundle properties to composite properties. In a bundle test, once a fiber has failed, it contributes nothing to the strength on the bundle, whereas in a composite, a broken fiber can carry load outside the shear lag region of a fiber break. Reported instances, however, of composite properties being predicted by rule of mixture calculations, based on fiber bundle strength, are possibly largely coincidental.

CONCLUSIONS

A computer model has been developed to predict the accuracy of Weibull modulus calculations based on a limited number of samples and to assess the quality of the available statistical estimators.

A second model was used to predict the properties of mixed bundles of fibers. This model showed that adding a small fraction of poor quality fibers to a bundle lowers the bundle strength more than is predicted by rule of mixture calculations.

REFERENCES

1. A.G. Evans and D.B. Marshall, Acta Metall., 37 , 2567 (1989).
2. A. Kelly and N. H. Nicholson, Strong Solids, 3rd ed. (Clarendon Press, Oxford, 1986), p. 274.
3. S. van der Zwaag, J. Test. and Eval., 17, 292 (1989).
4. S. L. Draper, P. K. Brindley, and M. V. Nathal, Met. Trans., in press.

FIBER PUSH-OUT TESTING OF INTERMETALLIC MATRIX COMPOSITES AT ELEVATED TEMPERATURES

JEFFREY I. ELDRIDGE
NASA Lewis Research Center, Cleveland, OH 44135

ABSTRACT

A newly developed apparatus for performing fiber push-out testing at elevated temperatures has been applied towards testing fiber-reinforced intermetallic and metal matrix composites. This new capability shows the effects of the relief of residual stresses and increased matrix ductility with increasing temperature on fiber debonding and sliding behavior.

INTRODUCTION

Fiber push-out testing has been successfully applied towards evaluating fiber debonding and sliding behavior in fiber-reinforced composite materials at room temperature. The benefits of extending these measurements to elevated temperatures include (1) generating data at composite service temperatures which could be used to optimize interfacial mechanical behavior at those temperatures and (2) evaluating the effects of residual stresses and matrix ductility on fiber debonding and sliding.

This paper describes a newly developed apparatus for performing elevated temperature fiber push-out tests and reports results for two intermetallic composites and one metal matrix composite. The results are discussed in terms of residual stresses, interfacial wear, matrix ductility, and changing modes of interfacial failure.

EXPERIMENT

Fiber Push-Out Apparatus

Extending fiber push-out testing to elevated temperatures added two basic requirements to the previously developed apparatus [1]: (1) heating the sample and indenter and (2) providing a nonoxidizing environment. Fig. 1 shows a schematic of the new elevated temperature fiber push-out apparatus. Controlled indenter displacement is performed using an Instron frame with a crosshead speed of 0.98 μm/s. The sample and indenter are located inside a cubical stainless steel chamber with conflat-flanged ports on each face. This test chamber is evacuated to a base pressure of 1×10^{-6} torr, which prevents significant sample oxidation. The indenter is a flat-bottomed cylindrical tungsten carbide punch. The indenter/load cell assembly is coupled to the Instron crosshead, moving inside a collapsible bellows. The specimen is mounted with a spring-loaded clamping device. The sample support block has a set of three 300 μm wide grooves underneath the sample which allow fibers to be pushed out without resistance from the support block.

Sample heating is achieved using a quartz halogen lamp inside an ellipsoidal reflector with the lamp at one focal point and the sample at the other where the radiation is focused. The reflector is bisected by the chamber's quartz window. Sample heating up to 1000°C takes less than 10 min. Sample temperature is monitored by a thermocouple attached to the Ta sample support block. A water-cooled plate below the sample support keeps the sample translation stage below 30°C.

Fiber/indenter alignment is performed remotely, using motorized translation stages to bring an individual fiber beneath the indenter. Alignment is attained by monitoring the video from a long-working-distance microscope. Video monitoring during the test is also used to identify the initiation of fiber movement and the generation of matrix cracks.

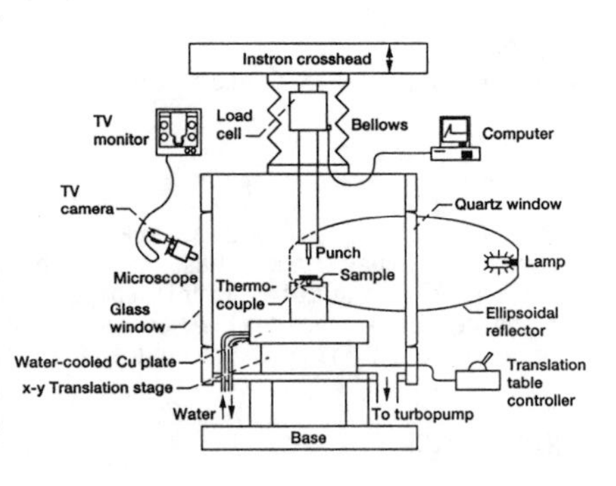

Figure 1. Schematic diagram of elevated temperature fiber push-out apparatus.

Materials

Three different composites were tested: SCS-6 SiC reinforced Ti-24Al-11Nb (Ti-24-11), SCS-6 SiC reinforced Ti-15V-3Cr-3Sn-3Al (Ti-15-3), and single crystal (Saphikon) sapphire-reinforced NiAl. The Ti-24-11 and NiAl matrix composites were prepared by a powder cloth process [2] at NASA Lewis, while the Ti-15-3 matrix composite was prepared by a foil consolidation process by Textron Specialty Materials. Thin specimens (0.28 to 0.48 mm) were sliced with a diamond saw perpendicular to the fiber axes and then mechanically polished down to a 1 μm finish.

RESULTS

Thickness Dependence

Fig. 2 shows a plot of average interfacial debond and frictional sliding stresses for push-out tests on different thicknesses of SCS-6 reinforced Ti-24-11 at room temperature. The average interfacial shear stress, τ_{av}, is taken as $\tau_{av}=P/(\pi d_{fiber}t)$, where P is the applied load, d_{fiber} is the fiber diameter, and t is the sample thickness. Fig. 2 shows that there is a range in thickness from about 0.28 to 0.48 mm for which τ_{av} has no significant thickness dependence. Based on this observation, testing was restricted to sample thicknesses in this range where τ_{av} offers a fair comparison between samples. Further work is required to validate this approach for all testing temperatures.

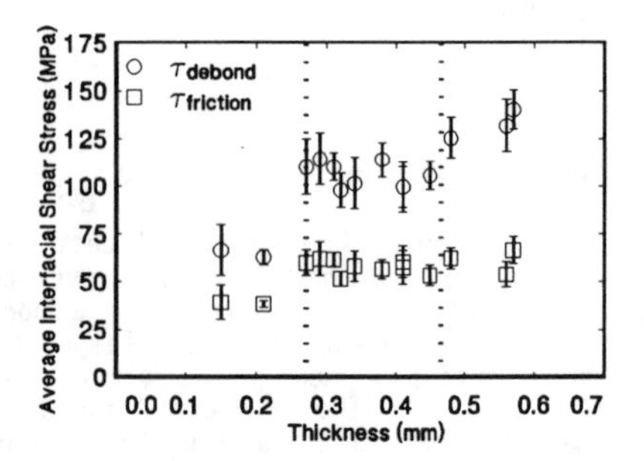

Figure 2. τ_{debond} and $\tau_{friction}$ vs. sample thickness for SCS-6/Ti-24-11 at room temperature.

Elevated Temperature Fiber Push-Out Data

Fig. 3 shows τ_{av} vs. crosshead displacement curves for representative push-out tests at different temperatures for the three composites being examined. Most of the curves consisted of an initial monotonically increasing portion followed by a sharp drop at debonding; the shear stresses after debonding corresponded to purely frictional resistance. Debond shear stresses, τ_{debond}, were taken to be the maximum values prior to debonding. The frictional shear

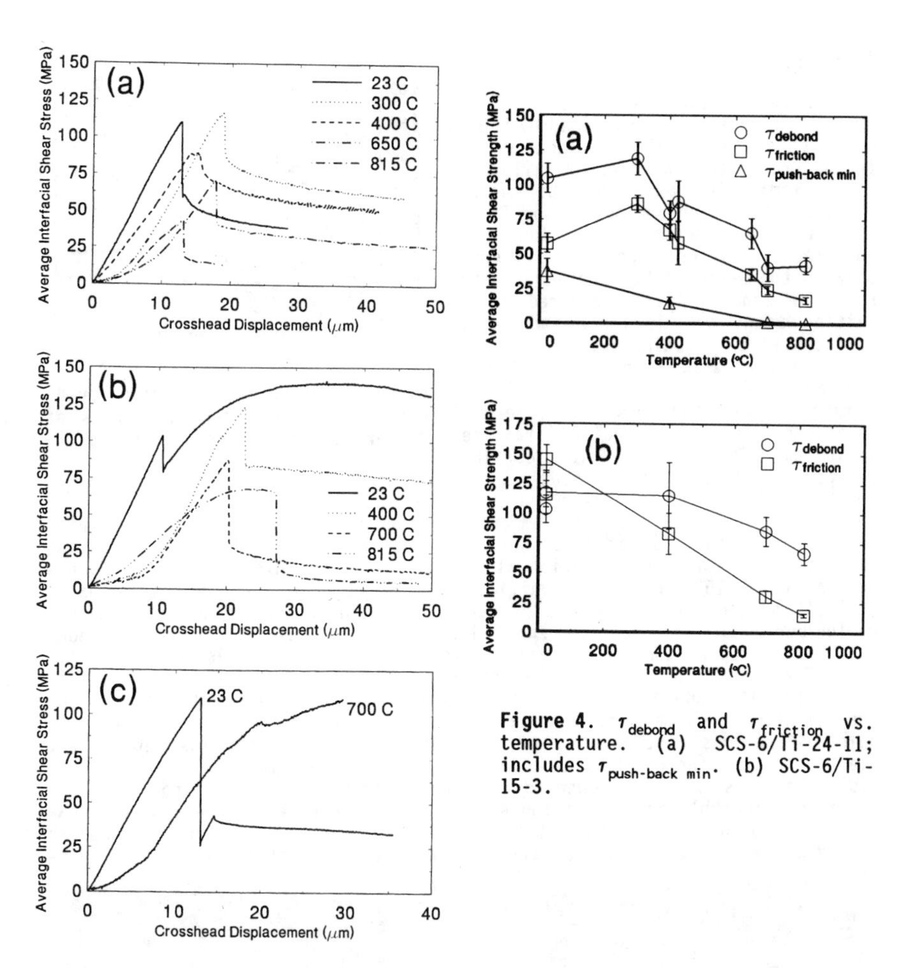

Figure 4. τ_{debond} and $\tau_{friction}$ vs. temperature. (a) SCS-6/Ti-24-11; includes $\tau_{push\text{-}back\ min}$. (b) SCS-6/Ti-15-3.

Figure 3. Representative fiber push-out τ_{av} vs. displacement curves at different temperatures. (a) SCS-6/Ti-24Al-11Nb. (b) SCS-6/Ti-15-3. (c) Sapphire/NiAl.

stress, $\tau_{friction}$, was taken to be the value of τ_{av} obtained immediately after the sharp drop at debonding.

Fig. 4 shows the variation of τ_{debond} and $\tau_{friction}$ as a function of temperature for the Ti-based matrix composites. The plotted results are average values from many tests. Also shown in Fig. 4a are results of push-back tests for the Ti-24-11 matrix composite. In these tests, all fibers are initially pushed out at room temperature, then pushed back at different temperatures. Representative τ_{av} vs. crosshead displacement push-back curves are shown in Fig. 5. A pronounced valley in these curves is observed as the fiber moves through its initial undisplaced position. This minimum occurs because the undisplaced position is the only fiber position where the contacting fiber and matrix surfaces are in registry.

Interfacial Failure

After testing, both ends of the pushed-out fibers were examined by SEM in order to evaluate both the location and the nature of the interfacial failure.

Examination of both ends of the SCS-6 SiC fibers pushed out of the Ti-24-11 matrix composites revealed that interfacial failure occurred primarily between the C-rich fiber coating and the reaction zone. There appeared to be little damage to the fiber and matrix surfaces at all testing temperatures.

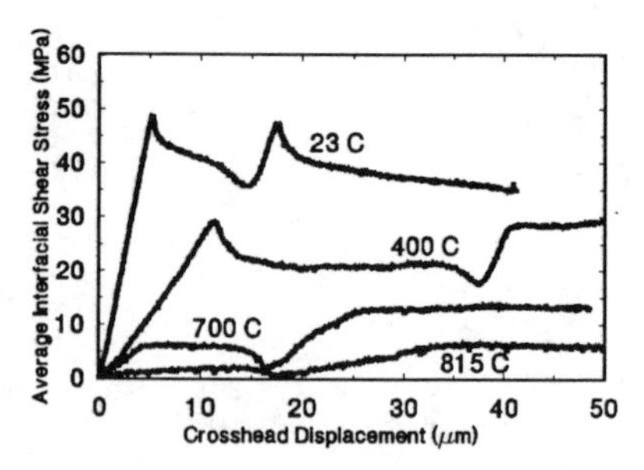

Figure 5. Representative fiber push-back τ_{av} vs. displacement curves at different temperatures for SCS-6/Ti-24-11.

In contrast, examination of both ends of SCS-6 fibers pushed out of Ti-15-3 matrix composites showed a variety of locations of interfacial failure, including within the C-rich coating, between the fiber and the coating, and between the coating and reaction zone. In addition, there was a marked temperature dependence for interfacial wear (Fig. 6). Fibers pushed out at 23°C showed severe fracturing of the C-rich coating with coating debris accumulation at the fiber/matrix interface, while fibers pushed out at 815°C showed no significant damage. In addition, the sample tested at 815°C showed varying amounts of matrix yielding on the topface in a narrow region surrounding the reaction zone. The amount of matrix yielding corresponded with the degree of nonlinearity of the load/displacement curves (Fig. 3b).

Examination of both ends of sapphire fibers pushed out of NiAl matrix composites also showed a marked temperature dependence. At 23°C there was no overall matrix deformation, only localized wear tracks on the side of the push-in hole as previously observed [3]. At 700°C, however, there was significant plastic deformation of the matrix on both sides of the sample. On the push-in side, there was extensive downward yielding of the matrix surrounding the fiber, and at considerably greater distances from the fiber than observed for the SCS-6/Ti-15-3 composite tested at 815°C. In addition the top ends of the sapphire fibers pushed out at 700°C typically displayed significant cracking. The push-

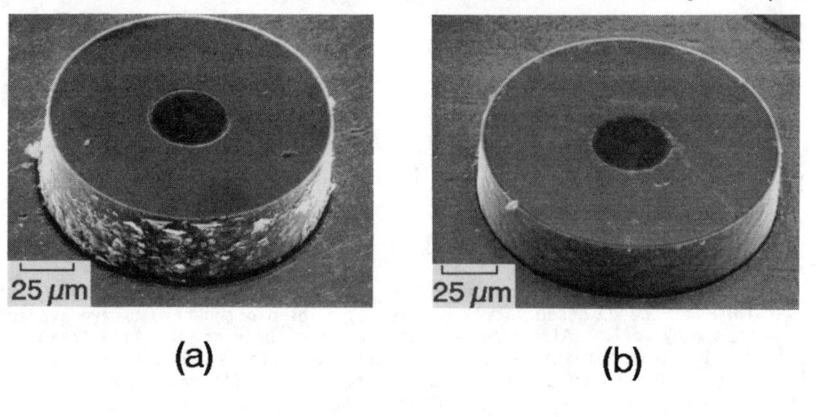

Figure 6. SEM micrographs of SCS-6 fibers pushed out of Ti-15-3 matrix. (a) 23°C. (b) 815°C.

out side of the fibers tested at 700°C showed a combination of outward matrix deformation as well as interfacial crack opening along some portion of the fiber circumference.

DISCUSSION

Fiber Debonding

One of the most important observations in the temperature dependence of τ_{debond} is that there is no significant decrease up to 300°C for SCS-6/Ti-24-11 (Figs. 3a and 4a) and up to 400°C for SCS-6/Ti-15-3 (Figs. 3b and 4b) even though residual stresses would be expected to be on the order of half their room temperature values. This is consistent with a chemically bonded interface, although a small decrease in τ_{debond} would be expected with a reduction of the residual shear stresses that oppose the shear stresses induced by the applied load near the topface [4,5].

At higher temperatures, there are two factors which may contribute to changing the mode of interfacial failure. One is the reduction of residual clamping stress, σ_r, with increasing temperature, allowing mode I debonding induced by bending stresses at the backface along with the mode II debonding at the topface which dominates at lower temperatures. The other factor is the increased ductility of the matrix. These factors may explain the significant decrease at higher temperatures in τ_{debond} (Figs. 3 and 4). The onset of a mixed mode interfacial failure may occur at lower loads than the simpler interfacial failure at lower temperature. Mixed mode failure has been previously observed at room temperature [6] for sapphire-reinforced Nb, which has only a very small fiber/matrix CTE mismatch and high ductility at room temperature. However, the only direct evidence for this mixed mode failure for the Ti-based matrix composites was for the 815°C tests of the Ti-15-3 matrix composite where there was obvious matrix yielding on the topface, slight interface circumferential cracking on the bottomface, and obvious yielding behavior in the push-out τ_{av} vs. displacement curves. While there is a suggestion of mixed mode failure for the Ti-based matrix composites, it was very clear for the sapphire-reinforced NiAl at 700°C, possibly because NiAl has a much more pronounced brittle to ductile transition. SEM examination clearly showed circumferential crack opening of the interface on the bottom face along with inward matrix shear on the topface. The τ_{av} vs. displacement curve at 700°C was further complicated by inflections in the curve due to cracking of the top end of the fiber.

Fiber Sliding

The frictional sliding stresses are presumed to follow the relationship, $\tau_{friction} = \mu\sigma_r$, where μ is the coefficient of friction between the sliding surfaces. For an undisplaced fiber, σ_r is due to residual stresses only; however, the push-out test determines $\tau_{friction}$ for a displaced fiber. A displaced fiber will experience an increased misfit strain due to the contacting irregular fiber and matrix surfaces being out of registry. A sliding fiber will also generate wear at the interface.

The data indeed show that $\tau_{friction}$ is not simply related to the reduction in σ_r with temperature. The results for SCS-6/Ti-24-11 (Figs. 3a and 4a) show an increase in $\tau_{friction}$ up to 300°C even though there should be a substantial decrease in the residual clamping stresses. This can only be explained by differences in interfacial wear. While there was no gross wear at any temperatures, the texture of fiber surfaces pushed out at lower temperatures indicated more significant wear on a submicron scale. The more severe fine-scale wear at lower temperatures may cause $\tau_{friction}$ to be lower than it would have been without any wear. The fiber push-back results (Figs. 4a and 5) give a better indication of residual clamping stresses at different temperatures, because the measurements are made at the initial undisplaced fiber position [7], and because the interfacial wear at different temperatures is equalized by performing all the initial forward pushes at room temperature where the wear is the most severe. Indeed, $\tau_{push\text{-}back\ min}$ shows a steady decrease with temperature all the way to zero, as one expects σ_r to behave.

The SCS-6/Ti-15-3 push-outs showed gross coating fracturing at room temperature (Fig. 6a), with evidence of coating debris accumulating at the interface during fiber sliding. This large-scale debris accumulation is responsible for the increase in $\tau_{friction}$ during continued fiber sliding observed at room temperature (Fig. 3b). The wear becomes less severe with increasing temperature. Unfortunately, the fiber coating damage is so severe at room temperature, that fiber push-back tests would have little meaning.

The sapphire/NiAl results have the additional consideration of the greatly increased matrix ductility at elevated temperatures which may produce large increases in μ. This may explain why $\tau_{friction}$ is higher at 700°C than at 23°C (Fig. 3c) even though σ_r is expected to be much lower at the higher temperature.

CONCLUSIONS

For the composites tested, τ_{debond} (τ_{av} at debonding) did not appear to depend on residual shear stresses since there was little temperature dependence up to 400°C. Subsequent decreases in τ_{debond} at higher temperatures may be due to a combination of reduced σ_r at the sample backface and increased matrix ductility bringing an onset of mixed mode interfacial failure.

Frictional sliding stresses were found to reflect reduced residual clamping stresses with increasing temperature, but were sometimes dominated by changes in interfacial wear. A fiber push-back technique was found in one case to provide a better indication of residual clamping stresses on undisplaced fibers.

ACKNOWLEDGMENTS

The author wishes to thank B. Ebihara and D. Dixon for assistance with the design and build-up of the testing apparatus. Thanks are also due to A. Korenyi-Both for SEM analysis, and to P. Kantzos, P. Brindley, and R. Bowman for providing material and helpful discussions.

REFERENCES

1. J.I. Eldridge, R.T. Bhatt, and J.D. Kiser, Ceram. Eng. Sci. Proc. 12, 1152 (1991).
2. J.W. Pickens, R.D. Noebe, G.K. Watson, P.K. Brindley, and S.L. Draper, NASA TM-102060, 1989.
3. C.A. Moose, D.A. Koss, and J.R. Hellmann in Intermetallic Matrix Composites, edited by D.L. Anton, P.L. Martin, D.B. Miracle, and R. McMeeking (Mater. Res. Soc. Proc. 195, Pittsburgh, PA 1990) pp. 293-299.
4. D.B. Marshall, M.C. Shaw, and W.L. Morris, Acta Metall. Mater., 40(3), 443 (1992).
5. R.J. Kerans and T.A. Parthasarathy, J. Am. Ceram. Soc., 74(7), 1585 (1991).
6. P.D. Jero, R.J. Kerans, and T.A. Parthasarathy, J. Am. Ceram. Soc., 74(11), 2793 (1991).
7. D.A. Koss, R.R. Petrich, J.R. Hellman, and M.N. Kallas, in HITEMP Review 1991, NASA CP 10082, pp. 27-1 to 27-13, 1991.

USE OF SHOCK WAVES TO MEASURE ADHESION AT INTERFACES

GERALD L. NUTT, and WAYNE E. KING
Lawrence Livermore National Laboratory, 7000 East Avenue, P.O.Box 808, Livermore, CA. 94550

ABSTRACT

The central problem in the study of composite materials is the adhesive strength of the electronic bond between reinforcement and matrix. We have introduced a unique method of measuring the interface bond strength of a wide variety of engineering interfaces (e.g. metal/ceramic, semiconductor/metal, metal/polymer). Specimens are composed of a relatively thin overlayer on a thick substrate. The specimens are shocked using a magnetic hammer which accelerates a thin metal flyer onto the substrate. The shock, upon reflection at the free surface, is incident on the bonded interface as a tensile wave, spalling the overlayer. The method is unique in using free surface velocity measurements to determine the interface stress at the instant of separation. The debonding process is sufficiently rapid (on the order of 1.0 ns) that debonding occurs by the simultaneous breaking of atomic bonds, rather than by propagation of cracks nucleating at stress concentrations near existing flaws.

INTRODUCTION

There are essentially two methods of determining the bond strength from the spall stress at the interface. The first, introduced by Snowden [1] determines the minimum shock strength required to spall a film from a substrate. The event is then modeled using a computer code to determine the maximum tensile stress at the interface generated by the known shock input. The tensile stress is taken as the ultimate strength of the bond. This method has been reintroduced by Gupta *et.al.*[5, 3, 4, 2], generating the shock by deposition of laser energy.

The second method, introduced by us [6, 7, 8, 9],generates a plane shock wave parallel to the interface by flyer impact. We do not require the measurement of the threshhold shock profile that produces the spall. All that is needed is that the shock wave produce a spall event. The stress at the bond is found from the free surface velocity history of the spalled film. We measure this velocity using a VISAR laser interferometer. We shall discuss the important features of both types of experiments, illustrating with numerical simulations.

There are advantages of this type of numerical study as well as limitations. One advantage is, a modeling calculation can show details of the experiment that cannot be observed experimentally. This allows a quick assessment of the effect of parameter changes on the outcome of the experiment, (i.e. sensitivity studies). As we shall see, we will also be able to evaluate assumptions and graphically display their effect on the outcome of the experiment. At the same time we must be cautious when applying a numerical model to an experiment. The model is always a simplification. Additionally, these techniques invariably introduce numerical artifacts which are unphysical, and which must be allowed for when interpreting the results.

Several conclusions can be drawn from our numerical studies. Generally, steps must be taken to make the shock wave incident on the interface, planar. Gupta assumed his laser generated shocks were planar, but we will show that in fact they were not. The consequence was a sizable error in his reported results. We further conclude that the method of shock generation is important. Snowden's experiment, as well as in our own, used flyer impact to produce the shock. It is possible to guarantee plane shocks with this approach.

We will also show that if the spall event is to be diagnosed using computer simulation, the shock profile must be known precisely. This is extraordinarily difficult at best, but especially so with laser generated shocks. We show how this leads to additional errors in Gupta's results.

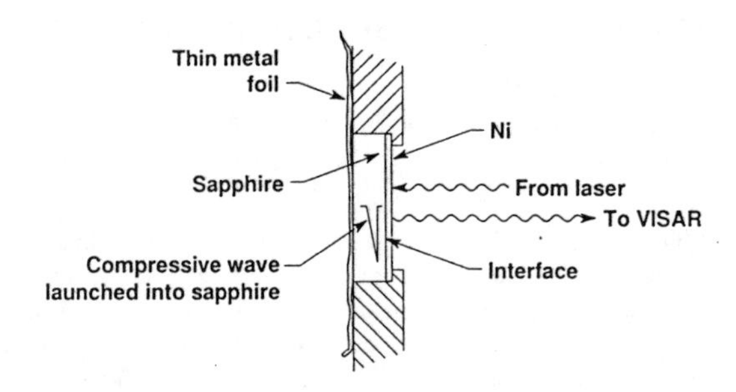

Figure 1: Schematic of flyer impacting specimen.

SHOCK WAVES USING FLYER IMPACT

Figure 1 shows a schematic of a flyer impacting a substrate in one of our sapphire/Ni experiments. If the flyer is flat on impact, and if the width of the substrate/film specimen is more than twice its thickness, the planarity of the shock in the central part of the specimen is guaranteed. A variation of the experiment could include a pad between the flyer and substrate for the purpose of shaping the shock wave.

The measurement is made by interferometry with the laser light reflected from the free surface of the film. The spall stress is calculated using the formula [9]

$$\sigma_s = \rho_0 \frac{c_l c_b}{c_l + c_b}(u_0 - u_k),$$

where $u_0 = 0.02175 \text{cm}/\mu\text{s}$, and $u_k = 0.02093$ cm/μs. $\rho = 8.90 \text{g/cm}^3$, and c_l and c_b are the longitudinal and bulk sound speeds respectively. The result is a bond strength of 190 MPa.

A typical surface velocity record is shown in Fig. 2. The sampling interval is 3.0 ns. Reading the indicated jumpoff velocity and the minimum at the first pullback, we are able to calculate the breaking stress at the interface. This record appears rather mesy when compared with the same velocities from the model calculations. A little interpretation helps to get oriented. The ringing period of the Ni film is 6.8 ns for a nominal thickness of 2.0×10^{-3}cm and a longitudinal sound speed of 0.587 cm/μs. The shortest period oscillation appearing in Fig. 2, a is identifiable as the Ni film period. Similarly, the sapphire substrate has a period of about 0.5μs - long compared to the shorter structure in Fig. 2,a. The flyer period is of the order of 21 ns and appears to be superimposed on the Ni vibration in the next four peaks in the free surface velocity.

Figure 2, b shows a one-dimensional simulation of the spall event superimposed on the data of Fig. 2, a. The time scale is expanded. The calculated signal is much cleaner than the VISAR measurement. For example, the flyer period is not present in the calculated free surface velocity. This is because the foil separates completely from the substrate on the first reflection of the shock from the free surface and is no longer in communication with the substrate. In the actual experiment, the foil does not completely separate but remains partially attached near the edge of the specimen.

We believe there are several reasons for incomplete separation of the film. Among these are nonuniform thickness of the Ni film. Upon examination of the batch of Ni/sapphire specimens we found as much as a 50% variation in thickness of the Ni across the diameter of the sample. We also found that the flyer wrinkled during acceleration due to a "pinch" effect during the conduction of high current. This caused a wrinkled impact on the base of the

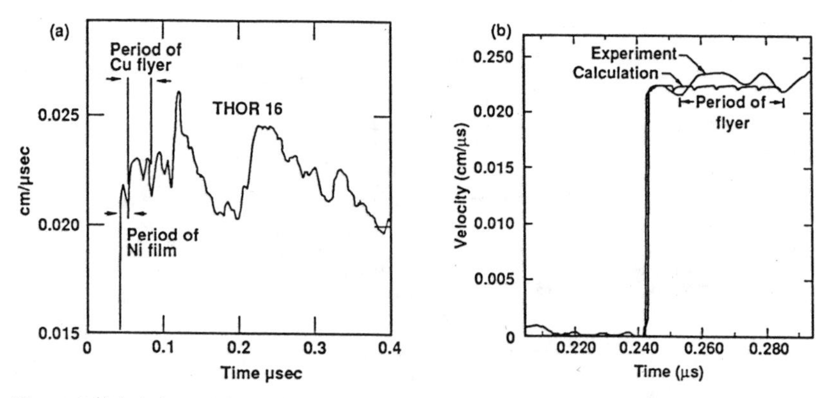

Figure 2: (a) A free surface velocity history for an exfoliation experiment. (b) Superposition of calculated and measured free surface velocities

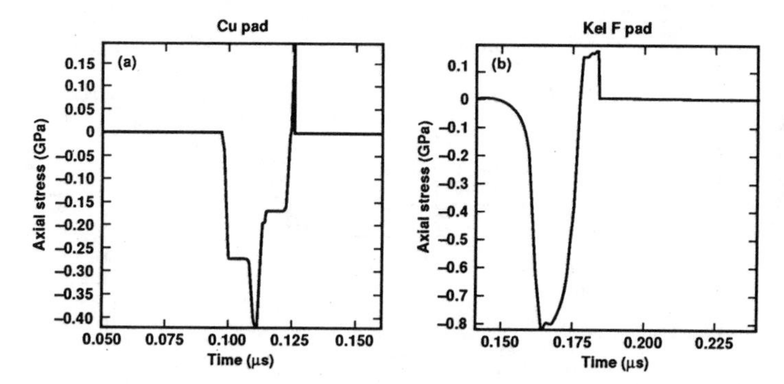

Figure 3: Comparison stress risers resulting from use of stiff (Cu) and compliant (Kel F) pad between flyer and substrate.

substrate. Of course, the shock wave is weakened near the edge of the sample due to lateral relief waves from the edge, with the consequence that the tensile wave is correspondingly weakend near the edge.

Several solutions to these problems present themselves. First, the substrate could be masked during deposition of the overlayer so as to confine the film to the central portion of the specimen. Second, a pad of material such as Cu, or Kel F (C_2F_3Cl) could be placed over the impact surface to smooth the shock front. The softer material would also have the effect of slowing the stress rise time at the bonded interface in order to improve the resolution of the spall event. This is illustrated in Fig. 3, comparing calculations of two specimens, one with a stiff Cu pad and a flyer velocity of 0.002 cm/μs, and a similar specimen with a Kel F pad, and flyer velocity of 0.011 cm/μs. The result is that the stress is applied over a time of 5 ns with the Kel F pad, as opposed to a time of less than 1.0 ns with the copper pad.

LASER GENERATION OF SHOCK WAVE

A schematic of Gupta's experiment is shown in Fig. 4. In this experiment a 1.0 μm film of SiC was deposited on one surface of a ≈ 2 mm thick Si substrate. A 1.0 μm thick layer of Au was deposited on the other Si surface. Several millimeters of quartz were laid over the Au film. A Nd-YAG laser ($\lambda = 1.06$ μm) illuminated a ≈ 1 mm^2 spot on the Au through the quartz, generating the shock wave.

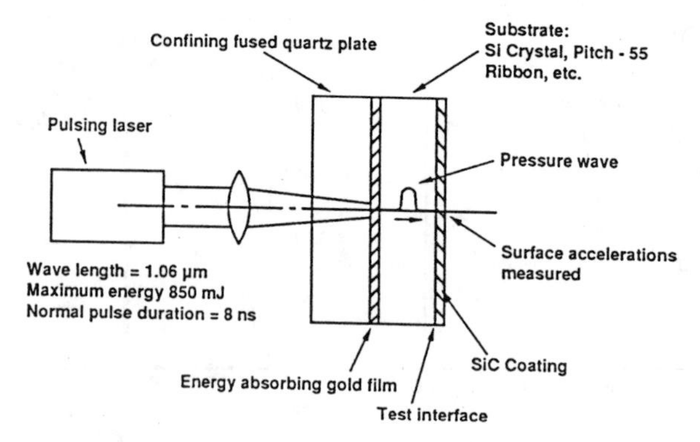

Figure 4: Schematic of the experiment using laser induced spall reported by Gupta et. al. [5]

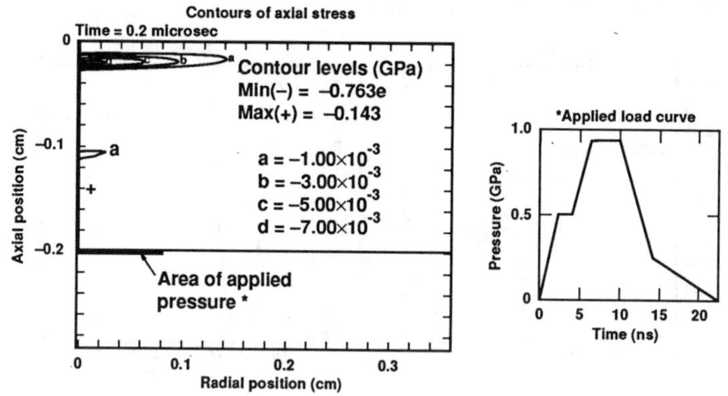

Figure 5: A two-dimensional simulation of the experiment. Contours are shown just as the shock reaches the interface. The stress history at the indicated surface approximates one of Gupta's measurements.

In order to model one of Gupta's experiments as closely as possible, we took the stress profile measured by the piezoelectric gauge corresponding to a dose of 3.64×10^4 J/m^2.[5] The pressure boundary condition was applied over a 2.0 mm^2 circular area of one face of the Si substrate as shown in the diagram of the two-dimensional problem in Fig. 4. The axial zoning was uniform with elements 4.0×10^{-5} cm thick.

The effect of lateral dispersion on the shock pulse is shown in Fig. 6 by the approximately 20% decrease in the amplitudes of the stress profiles as they propagate through the substrate. The amplitudes shown at 0.25 μsec are reduced even further by the desructive interference of the incident compressive wave and the tensile wave reflected at the free surface.

Finally, the stress history at the bond interface is shown in Fig. 7. The reduction in stress amplitude from an input stress of ≈ 1 GPa to a maximum tensile stress of ≈ 30 MPa is due to the input stress wave length relative to the thickness of the film. The stress amplitude at the bond is the difference between two points on the stress profile separated by twice the film thickness. To be able to control this experimentally requires an unrealistic precision in controlling the input shock as well as in the calculation of its motion through the specimen.

We conclude that Gupta's results are in serious error because the non-planarity of the shock wave requires at least a two-dimensional calculation to account for the geometric dispersion. In addition, we estimate that the problem caused by the relative thickness of the film to the length of the stress wave lead him to overestimate his measurements by a factor in the neighborhood of two orders of magnitude. The accuracy of this estimate is limited by the accuracy of our input pressure profile taken from Gupta's published data.

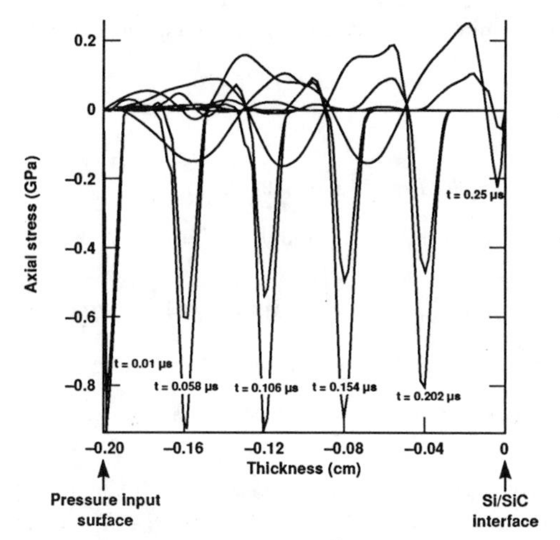

Figure 6: Stress wave profiles propagating through the substrate. The larger amplitudes are found along the axis while the smaller amplitudes are 0.7 mm from the axis.

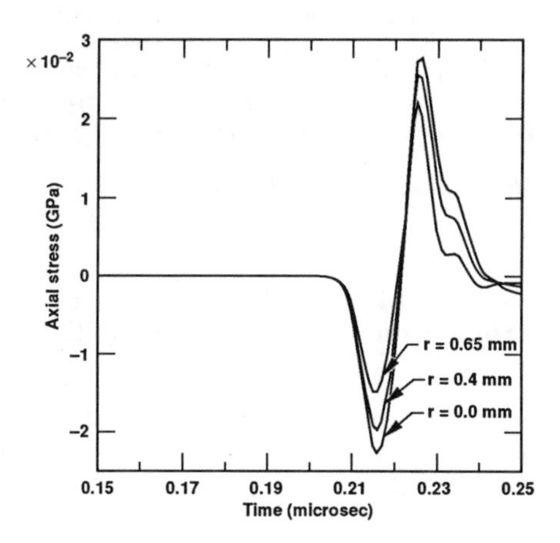

Figure 7: Stress history of interface in the neighborhood of the axis.

CONCLUSIONS

There are four conclusions to be drawn from this study. First, contrary to assumption, the shock wave in the experiments of Gupta *et. al.* is not planar. Rather, considerable dispersion occurs before it interacts with the bond. Second, when the length of the stress wave is much greater than the thickness of the film, as is the case in their experiment, the error in the calculated interface stress becomes unmanageable. Third, our simulations imply that Gupta's reported measurements are too high by more than an order of magnitude. Finally, laser interferometry appears to offer the best prospects for bond strength measurements since the measurement is direct, and accurate simulation of the experiment is unnecessary.

References

[1] W. E. Snowden. *Mater. Sci Res.*, 14:651, (1981).

[2] V. Gupta, A. S. Argon, D. M. Parks, and J. A. Cornie. *J. Mech. Phys. Solids*, to be published.

[3] J. A. Cornie, A. S. Argon, and V. Gupta. *MRS Bulletin*, XVI(4):32–38, (1991).

[4] V. Gupta. *MRS Buletin*, XVI(4):39–45, (1991).

[5] V. Gupta, A. S. Argon, J. A. Cornie. and D. M. Parks," *Mater. Sci. & Eng.*, A 126:105–117, (1990).

[6] Gerald L. Nutt, William Lai, Kenneth E. Froeschner, and Wayne E. King. In M. Rühle, A. G. Evans, M. F. Ashby, and J. P. Hirth, editors, *Metal - Ceramic Interfaces*, pages 307 – 312, Acta-Scripta Metallugica, (1989).

[7] Gerald L. Nutt, William Lai, Kenneth E. Froeschner, and Wayne E. King. In B.M. DeKoven, A.J. Gellman, and R. Rosenberg, editors, *Interfaces Between Polymers, Metals, and Ceramics*, pages 385–390, Materials research Society Proceedings **153**, 1989.

[8] Gerald L. Nutt. Numerical model of bond strength measurements. In R. D. Bringans, J. M. Gibson, and R. M. Feenstra, editors, *Atomic Scale Structure of Interfaces* **159**, pages 465–471, Materials Research Society Proceedings **159**, (1990).

[9] Gerald L. Nutt. *J. Mater. Res.*, 7(1):203–213, 1992.

---

* Work performed under the auspices of the U.S. Department of Energy by the Lawrence Livermore National Laboratory under contract number W-7405-ENG-48.

INTERFACIAL SHEAR BEHAVIOR OF TWO TITANIUM-BASED SCS-6 MODEL COMPOSITES

I. ROMAN* AND P. D. JERO

Wright Laboratory, Materials Directorate, WL/MLLM, WPAFB OH 45433
*Permanent address: Graduate School of Applied Science and Technology,
The Hebrew University of Jerusalem, JERUSALEM, ISRAEL 91904

ABSTRACT

Single fiber push-out and push-back tests combined with acoustic response monitoring were used to examine the interfacial behavior in two titanium alloy-SiC fiber composites. Distinctly different behaviors were observed in the two systems. The differences were attributed to the formation of a substantial interfacial reaction layer in one of the composites which changed the interfacial chemistry and the resulting debond topography. The reaction layer caused an increase in the interfacial bond strength and in the roughness of the debonded interface. The latter resulted in substantially increased sliding friction. Although both composite interfaces exhibited some roughness, only one showed a seating drop during fiber push-back. This is related to the fact that the reaction layer which formed in one of the composites was severely degraded during fiber push-out. Although this interface was still rough, the roughness correspondence between fiber and matrix was destroyed during sliding, such that seating was no longer possible.

INTRODUCTION

Titanium and titanium aluminide composites are emerging as a new class of materials for aerospace applications because of their attractive high temperature specific strength and stiffness, and due to their potential to provide properties not achievable in monolithic alloys. The interfaces between matrix and reinforcement play a critical role in determining the mechanical properties of these advanced composites, for reviews see [1,2]. Low interfacial shear strength, which enables debonding between the fiber and matrix at the crack front, leads to higher composite toughness, whereas high interfacial shear strength is required for efficient load transfer, which promotes composite strength. A substantial amount of information on the nature of the microstructure and chemical composition of interfacial zones in titanium-base composites, along with the kinetics of interfacial reactions is available in the literature, [e.g. 3]. However, quantitative information on the interfacial mechanical properties, which is essential to fully characterize a composite system, is scarce, as experimental testing methods capable of determining interfacial mechanical properties of metal matrix composites have not yet been sufficiently established. Fiber push-out (or indentation) testing [4], is one of the techniques that can be employed to experimentally determine interfacial properties in metal matrix composites. This technique, has recently motivated a large number of studies in ceramic matrix composites [5-11], and is just beginning to be employed in testing of metal and intermetallic matrix composites [12-15]. Push-out testing was employed by Yang, Jeng, and Yang [12], to determine interfacial shear strength in different titanium-matrix composites and recently by Roman and Aharonov [13], who studied aluminum-based composites. Both groups utilized simple push-out instrumentation consisting of a modified standard microhardness tester where the indenter (Vickers and Rockwell C, respectively) was loaded in steps by weights until first sliding of the pushed fiber was noticeable. The interfacial strength in SCS-6/Ti-24Al-11Nb (atomic percent) composites and in sapphire reinforced NiAl composites, was determined by Eldridge and Brindely [14], and by Moose, Koss and, Hellmann [8], respectively. Flat bottomed tungsten carbide probes (100-125 μm in diameter) loaded continuously, were employed in both studies and load-displacement behavior was recorded. Analysis of test results in these studies relied on peak load as a measure of debonding, and distinction between frictional resistance and actual bond strength was obtained by noting the difference between peak load and the load required for pushing back the debonded fiber.

EXPERIMENTAL PROCEDURE

The materials used in this study were two different titanium-based composites containing Textron's SCS-6 SiC fiber as the continuous reinforcement. The SCS-6 fiber is a 142 μm diameter SiC filament fabricated by chemical vapor deposition (CVD) on a 33 μm diameter carbon core. It has an outer, 3.6 μm thick, double pass silicon doped carbon coating added to preserve fiber strength and enhance chemical compatibility with titanium-based matrices.

Two different matrix compositions were employed. One was a conventional $\alpha + \beta$ titanium alloy having a nominal composition (weight percent) of Ti-6Al-4V (Ti-6-4). The other, having a nominal composition (atomic percent) of Ti-24Al-11Nb (Ti-24-11), was an $\alpha_2 + \beta$ alloy based on the Ti_3Al (α_2) intermetallic compound (DO_{19}) which exhibits an ordered hexagonal structure. Composite panels of 35% fiber volume fraction manufactured by Textron Speciality Materials were utilized for specimen preparation. The material was consolidated in a single stage process, and was made of eight unidirectional plies.

Transverse slices, measuring about 12 mm x 2 mm in cross-section and ranging in thickness from 0.25 to 0.75 mm were diamond saw cut from the samples, ground and polished to a surface finish of 0.06 μm RMS and utilized in fiber push-out tests. The samples were mounted on a steel support plate, with the fiber(s) to be pushed-out aligned over a 0.4 mm wide slot. Single fibers were pushed out of the matrix, one at a time, by either a 100 μm or a 125 μm diameter, flat tip tungsten carbide probe, mounted on a small load cell, and driven at a cross-head speed of 2.12 μm/s. Fiber displacement was monitored utilizing a capacitance gauge (Capacitec Inc., Ayer, MA). Data acquisition was computerized with a sampling rate of 20 Hz. Fibers were pushed a distance of about 100-200 μm, after which the sample was flipped, and the fiber was pushed back through its original position in order to examine the effect of interfacial roughness on the frictional sliding [9]. 5 to 10 fibers were pushed from each of the composites.

Acoustic emission (AE) activity during push-out was monitored employing a resonant transducer with a nominal center frequency of 250 KHz (Micro 30, Physical Acoustics Corporation (PAC), Princeton, NJ) which was coupled via high vacuum grease to the loading probe holder or to the support plate. Transducer outputs were amplified first by 40 dB using a preamplifier (Model 1220A, PAC) with a bandpass filter of 100-400 KHz and then by an additional 20 dB at the main amplifier (Model 4300, PAC). The AE parameter recorded by the computerized data acquisition system was the RMS voltage of the amplified transducer outputs, measured using an rms voltmeter (Model 3400A, HP).

RESULTS AND DISCUSSION

Load-displacement curves typical of the push-out tests are shown in Fig. 1. The results for both composites indicate that after settling (i.e. overcoming the "slop" in the system associated mainly with imperfect sample mounting), the load initially increased to a point where it dropped. The nature of the load drop and subsequent behavior differ for the two composite systems. It is not apparent from the measured load-displacement behavior whether the curves are linear up to the point where the load dropped, however, the ability to make such a determination is significant, as deviation from linearity implies a progressive debonding of the interface. Fig. 2 shows the same data from which the compliance of the load train has been removed. The compliance removal procedure has been described previously [16,17]. With the system compliance removed, the true response of the specimen is apparent. It is seen, for instance, that in the Ti-24-11 composite the displacement of the fiber actually begins before the abrupt drop. Note that elastic or plastic deformation of the matrix immediately adjacent to the fiber could also cause an apparent deviation from linearity. Therefore, in order to minimize bending, the slot under the specimen was kept as narrow as practically possible. In addition, in order to assess whether the matrix had deformed plastically around the pushed fibers, samples were subsequently examined using a laser interferometric microscope (model 5610, Zygo Corp., Middlefield, CT). Comparison of matrix areas immediately adjacent to both pushed and unpushed fibers showed no differences, indicating that there was no plastic deformation associated with pushing the fibers. It is thus ascertained that neither sample bending nor plasticity contributed significantly to the measured displacement.

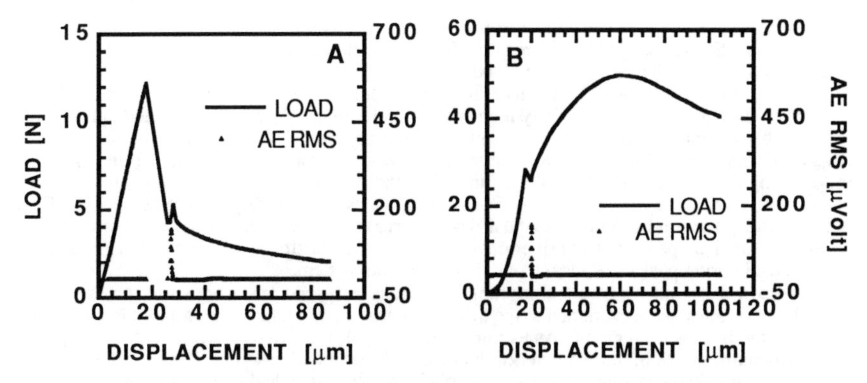

Figure 1. Typical load/acoustic emission-displacement curves for the (A) Ti-24-11 and (B) Ti-6-4 composites.

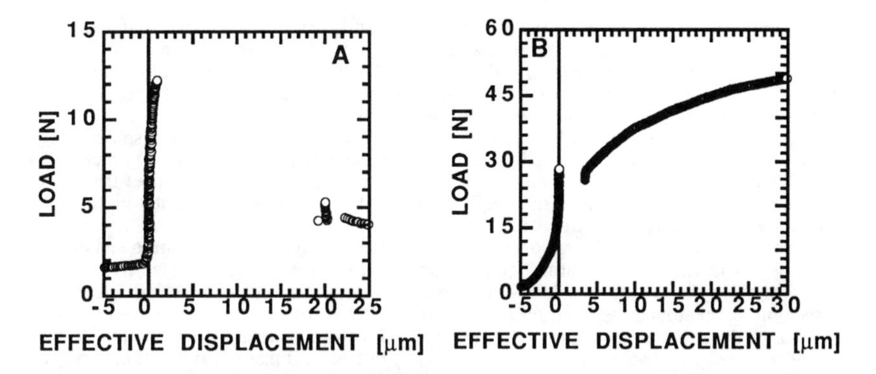

Figure 2. Typical load-corrected effective displacement curves for the (A) Ti-24-11 and (B) Ti-6-4 composites.

In the Ti-24-11 composite a small but distinct region of increasing displacement with increasing load was followed by a large, abrupt load drop. The initial deviation from linearity is believed to be due to stable, progressive debonding, whereas the abrupt load drop and the associated displacement are the result of instantaneous failure of the remaining intact ligament and rapid sliding of the entire fiber. This was followed by a slow, nonlinear decrease in load as the fiber continued to slide out. The acoustic emission (AE) activity during these tests was limited. No activity was detected during stable debonding, however, a sharp spike in the AE rms voltage level accompanied the abrupt load drop. This observation tends to support the notion of catastrophic crack extension through the remaining intact ligament, however, it is possible that the AE is produced as a result of either the cracking or a frictional process during the rapid sliding of the fiber.

In the Ti-6-4 composite, on the other hand, debonding appeared to be catastrophic; deviation from linearity began as a large, abrupt increase in displacement. The magnitude of this initial displacement was sufficient to ensure that the fiber was completely debonded. In some cases a relatively small load drop occurred with the abrupt increase in displacement. In these cases a burst of AE activity was associated with the abrupt displacement and indicates catastrophic crack extension, as discussed above. In all cases, the load subsequently increased in a non linear fashion to a peak, followed then by a slow non linear load decrease. The rate of this final load decrease, when the fiber was sliding out, was similar for both composites but the actual load levels were much higher for the Ti-6-4 system. For this system, in this region of decreasing load, random spikes in the AE rms voltage level were noted for most of the samples which did not exhibit an AE spike when the initial displacement occurred.

Fig. 3 shows typical push-back curves for the two composite systems tested. In the Ti-24-11 composites the load increased initially then peaked as the fiber began to slide. The load required for push-back remained almost constant until seating occurred [18]. The seating phenomenon began with an abrupt drop in the load level as the fiber began to traverse its origin. This was followed by a rise to a peak above the initial value and then a return to approximately the initial value. The push-back behavior of the Ti-6-4 composites was entirely different, after the fiber began sliding, the load required for push-back increased monotonically, no seating drop was observed. The load peaked as the fiber traversed its original position. No significant AE activity was detected during push-back testing of either composite system.

Pushed-out samples and metallographic sections prepared from both as-fabricated, and pushed-out samples, were examined in the SEM, to reveal topographic features and microstructural details of the interfacial regions. As exemplified in Fig. 4, the two composite systems display a very distinct difference in the character of their matrix/fiber interfaces. In the case of the Ti-24-11 composite, a relatively narrow and smooth interfacial reaction zone had formed whereas in the case of the Ti-6-4 system, a significant reaction zone with a very rough interface was produced due to preferential reaction between the β-phase and the outer carbon layer of the fiber. Much of the outer layer was consumed by the reaction.

Longitudinal sections of composites with pushed-out fibers, Fig. 5, indicate that when a fiber in the Ti-24-11 composite was pushed-out separation occurred either between the outer carbon-rich layer and matrix, or between the inner and outer carbon-rich layers. The longitudinal cross-section for the Ti-6-4 system indicates that the outer carbon-rich layers were mostly destroyed during fiber push-out . It is difficult to determine where the separation initiated.

The surfaces of the pushed fibers in the Ti-24-11 composites are relatively smooth, Fig. 6, with a topography that shows small dimples or nodules characteristic of the CVD process. This appearance is typical of SCS-6 fibers [11,19]. In this composite, the roughness of the interface appears to be controlled by the initial roughness of the fiber. This is consistent with the limited reaction between the fiber and matrix in this system.

Conversely, in the Ti-6-4 composites, substantial fiber/matrix reaction occurred and extensive cracking and flaking of the coating (and reaction product) are apparent. Once free of matrix constraint, this damaged layer could not be pushed back into the matrix with the fiber; debris could be seen piling up around the fiber during push-back. These observations can explain the seating behavior of the two systems examined if it is realized that the seating phenomenon is mainly an indication of relative mating between the topography of fiber and matrix cavity. If damage to the interfaces is not extreme, (as in the Ti-24-11 composite) then the seating drop is a measure of surface roughness amplitude. However, when damage is substantial, (as in the Ti-6-4 composite), then a seating drop is not expected and indeed is not observed. In this case, lack of seating does not mean lack of roughness but merely that the mating surfaces no longer resemble each other.

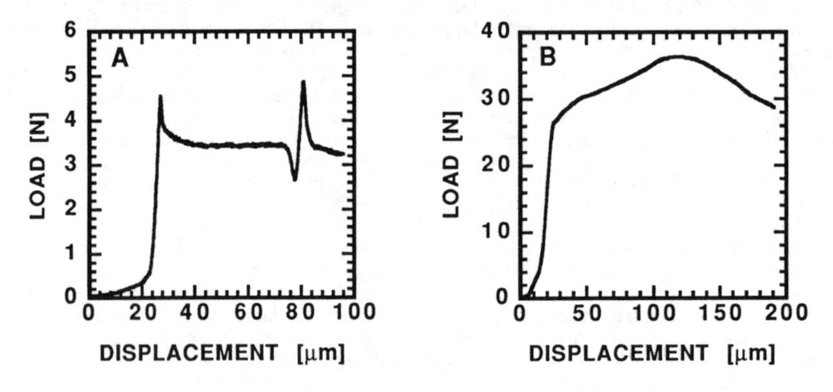

Figure 3. Typical push-back load-displacement curves for (A) Ti-24-11 and (B) Ti-6-4 Composites.

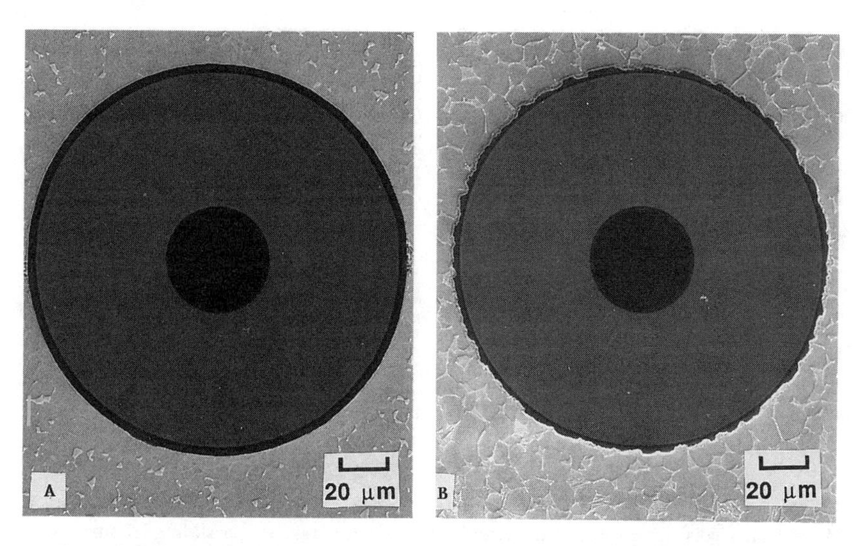

Figure 4. Micrographs displaying the interfacial microstructures of the (A) Ti-24-11 and (B) Ti-6-4 composites.

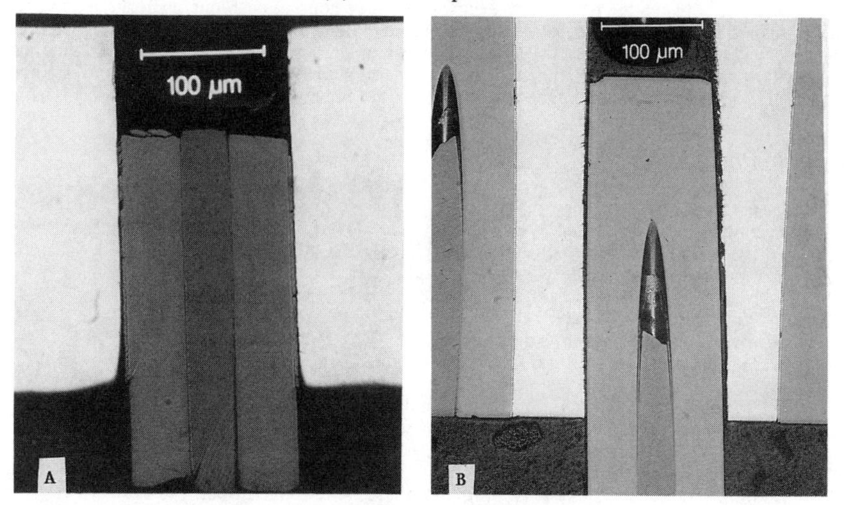

Figure 5. Micrographs showing longitudinal cross-sections of pushed-out fibers in the (A) Ti-24-11 and (B) Ti-6-4 composites.

The broad peak in the load when the fiber traverses its origin is due simply to the change in the sliding surface area.

CONCLUSION

Single fiber push-out tests were used to examine interfacial behavior in two commercial metal matrix composites, SCS-6/Ti-24Al-11Nb and SCS-6/Ti-6Al-4V. The formation of a significant

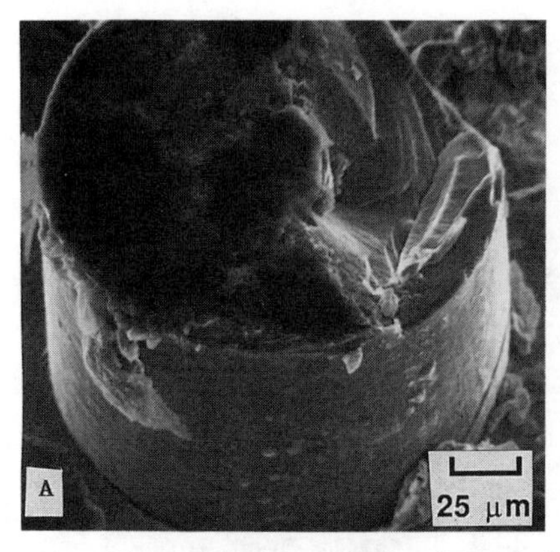

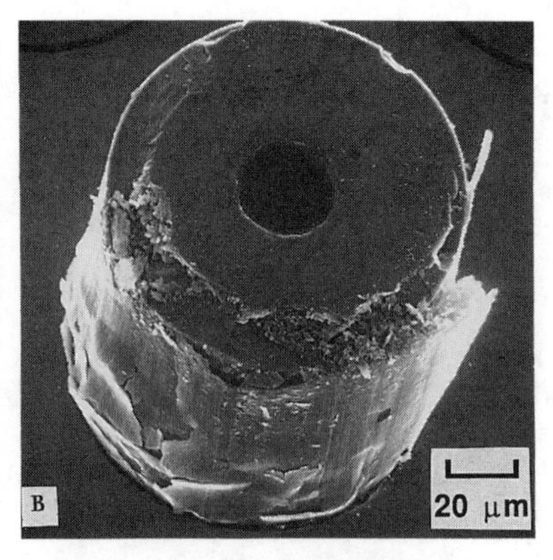

Figure 6. Fractographs showing typical topographies of pushed-out fiber in the (A) Ti-24-11 and (B) Ti-6-4 composites.

fiber/matrix reaction layer in the Ti-6-4 composite resulted in a slightly higher bond strength and an increased interfacial roughness. This interfacial roughness resulted in sliding friction which increased with sliding distance for a time. The reaction layer was, however, found to be readily degraded upon sliding, such that the friction rapidly decreased after peaking. The roughness correspondence between the fiber and matrix was destroyed during push-out, as illustrated by the lack of seating during fiber push-back. It is postulated that the high level of initial sliding friction observed in these tests is effectively diminshed by the decreased fiber strength seen in fiber fragmentation tests [20], such that tensile fiber failure may occur before substantial fiber debond/pull-out, seriously limiting the performance of the composite.

REFERENCES

1. S. G. Fishman, in Interfaces in Metal-Ceramics Composites, edited by R. Y. Lin, R. J. Arsenault, G. P. Martins, and S. G. Fishman (TMS, Warrendale, PA, 1990) p.3.
2. A. G. Evans and D.B. Marshall, Acta Met. 37(10), 2567 (1989).
3. S. F. Baumann, P. K. Brindely and S. D. Smith, Met. Trans. 21A , 1559 (1990).
4. D. B. Marshall, J. Amer. Ceram. Soc. 67(12), C259 (1984).
5. R. W. Goetler and K. T. Faber, Ceram. Eng. Sci. Proc., 9(7-8), 861 (1988).
6. M. K. Brun and R. N. Singh, Adv. Ceram. Mat., 3(5), 506 (1988).
7. A. J. G. Jurewicz, R. J. Kerans, and J. Wright, Ceram. Eng. Sci. Proc., 10(7-8), 925 (1989).
8. J. D. Bright, D. K. Shetty, C. W. Griffin, and S. Y. Limaye, J. Am. Ceram. Soc., 72(10), 1891 (1989).
9. G. Morscher, P. Pirouz, and A. H. Heuer, J. Am. Ceram. Soc., 73(3), 713 (1990).
10. J. I. Eldridge and F. S. Honecy, J. Vac. Sci. Technol., A, 8(3), 2101 (1990).
11. P. D. Jero, R. J. Kerans, and T. A. Parthasarathy, J. Am. Ceram. Soc., 74, 2793 (1991).
12. C. J. Yang, S. M. Jeng and J.-M. Yang, Scripta Met., 24(3), 469 (1990).
13. I. Roman and R. Aharonov, Acta Met., 40(3), 477 (1992).
14. J. I. Eldridge and P. K. Brindely, J. Mat. Sci. let., 8, 1451 (1989).
15. C. A. Moose, D. A. Koss and J. R. Hellmann, in Intermetallic Matrix Composites, edited by D. L. Anton, P. L. Martin, D. B. Miracle, and R. McMeeking (Mater. Res. Soc. Proc. 194, Pittsburgh, PA, 1986) pp. 293-299.
16. T. A. Parthasarathy, P. D. Jero, and R. J. Kerans, Scripta Met. et Mat., 25, 2457 (1991).
17. P. D. Jero, T. A. Parthasarathy, and R. J. Kerans, Ceram. Eng. & Sci. Proc., In press.
18. P. D. Jero and R. J. Kerans, Scripta Metall. Mat., 24, 2315 (1990).
19. P. D. Jero, T. A. Parthasarathy, and R. J. Kerans, Ceram. Eng. & Sci. Proc., In press.
20. I. Roman, S. Krishmurthy, and D. Miracle, to be published.

EVALUATION OF COATINGS FOR SAPPHIRE FIBERS IN TiAl USING THE FIBER PUSH-OUT TEST

T. J. Mackin and J. Yang
Materials Department, The University of California at Santa Barbara, Santa Barbara, CA 93106

ABSTRACT

Fiber push-out tests were used to evaluate several double coating systems for sapphire fibers in γ-TiAl. The double coatings consisted of an inner debond coating, followed by an outer coating of alumina that serves as a diffusion barrier. The inner coatings were of three types: carbon black, colloidal graphite and porous alumina. By comparing estimates of interface toughness computed from the push-out tests, each double coating is found to permit debonding and sliding of the sapphire fibers. Consequently, each of the double coatings would afford improved fracture toughness for sapphire reinforced γ-TiAl.

INTRODUCTION

The mechanical properties of fiber-reinforced materials are known to depend upon the mechanical properties of the fiber/matrix interface. [1-3] Optimum longitudinal properties, for both brittle and ductile matrix materials, require that the interface debond and slide with a relatively low shear resistance, τ, relative to fiber strength, S. A high composite fracture resistance requires small values of τ/S to achieve large fiber pull-out lengths and the consequent frictional dissipation that contributes to the composite's work of rupture.[3,4,] In many cases, fibers must be coated to permit debonding and sliding. A debonding criterion derived by He and Hutchinson[5] can then be used to screen candidate fiber coatings. This criterion states that the Mode II interfacial toughness must be less than 1/4 of the reinforcement toughness. Subsequent frictional pullout dissipation will depend upon the mechanical properties of the sliding interface. This paper summarizes a comparison of three double coating systems for sapphire fibers in γ-TiAl.

Sapphire fibers have a sinusoidal surface roughness that exerts an important influence on coating design.[6] Specifically, when the inner coating is chosen to have a thickness less than the peak-to-peak amplitude of the fiber roughness, the debond surfaces would mimic the underlying fiber geometry, leading to explicit roughness effects on τ. Conversely, coating thicknesses many times the fiber roughness amplitude would tend to obscure any effects of the underlying fiber geometry.

MATERIALS

Sapphire fibers [Sapphikon, R=67μm] were double coated before consolidation into a γ-TiAl matrix. In principle, the inner coating is used to assure debonding between the fiber and matrix during fracture of the composite and is chosen according to the aforementioned debonding criterion. Three different coatings were tested in this regard: carbon black, colloidal graphite, and porous alumina. Since carbon will react with TiAl during consolidation, a second coating must be placed around the inner carbon coating to prevent diffusional interactions. The choice for the outer coating was dense Al_2O_3.

The inner coating was of three types: carbon black, colloidal graphite and a mixture of graphite with 10% alumina sol to produce a porous alumina network. Carbon black was placed on the sapphire fibers by passage through an acetone flame. This results in a uniform, sub-micron soot layer on the fiber. Colloidal graphite coatings were placed on the fibers by dipping into a graphite slurry. Fibers are slowly immersed into a slurry of 1μm graphite particles mixed with a dispersing agent and polymer material in an organic solvent. After dipping, coated fibers are placed in a holder and allowed to air dry. Porous alumina coatings arise by dipping fibers into a mixture of graphite particles in an ageous alumina sol. The outer coating was produced by dipping into an alumina sol containing 40 wt % α-Al_2O_3 seeds,‡ air dried and then sintered at 1350°C in Argon. The coated fibers were mixed in a γ-TiAl slurry, vacuum canned and HIPed for four hours at 1066°C in a pressurized Ar atmosphere at 276 MPa. Typical cross sections of the coated fibers in the γ-TiAl matrix indicate that the inner coating of carbon black has sub-micron thickness, whereas both the colloidal graphite and C/alumina coatings are ~ 1 μm thick. The outer Al_2O_3 coating, in all cases, is several microns thick.

EXPERIMENTAL PROCEDURE

Push-out tests were conducted to determine the interfacial characteristics associated with these coatings. For testing purposes, thin wafers of the composite were cut perpendicular to the fiber axis. Each side of the wafer was polished to a 1 μm diamond finish, so that the sides were nearly parallel. The push-out apparatus[7] employs a cylindrical indenter to apply loads to the fibers. The fiber is pushed into a 220 μm diameter hole in the support base with a transducer used to monitor fiber displacement, Fig 1.

AKP-50, 0.18 mm diameter.

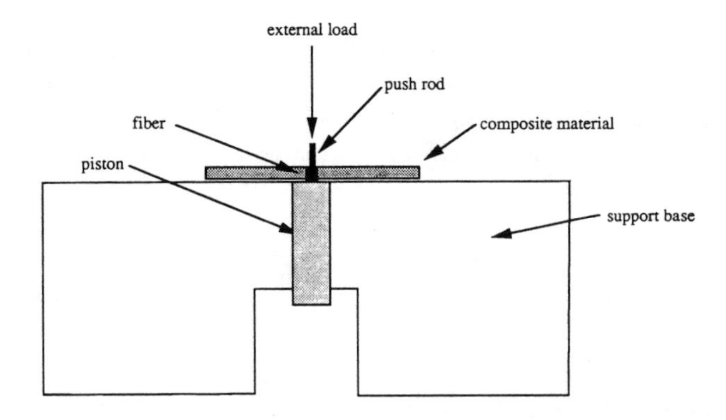

Figure 1. Schematic of the push-out apparatus used for testing. Load is applied through a 490N load cell to a SiC push-rod that sets on top of the chosen fiber. Displacement is monitored on the bottom side of the fiber using a displacement piston that deflects the end of a strain-gaged cantilever beam.

Push-out curves show a linear elastic loading region, followed by stable debond propagation down the fiber/matrix interface, Fig 2. Subsequently, the debond crack becomes unstable at ~ 1.5 fiber radii from the bottom side of the specimen, resulting in a load drop, ΔP. Thereafter, the fibers slide relative to the matrix. The debond load drop, ΔP, is related to the debond energy, Γ, by[8]

$$\Delta P = \left[2\sqrt{\frac{\Gamma E_f}{B_2 R}}e^{3\mu B_1} + \frac{(\tau_0 + \mu n_R)}{\mu B_1}\left(e^{-3\mu B_1} - 1\right) + p_R\right]e^{2\mu B_1 h/R} \tag{1}$$

where

$$B_1 = \frac{\nu_f E}{(1 - \nu_f)E + (1 + \nu)E_f}$$

and

$$B_2 = 1 - 2\nu_f B_1$$

E is the Young's modulus, ν is Poisson's ratio, subscripts f and m refer to fiber and matrix, respectively, h is the section thickness, μ is a coefficient of friction, R is the fiber radius, p_R is the residual axial compressive stress in the fiber, n_R is the residual radial compressive stress at the interface, τ_0 is the shear strength at zero normal stress (mechanistically related to an asperity pressure that arises during fiber sliding).[9] The nominal sliding strength, τ, at push-out displacement d is obtained from the instantaneous load, P, using

$$\tau = \frac{P}{2\pi R(h - d)} \tag{2}$$

In general, the sliding stress is not constant, because of asperity effects and wear processes.[9] Consequently, the sliding stress just after the debond load drop is used for comparing coatings. It is also useful to compare the coefficients of friction for each of the sliding interfaces. For this purpose, since the fiber/matrix asperities are in near perfect registry at the onset of sliding ($\tau_0 \approx 0$), μ is given by

$$\mu \approx \frac{R}{2B_1 h} \ln\left(\frac{P}{n_R - B_1 p_R}\right) \tag{3}$$

The stresses p_R and n_R are evaluated from the thermal expansion mismatch, using the formulae summarized in reference [10].

He and Hutchinson[5] derived a debond criterion based upon a comparison of the strain energy release rates for incremental crack propagation into the fiber/matrix interface or into the fiber. The debonding criterion depends upon the first Dundurs' parameter defined by

$$\alpha = \left[\frac{\bar{E}_f - \bar{E}_m}{\bar{E}_f + \bar{E}_m}\right] \tag{4}$$

where

$$\bar{E} = \frac{E}{(1-\nu^2)}$$

Using elastic properties of fiber and matrix (E_f=434GPa, ν_f=0.27, E_m=173GPa, ν_m=0.33), the first Dundurs' parameter is computed to be α=0.4. An estimate of the dedonding criterion for Γ_i/Γ_f at the computed value of α is taken from Fig 3, He and Hutchinson,[5] and is well approximated by $\Gamma_i/\Gamma_f \leq 0.25$. The fracture toughness of sapphire fibers loaded along the c-axis lies within the range [2.4-4.5MPa$\sqrt{m}$], or, equivalently [7-23 J/m^2][11]. Using the lower bound of this range, a conservative estimate of acceptable interface fracture energies is

$$\Gamma_i \leq 1.75 \text{ J/m}^2 \tag{5}$$

This value shall be used to evaluate the aforementioned coatings.

RESULTS

Typical push-out curves for the as-HIPed specimens are shown in Fig. 2. The *carbon black* interface (Fig. 2a) exhibits a very small debond load drop and a relatively high sliding stress. This is the *thinnest* of the three inner coating systems and will be most affected by the underlying fiber geometry. This feature is clearly demonstrated in Fig. 3, where the sinusoidal modulations in the load displacement trace have a wavelength identical to that of the fiber surface roughness. Push-outs for the *colloidal graphite* inner coatings (Fig. 2b) indicate a small debond load drop (though larger than that of the carbon black coating) and a low sliding stress. The low τ arises because the relatively thick C coating behaves as a compliant zone that mediates the mismatch pressure and also provides a debond propagation path that differs from that of the underlying fiber geometry. The C/Al_2O_3 inner coating (Fig. 2c) has the largest debond load drop, because the associated alumina network provides a bonded framework between the fiber and outer alumina coating . However, once sliding commences, the sliding stress has intermediate magnitude. Values of the debond and sliding parameters for each system, Γ, τ and μ, obtained from Eqns. (1–3) are summarized in Table I. Each of the coatings tested is found to satisfy the criterion of Eqn (5), and consequently, each coating should permit sapphire fiber debonding and sliding within the γ-TiAl matrix.

Table I
Comparison of interfacial properties for each coating

SAMPLE	Γ_i (J/m^2)	τ (MPa)	μ
Carbon black/Alumina	0.01	100	0.23
Colloidal graphite/Alumina	0.01	40	0.09
Porous Alumina/Alumina	0.05	75	0.09

CONCLUSIONS

Three double coating systems for sapphire fibers in γ -TiAl were compared using fiber push-out tests. These double coating systems employ an inner, debond coating of either carbon or porous alumina that provides a path for interfacial crack propagation and also protects the fiber during consolidation. These inner coatings are, in turn, protected from the matrix by a second, diffusion barrier of dense alumina. Each coating system was found to permit debonding and sliding, in accordance with accepted debonding criteria. Choosing between these coatings requires a more careful consideration of the service requirements of the composite, including high temperature oxidizing environments. Thus, the debonding and sliding behavior of these coatings must be evaluated after simulated in-service environmental exposures. Such a study is the subject of an upcoming paper.

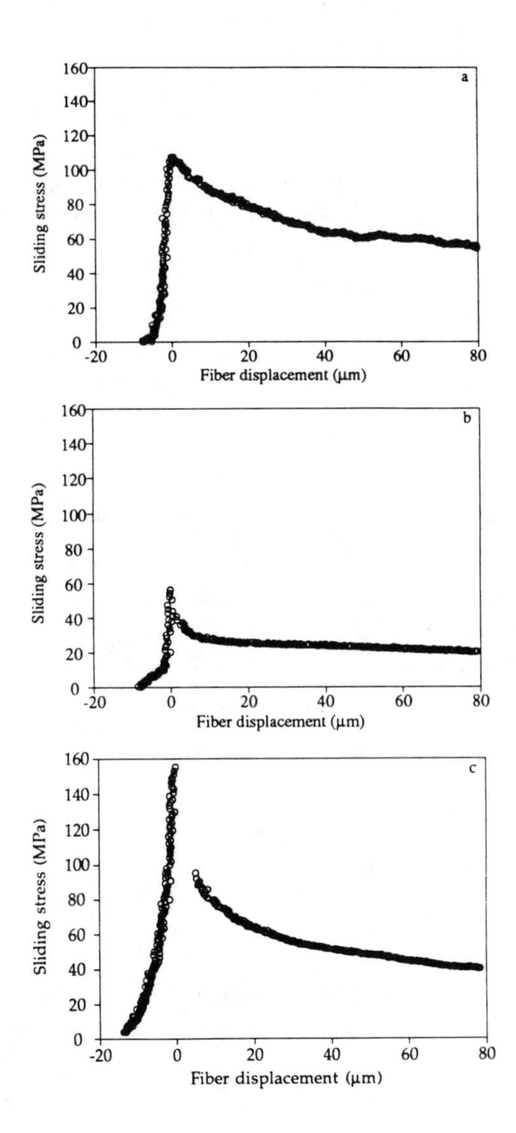

Fig. 2. A comparison of representative push-out curves with a) a carbon black inner coating, b) a colloidal graphite inner coating and c) a carbon/alumina inner coating.

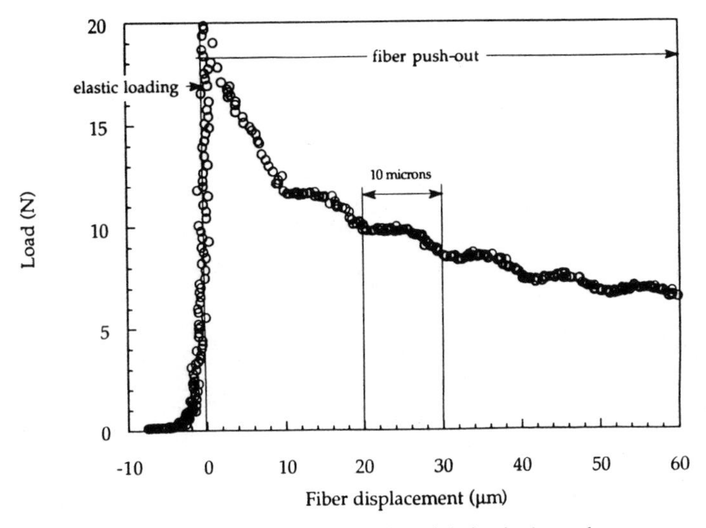

Fig. 3. Fiber roughness appears as a sinusoidal modulation in the push-out curve.

REFERENCES

1. A. G. Evans, F. W. Zok and J. Davis, Composites Science and Technology, **42**, 3-24, (1991).
2. R. J. Kerans, R. S. Hay, N. J. Pagano and T. A. Parasarathay, Cer Bull , Vol 68, No. 2, (1989).
3. W. Curtin, J. Am. Cer. Soc., 74(11), 2837 (1991).
4. S. Ochai and K. Osamura, Met Trans 21A, 971 (1991) .
5. M-Y He and J. W. Hutchinson, Int J. Solids Structures Vol. 25, No. 9, pp 1053-1067 (1989).
6. T. J. Mackin, J. Y. Yang and P. D. Warren, submitted to the J. Amer. Cer. Soc., April (1992).
7. P. D. Warren, T. J. Mackin and A. G. Evans, Acta metall. mater. Vol. 40, No. 6 pp 1243-1249 (1992).
8. C. Liang and J. W. Hutchinson, to be published.
9. T. J. Mackin, P. D. Warren and A. G. Evans, Acta metall. mater. Vol. 40, No. 6 pp 1251-1257 (1992).
10. B. Budiansky, J. W. Hutchinson and A. G. Evans, J Mech Phys Solids, 2 167(1986).
11. M. Isawa and R. C. Bradt, Structure and Properties of MgO and Al2O3 Ceramics, 767-779, ACS, Columbus OH.

CONSOLIDATION OF CONTINUOUS FIBER, INTERMETALLIC-MATRIX COMPOSITES

S.L. SEMIATIN*, R.L. GOETZ**, AND W.R. KERR*
* Materials Directorate, Wright Laboratory, WL/MLLN, Wright-Patterson Air Force Base OH 45433-6533
** UES, Inc., 4401 Dayton-Xenia Road, Dayton OH 45432

ABSTRACT

Processing routes for fabrication of continuous fiber, intermetallic-matrix composites are reviewed. These methods include conventional and isostatic hot pressing of layups of matrix material (e.g. foil or powder cloth) and fiber mats; consolidation of monotapes made by techniques such as arc, plasma, vapor, or electron beam deposition or tape casting; and liquid metal infiltration-base methods. The advantages and disadvantages of the various methods are discussed. Particular attention is focussed on HIP consolidation via foil-fiber-foil techniques. Process modeling techniques to assess the effects of pressure, temperature, and time on consolidation behavior are described. By this means, maps to delineate the interaction of process variables in such methods can be developed and applied for process optimization.

INTRODUCTION

Intermetallic matrix composites represent a relatively new area of engineered materials. The combination of an intermetallic matrix and ceramic reinforcements can provide exceptional levels of room and elevated temperature strength and stiffness as well as low density. Further, tailoring of the fiber-matrix interface properties can enhance composite toughness through control of fiber pullout during fracture.

For given matrix and reinforcement materials, the mechanical properties are usually best when the fiber is in a continuous rather than particulate or whisker form. Particulate and whisker reinforced composites can be consolidated and/or shaped via a large number of conventional powder, solidification, or deformation processes. By contrast, the manufacture of continuous-fiber reinforced metal-matrix composites is limited to relatively few methods. Three broad classes of fabrication of continuous-fiber, intermetallic matrix composites have been developed: (1) consolidation of the discrete composite components via hot pressing, (2) consolidation of composite monotapes via hot pressing, and (3) infiltration of liquid (matrix) metal under high pressure into fiber bundles or preforms. Specific methods in each class will be briefly summarized. Following this discussion, one of the more popular approaches under the first category, the foil-fiber-foil technique, will be described. Special emphasis will be placed on the application of process modeling techniques to select consolidation parameters to minimize reaction zone formation, fiber breakage, residual stresses, etc.

FABRICATION TECHNIQUES

Extensive research and development on fabrication of continuous fiber, intermetallic matrix composites has been in progress for approximately a decade. Unfortunately, much of this work is summarized in limited distribution reports. The

Mat. Res. Soc. Symp. Proc. Vol. 273. ©1992 Materials Research Society open literature, to which the discussion below is restricted, does provide a good overview of the overall direction of such efforts, however.

Consolidation from Discrete Components

Fabrication of continuous fiber, intermetallic matrix composites from matrix and fiber materials in discrete form represents the simplest and probably most common approach at present. The matrix is usually in the form of foil or powder cloth, and the fibers are assembled as mats held in place with cross-weave wires or ribbons or a fugitive binder. Foil matrix alloys are typically made by an ingot metallurgy approach in which large billets are converted into sheet by conventional and/or pack rolling. The final foil is then made via cold rolling with or without intermediate and final annealing treatments[1] or hot pack rolling[2]. Both of these processes can be very labor intensive and costly. However, they can provide matrix material in long continuous lengths of uniform gage and microstructure, provided that primary processing has been suitably designed and controlled.

Powder cloths are made by mixing prealloyed atomized or pulverized powder with a fugitive binder (typically an organic compound such as polytetrafluoroethylene) and a wetting agent (Stoddard solution, or a high grade kerosene). After most of the wetting agent has evaporated, the mixture has a dough-like consistency and can be rolled to relatively thin gage material from which the balance of the Stoddard solution is evaporated[3-8]. The primary advantage of the powder cloth method is that many matrix compositions are readily available in the form of powder, and the labor involved in going from ingot to foil in wrought products is avoided. The primary disadvantage with the use of powder cloth is related to the difficulty in insuring complete binder removal during consolidation and thus avoidance of matrix contamination and embrittlement. In addition, some binders can be toxic or explosive. A major challenge with the use of powder cloths lies in scaleup of the process especially with respect to removal of binder from large panels.

Fibers, often manufactured by chemical vapor deposition onto a filamentary substrate[9], are fabricated into mats via either drum winding or weaving type processes. In the former instance[4-6], the fiber is wound continuously on a drum mounted in a lathe. When the desired width has been achieved, the mat is coated with a binder such as polymethylmethacralate in a solvent base. The solvent is then evaporated and the mat removed from the drum and cut into pieces of the desired size. Woven fiber mats are made using conventional Rapier-or shuttle-type looms common in the textile industry[9]. In the Rapier-type loom, the filaments lie in the transverse direction and the crossweave wire or ribbon lies in the long direction of the fabric. The shuttle-loom product, on the other hand, weaves the reinforcing filaments in the longitudinal direction and the crossweave in the transverse, or "fill", direction.

After manufacture of the composite components, they are cut to size, stacked and assembled in some form of stainless or mild steel can and loaded into a vacuum hot press or HIP (hot isostatic pressing) vessel for consolidation[10]. In both cases, the vessel or can is evacuated, and a small positive pressure is applied via the ceramic plattens (hot press) or isostatic pressure (HIP) to hold the fibers in place. The heating cycle usually includes a dwell at temperatures of the order of 425 to 535°C to allow binder decomposition and outgassing when powder cloths or drum-wound mats are utilized. Subsequently, the temperature and pressure are raised to effect plastic flow/densification of the matrix around the fibers. Typical peak pressures and temperatures for titanium matrix composites are approximately 925°C and 105 MPa, respectively[10].

Intermetallic matrix composites made by foil-fiber-foil or powder cloth methods include Ti-24Al-11Nb (atomic percent), NiAl, and FeAl, all with silicon carbide fibers[8,11].

Consolidation from Monotapes

Manufacture of continuous-fiber, intermetallic matrix composites from monotapes is similar to the foil-fiber-foil and powder cloth processes in that layups are assembled, debinded if necessary, and consolidated via conventional or isostatic hot pressing. The principal difference lies in the use of pre-composited monolayers of the matrix and fiber components in a tape form. The advantage of these approaches, compared to the former, lie in the improved ability to control uniformity of fiber spacing by placing matrix material between the fibers prior to hot consolidation (Figure 1).

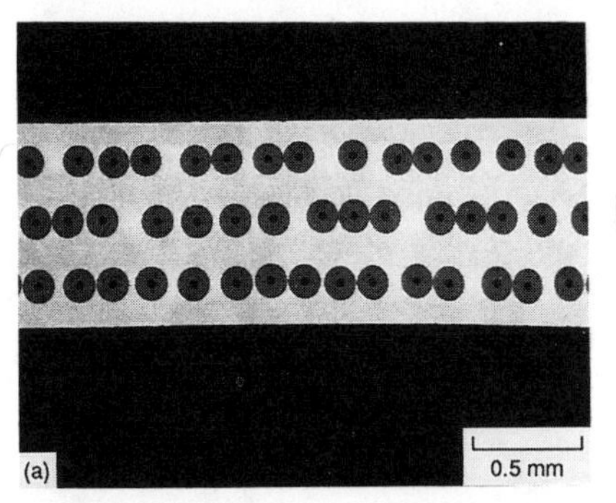

Figure 1. Comparison of fiber arrangements in multi-ply metal matrix composites fabricated via (a) powder cloth, (b) foil-fiber-foil, and (c) plasma sprayed monotape techniques[6].

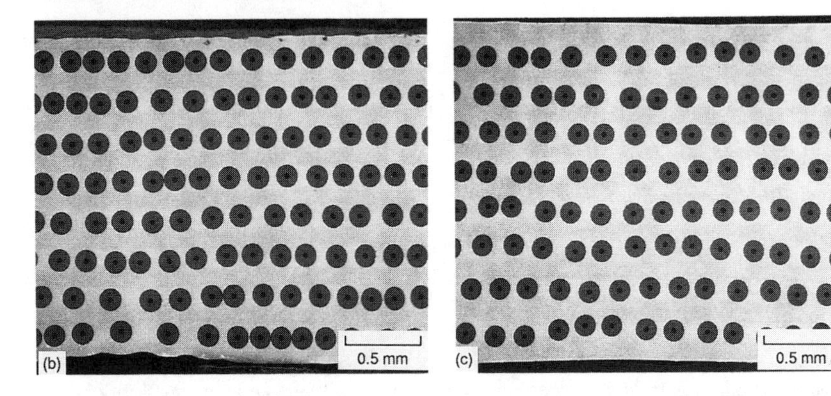

Monotape production techniques fall into the main categories of thermal spray, vapor deposition/plating, and tape casting. Thermal spray approaches include arc-spraying and induction-coupled plasma deposition[6,8,12,13]. In both cases, fiber is precisely wound on a drum or mandrel and sprayed with a layer of matrix material, thereby eliminating the need for organic binders. Thus, the possibility of matrix contamination mentioned above is eliminated, and potential for process scaleup is greatly enhanced. However, matrix contamination (e.g. interstitial pickup) during spraying must be controlled. In arc-spraying, the matrix material must be obtained in the form of wire, a product which can be difficult to produce for intermetallic materials. By contrast, the induction-coupled plasma technique makes use of more readily obtained prealloyed powder. Both spray techniques suffer from the tendency for spallation or exfoliation of the outer layers of the fiber due to thermal stresses generated during processing as well as the formation of substantial matrix-fiber reaction zones. Some of these problems are being addressed through the application of sophisticated material and process models, on the one hand, and sensor and control technologies, on the other, which are integrated through the intelligent processing of materials (IPM) methodology[12,13]. For example, processing maps can be developed to predict defects due to thermal shock (Figure 2), thereby providing information such as required drum/fiber preheat temperatures prior to thermal spray.

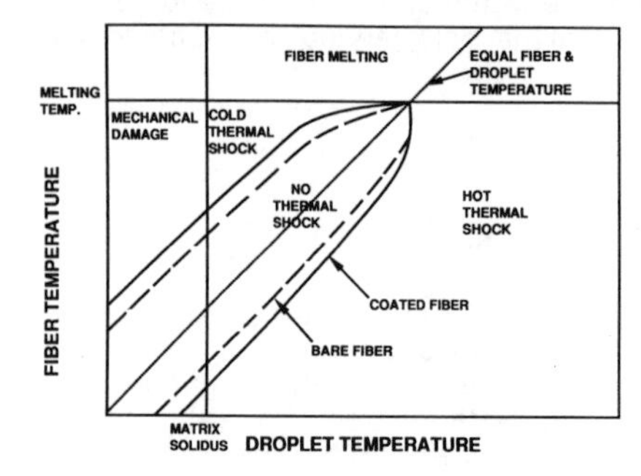

Figure 2. Processing map for the induction-coupled plasma deposition process showing the influence of fiber and droplet temperature on fiber thermal shock[13].

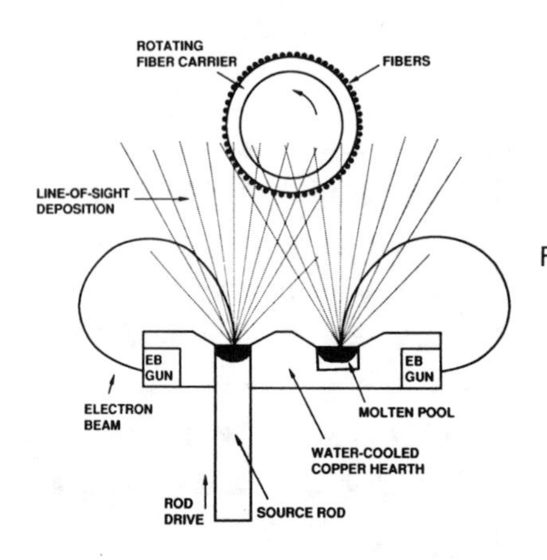

Figure 3. Schematic illustration of the electron beam evaporation technique for making continuous fiber, intermetallic matrix monotapes[15].

Vapor deposition and plating methods for producing monotapes, although slower, avoid many of the problems associated with thermal spray fabrication routes. Chou, et al.[14], Stoloff and Alman[4], and Ward-Close and Partridge[15] have summarized a number of the coating processes. Chemical vapor deposition (CVD) involves production of depositing elements through chemical reactions at temperatures much lower than those of the liquidus temperature of the matrix. Intermetallic compounds such as Nb_3Al, Nb_2Al, $NbAl_3$, and $MoSi_2$ have been produced by CVD. Often the coatings are applied to fiber moving through a reactor, after which the fiber (with an organic binder) is wound onto a mandrel. Monotapes are cut from this preform, stacked, and consolidated as in the powder-cloth process.

Physical vapor deposition (PVD) methods include sputter ion plating (SIP) and electron beam evaporation. SIP uses a high ion current to generate a flux of matrix atoms by physical sputtering. Although SIP is very slow, both protective interfacial layers and matrix material can be deposited with tight compositional control by this means. Electron beam evaporation can deposit matrix alloy onto drum-wound fibers (Figure 3) at considerably higher deposition rates viz., on the order of 5 to 10 micrometers per minute. Vanadiam-modified intermetallic matrices of Ti_3Al and TiAl have been produced on SiC fibers using two evaporation sources, one of Ti-6Al-4V and one of pure aluminum[15]. Although electron beam processing is conducted at relatively low temperatures, nonnegligible levels of oxygen pickup (order of 600 ppm, by weight) have been noted for titanium alloys.

Tape casting of continuous fiber monotapes draws on the slurry casting techniques developed by the ceramics industry. In brief, collimated fiber mats with an organic binder are first produced by drum winding. These mats are cut and placed flat on a glass plate in the casting machine. The principal component of this machine is a casting chamber (to hold a slurry mixture of powder plus binder) one side of which is comprised of a doctor blade positioned over the fiber mat. The height of the doctor blade is adjusted to control the thickness of the deposited slurry. The casting chamber is motor driven to allow continuous casting of monotapes[16].

Liquid Metal Infiltration

Liquid metal infiltration has been used to make a variety of intermetallic matrix composites with moderate to high volume percentages of continuous fibers. The most extensively investigated systems have been TiAl, Fe_3Al, Ni_3Al, and NiAl, all reinforced with alumina fibers[5,17,18]. Because many molten matrix materials do not wet fibers such as alumina, pressure is often superimposed during casting to ensure fill and matrix-fiber bonding. The actual process involves melting of the matrix under vacuum in a crucible placed on top a mold containing the fibers, which are heated to the same temperature. Upon reaching the desired temperature (usually approximately 100°C above the matrix liquidus temperature), the liquid metal is poured onto the fibers. Simultaneously, argon gas is introduced into the chamber to facilitate infiltration. The application of gas pressure to achieve infiltration allows the use of ceramic molds which have higher temperature capability and are less reactive than the metallic ones used to make low temperature composites. After infiltration is completed, cooling is usually rapid to minimize fiber degradation and fiber/matrix reaction zone formation.

The design of such casting systems for continuous-fiber, intermetallic matrix composites requires estimates of the minimum pressure for infiltration. As discussed by Nourbakhsh and Margolin, the resistance to infiltration arises from three sources - capillary, viscous drag, and gravity forces. The first two forces are most important. When sufficient pressure have been applied to overcome capillary forces, an additional force is required to overcome viscous drag and propagate molten metal through the fiber preform. The viscous drag can be quite large as the preform length and desired rate of infiltration increase.

MODELING OF FOIL-FIBER-FOIL CONSOLIDATION

At present, the foil-fiber-foil approach is probably the most widely used for consolidation of continuous-fiber, intermetallic matrix composites. Much of the research and developmental effort has focussed on making flat, unidirectionally reinforced panels via hot isostatic pressing (HIP) in order to obtain samples for mechanical property evaluation. Unfortunately, it appears that most of this processing work has relied on trial-and-error methods to select processing parameters. Hence, it may be surmised that current HIP cycles may not be optimized relative to the minimization of reaction zone formation, fiber breakage, residual stresses, etc. In the remainder of this paper, some initial work on modeling of the foil-fiber-foil process conducted at the Materials Directorate of Wright Laboratory is reviewed.

Model Formulation

The isothermal version of ALPID, a rigid, viscoplastic large strain finite element code originally developed by Battelle under Air Force sponsorship[19] was used to model the HIP consolidation process for a unidirectionally reinforced material. In the simulations, the consolidation was modeled as a plane strain deformation problem in which the foil matrix flows around rigid, stationary fibers. Examination of cross-sections of consolidated sheets, such as that shown in Figure 1b revealed that the analysis should focus on three fiber-matrix geometries which could be described in terms of so-called "unit cells" or representative volumes. These are shown schematically in Figure 4 and are described by the fiber arrangement characterizing each: "triangle", "cusp or trapezoidal", and "rectangle". Consideration of fiber alignment in adjacent layers and the bending in the foil suggests that fibers in a given layer tend to move into the interfiber positions of the adjacent layer during initial loading; this results in the triangular pattern or an array thought to be most stable during metal flow involving multiple layers of foil and fiber mats. However, in some cases, not all the fibers in a given layer move in the same direction resulting in either more or less than the nominal space between fibers. Continued loading forces fibers together in one layer and further apart in the other leading to the cusp arrangement. The rectangular fiber array represents a "metastable" arrangement between the triangular and cusp geometries. In the finite element analyses, it was assumed that the above fiber arrangements occurred before foil deformation around the fibers.

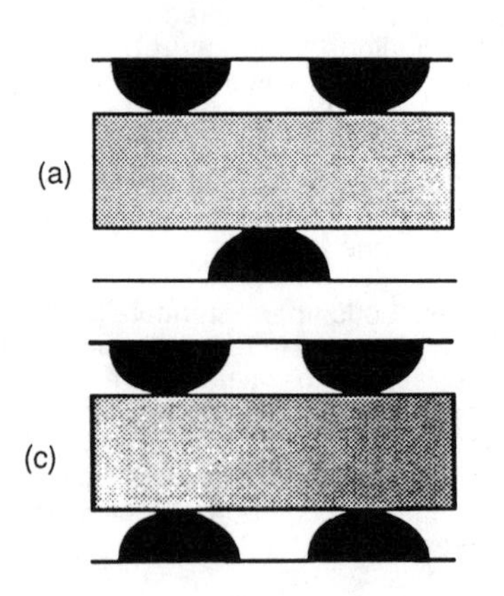

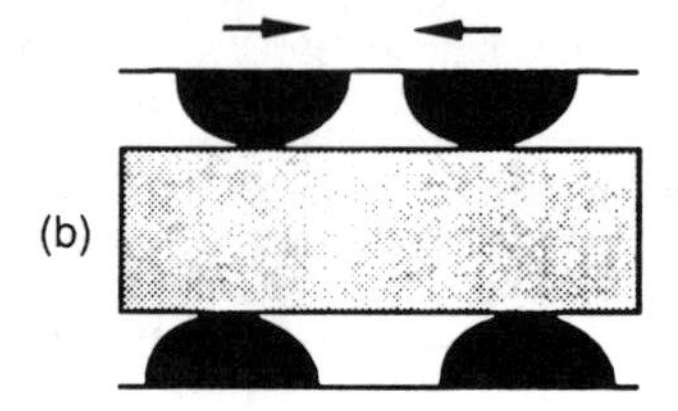

Figure 4. Representative "unit cells" used in the modeling of foil-fiber-foil composite consolidation: (a) "triangle", (b) "cusp", and (c) "rectangle".

The finite element method results below deal solely with the triangular fiber array, although those for the other arrangements have been conducted and found to be qualitatively similar[20]. Figure 5 shows the geometry used in simulations involving the triangular array. The rigid fibers were taken to be 140 μm in diameter; the foil thickness was 115 μm. Because of symmetry, only one-fourth of the top two fibers and one half of the bottom fiber was used in the simulation. The horizontal line joining the top two fibers and that extending in an equatorial fashion from the bottom fiber can be considered as neutral planes. Therefore, the equivalent of top and bottom rigid dies can be defined by the fibers. Similar symmetry considerations define the vertical neutral surfaces and thus the volume into which the foil must flow for this fiber arrangement. The foil itself was divided into 544 quadrilateral elements with a total of 595 nodes.

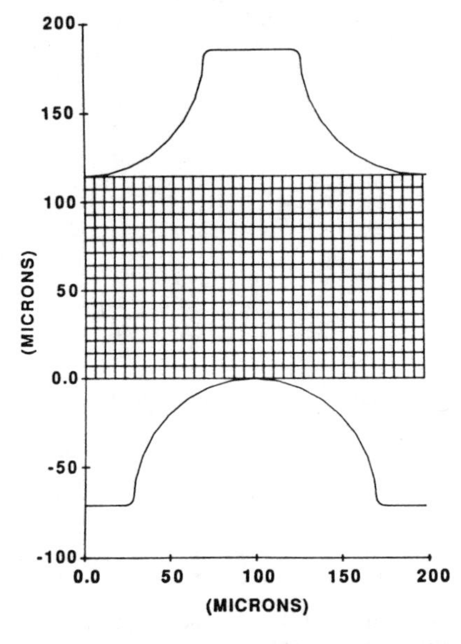

Figure 5. Foil and fiber geometry for the triangle unit cell used in the finite element simulation of foil-fiber-foil composite consolidation.

Boundary conditions consisted of specification of the friction conditions between the fiber and foil and the applied traction. Friction was characterized by a constant friction shear factor M whose value was taken to be between 0.1 to 1.0. The value of 1.0 indicates sticking friction or a level that would pertain to either a rough fiber surface or the development of a reaction zone between the fiber and the foil. The traction boundary condition consisted of the application of a fixed, prespecified pressure, or, in the two-dimensional plane-strain problem here, a constant force. This force was assumed to be uniformly distributed over the top and bottom horizontal surfaces defined by the fibers and neutral surfaces. In the actual simulations, the relative displacement rate of the top fibers relative to the bottom was specified, and corrected during subsequent steps to ensure near-constancy of load. This correction was based on the constitutive equation of the foil alloy, or more specifically the strain rate sensitivity m, according to the relation:

$$v_u/v_c = (F^*/F_c)^{1/m} \tag{1}$$

where v_c, v_u ≡ current and "updated" displacement rate, F^* is the desired (constant) force, and F_c is the (calculated) force applied during the current deformation step.

The foil constitutive relation was taken to be of a nonhardening, power law creep form, viz.,

$$\dot{\bar{\varepsilon}} = A\bar{\sigma}^{n} \exp(-Q/RT), \qquad (2)$$

where $\bar{\sigma}$ and $\dot{\bar{\varepsilon}}$ denote the effective stress and strain rate; A is a constant; n is the stress exponent (= 1/m), which can be a function of temperature and strain rate; and Q, R, and T are an apparent activation energy for the deformation process involved, the gas constant, and temperature, respectively. For the isothermal simulations discussed below, the constitutive relation reduces to the simple form $\dot{\bar{\varepsilon}} = A'\bar{\sigma}^{n}$. The matrix material used in the present work was Ti-24Al-11Nb foil whose constitutive relation was determined by step strain-rate change tests[21]. The flow stress dependence on temperature and strain rate is shown in Figure 6.

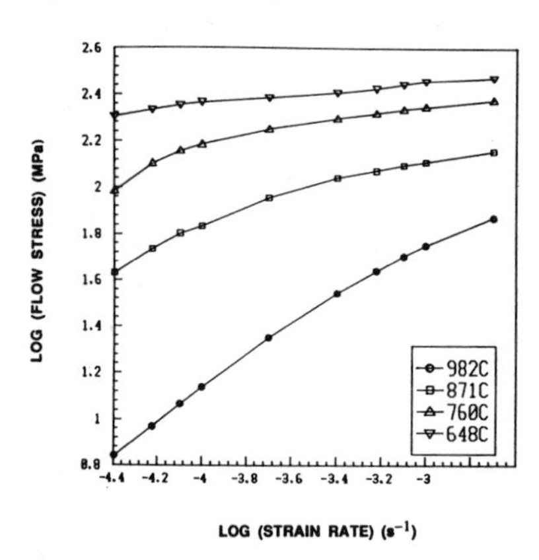

Figure 6. Flow stress versus strain rate data for Ti-24Al-11Nb.

Simulation Results

Metal flow predictions for the consolidation of Ti-24Al-11Nb matrix composites at 648°C/414 MPa and 980°C/70MPa are summarized in Figures 7 and 8, respectively. In both cases, the assumed friction factor was 0.1. The computer predictions are in terms of grid distortions and strain contours at the near final fill stage. Both simulations show some nonuniformity of flow, which is as expected in view of the complexity of the geometry. Careful examination does reveal a degree of difference, however. Consolidation at the lower temperature and higher pressure shows a somewhat higher degree of strain nonuniformity and a tendency to form shear bands in the deforming foil. This is not surprising in view of the flow stress behavior. Replotting the data in Figure 6 in terms of the stress exponent as a function of temperature and strain rate reveals that the low temperature, high pressure conditions correspond to a high n regime, i.e., $n \approx 10$, or m = 0.10 (Figure 9). Conversely, the n at the high temperature and low pressure is approximately 1.6, or $m \approx 0.63$. The latter regime corresponds to conditions representative of superplastic flow under which flow localization tendencies would be minimum. Despite this difference, the <u>average</u> effective strain in the foil in both cases is approximately 0.6, which is principally controlled by the geometry of this pseudo closed-die forging.

The simulation results also indicate that the tensile strain at the surfaces where the foils bond is approximately 0.4, or a deformation of the magnitude at which surface oxides might be expected to break up and not interfere with bonding. In addition, the deformation at the foil-fiber interfaces is somewhat larger, being of the order of 2.0. Such a large deformation would greatly enhance oxide breakup and enhance the tendency for reaction zone formation.

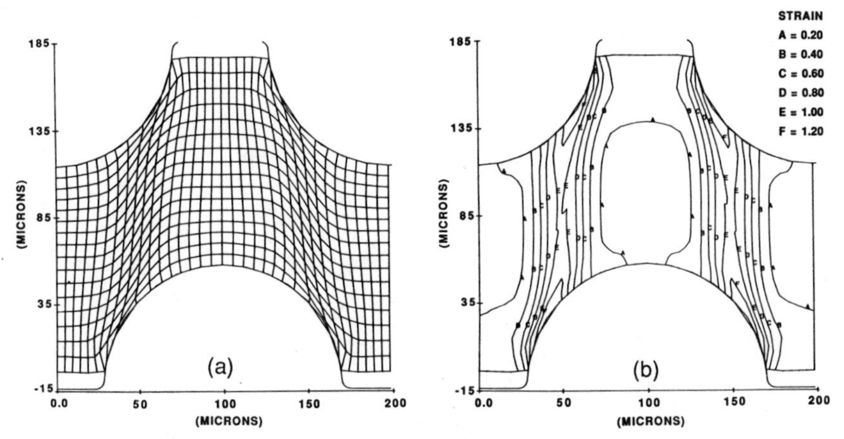

Figure 7. Computer model prediction of (a) grid distortions and (b) strain contours in the consolidation of Ti-24Al-11Nb matrix composite via HIP at 648°C/414 MPa. The friction factor M was taken to be 0.1.

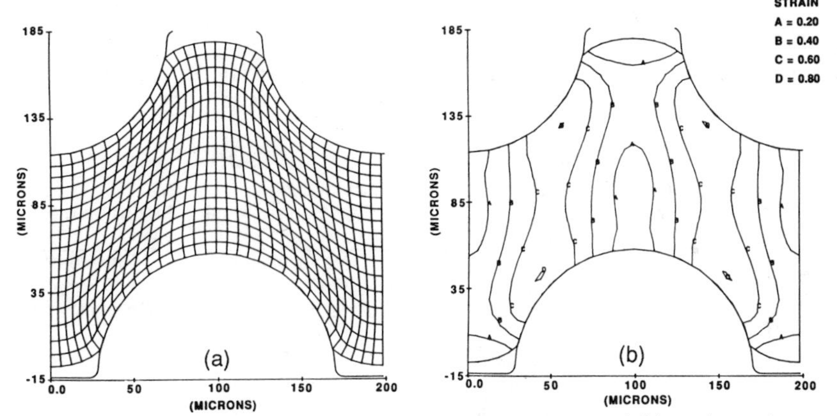

Figure 8. Computer model prediction of (a) grid distortions and (b) strain contours in the consolidation of Ti-24Al-11Nb matrix composite via HIP at 980°C/70 MPa. The friction factor M was taken to be 0.1.

Closer examination of the simulation results gives a more in-depth idea of the details of the actual consolidation. For example, plots of displacement rate versus time (or displacement), Figure 10, reveal that the initial stages of consolidation occur quite rapidly. Under constant load conditions, the displacement rate decreases very rapidly as the final filling stage occurs. In other words, final pore closure around the fibers requires substantial time. This effect is summarized in Table I, which contains simulation predictions of the time to complete the initial 97 percent of the stroke versus that for the final 3 percent (i.e., the final pore closure stage). The finite element simulations suggest that the times for the initial 97 percent and the final 3 percent are approximately equal, in agreement with experimental observations[22]. The simulations also reveal that the deformation involved in final pore closure is a strong function of material properties[20]. The details of the final deformation stage may therefore affect foil bond formation in the vicinity of the fibers and thus final composite properties.

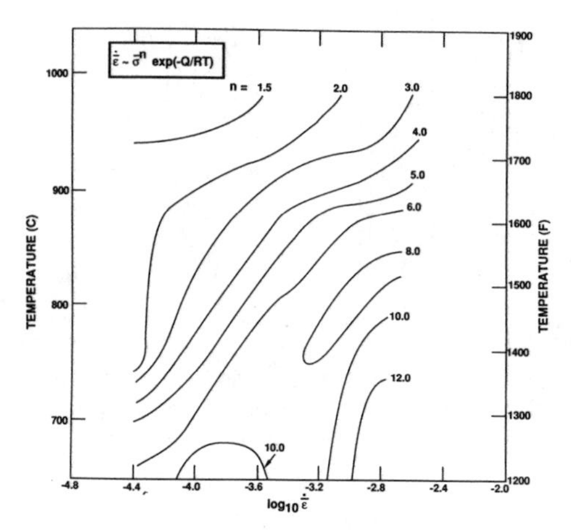

Figure 9. Stress exponent n versus temperature and strain rate for Ti-24Al-11Nb.

Table I. Predicted Time for Partial or Full Consolidation of Ti-24Al-11Nb Matrix Composites Fabricated Via Foil-Fiber-Foil Techniques (Friction Factor M=0.1)

Temperature, °C (°F)	Pressure, MPa (ksi)	Time for Initial 97 Percent Consolidation,s	Time for 99.7 Percent Consolidation,s
648 (1200F)	310 (45)	3281	4142
" "	345 (50)	1098	1348
" "	379 (55)	417	505
" "	414 (60)	165	199
760 (1400F)	138 (20)	17700	44954
" "	207 (30)	2475	6000
" "	275 (40)	610	1500
" "	345 (50)	207	500
871 (1600F)	69 (10)	8942	19185
" "	103 (15)	2366	5077
" "	138 (20)	921	1977
" "	207 (30)	244	523
982 (1800F)	11 (1.6)	10035	16995
" "	35 (5)	1611	2700
" "	69 (10)	527	892
" "	103 (15)	274	465

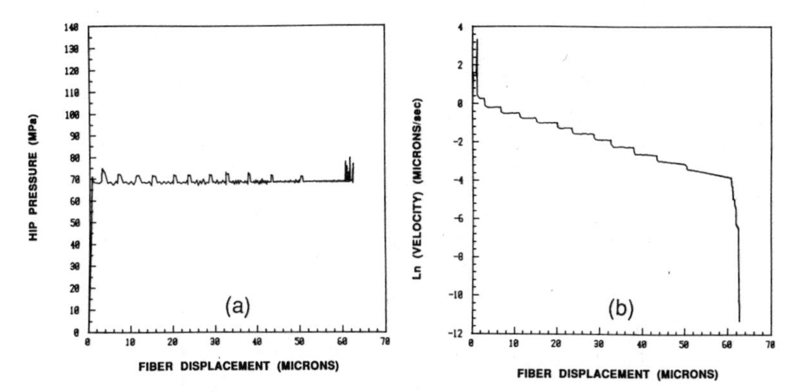

Figure 10. Plots of pressure (a) and velocity (b) versus total displacement from a computer simulation of HIP consolidation of a Ti-24Al-11Nb matrix composite at 980°C/70 MPa.

Application of Modeling Results

The results of finite element simulations offer a powerful tool for design of foil-fiber-foil consolidation processes. For example, results such as those obtained above can be utilized to generate HIP diagrams analogous to those used to predict HIP consolidation of porous metals and ceramics[23]. By analogy, HIP diagrams for foil-fiber-foil processing yield information on consolidation time as a function of pressure and temperature. An example of such a diagram for a Ti-24Al-11Nb unidirectionally reinforced composite is given in Figure 11; these results have been derived from "triangle" fiber array simulations assuming M = 0.1. Modified diagrams taking other friction factors and fiber geometries (or combinations of geometries) into account are under development[20].

The effect of nonisothermal, nonisobaric conditions (that pertain to actual HIP operations) on consolidation of composites may be ascertained using some form of numerical integration scheme as has been used for HIP'ing of monolithic powders[24,25]. However, some insight into such effects can be obtained if the material flow behavior is assumed to follow the power law creep formulation, Equation (2). Taking the effective stress $\bar{\sigma}$ to scale with the imposed pressure, the ratio of the effective strain rate $\dot{\bar{\varepsilon}}_i$ at a given temperature T_i and pressure P_i to that at the peak values in the cycle, T^* and P^*, $\dot{\bar{\varepsilon}}^*$, is given by:

$$\dot{\bar{\varepsilon}}_i/\dot{\bar{\varepsilon}}^* \sim (P_i/P^*)^n \exp[(-Q/R)(T_i^{-1} - T^{*\,-1})], \tag{3}$$

For typical values of n (~ 2 to 10) and Q/R (~ 25,000°C), $\dot{\bar{\varepsilon}}_i/\dot{\bar{\varepsilon}}^*$ is very small except for $P_i \rightarrow P^*$ and $T_i \rightarrow T^*$. Thus, the deformation occurring during heating and pressurization cycles whose duration is of the same magnitude as the time required for consolidation under given isothermal/isobaric conditions may be expected to be small typically.

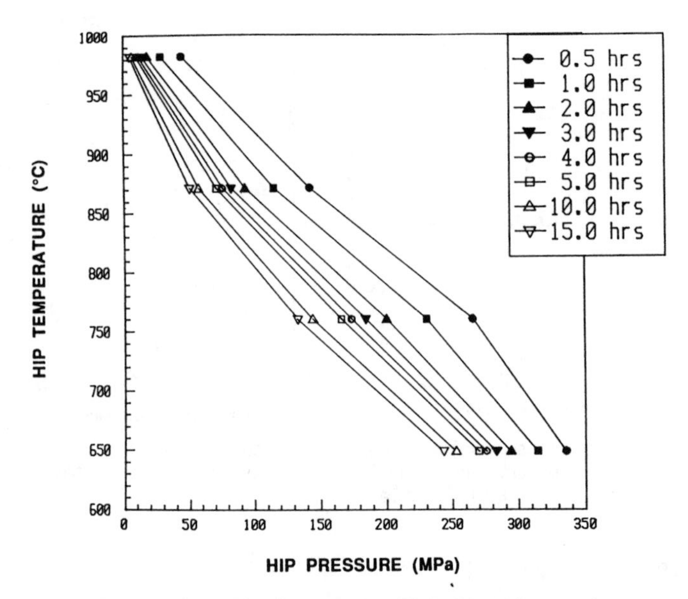

Figure 11. HIP consolidation diagram for Ti-24Al-11Nb matrix composites fabricated via foil-fiber-foil techniques; triangle unit cell, M=0.1.

HIP diagrams may be used to optimize foil-fiber-foil consolidation in terms of minimizing reaction zone formation (minimize temperature and time) and residual stresses developed due to fiber-matrix CTE mismatch (minimize temperature)[26,27]. Such optimization is subject to the pressure capabilities of the HIP vessel, and the need to control pressurization rates to avoid fiber breakage. Despite the fact that foil-fiber contact during processing is initially line contact in form, it is unlikely that Hertzian stresses would be sufficient to damage the fibers. This is because the substantially lower flow stress of most matrix alloys relative to the yield strength of typical fibers would lead to matrix flow and reduction of the contact stresses. A more serious concern with regard to fiber damage and consolidation in general relates to nonuniform temperatures throughout multi-ply layups during the heating cycle that could induce stresses in the fibers and lead to bonding of the various composite layers during different periods and hence component distortion. Such effects are well known in the HIP'ing of powder compacts[28,29]. Also, fiber damage due to crossweave wires may be of importance.

SUMMARY AND FUTURE OUTLOOK

Continuous-fiber, intermetallic matrix composites are attractive candidates for high strength, high temperature applications. A number of processing techniques for these materials have been proven to be technically feasible. However, no intermetallic matrix composite has been produced in a sufficient quantity, let alone in a production environment, to allow an assessment of the commercial viability of these materials. Thus, there are a number of issues which must be addressed, including the following:

- Quality control. This includes quality control for the starting materials (e.g. foil, powder precursors) as well as process control during consolidation. This will require development of process specifications and process control technologies which most likely will be more stringent than for conventional materials because of the larger effect of microstructural and consolidation defects on degradation of performance.

- Process modeling. Process modeling is a useful tool in establishing processing windows and the effects of "nonoptimal" practice on final microstructure. Moreover, process models can be an invaluable tool in scaleup from laboratory to production. For example, process modeling may prove useful to understand and overcome the unavoidable temperature nonuniformities that are developed during composite processing in large, heavily loaded HIP vessels. The success of process models will rely of course on a close interaction between equipment designers, materials engineers, and the modelers.

- Cost/performance tradeoffs. Very little effort has been put into estimating the cost of final intermetallic matrix composite products. The cost of such products is principally that associated with processing. Such processing costs must be weighed against the properties that can be obtained to justify pursuing such technologies.

ACKNOWLEDGEMENTS

This work was conducted as part of the in-house research activities of the Metals and Ceramics Division, Materials Directorate, Wright Laboratory. The authors thank P. Brindley for supplying Figure 1 and L. Farmer for careful preparation of the manuscript.

REFERENCES

[1] S.C. Jha, J.A. Forster, A.K. Pandey, and R.G. Delagi, Advanced Materials and Processes, 139, 87 (April, 1991).
[2] C. Bassi, J.A. Peters, and J. Wittenauer, JOM, 41, 18 (September, 1989).
[3] P.K. Brindley in High Temperature Ordered Intermetallic Alloys II, edited by N.S. Stoloff, et al. (Mater. Res. Soc. Proc. 81, Pittsburgh, PA, 1987) pp. 419-424.
[4] N.S. Stoloff and D.E. Alman in Intermetallic Matrix Composites, edited by D.L. Anton, et al. (Mater. Res. Soc. Proc. 194, Pittsburgh, PA, 1990) pp. 31-43.
[5] S. Nourbakhsh and H. Margolin in Metal and Ceramic Matrix Composites: Processing, Modeling, and Mechanical Behavior, edited by R.B. Bhagat, et al. (TMS, Warrendale, PA, 1991) pp. 75-89.
[6] R.A. Mackay, P.K. Brindley, and F.H. Froes, JOM, 43, 23 (May, 1991).
[7] E.A. Feest and J.H. Tweed in High Temperature Intermetallics (Institute of Metals, London, 1991) pp. 30-42.
[8] R. Bowman and R. Noebe, Advanced Materials and Processes, 136, 35 (August, 1989).
[9] M.A. Mittnick in Metal and Ceramic Matrix Composites: Processing, Modeling, and Mechanical Behavior, edited by R.B. Bhagat, et al. (TMS, Warrendale, PA, 1991) pp. 605-617.
[10] P.R. Smith and F.H. Froes, Journal of Metals, 36, 19 (March, 1984).
[11] J.-M. Yang, W.H. Kao, and C.T. Liu, Mater. Sci. Eng., A107, 81 (1989).

[12] H.P. Wang, E.M. Perry, R.D. Lillquist, and J.H. Taylor, JOM, 43, 22 (January, 1991).
[13] D. Backman, JOM, 42, 17 (July, 1990).
[14] T.W. Chou, A. Kelly, and A. Okura, Composites, 16, 187 (July, 1985).
[15] C.M. Ward-Close and P.G. Partridge, Journal of Mat. Sci., 25, 4315 (1990).
[16] M.A. Kelley and M.F. Amateau in Metal and Ceramic Matrix Composites: Processing, Modeling, and Mechanical Behavior, edited by R.B. Bhagat, et al. (TMS, Warrendale, PA, 1991) pp. 23-30.
[17] S. Nourbakhsh and H. Margolin, Metall. Trans. A, 21A, 213 (1990).
[18] S. Nourbakhsh, O. Sahin, W.H. Rhee, and H. Margolin, Metall. Trans. A, 22A, 3059 (1991).
[19] S.I. Oh, Inter. J. Mech. Sci, 24, 479 (1982).
[20] R.L. Goetz, W.R. Kerr, and S.L. Semiatin, unpublished research, Materials Directorate, Wright Laboratory, Wright-Patterson Air Force Base, OH (1991).
[21] C.C. Bampton, unpublished research, Rockwell International Science Center, Thousand Oaks, CA (1991).
[22] P.D. Nicolaou, H.R. Piehler, and M.A. Kuhni, unpublished research, Carnegie-Mellon University, Pittsburgh, PA (1991).
[23] A.S. Helle, K.E. Easterling, and M.F. Ashby, Acta Metall., 33, 2163 (1985).
[24] H.N.G. Wadley, R.J. Schaefer, A.H. Kahn, R.B. Clough, M.F. Ashby, Y. Geffen, and J.J. Wlassich in Hot Isostatic Pressing: Theory and Applications, edited by R.J. Schaefer and M. Linzer (ASM International, Materials Park, OH, 1991) pp. 91-95.
[25] M. Dietz, H.P. Buckremer, and D. Stover in Proc. Fourth Inter. Conf. on Isostatic Pressing, Stratford-upon-Avon, United Kingdom (November, 1990).
[26] D.B. Gundel and F.E. Wawner, Scripta Metall. et Mater., 25, 437 (1991).
[27] D.A. Saravanos, P.L.N. Murthy, and N. Morel in Proc. 35th Inter. SAMPE Symposium, SAMPE, Covina, CA, 1990, pp. 506-519.
[28] W.A. Kaysser in Hot Isostatic Pressing: Theory and Applications, edited by R.J. Schaefer and M. Linzer (ASM International, Materials Park, OH, 1991) pp. 1-13.
[29] J. Besson and M. Abouaf in Hot Isostatic Pressing: Theory and Applications, edited by R.J. Schaefer and M. Linzer (ASM International, Materials Park, OH, 1991) pp. 73-82.

PROCESS MODELING FOR TITANIUM ALUMINIDE MATRIX COMPOSITES

C.C. BAMPTON AND J.A. GRAVES
Science Center, Rockwell International Corporation, 1049 Camino Dos Rios, Thousand Oaks, CA 91360
K.J. NEWELL, R.H. LORENZ
North American Aircraft, Rockwell International Corporation, 201 North Douglas Street, El Segundo, CA 90245

ABSTRACT

Consolidation of continuous fiber-reinforced titanium aluminide matrix composites (TMC) by the foil/fiber/foil method has traditionally taken an empirical approach utilizing processing cycles derived by simple trial and error. In an effort to reduce the empirical nature of producing TMC, a simple but effective analytical approach is employed. This approach analyzes the effect of fiber and foil geometries on consolidation parameters by combining a physical constitutive creep model with computational methods of interpreting raw materials characterization data. Examples of SCS-6/super α_2(Ti-25Al-10Nb-3Mo-1V) and Saphikon/γ-TiAl composites consolidation are discussed by comparing the model predictions with equivalent validation specimen microstructures.

1. INTRODUCTION

In the foil/fiber/foil consolidation process, heat and pressure are applied to a composite lay-up of metal foils and fiber mats usually by means of a hot isostatic press (HIP) for the purpose of: (1) embedding the fibers in the metal matrix by creep forming the metal around the fibers, and (2) providing suitable conditions for diffusion bonding the metal to metal and fibers to metal surfaces as they make contact. Readily apparent is that the selection of bonding parameters for a given combination of fiber and matrix is a principal criterion for the achievement of well consolidated titanium matrix composite. Equally important however, is that the consolidation take place in a reasonable amount of time, at practical temperatures and pressures while taking into account the effect of these parameters on the composite constituents and residual stresses.

Issues which are specific to processing of TMC's are: (1) Consolidation temperatures which are close to or above phase transition temperatures are often used in attempts to facilitate metal flow and avoid fiber mechanical damage or displacement. This is not, however, generally desirable with regard to the final composite microstructure; (2) Titanium-based matrices have the ability to absorb (interstitially) surface oxides. This enhances foil-to-foil diffusion bonding but may adversely affect phase stabilities and mechanical properties in the oxygen-rich areas; (3) All titanium-based alloys are known to be reactive with currently available structural fibers. This may degrade the fiber integrity and embrittle the near-fiber matrix. Minimized temperatures and exposure times are therefore generally desirable.

Conventional titanium alloys, such as β-21S, are generally "well-behaved" as TMC matrices, consolidating easily and predictably. Even in these cases, however, we have found successful consolidation cycles, using a new predictive model, which involve significantly less thermal exposure than typically used in practice. Titanium aluminides present more challenges for composite consolidation due to (1) higher flow stresses at elevated temperatures; (2) higher strain rate sensitivities; (3) slower diffusion rates; (4) more severe localized oxygen effects (embrittlement and phase stability); and (5) lower ambient temperature ductilities and fracture toughnesses.

In an effort to reduce the empirical nature of producing TMC and to facilitate investigation of alternate HIPing parameters, the Consolidation Analysis Program for Process Selection (CAPPS) is being developed. The initial goal of the CAPPS program is to provide a means of predicting densification times for various combinations of foil/fiber/foil constituent geometries while employing easily obtainable characterization data. The approach emphasizes convenient and rapid simulation of consolidation processing cycles so that more complex cycles, such as those including temperature and pressure "spikes", are easily evaluated. This may provide both quality and cost benefits in more sophisticated processing cycles. The CAPPS program is also intended to provide the framework for implementing methods of predicting diffusion bonding rates, interfacial reactions and microstructural damage.

While the CAPPS program is already proving useful as a research tool in the development of new composite materials, an ulterior goal is to support an automated, intelligent manufacturing system for lower cost, higher quality MMC's.

A complete composite process modeling package will address a large number of issues, including encapsulation, out-gassing, tooling thermal effects, consolidation; fiber "swimming", diffusion bonding, matrix microstructure evolution, CTE mismatch residual stresses, and interface reactions and damage. The approach described here focuses on the consolidation stage with a brief practical demonstration of the interaction and implications of the diffusion bonding stage.

2. CAPPS METHODOLOGY

We have developed and utilized a computer program for predicting processing parameters for foil/fiber/foil consolidation of metal matrix composites. This program is called "Consolidation Analysis for Processing Parameter Selection" (CAPPS).

The CAPPS program has proven to be a useful tool in determining load profiles necessary to consolidate SCS-6/β21S and SCS-9/β21S [1], SCS-6/Super α_2, SCS-6/O-Ti_2AlNb and Saphikon/γ-TiAl foil/fiber/foil composite systems.

The analytical approach used by the CAPPS program to model the consolidation of titanium matrix composites centers on the application of a physical constitutive equation for creep. This simplified form was shown by Feltham [2] to be applicable for relatively high applied stresses. This is pertinent to this case since the local stresses causing metal flow are high. This formulation was chosen as a baseline equation for use as a phenomological transformation of data on which flow stress strain rate data is curve fitted. The equation takes the form:

$$\frac{d\varepsilon}{dt} = A \exp\left(\frac{\beta\sigma}{RT} - \frac{Q}{RT}\right) \quad (1)$$

where
ε = strain
σ = stress
t = time
T = temperature
A= frequency factor
β = stress factor
Q = activation energy
R = gas constant

Equation (1) was then expanded while adding two product terms:

$$y = \ln(\dot{\varepsilon})$$
$$x = C_1 + C_2 \ln(\sigma) + C_3 \ln(T) + C_4\left(\frac{\sigma}{RT}\right) + C_5\left(\frac{-Q}{RT}\right) \quad (2)$$

In this form (Eq. 2), flow stress-strain rate data like that shown in Figure 1 are transformed into coefficients from multilinear least squares regression accurately describing the materials creep flow characteristics over a wide range of temperatures and pressures.

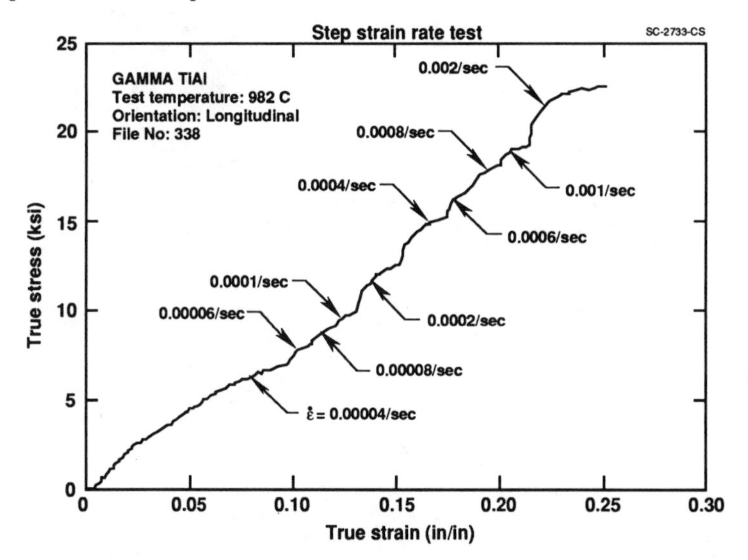

Fig. 1 Step-strain rate tensile flow stress data for γ-TiAl (Ti-48Al-2Cr-2Nb-1Ta[at.%]).

To apply the above formulation, a means of calculating an effective stress was established. Rather than employ a finite element model, an alternate, more convenient method which analyzes the change in unit cell geometry based on an incremental creep deflection induced by an average effective stress was chosen. Figure 2 illustrates the unit cell and how these

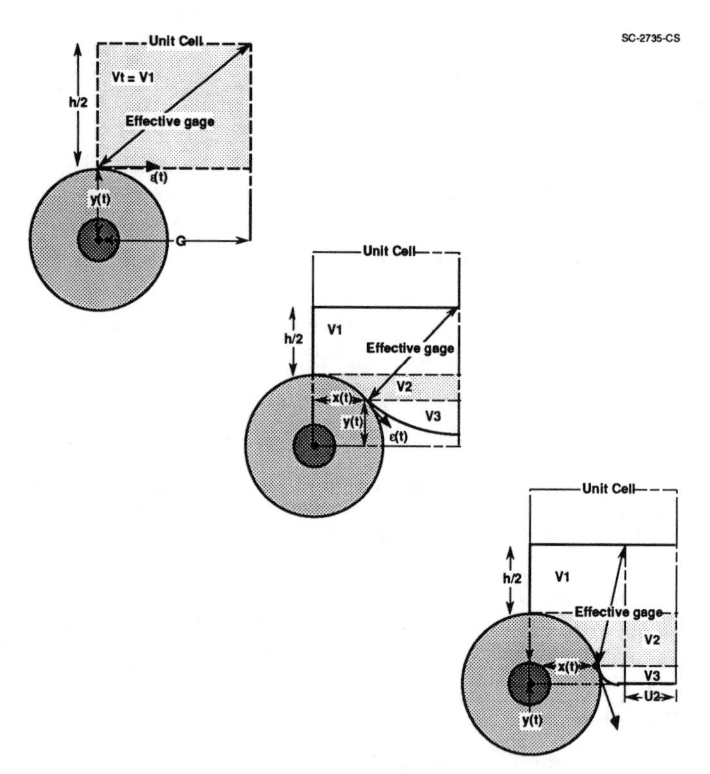

Fig. 2 Illustration of the incrementally changing volumes of the unit cell used by the CAPPS program to track the movement of the foil material into the voids between fibers.

incremental deflections are applied tangent to the point of contact at the fiber-matrix interface.

The unit cell is composed of a volume of matrix and a rigid quarter section of the fiber. At the onset of the analysis, the initial volume of matrix is measured in terms of:

$$V_1 = 2G(h/2)$$

where G is half the core-to-core spacing of the fibers and h/2 is the location of the neutral axis. The neutral axis is defined as the line from which matrix material will either flow into the unit cell or out into an adjacent void. The volume V_1 is taken to be a constant in the analysis and is set equal to the total volume V_t. Therefore, the assumption is that no transport of material is made in or out of the cell boundaries. For modeling purposes, the geometry and volumes are calculated for two identical unit cells with a mirror symmetry about the vertical line at G. A simple square array of fibers in adjacent layers is assumed. This is a rate-limiting case for time to consolidate. Although the computation is not very sensitive to changes in fiber arrays, future versions of the model are planned to incorporate such variables.

As the analysis proceeds, calculation of the average effective stress is performed at two boundary points. The first boundary point is located at the fiber-matrix contact point (henceforth referred to as the contact point) and the other is located at the intersection of the neutral axis and the midpoint G (henceforth referred to as the neutral point). The initial boundary conditions at these points are assumed to be equal to the principal stress state having the principal axis parallel to the plane of the foil. By utilizing finite element analysis and studies performed by Fleck et. al [3], on yield surfaces, the stress state at the contact point is assumed to have a concentration factor (contact stress/applied stress) of approximately = 2 at the onset of loading. This factor rapidly decreases as the matrix forms around the fiber. The stress state at the neutral point remains constant until the geometry of the unit cell predicts that the flowing matrix has come into contact with the matrix flowing from the opposite side of the fiber. This contact will produce a bondline interface whereby the stress state at the neutral point will approach isostatic.

Throughout the analysis, the geometry is recalculated after every time-step. The model utilizes two basic geometries for the analysis. The first geometry follows the flow of material downward between two adjacent fibers while the second geometry formulates the lateral flow of matrix once a bondline contact has occurred. Figure 3 shows both of these geometries with the corresponding variables which are calculated after each time-step. If creep deformation is predicted, the resulting change in the location of the contact point is used to establish the geometry for the next increment of time.

After each iteration the geometry is checked for the occurrence of contact at the bondline located along the line joining the centers of the fibers. The analysis is performed incrementally based on a time step of 0.01 seconds. It continues until (V_2 + V_3) is equal to the void volume at which time the analysis ends (V_2 is the volume of matrix displaced by the fiber and V_3 is the volume of matrix within an assumed meniscus between fibers).

3. EXPERIMENTAL VALIDATION OF CAPPS

Development and validation of the CAPPS model required concurrent experimentation and model analysis. The experimentation centered on the small scale HIPing of 1" x 1" test coupons made up of unidirectional composite lay-ups. Each HIP run was performed under the guidance of CAPPS which provided specific heat-up and pressure ramps with predicted hold times.

This study employed a partial consolidation technique. The consolidation process is stopped short of full density, thus leaving voids which were used to establish and validate methodology. The voids provided a means of measurement for the amount of fill taking place for a given profile of time, temperature and load. Prior to the actual HIP of the material, numerous computational predictions were made for core-to-core fiber spacings ranging in magnitude from touching to several fiber diameters apart. Post-processing files of each of these runs were made for comparison to the metallography obtained from the actual run. Once the actual HIP run was completed and metallography was obtained, measurements of the core-to-core spacings

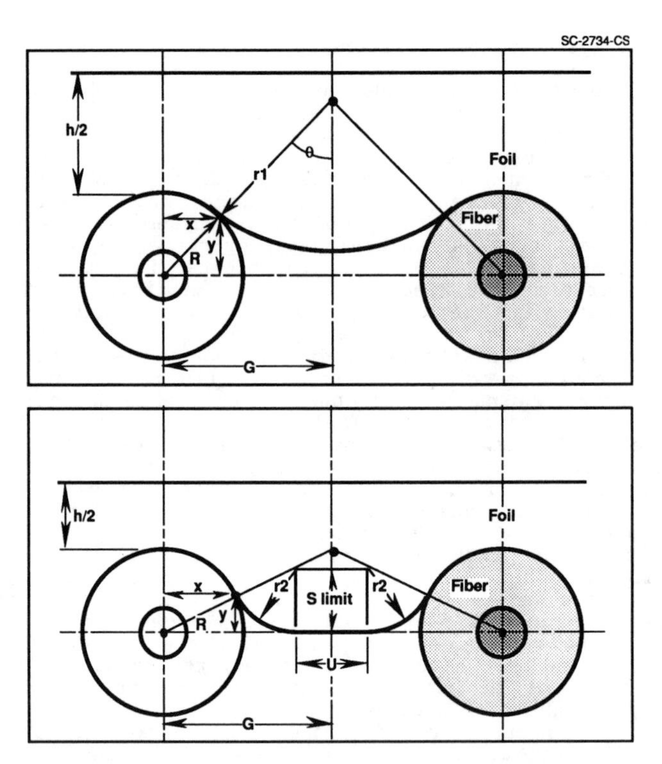

Fig. 3 Illustration of two fiber unit cell and the geometry implemented to track foil contact and final void closure.

were taken and matched with predicted results for the same spacing.

SCS-6/Super α_2

Figures 4 and 5 show two frames taken at the same instant for a simulated HIP run with a SCS-6/Super α_2 (Ti-25Al-10Nb-3V-1Mo) composite with two different fiber spacings. Figure 4 represents core-to-core spacing of 155 μm and Figure 5 core-to-core spacing of 228 μm. The HIP cycle in this case was a linear ramp in temperature to 913°C (1675°F) with two linear stages of pressure ramp to 70 MPa (10 ksi), with 1.5 hours hold at this temperature and pressure. This was followed by a second linear ramp in temperature to 982°C (1800°F) and a drop in pressure to 14 MPa (2 ksi) with a 3 hour hold at this temperature and pressure. The relatively low maximum pressure was used to reduce fiber movement during load-up. The high strain-rate sensitivity of the super α_2 matrix alloy tended to cause juxtaposed fibers in adjacent laminates to displace each other during an early stage of consolidation if high loading rates were used. The second (high temperature, low pressure) stage of this HIP cycle was necessary to diffusion-strengthen the foil-foil bonds in the few

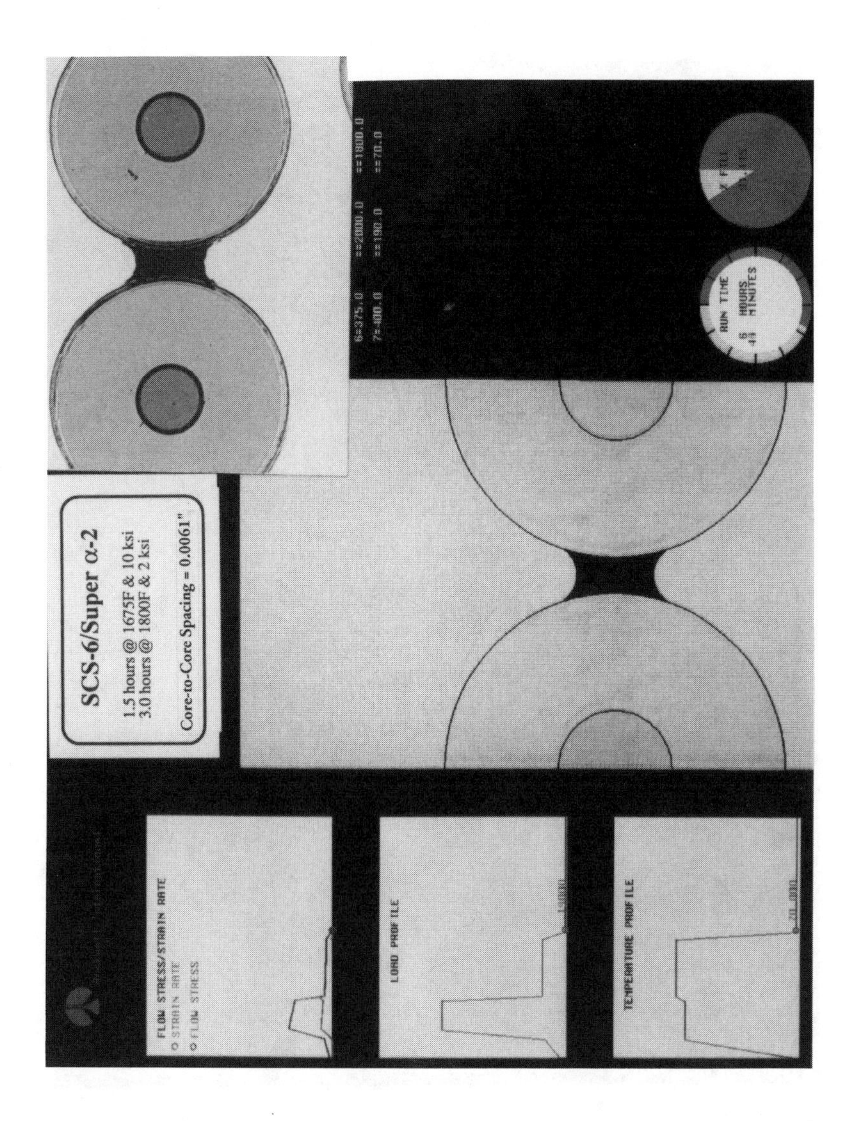

Fig. 4 A stop-run frame taken from a CAPPS simulation of SCS-6/Super α_2 partial consolidation with micrographs of similar spacing from an equivalent validation experiment superimposed in the top right corner.

areas where contact was made (between widely spaced fibers). In the absence of this diffusion-strengthening, the thermal expansion mismatch residual stresses between the fibers and matrix caused delamination of the specimens along the weak diffusion bonds during the cool-down stage. The HIP cycle processing conditions are represented graphically in the windows on the left side of the Figures 4 and 5. Cross section micrographs are

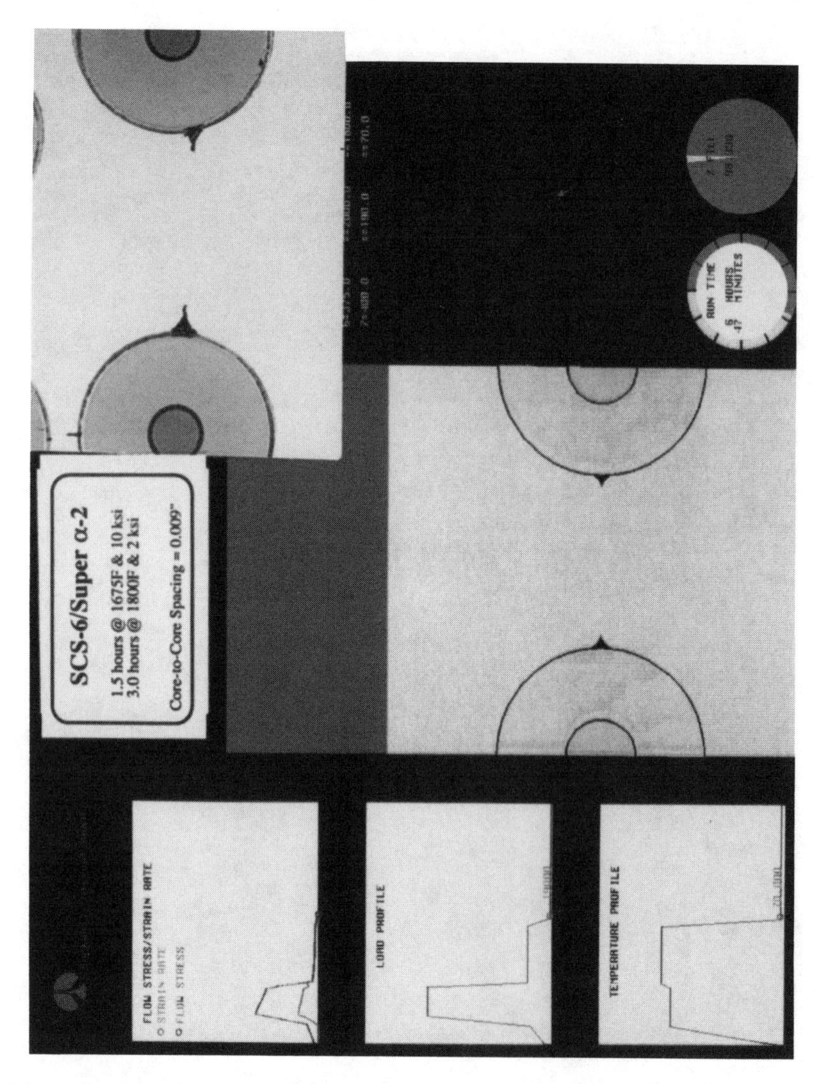

Fig. 5 A stop-run frame taken from a CAPPS simulation of SSCS-6/Super α_2 partial consolidation with micrographs of similar spacing from an equivalent validation experiment superimposed in the top right corner.

superimposed on the top right of each of the model graphical displays. The micrographs were selected from the validation samples run with nominally the same HIP conditions with closely similar fiber spacings to those used in the model. Comparison of the validation specimen micrographs with the model simulation graphics shows the model predictions to be accurate and to require no parametric adjustment. This was true of all intermediate fiber spacings examined in this HIP cycle (not shown here).

Saphikon/γ-TiAl.

Figure 6 shows optical micrographs of a Saphikon (150 μm diameter sapphire fibers)/γ-TiAl (Ti-48Al-2Cr-2Nb-1Ta[at.%]) composite consolidated according to a CAPPS processing cycle which predicted just-complete consolidation for touching fibers, (105 MPa for 4 hours @ 1088°C). As seen in Figure 6, the equivalent experiment validated the CAPPS prediction. Only very small (< 1 μm) voids were found adjacent to some of the fibers (not shown).

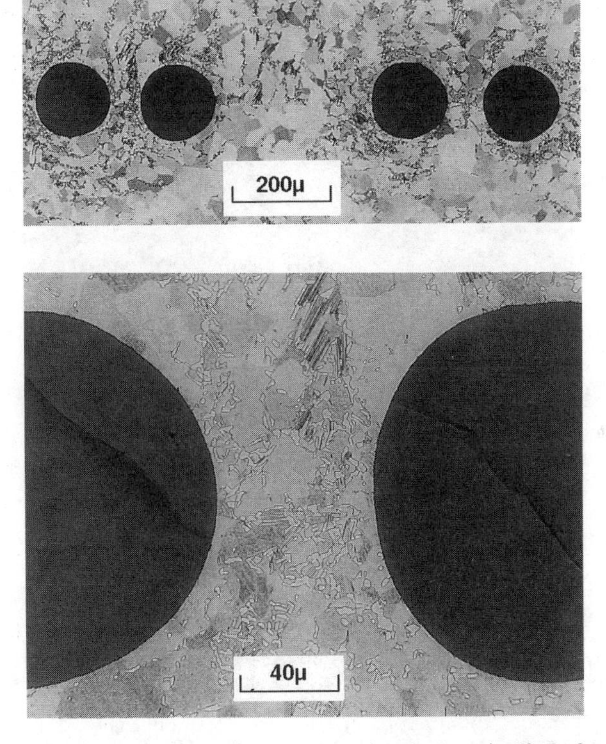

Fig. 6 Fully consolidated Saphikon/γ-TiAl. Ti-48Al-2Cr-2Nb-1Ta (at.%), 105 MPa (15 ksi) @ 1088°C (1990°F) for 4 h.

Figure 7 shows a slice from the same specimen as shown in Figure 6 after a post-consolidation exposure for 3 hours at 1204°C. This micrograph illustrates the significant increase in fiber/matrix reaction with only a modest increase in temperature. (Figure 7 shows a single-phase "halo" around the fibers which is not present in Figure 6.)

SCP-0267-E

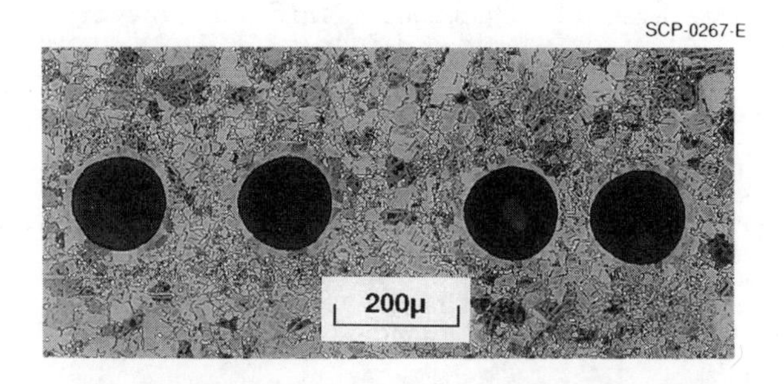

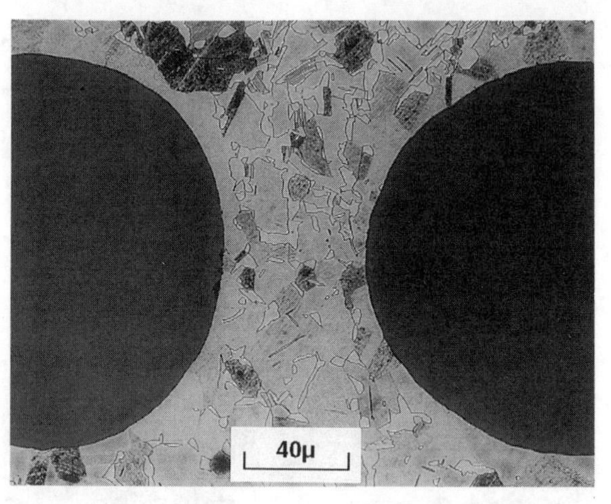

Fig. 7 Fully consolidated Saphikon/γ-TiAl. Ti-48Al-2Cr-2Nb-1Ta (at.%), with post-consolidation exposure @ 1204°C (2200°F) for 3 h.

Figure 8 shows a partially consolidated composite with the same constituents as in Figures 6 and 7. The temperature was lowered by only 50°C and the time at maximum temperature and pressure reduced from 4 to 3 hours compared to the fully consolidated case but consolidation was clearly far from complete. The damage incurred by the fibers which is visible in Figures 8 and 9 (profuse cracking and chipping) was due to cutting and polishing for metallography with poorly supported fibers in the partially consolidated composite.

Figure 9 shows three examples of different fiber spacing comparisons from the experimental sample and from the graphical output of the CAPPS model. As in the case of the SCS-6/Super α_2 case, the model simulation graphics show the model predictions to be accurate and requiring no parametric adjustment.

SCP-0265-E

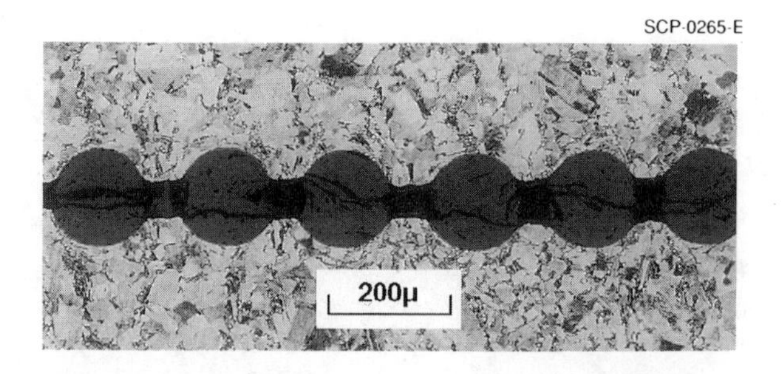

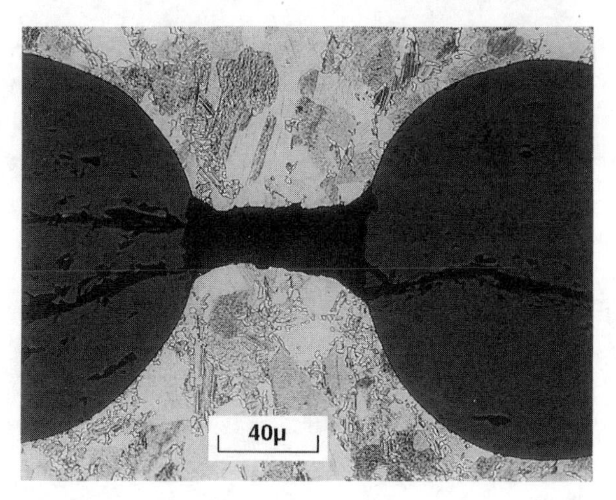

Fig. 8 Partially consolidated Saphikon/γ-TiAl. Ti-48Al-2Cr-2Nb-1Ta (at.%), 105 MPa (15 ksi) @ 1038°C (1900°F) for 3 h.

4. CONCLUSIONS

The embedment of continuous fiber reinforcement in a titanium aluminide matrix is readily amenable to analytical treatment by defining a unit cell geometry aiding in the formulation of boundary conditions to determine an average effective stress and a resulting creep deflection. The model is also general enough in nature that it can be applied to a wide range of titanium-based fiber-matrix systems producing reasonably accurate results. The effectiveness of the CAPPS model as a tool to optimize consolidation cycles for difficult-to-process alloys such as super-α_2 and a γ-TiAl alloy has been clearly demonstrated. The general nature of the model is a result of the transformation of flow stress strain rate data for the various matrix materials into constants of an expanded creep rate equation. As the model stands, additional work will be necessary to better predict the complete composite component fabrication process. Such additions would include the effects of multi-directional ply lay-ups and formulation for the diffusion rate of atoms across the bondline and the resulting grain boundary migrations. Other useful

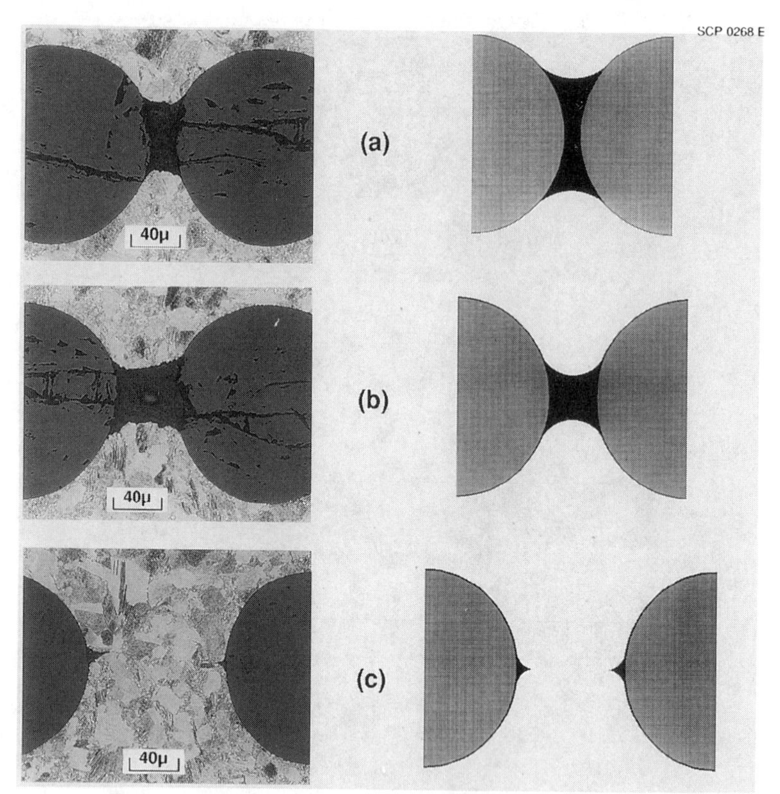

Fig. 9 Saphikon/γ-TiAl composite consolidation, experiment (left) versus model (right). Ti-48Al-2Cr-2Nb-1Ta (at.%), 105 MPa (15 ksi) @ 1038°C (1900°F) for 3 h.

additions include models for reaction rates at the fiber-matrix and foil/environment interfaces. At this time, the model is very accurate and extremely useful in predicting the densification envelopes for SCS-6/β21S and SCS-9/β21S, SCS-6/Super α_2, SCS-6/O-Ti_2AlNb and Saphikon/γ-TiAl foil/fiber/foil composite systems.

5. REFERENCES

1. NASP-MASAP "Titanium Matrix Composites," Final Report.

2. Feltham, P., "Creep in Face-Centered-Cubic Metals with Special Reference to Copper," Acta Met. 7, No. 9, 614-27, 1959.

3. N.A. Fleck, L.T. Kuhn and R.M. McMeeking, "Yielding of Metal Powder Bonded by Isolated Contacts," submitted to Journal of Mechanics and Physics of Solids, 1992.

PARTICULATE INTERMETALLIC MATRIX COMPOSITE MADE BY ONE-STEP PULSED LASER DEPOSITION

LI-CHYONG CHEN, ERNEST L. HALL, and KAREN A. LOU
General Electric Corporate Research and Development, Schenectady, NY12301

ABSTRACT

A novel way of producing particulate intermetallic matrix composites based on Nb-Al in one step using pulsed laser deposition (PLD) has been investigated. One unique characteristic, inherent to laser ablation, is the generation of particulates. These particulates condense on the substrate and become part of the film. In some cases, such as in high T_C superconducting or optical films applications, it is believed that the presence of particulates diminishes the quality of the films. In this work, we demonstrate that it can be advantageous in some applications to incorporate these particulates into the films.

Nb-Al films were prepared by laser ablation from a Nb_3Al target using a 248nm KrF excimer laser at various fluences and substrate temperatures. Transmission (TEM) and scanning electron microscopy (SEM), energy dispersive X-ray analysis (EDX), and Auger electron spectroscopy (AES) were used to characterize the samples. The resultant films consist of a homogeneous matrix with uniformly distributed inclusions. The size of the particles generated by PLD is in the range of a few tens of nanometers to a few microns. EDX shows that the matrix has a composition of Nb_7Al_3 and the particulates contain barely detectable Al. The mechanism of the depletion of Al in the particulates will be discussed. The merit of having a ductile Nb phase embedded in the intermetallic matrix is evident as the cracks generated during TEM sample thinning process propagate in the brittle matrix and finally are arrested at the ductile inclusions.

INTRODUCTION

Ductile phase reinforced intermetallic matrix composites have been recognized for several years as a new class of materials that would offer low density and high-temperature strength from the intermetallic matrix while the intervening ductile reinforcements provide low-temperature fracture toughness. Because interdiffusion and reaction can be significant in the potential service temperature range of 1000 °C to 1600 °C, thermal stability is one of the essential factors in the selection of an intermetallic-reinforcement system. A system consisting of intermetallic and adjacent ductile terminal solid solutions that are in chemical equilibrium such as the Nb_3Al/Nb composite, for example, is desirable and is being extensively investigated [1, 2].

Many innovative techniques have been developed to produce intermetallic matrix composites, of which several representative ones have been reviewed by Stoloff and Alman [3]. Because of the relatively high melting temperatures and extreme brittleness of most intermetallic compounds, the majority of processes involve the use of powders. Vapor deposition has also received increasing attention. For example, high-rate sputter deposition [2, 4] as well as electron beam evaporation [4, 5] have been employed to produce in-situ intermetallic matrix composites.

In this work, we demonstrate for the first time that pulsed laser deposition can be used to synthesize a particulate intermetallic matrix composite in one step. PLD is rapidly emerging as the technique of choice for producing high temperature superconducting (HT_cS) thin films. Specific advantages of PLD include greater composition control and lower substrate temperature as well as greater ambient atmosphere compatibility than competing techniques. In addition to the HT_cS application, a rather wide application of the PLD technique has been reported in the literature and the tremendous potential of PLD technique is being extensively investigated in many laboratories[6, 7].

In PLD, the delivery of an extremely high level of energy to a target surface from a pulsed laser causes the removal of a volume of target material. The ejected flux often consists of neutrals, ions, electronically excited species, electrons, soft x-ray emission, and both liquid droplets and solid particles in a variety of concentrations. Although a number of models have provided some grasp of the physical picture [8, 9], many questions remain unanswered. The generation of particulates, formed by the splashing of the molten layer from the recoil pressure or by dislodging from protruding surface features, is one of the unique characteristic observed in PLD. For superconducting and optical applications, the presence of such particles is generally deemed to be adverse to the quality of the deposited film. For other applications, the problem seems less acute and the presence of particles may even be desirable. This paper demonstrates that, for an intermetallic containing system such as Nb-Al, it can be advantageous to incorporate these particulates into the films.

EXPERIMENTAL

The deposition experiments were performed in a vacuum chamber with a base pressure of 10^{-7} Torr. The laser employed in this work was a 248 nm KrF excimer laser (Lambda Physik EMG-103 MSC). A shorter wavelength is preferred because of smaller reflectivity of 248 nm laser beam in metals and hence the higher efficiency of photon absorption. Laser pulses can be generated at a rate as high as 200 Hz. For each pulse, typical FWHM of the temperal duration is 30 nanoseconds. The Nb_3Al target was a 1 inch diameter pellet, sintered from commercially available Nb_3Al powder purchased from CERAC. The target was rotated at 20 rpm during the ablation. The laser was incident on the target at 45° and the focused spot size of the laser at the target was 2 mm × 0.2 mm. Fluence was varied by varying the pulsed laser energy which was measured in front of the processing chamber window by a powermeter. Prior to film deposition, the deposition rate as a function of fluence was measured by a quartz crystal microbalance located in the same position as the substrate for future film deposition. Oxygen-free high conductance copper substrates were used. The substrate holder consisted of an Inconel heater with a chromel-alumel thermocouple welded to the back side of the substrate. An Omega CN-2012 temperature controller was used to control the substrate temperature to within ± 2 °C. The substrate-to-target distance was 5.5 cm.

Scanning electron microscopy was used to examine the film morphology and the image analyses of particle density were made from the SEM micrographs. Additional microstructural examination was performed using TEM, and the chemical compositions of the films were determined by EDX. Point specific chemical information was obtained from Auger electron spectroscopy.

RESULTS AND DISCUSSION

PLD is a highly directional deposition process in which the cone-shaped plume of ejected material is characterized by $\cos^n(\theta)$, with $8<n<12$ [9]. The morphology and composition of the film can depend on the position relative to the center of the plume of the specific area under investigation. In this report, we compare only the portion of the film within approximately 1 cm^2 near the center of the plume for convenience. Typical SEM micrographs of thin films deposited at two different laser fluences are shown in Figure 1. The film consists of randomly distributed particles embedded in a relatively uniform and featureless matrix. Notice that the majority of the particles are spherical and irregular shaped fragments are only occasionly found. The fact that the particles are spherical in shape suggests that most of the particles are solidified from molten droplets rather than broken loose from the target. The molten droplets are likely to be formed by the splashing of the molten layer from the recoil pressure induced by the high power pulse laser. Typical particle size distributions for films deposited at various fluences are shown in Figure 2. For the sample prepared with a laser fluence of 30 J/cm^2, a few particles with diameters up to 8.5 μm were found, but only the portion that contains data up to 5 μm is plotted here for convenience. For the samples prepared with other laser fluences, no particle was found with a diameter greater than 5.5 μm. The particle number density generated per pulse and the average particle size at various laser fluences are summarized in Table I. The particle number density as well as the average particle size appears to increase with increasing fluence. For comparison, the deposition rate derived from the thickness of the uniform matrix alone is also listed in Table I.

For the film deposited at a substrate temperature below 300 °C, TEM investigation showed that the uniform matrix exhibits a very fine grain randomly-oriented bcc structure with an average grain size estimated at 2 nm. EDX spectra from the matrix gave an Al content of 30.5±2 % using estimated calibration factors. In contrast, the large embedded particles, which are 2-6 μm in diameter, are polycrystalline with grain size on the order of 0.2 μm, while many of the smaller particles (< 1 μm) are single crystals. The electron diffraction patterns from the particles

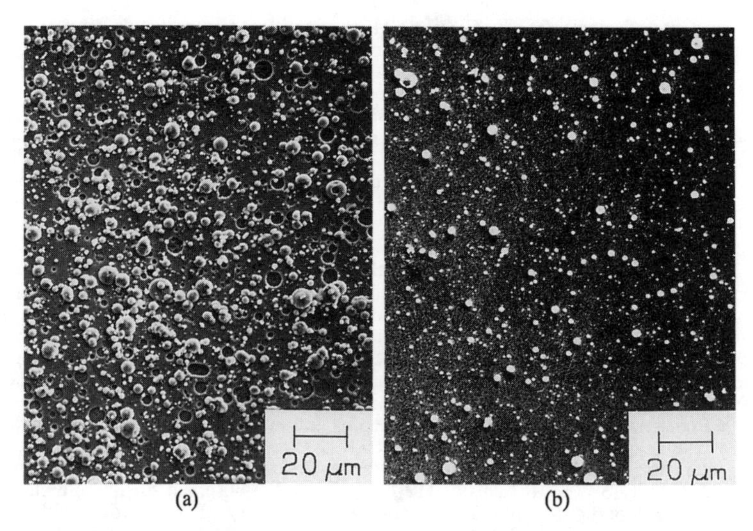

Figure 1. SEM micrographs showing the surface morphology of Nb-Al films deposited by 248 nm KrF excimer laser pulses with different powder density. The laser fluence of (a) is 30 J/cm^2 and (b) is 25 J/cm^2.

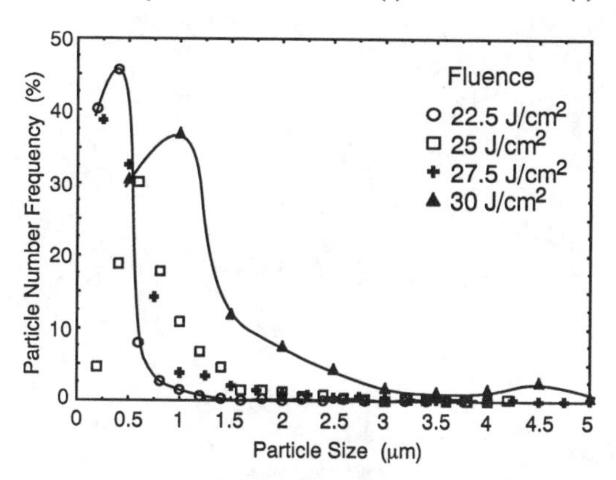

Figure 2. The particle number distribution for the size of the particles produced by laser ablation of a Nb_3Al target at various laser fluences. The curves drawn through the data points are for guiding the eyes. Only the portion that contains data up to 5 μm is plotted here for convenience.

Table I. Image analyses for PLD films at various laser fluences

Fluence (J/cm^2)	Particle generation rate (number/mm^2/pulse)	Average particle size[(a)] (μm)	Matrix deposition rate (nm/100 pulses)
22.5	0.86	0.3 (2.6)	0.062
25	1.65	0.7 (1.9)	0.13
27.5	2.5	0.5 (3.4)	0.23
30	3.01	1.1 (5.1)	0.58

(a): Average particle size based on number statistics. The average particle size based on volume statistics is shown in the parentheses.

are consistent with elemental Nb. Closer TEM examination indicates that the smaller particles also contain a higher density of defects than the larger particles. This provides additional evidence that the particles are indeed formed from molten droplets. The smaller particles often offer a higher cooling rate and a larger undercooling and the resultant microstructure contains more defects than the larger ones. The presence of single crystal particles also suggests a homogeneous nucleation from the melt. Figure 3 is a TEM micrograph which gives the best representation of the overall film. Figure 4 is a TEM micrograph taken at a higher magnification showing the highly defective single crystal particle embedded in the matrix. The corresponding EDX spectra taken from the matrix and a particle are shown in Figure 5. Notice that the EDX spectra taken from the particles contain no detectable Al.

The observation that the particles are deficient in Al is quite intriguing. The possibility that it is due to segregation of target material is ruled out since the target is a single phase material. We believe that the composition of the ejected particles right above the surface of the target was quite close to that of the target, and the loss of Al occurred mainly during the flight to the substrate. Dupendant *et. al.* [10] and Gavigan [11] have shown that, for PLD processes in a number of metallic systems with considerably different atomic masses and densities, the particle velocity distribution maxima were found to vary only between 100 and 200 meters per second, an order of magnitude smaller than those of atomic and ionic species. The time it takes for a particle to traverse a 5.5 cm distance between target and substrate as used in this work, with an estimated velocity of 200 m/s, is 275 microseconds. Assuming the particle remains in the liquid phase for that same period of time, the diffusion distance is estimated to be 1.17 μm with an atomic diffusivity of 5×10^{-5} cm^2s^{-1} [12], a typical value observed for liquid metals. For submicron-sized particles, complete depletion in Al is likely. For a particle with a larger diameter, one would expect that the depletion may be incomplete. Core regions of the big particles may still contain an Al content higher than that in the outer regions of the particles.

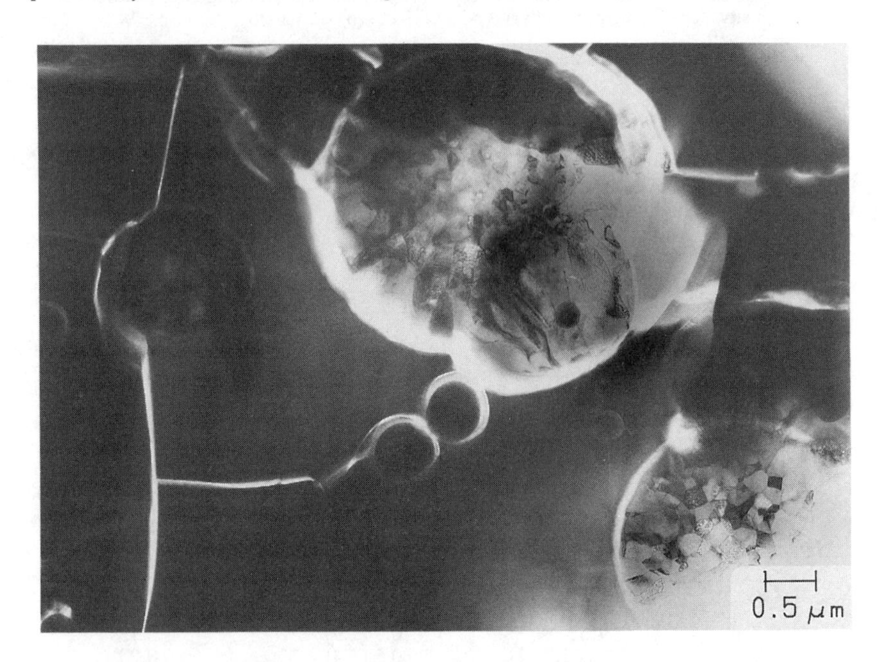

Figure 3. TEM micrograph of a Nb-Al film deposited by 248 nm KrF excimer laser pulses at room temperature and at 30 J/cm^2 laser fluence. Notice that the cracks generated during TEM sample thinning process propagate in the brittle matrix and finally are arrested at the ductile inclusions.

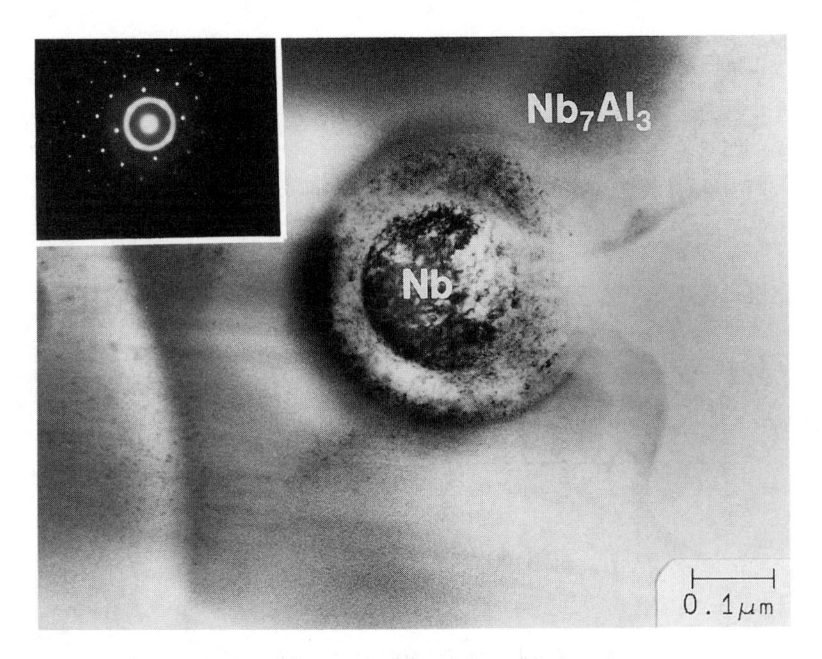

Figure 4. TEM micrograph of the same film as shown in Figure 3 except at a higher magnification. While the larger particles are polycrystals, the smaller particles are single crystals. Notice the single crystal particle is highly defective.

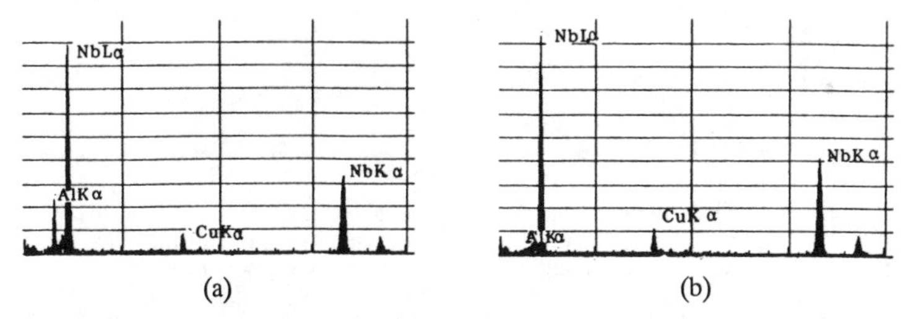

Figure 5. The corresponding EDX spectra taken from (a) the featureless matrix, and (b) a particle in the film shown in Figure 3 and 4. The particle is Nb and the matrix has a composition of Nb_7Al_3.

Figure 6 is a SEM micrograph showing the particle-containing cross-section of the film. The corresponding AES spectra for different regions are shown in Figure 7. While both Nb and Al were present in the matrix, marked as feature A in Figure 6, the AES spectra taken from the region at the outer edge but well under the surface of the particle, marked as feature B in Figure 6, show no detectable Al. The Al/Nb ratio derived from the Al:Nb peaks after correcting the sensitivity factor is 0.83 for the matrix. In the particle interior, marked as feature C in Figure 6, Al does appear but is still quite small. The Al/Nb ratio is reduced to 0.11 for the particle interior. That a trace amount of Al was observed in the interior but not in the edge of particle provides the evidence of the diffusion of Al in the liquid droplet as hypothesized above. The particle surface gave an AES spectra similar to that of the matrix. The surface Al is believed to be condensed from the gaseous component.

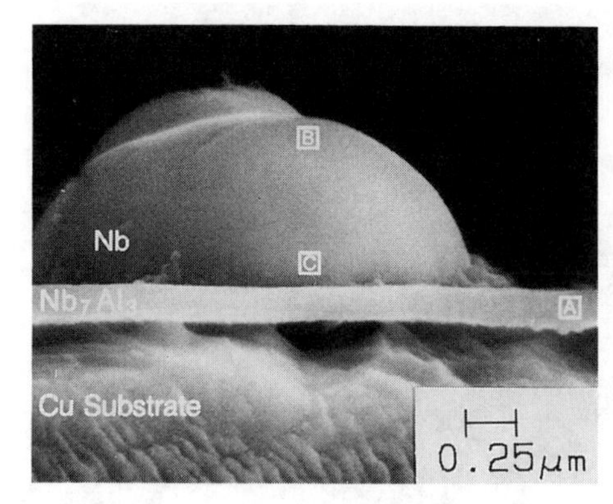

Figure 6. SEM micrograph of the particle-containing cross-section of the pulsed laser deposited Nb-Al film.

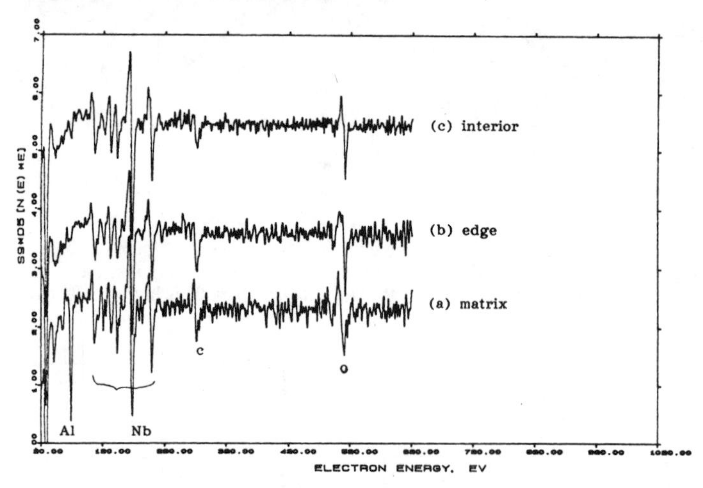

Figure 7. The corresponding AES spectra taken from (a) the matrix, (b) the outer edge of the particle, and (c) the interior of the particle in the film shown in Figure 6. Notice the trace amount of Al was observed in (c) but not in (b).

As shown in Figure 3, the merit of having a ductile Nb phase embedded in the intermetallic matrix is evident as the cracks generated during TEM sample thinning process propagate in the brittle matrix and finally are arrested at the ductile inclusions. No other mechanical testing has been done at present.

SUMMARY

We have demonstrated that films made by excimer pulsed laser ablation from a Nb_3Al target consist of micron and submicron sized Nb particulates embedded in a uniform Nb_7Al_3 matrix. Such a composite structure may be useful for high temperature applications. A plausible mechanism whereby highly Al-deficient particulates are formed is discussed. The Al-deficient particle size is consistent with the diffusion length of Al in the molten droplet with an estimated time of flight and an estimated atomic diffusivity. Only a trace amount of Al is found in the core of larger particles with size greater than 1 μm.

The PLD technique also has the flexibility of varying volume fractions of the reinforcing phase intervened within the matrix phase. The particulate generation rate and the size distribution depend on a number of process parameters. The laser fluence, in particular, strongly affects the particle number density and the average particle size. The deposition rate appears low; however, since the particulates generated by PLD are much finer than those by other conventional techniques, further development of a higher power pulsed laser would make this novel technique an effective alternative for fabrication of intermetallic matrix composites with fine dispersion.

ACKNOWLEDGEMENTS

The authors are very grateful to Ms. K. Denike and Mr. C. Erikson for technical assistance with pulsed laser deposition, Ms. E. Whittemore for the SEM, and Mr. J. Grande for the image analyses. L. C. Chen would also like to thank Prof. H. Lipsitt of Wright State University, Dr. G. Rowe, Dr. D. Skelly, and Dr. A. Taub for many valuable discussions.

REFERENCES

1. D. L. Anton and D. M. Shah, in *Intermetallic Matrix Composites*, D. L. Anton *et. al.* eds., MRS **194**, 45, Pittsburgh, PA (1990).
2. R. G. Rowe and D. W. Skelly, General Electric, this symposium (1992).
3. N. S. Stoloff and D. E. Alman, in *Intermetallic Matrix Composites*, D. L. Anton *et. al.* eds., MRS **194**, 31, Pittsburgh, PA (1990).
4. D. A. Hardwick and R. C. Cordi, in *Intermetallic Matrix Composites*, D. L. Anton *et. al.* eds., MRS **194**, 65, Pittsburgh, PA (1990).
5. R. S. Bhattacharya, A. K. Rai and M. G. Mendiratta, in *Intermetallic Matrix Composites*, D. L. Anton *et. al.* eds., MRS **194**, 71, Pittsburgh, PA (1990).
6. J. T. Cheung and H. Sankur, CRC Critical Reviews in *Solid State and Materials Sciences*, **15** (1), 63 (1988).
7. MRS Bulletin Vol. **XVII** (2), February issue (1992).
8. C. L. Chan and J. Mazumder, *J. Appl. Phys.* **62**, 4579 (1987).
9. R. Singh and J. Narayan, *Phys. Rev. B* **41**, 8843 (1990).
10. H. Dupendant, J. P. Gavigan, D. Givord, A. Lienard, J. P. Rebouillat and Y. Souche, *Applied Surface Science* **43**, 369 (1989).
11. J. P. Gavigan, in *The Science and Technology of Nanostructured Magnetic Materials*, Proceedings of NATO ASI series, Plenum Publishing Co. (1990).
12. N. H. Nachtrieb, in *Liquid Metals*, S. Z. Beer ed., Marcel Dekker Inc., New York (1972).

IN-SITU SYNTHESIS OF INTERMETALLIC MATRIX COMPOSITES

DILIP M. SHAH* AND DONALD L. ANTON**
* Pratt & Whitney, 400 Main St., E. Hartford, CT. 06108
**United Technologies Research Center, Silver Lane, E. Hartford, CT. 06108

ABSTRACT

In pursuing the development of intermetallics as high temperature structural materials, a composite approach is considered necessary for achieving room temperature toughness. If not further qualified, the term "composite" generally implies the application of mechanical processes by which a strong reinforcing phase is dispersed in the matrix, often as aligned continuous fibers. While this approach appears simple and promising in principle, in practice it is limited by the availability of compatible fibers, controlled processing, and microstructural homogeneity and reproducibility. Alternatively, the composite microstructures may be created in-situ either synthetically or naturally. In synthetically derived composites, the desired phases may be deposited layer by layer using such techniques as chemical vapor deposition (CVD), and potentially a variety of lithographic techniques may be employed to control the microstructure. However, such techniques are currently rate limited and not well developed for the large dimensions required for structural composites. In contrast, the in-situ composites, which rely on phase separation by either eutectic solidification or solid state precipitation, are economical and especially well suited for generating naturally compatible ductile phase toughened composites with uniform fine scale microstructures. This paper attempts to classify these approaches in perspective, discuss the benefit of in-situ composites relying on the natural phase separation mechanisms, and review the current activities with emphasis on the concept of ductile phase toughening.

INTRODUCTION

The need for high temperature materials beyond the current generation of single crystal superalloys for jet propulsion is a given, in spite of the changing economic and political climate [1]. If anything, for the next generation of materials, the emphasis is likely to shift toward greater reliability, durability and low cost. For fulfilling these demanding requirements of high temperature structural materials for dynamic applications, the development potential of refractory intermetallics has been explored[2-4]. With the current design philosophy, the relatively low modulus, high thermal conductivity and ductile behavior at high temperatures associated with the metallic characteristics of intermetallics are considered beneficial. The potential for improving the ultimate strength, high temperature creep resistance and oxidation resistance, to a realistic level, has been demonstrated for various refractory intermetallics with ternary additions. The problem at hand is to optimize all these properties for one system with acceptable room temperature toughness - or perhaps ductility. To achieve this goal, several composite approaches are being evaluated. Among these, in-situ synthesis of composites as opposed to conventional processing of fiber reinforced composites, appears more attractive, with its potential for alloying to achieve a balance of engineering properties.

REFRACTORY INTERMETALLICS FOR STRUCTURAL APPLICATIONS

In general, as depicted in Fig. 1, the intermetallics may be used in principle as single phase monolithic materials or may be used as a principal phase of two phase composites. The composite may be produced either in-situ or artificially by judiciously juxtaposing, individually identifiable, matrix and reinforcement materials with optimized, intrinsic properties. $MoSi_2$ reinforced with single crystal alumina monofilaments (Saphikon ®) may be cited as an ideal example of the latter. Conceptually, this is attractive on the assumption that the balance of all other properties will be achieved once one of the most oxidation resistant refractory intermetallics is reinforced with strong fibers. The in-situ composite approach seeks to produce a two-phase composite microstructure in a controlled, nonmechanical fashion. The composites produced in-situ may be further divided into natural and synthetic in-situ composites; the former relying largely on the natural phase separation mechanisms versus the latter employing techniques primarily developed for coatings such as sputtering, physical vapor deposition (PVD), plasma spray or chemical vapor deposition (CVD), and many variants thereof. Even though the natural and the synthetic multiphase materials are in-situ composites, there are sufficient differences to compare them separately with the artificial or largely fiber reinforced composites as presented in Table I.

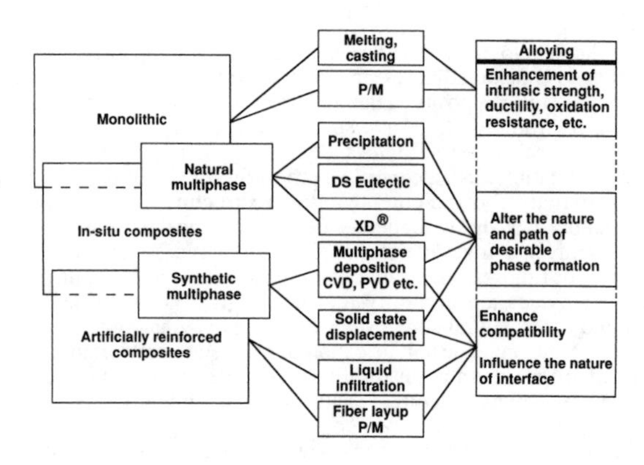

Fig. 1. Schematic overview of strategies for the development of intermetallic structural materials.

Table I
Comparison of Composite Approaches

	IN-SITU COMPOSITES		ARTIFICIAL COMPOSITES
	Natural	Synthetic	(Fiber Reinforced)
CONCEPT	Complex	Simple	Simple
MAT. SELECTION	Constrained	Flexible	Flexible
DEV. EFFORT	Specific	Specific	Generic
MICROSTRUCTURE			
Morphology	Constrained	Partially Const	Flexible
Interface			
Stability	Compatible	Critical	Critical
Scale			
nm	Precipitate	Nanocrystal	Whiskers
µm	Eutectic	Microlaminate	Fiber tow
mm		Macrolaminate	Monofilaments
Control	Excellent	Good	Poor
STRENGTH INCREMENT	Nonlinear	Nonlinear	Rule of Mixture
PROCESSING			
Cleanliness	High	Controllable	Problematic
Reliability	High	Acceptable	Unacceptable
Cost	Low	Acceptable	High

COMPARISON OF COMPOSITE APPROACHES

Conceptually, the artificial composites and synthetic in-situ composites are easily understood, modeled and appear to offer a limitless possibility in terms of the selection of materials. However, a careful consideration of thermodynamic and mechanical compatibility must be exercised in this selection process. For the natural in-situ composites, the compatibility naturally exists unless the atomic bonding characteristics of the two phases are vastly different. Development effort for both the in-situ approaches has to be rather specific, requiring different processing routes with different alloy chemistry. For the fiber reinforced composites, generic processing may be applied.

Scale and Control of Microstructures

The real differences between in-situ and artificial composites is in their morphology, scale, and control of microstructure, as is illustrated in Fig. 2. In principle, any morphology can be achieved artificially but at a very labor intensive cost if mechanical means are employed. Fig. 2(c) represents the best case of NiAl reinforced with FP alumina fibers prepared using fine powders to improve infiltration [5]. In contrast, the selection of morphology is constrained if natural phase separation mechanisms are used. As shown in Fig. 2(a) for the Nb_3Al/Nb in-situ composite, strings of Nb-Al solid solution are precipitated in the intermetallic matrix. Similarly, if synthetic or nonmechanical means are employed, then with ingenious processing, the constraint on morphology may be removed; but at the present laminate structures are the most viable microstructure producible by such means with good control.

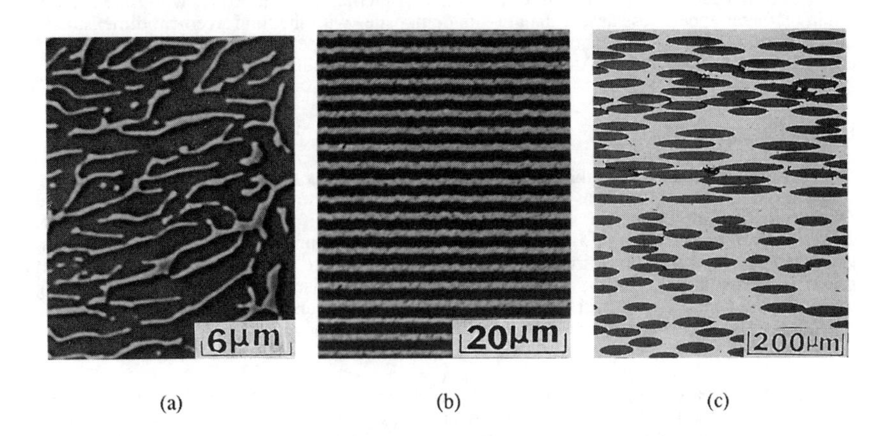

Fig. 2. Comparison of in-situ versus artificial composite microstructures: (a) Nb precipitation in Nb_3Al by phase separation, (b) microlaminate of Nb/Nb_3Al produced by magnetron sputtering, (c) powder infiltrated NiAl/aligned FP alumina composite.

As a direct comparison to the naturally produced Nb_3Al/Nb composite in Fig. 2(a), a microlaminate microstructure of the same composite produced by magnetron-sputtering is presented in Fig. 2(b)[6]. Note that the scale of microstructures in Fig. 2(a) and (b) are within a factor of three. However, the scale of microstructure in Fig. 2(c) is a factor of ten coarser using the finest continuous reinforcement produced. In all cases, nano-scale microstructural features may be introduced, but the scale of mean free distance between features is difficult to control if mechanical means are employed. For example, very fine whiskers may be artificially blended in a composite but intergrannular clumping of such whiskers is difficult to avoid. In synthetically produced microlaminates, the volume fraction of the phases can be varied by processing, which is difficult to achieve by natural phase separation without altering the size and morphology of the microstructure as well.

Strength Increment

When the scale of both the second phase and the mean free distance are comparable to slip distances, generally a nonlinear increment in strength occurs as in most precipitation hardening systems. This is also the underlying motivation in pursuing the nanocrystalline and microlaminate materials with the promise of unusual physical and mechanical properties. When the scale of microstructure is too coarse, each individual phase behaves as it does in bulk form and the rule of mixtures applies. The rule of mixtures is always easy to understand because no atomistic details need be invoked. Intrinsic nonlinear plasticity in solids is complex since the dislocations and twin boundaries, as carriers of strain, are one- and two-dimensional defects, respectively. Notwithstanding the complexity, however, it is clear that a nonlinear increment in strength occurs when the scales of microstructures are measurable in finite multiples of Burgers vector, a characteristic dimension of dislocation type defects[7].

Besides the nonlinear strength increment, a fine scale microstructure is desirable from a design standpoint. In current practice, considerable technological effort is expended in producing very complex cooling passages in investment cast superalloy turbine blades[8]. The dimensions of some of the structural features, such as cooling holes, are in the range of 0.5-1 mm. For the design system to treat the properties of the material to be truly intrinsic with the use of such features, the scale of the microstructure has to be several orders of magnitude smaller. The scale of acceptable microstructures is difficult to predict, but composites with fibers of 100 μm in diameter appear less usable than in-situ composites with the reinforcement dimension in the range of 1-2 μm. As an example, the microstructure of superalloys consist of typically 0.5 μm precipitates of ordered Ni_3Al in a disordered nickel-base solid solution matrix.

Process Control

Ultimately, what favors the natural in-situ composites is the high degree of processing cleanliness, reliability and low cost of production. By allowing nature to separate the reinforcement from the matrix, we generally obtain a very uniform distribution if the process conditions are controlled. In a sense, the same thrust of technology that has taken us from vacuum tubes to transistors, mechanical relays to solid state transducers, and daisy wheel printers to laser printers, can be applied here. The lesser mechanical handling at the processing stage and the fewer the discreet components with relative mechanical motion, the greater the reliability. These are the underlying factors motivating the development of synthetic in-situ processes.

Application of Lithography: We note in passing, that the potential for generating more complex, pre-designed synthetic in-situ composite microstructures exists, if sufficient economic motivation is found to exploit the lithographic technology developed for producing integrated circuits (IC) [9]. To demonstrate feasibility, aluminum was sputter deposited on a niobium foil using conventional microlithography [10]. The goal was to create fibers of Nb_3Al, but inadvertently, owing to a modified experimental condition aimed at achieving a high deposition rate, Al_2O_3 fibers were produced as shown in Fig. 3. Similar attempts to produce Si_3N_4 fibers using low pressure chemical vapor deposition (LPCVD) and plasma enhanced chemical vapor deposition (PECVD) have been reported [11]. Besides the conventional photolithography, many variations of photo-assisted etching and deposition techniques can be used. Such techniques rely on the use of starting materials such as metallorganic gas or organometallic ink with low decomposition temperatures. Conceptually, the process is similar to stereolithography in which a photoresist liquid is polymerized using a UV laser to build models for investment casting[12].

Fig. 3. Deposition of uniformly spaced alumina fibers(white) on a niobium foil prepared using microlithography.

DUCTILE PHASE VERSUS BRITTLE PHASE TOUGHENING

For improving the fracture toughness, the reinforcing phase may be brittle or ductile. However, the brittle-brittle in-situ systems do not appear attractive since the strength of the reinforcing phase is not easily optimized as with strong fibers in conventional ductile matrices. Similarly, the ductile phase toughened composites are best made in-situ since a fine scale microstructure is difficult to achieve with commercially available metal wires and if fiber coating is required, this approach is not cost effective. Nonetheless, such attempts have been made and provide useful mechanistic insight [13-15]. We shall return to further discussion of ductile phase toughening in the context of natural in-situ composites.

NATURAL IN-SITU COMPOSITES

In forming natural in-situ composites, stable or metastable phase separation mechanisms are exploited. The selection of the processing route is dependent on the phase separation mechanism. The phase relationship, in turn, is very system specific and requires a judicious sorting of potential two-phase systems, satisfying other critical requirements such as environmental resistance. Since the ultimate goal of compositing for high temperature structural applications is to enhance room temperature fracture toughness, a better understanding of the factors affecting the preferred ductile phase toughening approach is necessary. Finally, as summarized in Fig. 1, alloying must be imaginatively used affecting the choice of processing as well as the ultimate performance.

Mechanisms of Phase Separation

To produce ductile phase toughened in-situ refractory intermetallic matrix composites, at least three kinds of binary phase diagrams, nearing a terminal solid solution may be considered, as schematically shown in Fig. 4. From the liquid state, one can use eutectic phase separation as shown in Fig. 4(a). In this case, directional solidification (DS) can be employed to produce an aligned microstructure. Solid state precipitation has been one of the most successful natural processes used, for example, in aluminum alloys and nickel-base superalloys. Here, one relies on the sloping phase boundaries with temperature to precipitate the intermetallic phase, as schematically shown in Fig. 4(b). In the simplest situation, the processing involves low temperature aging to precipitate the intermetallic phase, following a high temperature solution heat treatment. As shown in Fig. 4(c), in some cases, retrograde solubility may be used to disorder an ordered phase to precipitate the ductile phase. Many complex variations of the mechanisms cited here occur and increased complexity in multicomponent alloys can be expected.

Generally, the intermetallics of similar structural type are expected to be formed in similar manner and, hence, a specific phase separation mechanism is expected. However, the larger the occurrence of a structure type, the wider is the variation in the nature of phase formation and separation mechanisms. Refractory A_3B type intermetallics with A15 structure is one such example, where a systematic variation in the nature of phase formation has been observed with the types of A and B elements [16]. Among the binary systems of some practical applicability, examples of Re-V, Nb-Al, V-Ga and Cr-Si may be cited as forming an A15 compound by peritectoid, peritectic, congruent from solid and congruent from liquid transformations, respectively.

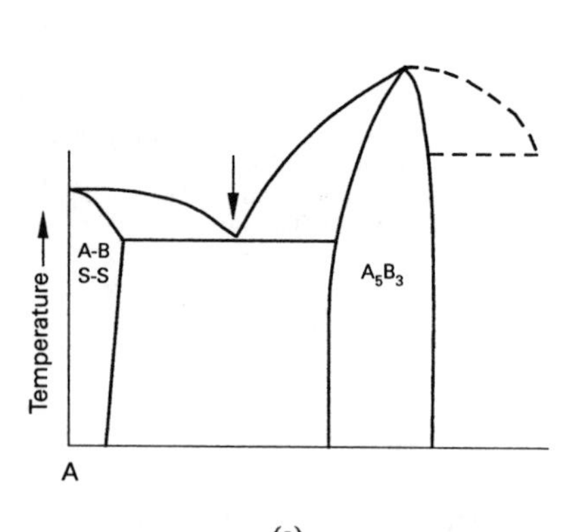

(a)

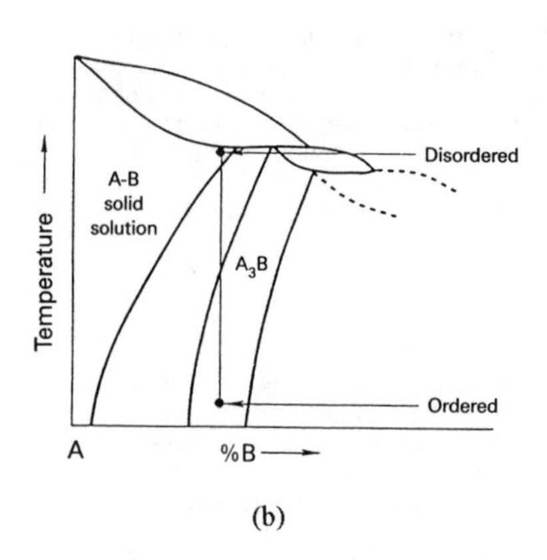

(b)

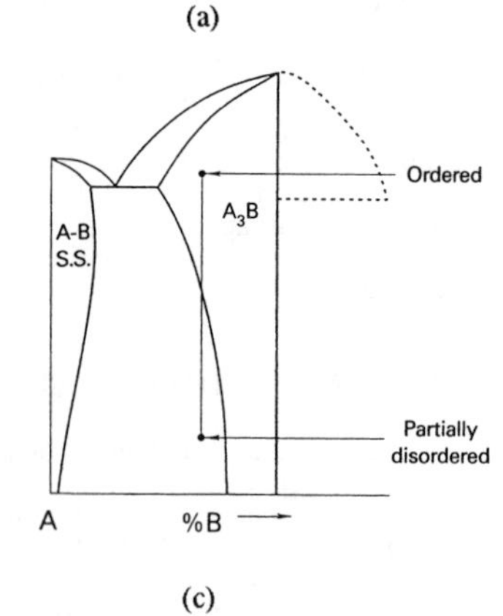

(c)

Fig.4. Mechanisms of Phase Separation: (a) eutectic, (b) precipitation, (c) system with retrograde solubility.

Processing Options

In-situ composites relying on solid state phase separation mechanisms need no special processing, though further improvement in mechanical properties may be gained by forming columnar grain structures or single crystals by directional solidification(DS). For the eutectic in-situ composite, DS becomes a critical step in producing aligned, fine scale microstructures. The various solidification and DS processes are compared in Table II. It must be emphasized that this comparison is presented from the perspective of experimental processing of refractory intermetallics with a melting point higher than 1600°C.

Thus the formation of near net shape components is not a near-term issue, but the ability to produce test pieces of large enough dimensions for the evaluation of mechanical properties is a legitimate concern. For most preliminary evaluations, conventional non-consumable arc melting provides good insight[17]. If rapid solidification rate is required, one can resort to splat quenching [18]. For DS, at least five different techniques are available. All of these techniques, with the exception of the conventional process, are containerless and suitable for refractory in-situ composites [19-20]. We have successfully used an optical float zone furnace and demonstrated the feasibility of producing Cr_2Nb/Nb eutectic, as shown in the microstructure presented in Fig.5 [21]. In this case, further iteration of processing parameters is required to obtain aligned fibers of Nb-Cr solid solution rather than the cellular morphology.

Table II
Comparison of In-situ Processing Options

	ADVANTAGES	DISADVANTAGES
Conventional Casting/ Arc Melting	Simplicity	Uncontrolled Microstructure
Splat Quenching	Rapid Solidification Rate	Small Volume, Thin Samples
High Gradient Conventional DS	Near Net Shape	Crucible Required
Induction Float Zone	Large Size	Difficult to Efficiently Couple
Electron Beam Float Zone	Simplicity	In Vacuum, Limited Size
Optical Float Zone	Simplicity	Limited Size
Cold Crucible Induction	Simplicity	Melt Chemistry Control

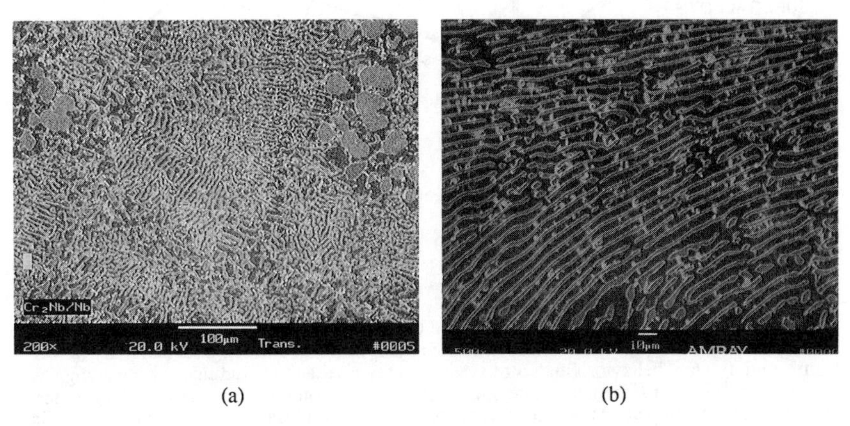

(a) (b)

Fig. 5. (a) Transverse, and (b) longitudinal microstructure of directionally solidified Cr_2Nb/Nb eutectic using optical float zone furnace at 20 mm/hr.

Potential Two-Phase Systems

The wide spectrum of the published work is best summarized with reference to a pseudo-quarternary phase diagram presented in Fig. 6. In this diagram, based on certain semi-theoretical arguments, elements are divided into four groups consisting of Ni and Co; Cr, Fe and Mn; refractory elements Nb,Ta and Mo; and metalloids Al and Si. Note that, with reference to this diagram, nickel-base superalloys belong to the two-phase region between the nickel-base solid solution and the intermetallic Ni_3Al. The earlier work on refractory metal was limited to the solid solution apex of the diagram. The current activity on the single crystal NiAl is limited to the narrow phase field between NiAl and the Heusler phase[22]. In the same manner, all other in-situ composite systems seek to exploit a stable two-phase region between the refractory metal solid solution and the intermetallic of choice.

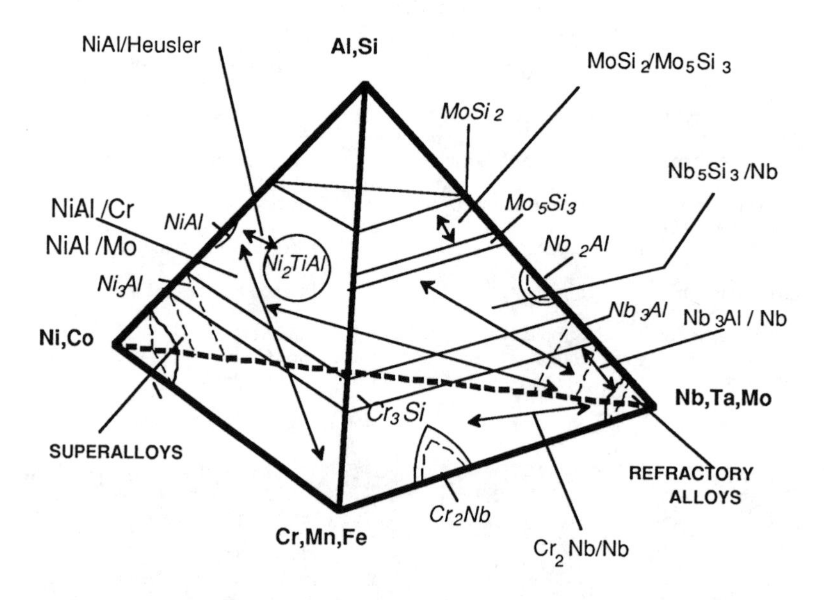

Fig.6. Pseudo-quarternary phase diagram representing in-situ composite systems, currently pursued.

The current activities on the ductile phase toughened in-situ composite systems is further summarized in Table III, where crystal structure type is also listed with a brief summary of room temperature fracture behavior. Studies of systems based on titanium aluminides have recognized the ability of thermo-mechanical processing of stable multi-component systems such as α_2/β and γ/α_2 which are in-situ composite to obtain enhanced strength and fracture capabilities. Excluding the NiAl based composites, the interest in which has been recently reviewed [23], the other refractory intermetallic composites may be grouped into three crystal structure types. As listed in Table III, these are: 5-3 silicides with $D8_x$ structure, cubic aluminides and silicides with the A15 structure, and Laves phases with C14/C15 structure.

Note that in the case of all brittle-brittle in-situ composites, the room temperature fracture behavior is questionable. However, for all ductile phase toughened systems, material is handleable at room temperature with ample evidence of crack blunting. In at least three cases, the fracture toughness has been measured to be greater than 10 MPa $\sqrt{m}$, which is encouraging. In the case of two systems, Nb/Nb_5Si_3 [24] and Ti/Ti_5Si_3[25], a high fracture toughness has been demonstrated for extruded and directionally solidified material, respectively. In both cases, the origin of the high fracture toughness may be at least partially attributed to the aligned and homogeneous microstructure.

Table III
Summary of current activities on in-situ intermetallic matrix composites.

MATRIX		REINFORCING PHASE		TOUGHNESS	SOURCE
ALLOY SYSTEM	CRYSTAL STRUCTURE	ALLOY SYSTEM	CRYSTAL STRUCTURE	INDICATOR ROOM TEMP.	
$\alpha_2(Ti_3Al)$	DO_{19}	α-Ti	A3	Ductility	
$(Ti,Nb)_3(Al,Si)$		$(Ti,Nb)_5(Si,Al)_3$	$D8_x$	?	Es-Souni et al.
NiAl	B2	Mo,Cr	A2	K_{IC} =14 MPa$\sqrt{m}$	Subramanian et al.
		Al_3Nb	DO_{22}	?	Sauthoff
		NbNiAl	Laves,C14	?	Sauthoff
		TaNiAl		?	Sauthoff
Nb_5Si_3	$D8_m$	Nb	Al	K_Q=21 MPa$\sqrt{m}$	Mendiratta et al. Bertero et al.
Ti_5Si_3	$D8_b$	Ti	A3	K_{IC}=11 MPa$\sqrt{m}$	Frommeyer et al.
Mo_5Si_3	$D8_b$	$MoSi_2$	$C11_b$	?	Mason & Aken
Nb_3Al	A15	Nb	A2	*No Cracking K_{IC}=5 MPa$\sqrt{m}$	Anton & Shah
Cr_3Si	A15	Cr	A2	*Crack blunting	Mazdiyasni & Miracle
Cr_2Nb	C14/C15	Nb	A2	Handleable K_{IC}=3.5 MPa$\sqrt{m}$	Anton & Shah
$Cr_2(Nb,Ti)$	C15	(Nb,Ti)	A2	Handleable	Thoma & Peripezko
Cr_2Ta		Ta	A2	*No Cracking	
Cr_2Ta		Cr	A2	*No Cracking	Mazdiyasni
Cr_2Hf		Cr	A2	*No Cracking	& Miracle
Cr_2Zr		Cr	A2	*Crack Blunting	

*Micro hardness

Even without the processing, considerable insight can be gained from an as-cast in-situ composite. It has been observed that, for the conventionally cast Cr_2Nb/Nb eutectic, the maximum bend strength varies over a wide range but it can always be correlated to fracture strain elastically. For the specimens machined from contiguous volume and with comparable microstructure; bend strength, failure strain and elastic modulus can be measured as a function of temperature and can be rationalized to be indicative of strengthening and toughening. In spite of the low value of K_{IC} of 3.5 MPa$\sqrt{m}$ for this material with uncontrolled microstructure, the fractograph presented in Fig. 7 shows plastic-stretching of ductile niobium phase as has been observed for Nb/Nb_5Si_3 [24]. Finally, a comparison of minimum creep rate versus stress for the monolithic Cr_2Nb and the in-situ composite Cr_2Nb/Nb in Fig. 8 shows an order of magnitude improvement in the creep resistance. Such improvement may not be predicted based on a simple rule of mixtures.

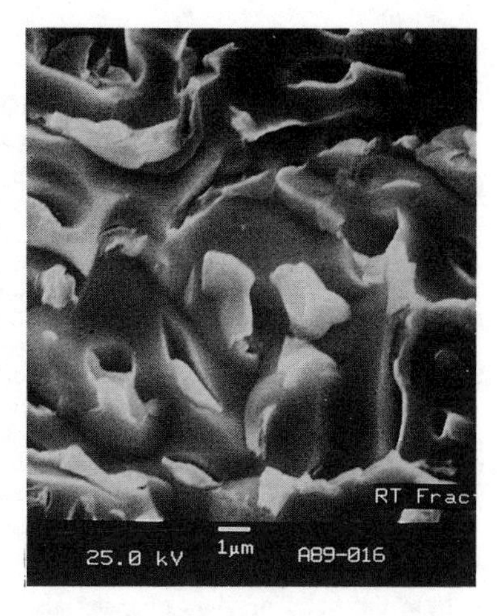

Fig. 7. Fractograph of Cr_2Nb/Nb showing plastic stretching of niobium particles.

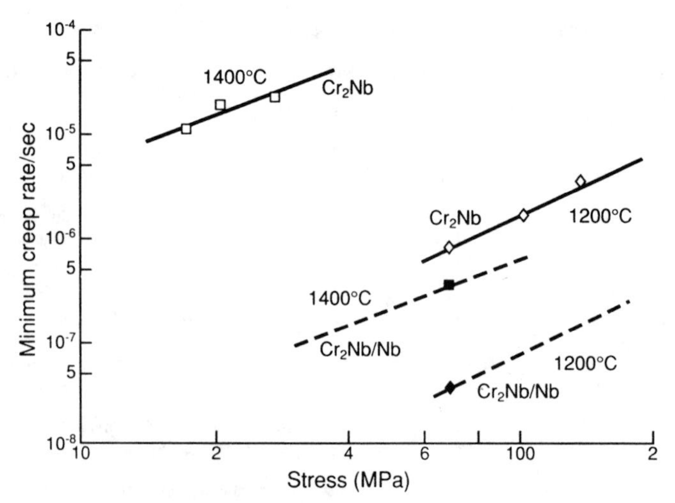

Fig. 8. Comparison of minimum creep rate versus stress.

Factors Controlling Ductile Phase Toughening

In a classical analysis of experimental work involving ductile lead wires constrained by thick-walled pyrex capillary tubing, Ashby et al. [13] have shown that a varying degree of constraint leads to a varying degree of toughening. Following this analysis, the increment of toughness over that of the matrix, ΔK_C, is given by:

$$\Delta K_C = E\,(\,C\,V_f\,(\sigma_o/E)\,a_o)^{1/2}$$

where E, V_f, σ_o and a_o are Young's modulus, volume fraction, tensile yield strength and radius of the ductile phase, respectively. The parameter, C, denotes the reciprocal of the degree of constraint. According to the relationship, the toughness increases with increasing E, V_f,σ_o and a_o.

As discussed by Mendiratta et al. [24], the in-situ composites, such as Nb\Nb_5Si_3, are much more complex than the model composite. In such systems, the intrinsic metallurgical changes such as recovery/recrystallization, solid solution strengthening and metastable phases affect the strength of the ductile phase. Secondly, the geometry/morphology of the microstructure can be significantly altered with minor variation in composition as well as heat treatment. Consider the micrographs for Nb/Nb_5Si_3 eutectic systems presented in Fig. 9 [24], where the system is intervened by a metastable Nb_3Si intermetallic. The as-cast microstructure in Fig. 9(a) consisting of primary-Nb particles(bright phase) and fine two-phase eutetic mixture of Nb_3Si(dark matrix) and Nb rods (bright phase), coarsens after a 100-hour heat treatment at 1500°C, as shown in Fig. 9(b). The high magnification examination of the structure presented in Fig. 9(c) shows the Nb_3Si phase completely transformed to the equilibrium Nb_5Si_3 and Nb phases via a eutectoid transformation.

One drawback of the elasto-plastic model is that it provides only partial mechanistic insight from the point of view of materials development. The above cited model provides no fundamental scale factor for the microstructure, as Burgers vector, which scales dispersion hardening. It is true that large ductile particles more effectively blunt the crack as we have observed in the Cr_2Nb/Nb composite [30]. However, in the same cited work it has also been observed that coarsening of the microstructure in Nb_3Al/Nb leads to increased propensity of cracking in the large region of the brittle intermetallic. As analyzed by Deve and Maloney [14], the matrix cracking in composite originates from existing flaw under a combination of applied external stresses and residual stresses. Residual stresses result from elastic and coefficient of thermal expansion (CTE) mismatch, which fortunately is expected to be low in in-situ composites. Nonetheless, the propensity for spontaneous cracking in the matrix presents a mitigating influence to enhanced fracture toughness. The success of the in-situ composite is likely to lie in the nonlinear effects associated with micron scale of their microstructures.

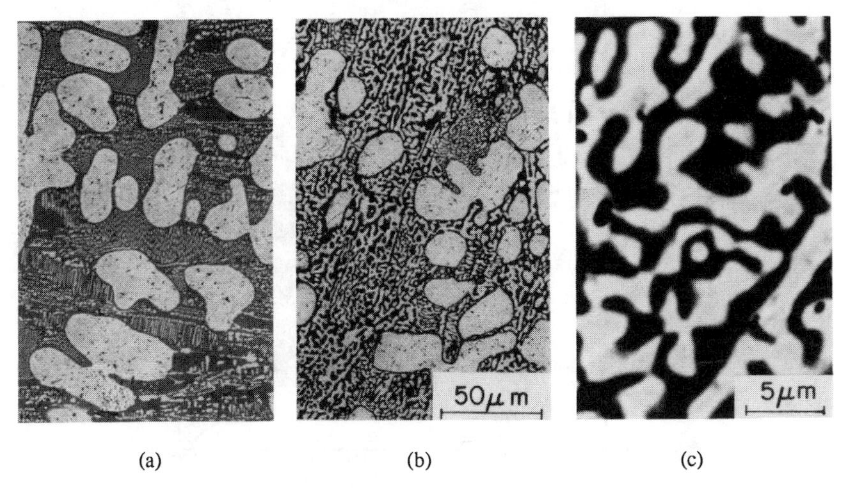

Fig. 9. Microstructure of the cast Nb-10Si alloy: (a) as cast; and 1500°C/100 hours at (b) low, and (c) high magnification.

Another factor that improves fracture toughness, as the model suggests, is 'limited decohesion' of the ductile phase. While this issue may appear conceptually simple in artificial composites where fiber coating is considered the panacea, it may indeed be better controlled in in-situ composites. The nature of interface may be varied from coherent to incoherent depending on the phase separation mechanisms, alloying and minor element additions. Obviously, the mechanistic models are not expected to provide any insight into these aspects.

Alloyed In-situ Composites: The importance of alloying in-situ composites for achieving a balance of engineering properties cannot be overemphasized. To achieve this, there is a clear need to identify a robust two-phase region. With reference to Fig. 6, while considerable work has been done on solid solution refractory alloys, only preliminary ternary diagrams exist covering two-phase regions with intermetallics.

To grasp the complexity of phase relationships, consider the qualitative attempt to construct the Nb-Mo-Si ternary diagram in Fig. 10, near the refractory metal terminal solid solution. The known equilibrium ternary phase diagram at 1200°C suggests that, in excess of 20 a/o Nb, the A15 phase is not stable. This is consistent with the known Nb-Si binary phase diagram. However, while the Mo_3Si(A15) phase forms in the hypo-eutectic region, the metastable Nb_3Si forms in the hyper-eutectic region. As shown in Fig. 10, the combination of the known data suggests that the intermetallic $(Mo,Nb)_3Si$ with A15 structure must intersect the eutectic trough between the Nb-Mo solid solution and the 5-3 silicide. If the qualitative analysis is correct, it is interesting to consider the possibility of a eutectic composition being an intermetallic, subsequently transforming to a ductile phase toughened composite by eutectoid transformation. Once again , the dual opportunity to control the microstructure through solidification as well as in solid state demonstrate the potential for generating fine scale microstructure and an associated balance of engineering properties.

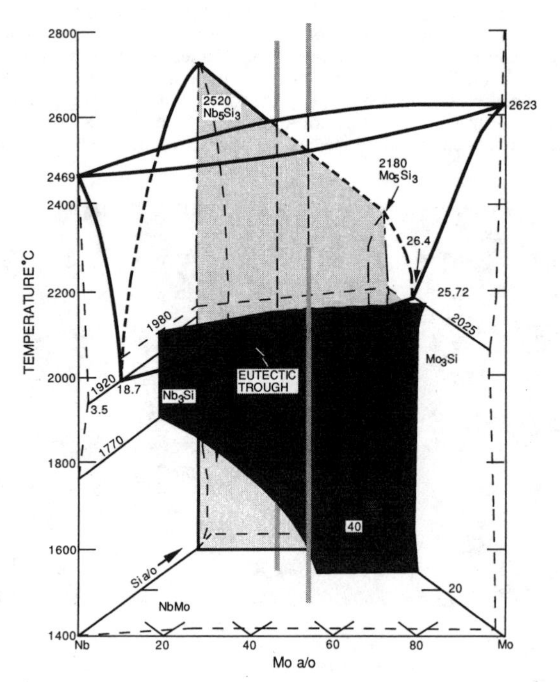

Fig. 10. A qualitative construction of Nb-Mo-Si ternary phase diagram.

SUMMARY

1. The approaches for in-situ synthesis of intermetallic matrix composites broadly fall in to two groups. The desirable two phase microstructure is either produced using natural phase separation mechanisms or synthetic means by which typically laminate structures are produced. Unlike artificial - typically fiber reinforced composites- the in-situ approaches provide a finer scale microstructure with significantly better control. Though conceptually complex, the natural in-situ composites once developed are easily transferred to practice at low cost and high reliability.
2. Since the natural in-situ processing relies on phase separation mechanisms, such as precipitation due to sloping phase boundaries and eutectic solidification, a specific knowledge of phase relationships in the context of refractory metals and high melting intermetallics is required to develop high temperature structural materials beyond nickel-base superalloys. The knowledge base of nickel-base superalloys and nickel-base eutectics provides a useful guide, but the new challenge of containerless processing of materials at high temperature must be undertaken.
3. Between the brittle-brittle and ductile-brittle combination of intermetallic matrix composites, the latter show potential for significantly improving the room temperature fracture toughness. The current theoretical understanding of the ductile phase toughening provides no absolute scale factor for the microstructure.
4. To achieve a balance of engineering properties, a better understanding of the multiphase alloys based on refractory intermetallics and refractory metals is required.

ACKNOWLEDGEMENT

Part of this work was supported by U.S. Air Force Contract F33615-90-C-5957. Thanks are also due to Drs. R. G. Rowe and M. Mendiratta for providing the original micrographs of microlaminate and in-situ composite, respectively.

REFERENCES

1. J. H. Brahney, Aerospace Engineering, August 1990, 17.
2. D. L. Anton and D. M. Shah, in High Temperature Ordered Intermetallic Alloys IV, edited by L. A. Johnson, D. P. Pope and J. O. Stiegler (Mater. Res. Soc. Proc. 213, Pittsburgh, PA 1991)p.733.
3. D. M. Shah and D. L. Anton in High Temperature Ordered Intermetallic Alloys IV, edited by L. A. Johnson, D. P. Pope and J. O. Stiegler (Mater. Res. Soc. Proc. 213, Pittsburgh, PA 1991)p.63.
4. R.L.Fleisher, C..L. Briant and R. D. Field in High Temperature Ordered Intermetallic Alloys IV, edited by L. A.Johnson, D. P. Pope and J. O. Stiegler (Mater. Res. Soc. Proc. 213, Pittsburgh, PA 1991)p.463.
5. D. L. Anton and D. M. Shah, This Proceedings.
6. R. G. Rowe and D. W. Skelly, This Proceedings.
7 M. F. Ashby , in Strengthening Methods in Crystals, edited by A. Kelly and R. B. Nicholson (Elsvier, New York, 1971)
8. G. K. Bouse and J. R. Mihalisin in Superalloys, Supercomposites and Superceramics, edited by J.K. Tien and T. Caulfield (Academic Press, Boston, 1989), p.105.
9. David J. Elliott, Microlithography Process Technology for IC Fabrication, McGraw-Hill, New York, 1986.
10. Faquir C. Jain, University of Connecticut, (Private Communication).
11. J. Koskinen and H. H. Johnson, Material Research Society Symposium Proceedings, 130,1989.
12 Ken M. Gettelman, "Stereolithography: Fast Model Making," Modern Machine Shop, October 1989, pp. 100-107.
13. M. F. Ashby, F. J. Blunt and M. Bannister, Acta Metall., 37, 1847-1857, (1989).
14. H. E. Deve and M. J. Maloney, Acta Metall. Mater., 39, 2275-2284, (1991).
15. H. C. Cao et al., Acta Metall., 37, 2969-2977, (1989).
16. D. M. Shah and D. L. Anton, Materials Science and Engineering A, A153, 402-409,(1992)
17. S. Mazdiyasni and D. B. Miracle in Intermetallic Matrix Composites edited by D. L. Anton, P. L. Martin, D. B. Miracle and R. McMeeking, (Mater. Res. Soc. Proc. 194, Pittsburgh, PA 1990)p.155.
18. G. Bertero, W. H. Hofmeister, M. B. Robinson and R. J. Bayuzick, Met.Trans. A, 22, 2713(1991).
19. D. Johnson, S. Joslin and Ben F. Oliver, This Proceedings.
20. Keh-Minn Chang, This Proceedings.
21. D. P. Pope and W. J. Romanow (Private Communication).
22. R. Darolia, JOM, March 1991, 44.
23. P. R. Subramanian et al. in Intermetallic Matrix Composites edited by D. L. Anton , P. L. Martin, D. B. Miracle and R. McMeeking, (Mater. Res. Soc. Proc. 194, Pittsburgh, PA 1990)p.147.
24. M. G. Mendiratta, J. J. Levandowski and D. M. Dimiduk, Metall. Trans. A, 22, 1573(1991).
25. G. Frommeyer, R. Rosenkranz and C. Ludecke, Z. Metallkde, 81, 30(1990).
26. M. Es-Souni et al. in Prc. Int. Symp. on Intermetallic Compounds (JIMIS-6), The Japan Institute of Metals, Sendai, 1991, p.525.
27. G. Sauthoff, Materials Science and Engineering, 1992, To be published.
28. D. Mason and D. Van Aken, in "High Temperature Intermetallic Matrix Composites," DOD-G-AFOSR-90-0141 Annual Progress Report, March 1992, p.15.
29. D. J. Thoma and J. H. Perepezko, Material Science and Engineering, A155,1992, (To be published).
30. D. L. Anton and D. M. Shah in Intermetallic Matrix Composites edited by D. L. Anton , P. L. Martin, D. B. Miracle and R. McMeeking, (Mater. Res. Soc. Proc. 194, Pittsburgh, PA 1990)p.45.

EVALUATION OF VARIOUS INTERMETALLIC MATRIX - CERAMIC PARTICLE SYSTEMS FOR MELT PROCESSING OF METAL MATRIX COMPOSITES

S. Sen, B. K. Dhindaw and D. M. Stefanescu
Solidification Laboratory, Metallurgical and Materials Engineering, The University of Alabama, Tuscaloosa, Al 35487

ABSTRACT

The possibility of incorporating SiC or TiC particulate reinforcements in a Ni_3Al matrix through the melt processing route was examined. Interface bonding and thermal stability of SiC and surface treated SiC in Ni_3Al has been investigated through sessile drop and DTA experiments. Feasibility of various processing techniques such as the mechanical mixing method and the vacuum infiltration method was also explored.

INTRODUCTION

Aluminum based intermetallic compounds inherently possess low density, high melting temperature, superior high temperature strength, and oxidation resistance. Consequently they are potential materials for high temperature structural and propulsion applications. The intergranular brittleness found in the majority of the polycrystalline aluminides has to a certain extent been overcome by micro and macro alloying, and grain refinement [1,2]. A typical example is the Ni_3Al system. Addition of 0.08 wt.% B to an alloy of Ni - 24 at. % Al yielded tensile elongation of up to 50% [3]. Hence it can be envisioned that introduction of ceramic reinforcements in such intermetallic systems would produce a composite with lower density, improved tensile strength, and stiffness. In this respect intermetallic matrix strengthened with fibrous or particulate reinforcements has recently been an area of considerable interest.

The majority of the intermetallic matrix composite systems that have been investigated to date were fabricated by conventional powder metallurgy routes involving mechanical alloying, sintering, HIPing, and extrusion[4]. Limited efforts have also been directed towards production of intermetallic matrix composites by the pressure casting route [5]. While developing a melting and casting route for such composites, one has to consider the physico-chemical characteristics of the particles in relation to the matrix for evaluating the processing feasibility, i.e., the ease with which the reinforcements can be incorporated in the liquid metal, their subsequent distribution in the matrix and the reinforcement - matrix bonding. The ease of incorporation and bonding is both related to the wettability and the extent of reaction taking place between the matrix material and the reinforcement. Wettability is quantitatively described by the contact angle between the molten metal and the reinforcements [6]. A reaction zone between the matrix and the reinforcement is desirable to a certain extent to ensure superior bonding between the two. However, excessive chemical reaction between the reinforcements and the matrix may result in detrimental effects on the mechanical properties of the composite. Previous works have reported the characteristics of the reaction between certain ceramic reinforcements and Ni_3Al matrix mostly at temperatures up to 1000°C [7]. Data on the behavior expected at the melt temperatures of Ni_3Al are not available. These data are essential if normal casting routes are to be explored for the manufacturing of these composites.

In the present work the interface behavior has been characterized by contact angle measurements in a sessile drop apparatus, and by the differential thermal analysis technique. Both mechanical mixing and in-situ techniques have been used for the preparation of the composites. The matrix alloy was Ni_3Al and the reinforcements studied were SiC and TiC. The choice of reinforcements was based on cost and on the strength to weight ratio.

EXPERIMENTAL PROCEDURE

The matrix - ceramic systems that were investigated are summarized in Table 1.

Table 1. Matrix -ceramic systems used for sessile drop experiments and for composite preparation

Matrix	Reinforcements	Substrate type for wettability study	Particle size for wettability study µm	Particle size for composite preparation µm
Ni_3Al	$SiC^{a,b,}$	HIPed	2 - 5	30 - 50
Ni_3Al	TiC	HIPed	5 - 7	5 - 7

a: substrate without any treatment;
b: substrate surface oxidized at 1000 °C for 20 hours

Contact angle measurements:

The Ni_3Al alloy used for the experimentation had a nominal composition of 87.5 wt. % Ni, 12.42 wt %. Al, and .08 wt % B. This alloy was produced in a Marko 200T vacuum arc furnace in the form of buttons. Cylindrical samples having a diameter and a height of 5mm were cut out of these buttons. Finally, these samples were polished to a mirror finish and used for contact angle measurements.

The ceramic substrates used for the contact angle measurements were placed in alumina crucibles having a height of 6mm and a diameter of 30mm. The alumina crucibles containing the substrates were positioned inside a sessile drop furnace and the cylindrical aluminide samples were placed on top of the substrates. The furnace chamber was evacuated to 5×10^{-4} Torr and then flushed with high purity argon for 5 minutes. This process was repeated twice, and after the third evacuation the furnace chamber was backfilled with argon up to a pressure of 1psi. The furnace was then heated to the melting temperature of Ni_3Al in 25 minutes, and as soon as the aluminide started melting five pictures were taken at intervals of 3 minutes. Similar experiments were performed with superheats of 100°C and 200°C to investigate the effect of temperature on wettability. These pictures were then projected on a flat screen TV monitor and the contact angles were measured using an Olympus image analyzer.

SEM and EDAX analysis were performed at the interface between the metal and substrate to obtain an indication of the extent of reaction between the two.

Differential thermal analysis (DTA):

DTA studies were performed on a DuPont 2000 instrument. The samples consisted of pulverized Ni_3Al ribbons mixed with SiC particulates in the as received form, or with SiC particulates surface oxidized at 1000°C for a period of 20 hours. The samples were run against alumina at a heating and cooling rate of 5°C/min between 1100°C and 1500°C.

Preparation of Composites:

Composites were prepared by two different methods, namely liquid mixing to create a vortex and vacuum infiltration.

a) Liquid Mixing Method

Ni_3Al with 0.08 wt. % B was prepared in a vacuum induction melting unit as shown in figure 1. Once the temperature of the melt was stabilized at 1500 - 1550°C the stirrer was lowered into the melt and the ceramic particulates were added through a steady flow of argon.

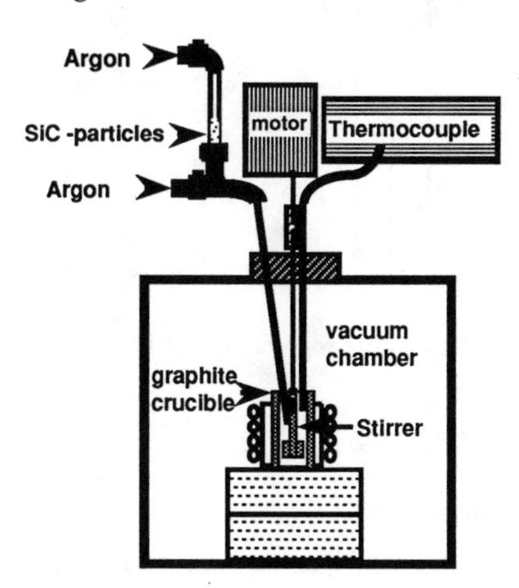

Figure 1. Schematic of vacuum induction melting unit used for preparation of composites.

The ceramic particulate additions were SiC with or without surface treatments. Their size range was typically 30 - 50 μm. The stirrer, rotating at approximately 350 rpm, created a vortex to ensure incorporation of the particles in the melt. This stirring action was carried out for a period of 1.5 - 3.0 minutes following which the melt was poured in a cast iron mold.

b) Infiltration Method

The melt infiltration method was used as an alternative approach for fabrication of the SiC reinforced composites. Figure 2 shows a schematic of the liquid metal infiltration process that was used. SiC particles 300 μm in size were used either in the as received form or after being surface oxidized. Typical castings produced by this method were 12mm in diameter and 6mm in thickness. Sectioned surfaces of these castings were examined through optical microscopy, SEM, and EDAX.

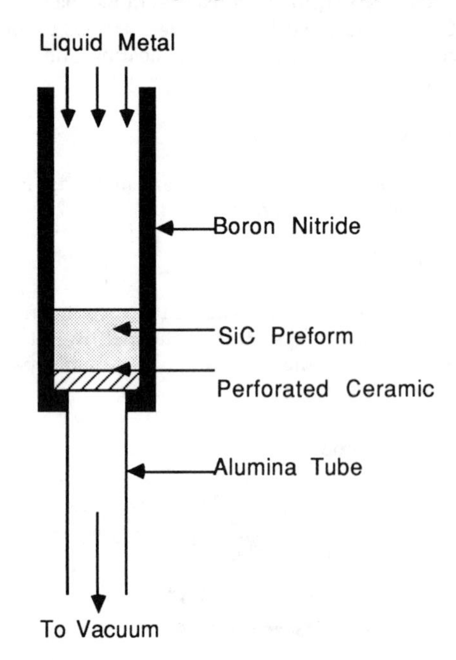

Figure 2. Schematic representation of the infiltration set up.

RESULTS AND DISCUSSIONS

Contact Angle Measurements

Contact angle is used as a measure of wettability between a molten metal or alloy and a ceramic substrate. Systems with $\theta \leq 90^{o}$ are considered to be wetting and systems with $\theta > 90^{o}$ are considered to be non - wetting.

Figures 3 and 4 show the variation of contact angle of Ni_3Al on SiC and surface oxidized SiC as a function of time and temperature.

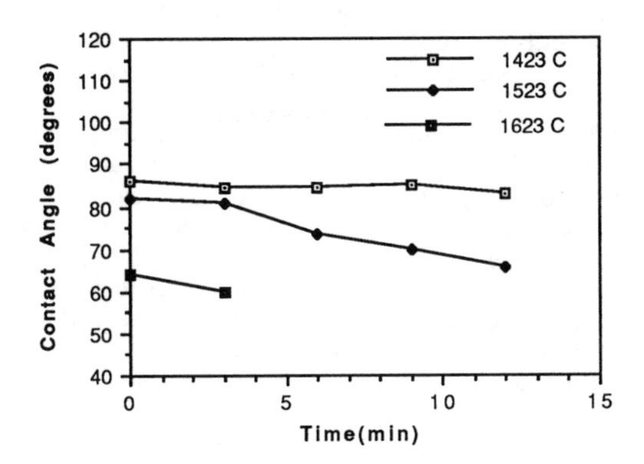

Figure 3. Influence of temperature and time on the contact angle of Ni_3Al on SiC.

As soon as the sessile drop is formed at 1423°C a contact angle of approximately 88° is obtained indicating a wetting behavior. At 1523°C the contact angle decreases to almost 70° over a period of 12 minutes. Further, as the furnace temperature was increased from 1423°C to 1523°C a gradual decrease in the height of the sessile drop was noticed indicating a progressive reaction between Ni_3Al and the SiC substrate. Finally, when raising the temperature of the furnace to 1623°C, the molten metal completely sank into the substrate after a period of 3 minutes as a result of extensive reaction between Ni_3Al and the substrate.

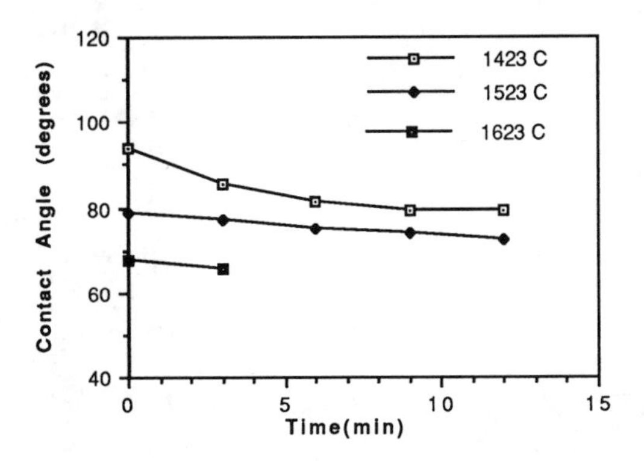

Figure 4. Influence of temperature and time on the contact angle of Ni_3Al on surface oxidized SiC.

When oxidized SiC was used as a substrate the contact angle was a few degrees higher than that on SiC without any surface treatment (Fig.4). This difference in behavior is probably due to the protective layer formed on SiC during oxidation which in the initial stages prevents any extensive reaction between the molten metal and SiC. However, at a higher temperature of 1623°C the reaction kinetics between Ni_3Al and SiC is very rapid. There is complete reaction between the molten metal and the substrate after approximately 3 minutes and no contact angle measurement was possible beyond this time as indicated in figure 4.

Figures 5(a) and (b) show the interfacial reaction between Ni_3Al and SiC, and surface oxidized SiC, respectively. The areas marked (1), (2), and (3) in each of the micrographs corresponds to the Ni_3Al, the reaction zone, and the substrate, respectively. EDAX analysis on these reaction zones indicated the formation of Ni_2Si on the intermetallic side of the interface and NiSi on the substrate side of the interface. However, it is interesting to note that there is a significant difference in the width of the reaction zone for Ni_3Al on SiC and surface oxidized SiC, as further presented in figure 6. Figure 6 supports the conclusion that the surface oxidation of SiC is instrumental in slowing down the intensity of the chemical reaction to a certain extent.

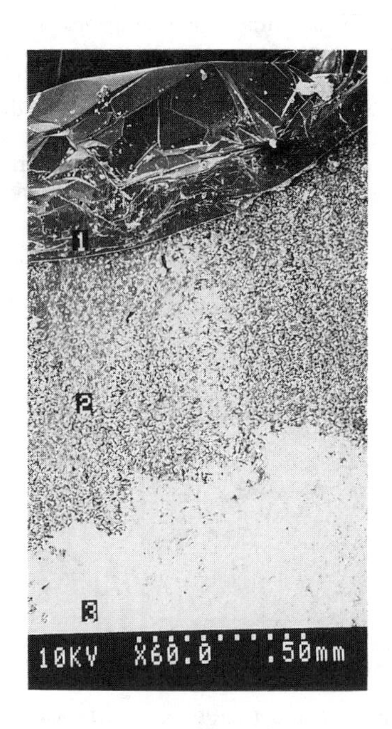

a)

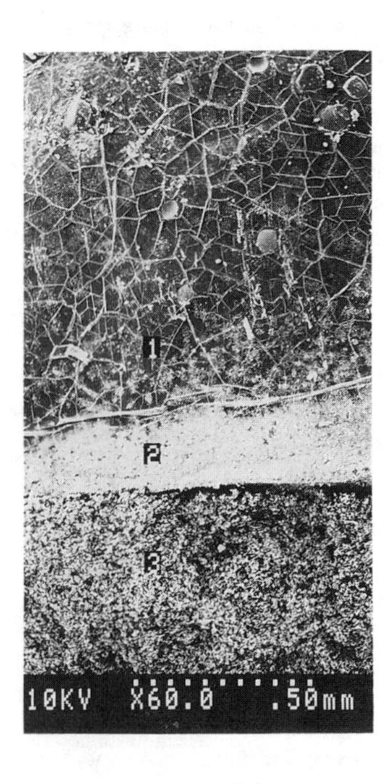

b)

Figure 5. SEM micrographs of interfacial reaction between Ni_3Al and SiC during sessile drop experiments at 1523°C: a) on as - received SiC, b) on surface oxidized SiC.

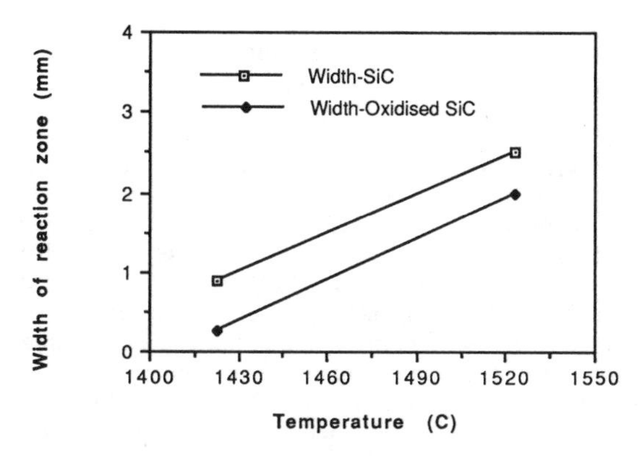

Figure 6. Comparison of width of reaction zone for Ni_3Al on SiC and surface oxidized SiC.

Ni_3Al reacted completely with the TiC substrate even at the low temperature of 1423 °C and consequently no contact angle measurement was possible. As shown in figure 7 there is complete reaction between the metal and the substrate. Accordingly, it can be concluded that it is impossible to incorporate TiC particles in a Ni_3Al matrix through the melt processing route.

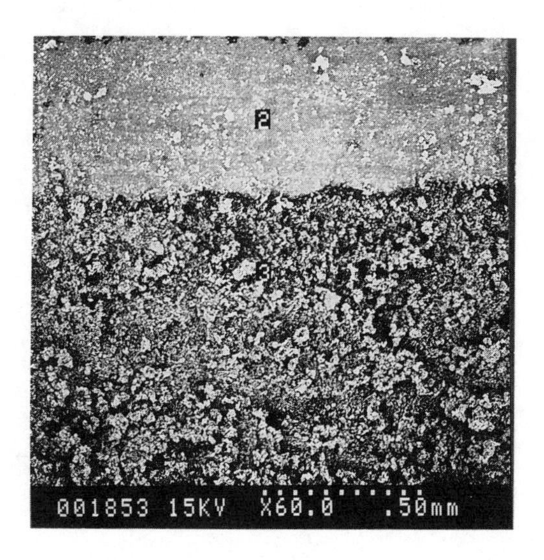

Figure 7. SEM micrographs of interfacial reaction between Ni_3Al and TiC during sessile drop experiments at 1423°C.

Differential Thermal Analysis (DTA) Results

To further confirm the beneficial effect of surface oxidation of SiC, DTA experiments were performed. Figure 8 shows the DTA curves for Ni_3Al and SiC, and for Ni_3Al and surface oxidized SiC, respectively. The heating curve for the Ni_3Al - SiC system exhibits an endothermic peak at 1377°C which corresponds to the melting of the intermetallic (Fig.8a). After the end of melting another series of exothermic events can be seen on the heating curve. They can be attributed to an extensive reaction between Ni_3Al and SiC. The cooling curve is rather complex, and does not show a clear melting point. A number of reaction peaks are observed. They indicate that several reaction products between Ni_3Al and SiC are precipitated during cooling. On the contrary, the heating/cooling curve for the system containing the oxidized SiC shows clear melting and solidification events (Fig.8b).

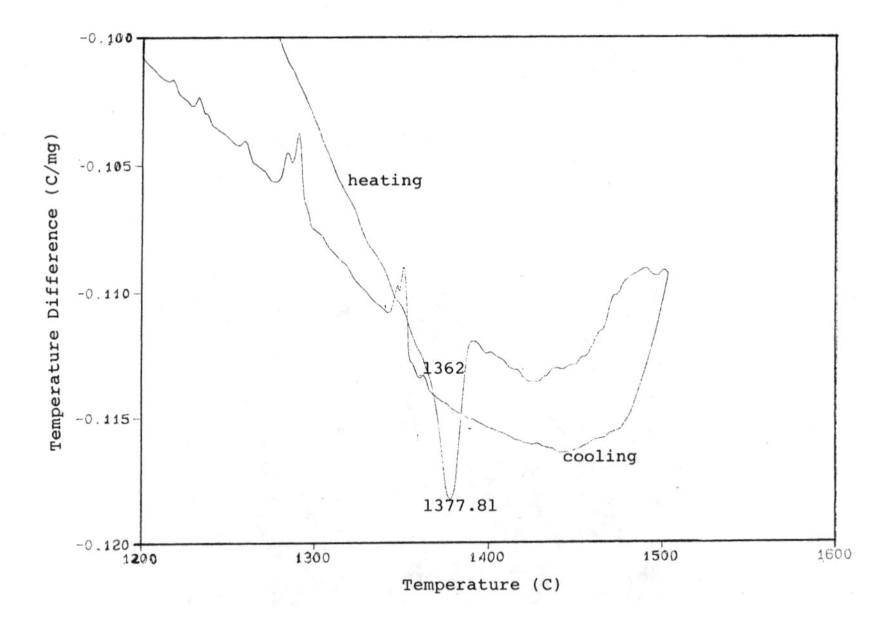

Figure 8 (a). DTA result of Ni_3Al mixed with as - received SiC.

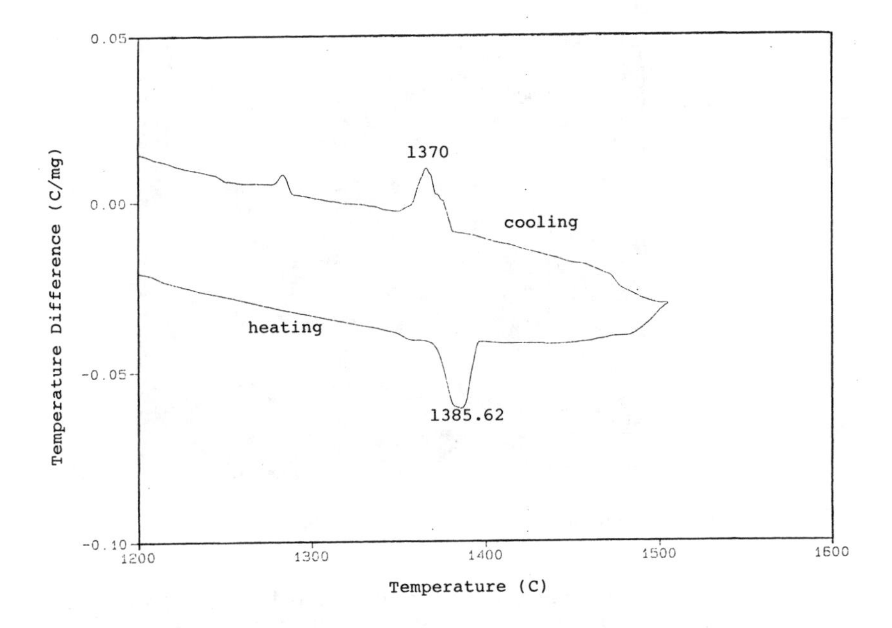

Figure 8(b). DTA result of Ni_3Al mixed with surface oxidized SiC.

Microstructural Evaluation of Composites

Ni_3Al + SiC

Typical castings were in the form of rods 12.5 mm in diameter and 20 mm in length. Longitudinal sections from different parts of these rods were prepared for metallographic examination using an image analyzer, and also for SEM, and EDAX analysis.

SiC particulates 30 - 50 μm in size were dispersed in the Ni_3Al matrix using the liquid mixing method. Although very short mixing time of 1.5 and 3.0 minutes were used, there was complete dissociation of the SiC particles, irrespective of the surface treatment given to the particles. The dissociation of SiC resulted in graphite nodules as shown in figure 9. EDAX analysis revealed 1.5 wt.% of Si in the matrix.. Hence it can be concluded that at melt temperatures of 1500°C a 50 μm SiC particle completely dissociates into Si and C. Therefore, under the present experimental conditions, liquid mixing cannot be used to produce Ni_3Al - SiC composites, even when using oxidized particles. It is necessary to employ a process which will allow only short contact time between the particles and the molten metal.

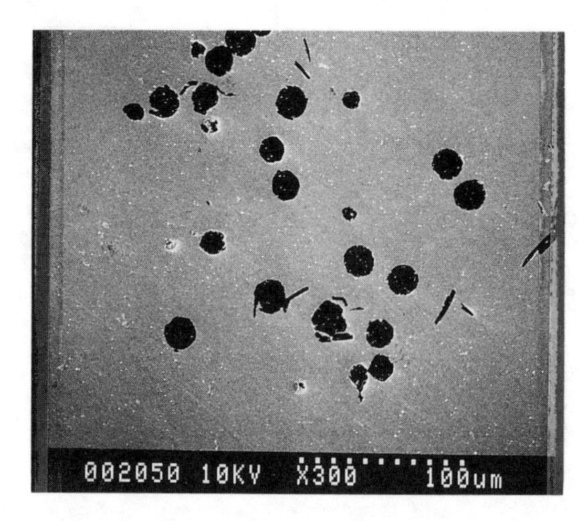

Figure 9. Graphite nodules formed in molten Ni_3Al after decomposition of SiC.

As an alternative approach the vacuum infiltration method was used. Figures 10(a) and (b) are SEM micrographs showing the interface between Ni_3Al and SiC, and Ni_3Al and oxidized SiC, respectively, for the vacuum infiltrated samples. It can be seen that a smoother interface is obtained between Ni_3Al and oxidized SiC as compared to Ni_3Al and SiC.

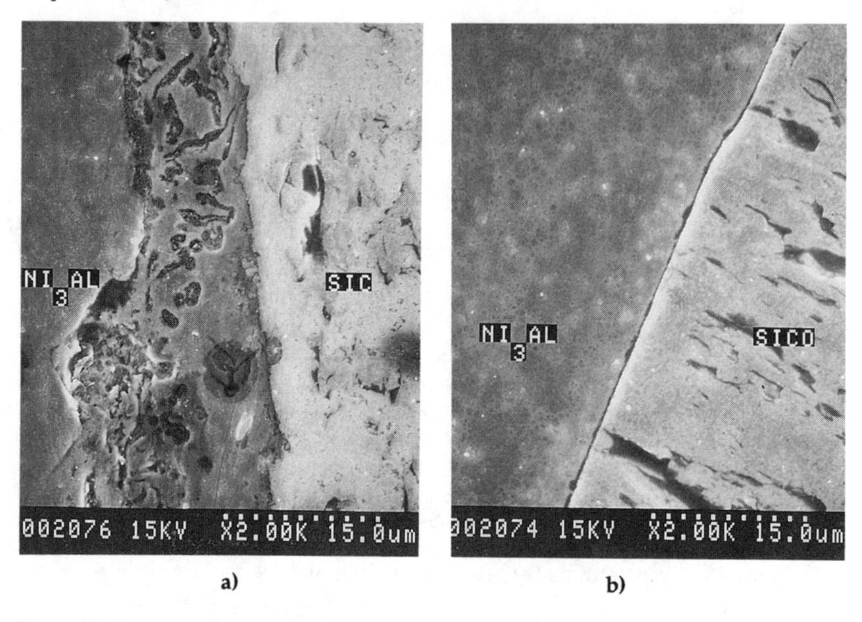

Figure 10. Interface between Ni_3Al and: a) as - received SiC, b) surface oxidized SiC, in vacuum infiltrated samples

The above results can be rationalized as follows. The thermodynamic driving force for the Ni_3Al/SiC reaction is the highly negative Gibbs free energy of the Ni-Al-Si ternary phase. However the overall Ni_3Al/SiC reaction is apparently limited by the decomposition of SiC [7]. A barrier on the SiC surface, as for example SiO_2, can retard Ni diffusion into SiC, and thus slow down the reaction. This explains the behavior of oxidized SiC during sessile drop and DTA experiments, where reduced reaction was observed between SiC and Ni_3Al. However, when the experimental conditions favored faster kinetics, as during liquid mixing experiments, even oxidized SiC particles ultimately reacted.

SUMMARY

Both sessile drop and DTA experiments have illustrated that compared to SiC, surface oxidized SiC particles have higher stability in contact with Ni_3Al matrix at temperatures of up to 1523°C. However, the reaction kinetics are still so rapid that even oxidized 50 μm SiC particulates are completely decomposed in the Ni_3Al melt during liquid mixing. Consequently, apart from oxidizing the SiC particulates, it is also necessary to employ a process reduces the particle - melt contact time. In this respect the melt infiltration technique appears to be the most feasible process.

The reaction between TiC and molten Ni_3Al was so rapid that it precludes fabrication of Ni_3Al/TiC particulates composites through liquid mixing.

REFERENCES

1. I. Baker and P.R. Munroe: *J. of Metals.*, 40, 28 (1988).

2. C.T. Liu, C.L. White, and J.A. Horton: *Acta Metall.*, 23, 213 (1985).

3. C.T. Liu and J.O. Stiegler: *Science*, 226, 226 (1984).

4. D.E. Alman and N.S. Stoloff: *High Temperature Ordered Intermetallic Alloys IV, MRS Symp Proc.*, L.A. Johnson, D.P. Pope, and J.O. Stiegler, eds., 213, 989 (1991).

5. S. Nourbakhsh, F.L. Liang, and H. Margolin: *Metall. Trans. A.*, 21A, 213 (1990).

6. J.T. Davies and E.K. Rideal: *Interfacial Phenomena*, (Academic Press, New York & London, 1961), p. 35.

7. T. C. Chou and T.G. Neih: *J. Mater. Res.*, 5 (9), 1985 (Sept 1990).

8. J.M. Yang, W.H. Kao, and C.T. Liu: *Metall. Trans. A.*, 20A, 2459 (1989).

THE SYNTHESIS AND EVALUATION OF Nb_3Al-Nb LAMINATED COMPOSITES

R.G. ROWE and D.W. SKELLY

GE Corporate Research and Development, Schenectady, NY 12309

ABSTRACT

Microlaminates of Nb_3Al and Nb were synthesized in-situ by high rate magnetron sputtering. Three composites were fabricated. They had thicknesses ranging from 19 to 146 μm (5.5 mils) with 11 to 91 layers. The volume fraction of the Nb lamellae was approximately 0.4 for two and 0.5 for the third. Room temperature fracture of the composites revealed that some cleavage cracking in the Nb_3Al intermetallic layers was arrested by the ductile Nb layer. The Nb layers failed by chisel point fracture and shear.

INTRODUCTION

High temperature intermetallic systems, which have strength and resist creep deformation at temperatures above 1000°C are brittle at low temperatures. [1] One approach to the utilization of these high temperature intermetallics is to combine them with ductile alloys for toughening at low temperatures. The toughening of a brittle material by ductile filaments or sheets has been demonstrated theoretically and experimentally for a number of systems. [2-5] In addition to toughening, the low temperature strength of a composite with a given fracture toughness can be increased by reducing the size of the largest cracks or defects. Since cracking across the intermetallic layer is the most likely defect, this can be accomplished by reducing the thickness of the intermetallic lamellae. The disadvantage of thin intermetallic layers is high interstitial element contamination for fabrication processes that begin with thin laminates or fine powder. Interstitial contamination causes higher ductile to brittle transition temperatures in both the toughening and the intermetallic phases. [6-8]

In-situ fabrication is one technique that can be used to produce low interstitial high temperature intermetallic laminated composites with thin intermetallic layers. Vapor phase deposition by high rate magnetron sputtering provides a means of producing low interstitial laminates with the freedom to control ductile and intermetallic phase compositions, volume fraction and laminate dimensions independently. This paper describes the synthesis and evaluation of multi-layer laminates of the intermetallic composite system Nb_3Al-Nb.

EXPERIMENTAL PROCEDURE

Figure 1. is a schematic of the magnetron sputtering process that was used to produce intermetallic Nb_3Al plus ductile Nb laminates. The two layers were produced by alternately sputtering from two different targets. An MRC 603 magnetron sputtering system, with Inset® magnetron cathodes was used. The sputtering target was produced by inserting segments of aluminum with bars of pure niobium as in Figure 2. The target composition was adjusted by changing the width of the aluminum segments. A target with 20 vol.% Al gave a film composition of Nb-21 at.% Al.

An 11-layer composite of Nb and Nb-20Al (Nb_3Al) was fabricated by sputtering onto a sapphire substrate. The composite consisted of six layers of Nb and five layers of Nb-20Al. The thickness of the Nb layers was 1.6 μm, and the Nb-20Al layers were 2.2 μm thick. The volume fraction of Nb-20Al was 58%, and the total sheet thickness was 19 μm. A thicker 91-layer Nb-(Nb-21Al) composite was produced by deposition onto a stainless steel substrate coated with a three-layer titanium-copper-titanium release layer. The 91-layer composite was applied with Nb and Nb-21Al layer thicknesses of 1.2 and 1.8 μm, respectively. It was

0.14 μm thick and exhibited no sign of cracking or buckling. A third run produced composites of 32 and 67 layers with equal Nb and Nb_3Al layer thicknesses of 2.3 μm.

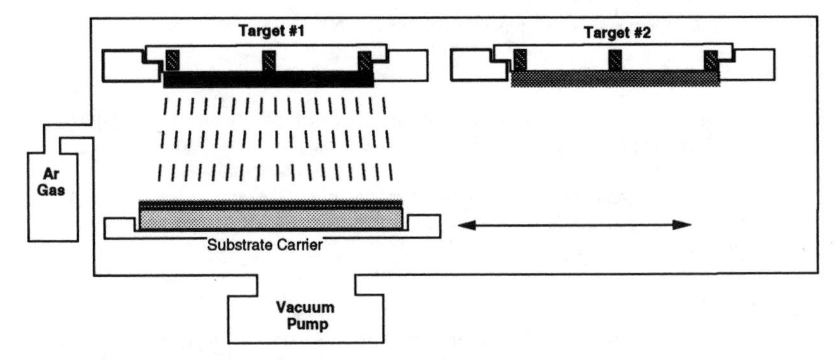

Figure 1. Schematic of sputtering system for synthesizing laminated metal-intermetallic composites by alternate sputtering from two targets.

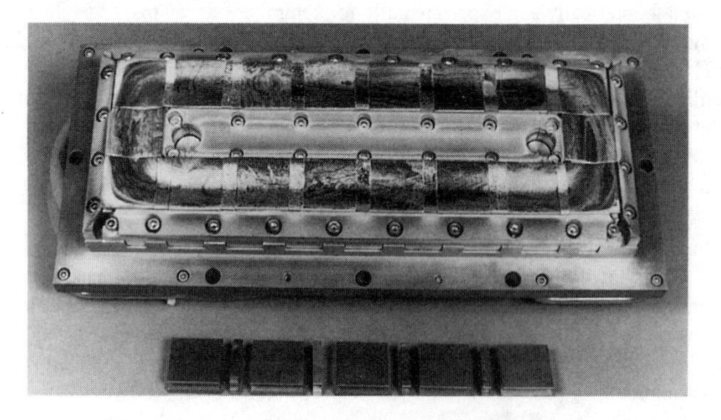

Figure 2. Racetrack sputtering target configuration

Flat free standing composite tensile samples were produced by EDM machining of the laminate on the stainless steel substrate followed by laminate removal by acid dissolution of the copper release layer. Tensile samples with a gage section of 13 mm long by 4.25 mm wide were tested by bonding them between tantalum sheet strips that were gripped in conventional tensile grips. The tantalum strips were used to accommodate misalignment strain in the tensile testing fixture.

EXPERIMENTAL RESULTS

Oxygen affects the low-temperature ductility of the metal phase and the DBTT of the intermetallic phase. An oxygen content of 1 at.% reduces the ductility of commercially pure Nb by a factor of two. Two at.% oxygen reduces it by another factor of two (to less than 5% elongation). [6] The residual vapor pressure of oxygen in vapor deposition systems (in the form of H_2O and CO) and the rate of deposition determine the level of oxygen contamination of the deposited material. It has been reported that the oxygen content of sputter deposited Nb which was deposited in a system with a background pressure of $1x10^{-6}$ Torr is as high as

8-12 at.%. [8, 9] This is consistent with an estimate of 12 at.% based upon the oxygen and metal atom impingement rates for a deposition rate of 10 Å/sec and a residual background pressure of 10^{-6} Torr. The background pressure of oxygen in the Magnetron sputtering system employed in the current study was low enough to produce low oxygen content $Nb-Nb_3Al$ multilayer composites. The oxygen contents of several laminated composites which were determined by neutron activation analysis are shown in Table 1.

Table 1. Oxygen Content of Sputtered Laminates

Metal Composition at.%	Intermetallic Composition at.%	Oxygen Content wt. ppm
Nb	Nb-18Al	730
Nb-35Ti-7Al	Nb-17Ti-23Al	200
Nb-27Ti-8Al		530

Figure 3(a) is a metallographic section of a 91-layer Nb-(Nb-21Al) composite laminate. The microstructure consisted of Nb and Nb-21Al layers with thicknesses of 1.2 μm and 1.8 μm, respectively. Both layers had fine columnar grain structures. Figure 3(b) is a fractograph of this laminated composite which shows conical growth defect structures. Growth of successive layers followed the surface contour of the previous layers, with the Nb layer filling most defects from previous layers. The Nb-21Al layers did not bridge defects above a minimum size, however, and conical defects were produced.

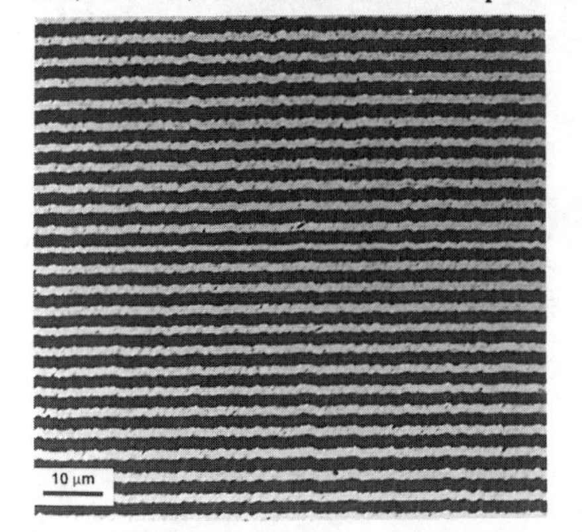

Figure 3(a) Scanning electron micrograph of the 91-layer Nb-Nb3Al laminated composite.

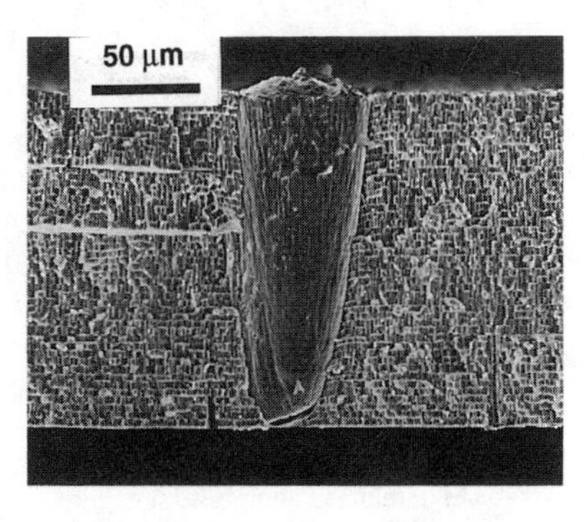

Figure 3(a) Scanning electron (negative backscattered electron) fractograph of bend fractured 91-layer composite revealing cone-shaped growth defects.

A sample of the 11-layer composite deposited on a sapphire substrate was fractured from the laminate side by bending. Figure 4(a) is a scanning electron fractograph of a region of this fracture surface. The Nb toughening layer appeared to be locally ductile, fracturing by shear (points denoted by A) and chisel-point separation (points denoted by B) between regions of intermetallic cleavage fracture. Cracking in the Nb-21Al lamellae was arrested by the Nb lamellae at points C in Figure 4(a). Bend fracture of the 32-layer composite that was fabricated with equal Nb and Nb-21Al layer thicknesses is shown in Figure 4(b). Cleavage fracture of the Nb-21Al layer was flat, and ductile Nb lamellae had chisel-point fracture profiles. There was no evidence of debonding between Nb-21Al and the ductile Nb lamellae.

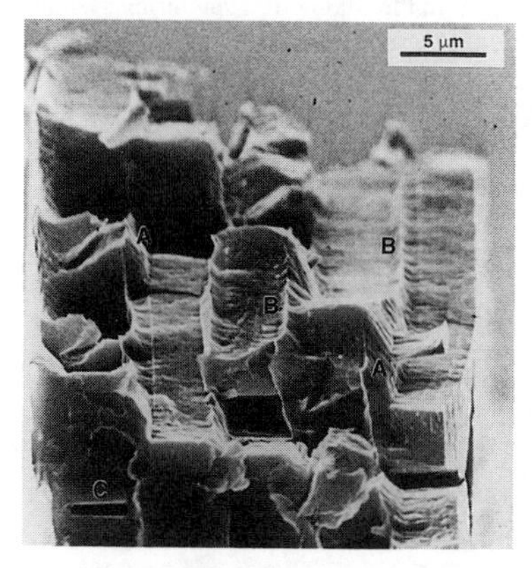

Figure 4(a) Fractography of the 11-layer Nb-Nb3Al laminated composite fractured by bending while attached to a sapphire substrate.

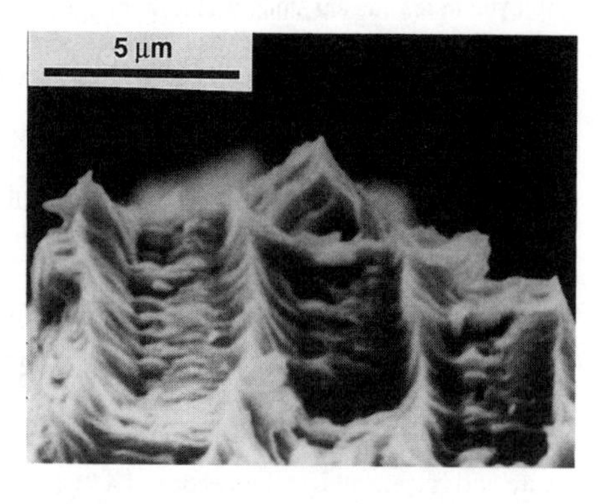

Figure 3(a) Scanning electron (negative backscattered electron) fractograph of bend fractured 91-layer composite revealing cone-shaped growth defects.

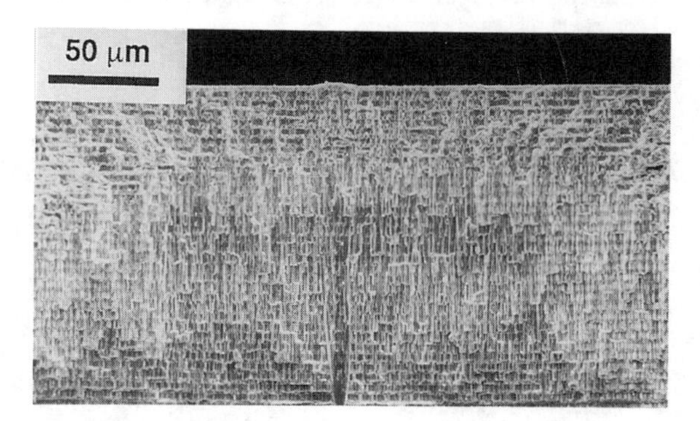

Figure 5. Scanning electron Fractograph of the tensile fracture surface of a 67-layer Nb-Nb3Al laminated composite showing the region of fracture initiation at the root of a cone-shaped growth defect.

The 67-layer composite was annealed one hour at 800°C to coarsen the as-grown columnar grain structure then tensile tested at room temperature. The sample failed with no indication of plasticity or subcritical fracture before failure. The sample failed at a mean stress of 335 MPa. A fractograph of the failed sample is shown in Figure 5. Initiation began at the center of a semicircular or thumbnail-shaped fracture region in the center of the gage section width. Initiation was at the base of a conical growth defect at the bottom growth surface. The initial semicircular fracture zone was regularly stepped from one lamella to the next indicating a macroscopic fracture path at a large angle to the gage section perpendicular. Beyond this semicircular initial crack, the fracture surface was macroscopically flat, lying in a plane that was close to the gage section perpendicular.

DISCUSSION OF RESULTS

Local ductility of the Nb lamellae was indicated by ductile dimpling and chisel-point fracture profiles in Figures 4(a) and 4(b) and the low oxygen content of the deposits was confirmed by analysis. This suggested that for the vacuum conditions of the system, interstitial contamination was negligible. Laminate structures were produced with good control of layer composition once the target configuration and the sputtering conditions were defined. The compositions of metal and intermetallic lamellae remained constant throughout multiple synthesis runs. This indicated that the technique of high rate sputtering was useful for synthesizing low interstitial intermetallic -metal laminates to evaluate the effect of laminate spacing and composition on composite toughening. Thick 91-layer laminates were produced in less than 16 hours of process time.

The growth defects that were observed were similar to defects associated with gap-filling failures in other vapor deposition experiments. [10] Particulate contamination internal to the sputtering chamber may be reduced by process changes to minimize spalling of previously deposited material onto the substrate.

Tensile and bend failure of the laminated composites was by defect-initiated fracture. The chisel-point fracture profile of the Nb lamellae was similar to that observed for metal films fractured under conditions of restrained plastic flow for other systems. [11] Examination of Figure 4 showed that there was no debonding between the metal and intermetallic lamellae or necking of the metal lamellae beyond the chisel-point area. It has been shown that debonding between intermetallic and metal lamellae increases the work of fracture and crack growth resistance. [2, 4, 11]

Failure of the Nb lamellae without debonding indicates that the strength of the Nb layer is low relative to the stress required for interface or transverse intermetallic fracture,or alternatively, the interface fracture energy between Nb and Nb_3Al is high.[12] The interfacial bond strength may be reduced to promote debonding or the strength of the ductile phase may be increased to impart a higher stress on the interface during fracture. Both may lead to an increase in debond length and necking of the ductile lamellae. Experiments to evaluate both modifications are underway. Deposition of a third layer between the Nb and Nb_3Al lamellae will be used to control the interfacial bond strength and alloying will be used to increase the strength of the ductile metal phase.

The laminated composites produced in this study had intermetallic and metal lamellar thicknesses ranging from 1.2 to 2.3 microns. Since it was expected that any cracks in the intermetallic layer would propagate entirely across the layer thickness, a small intermetallic laminate dimension was desired. On the other hand, optimization of elevated temperature properties would suggest a higher intermetallic volume fraction and greater lamellar thickness. It will be necessary to determine the effect of lamellar thicknesses and volume fraction on both high and low temperature properties to obtain a balance of properties. The technique of magnetron sputtering of laminated composites has the flexibility to control both intermetallic and ductile lamellar thicknesses without variations in composition and microstructure.

SUMMARY AND CONCLUSIONS

We have shown the high rate magnetron sputtering is a useful technique for in-situ fabrication of of low interstitial laminated metal-intermetallic composites. The interstitial level was below 730 wppm and the metal toughening films had ductile behavior at room temperature. The Nb-Nb_3Al laminated composite fractured with no debonding between the metal and intermetallic lamellae, and therefore did not have a high tensile work of fracture. Modification of the strength of the metal lamellae and the fracture energy of the interface between the metal and the intermetallic may be required to increase the room temperature fracture resistance. Although vapor phase growth defects were present in the laminates, it is expected that process refinement will minimize their occurrence. Their presence in these first laminates dominated the fracture behavior of both tensile and bend samples, however. Characterization of the fracture resistance may be more instructive than tests on smooth tensile samples to get the limit of fracture stress for small defect sizes.

REFERENCES

1. R.L. Fleischer, "Mechanical Properties of Diverse High-Temperature Compounds -- Thermal Variation of Microhardness and Crack Formation", Mat. Res. Soc. Symp. Proc., 133, 305-310, (1989).

2. H.E. Déve, A.G. Evans, G.R. Odette, R. Mehrabian, M.L. Emiliani and R.J. Hecht, "Ductile Reinforcement Toughening of Gamma-TiAl: Effects of Debonding and Ductility", Acta Metall. Mater, 38, 1491-1502, (1990).

3. L.S. Sigl, P.A. Mataga, B.J. Dalgliesh, R.M. McMeeking and A.G. Evans, "On the Toughness of Brittle Materials Reinforced with a Ductile Phase", Acta Metall. Mater, 36, 945-953, (1988).

4. H.C. Cao and A. Evans G,, "On Crack Extension in Ductile/Brittle Laminates", Acta Metall. Mater, 39, 2997-3005, (1991).

5. D.L. Anton and D.M. Shah, "Ductile Phase Toughening of Brittle Intermetallics", Mat. Res. Soc. Symp. Proc., 194, 45-52, (1990).

6. Anon., "Commercially Pure Nb", in Aerospace Structural Metals Handbook., MCIC, Battelle Columbus Laboratories: Columbus, OH, 43201. 1985, p. 4.

7. H. Conrad, "Effect of Interstitial Solutes on the Strength and Ductility of Titanium", Progress in Materials Science, 26, 123-403, (1981).

8. R.S. Bhattacharya, A.K. Rai, and M.G. Mendiratta, "Tailored Microstructures of Niobium-Niobium Silicides by Physical Vapor Deposition", Mat. Res. Soc. Symp. Proc., 194, 71-78, (1990).

9. D.A. Hardwick and R.C. Cordi, "Intermetallic Matrix Composites by Physical Vapor Deposition", Mat. Res. Soc. Symp. Proc., 194, 65-70, (1990).

10. R.F. Bunshah, Deposition Technologies for Films and Coatings, pp. 140-143. 1982, Park Ridge, NJ, Noyes Publications.

11. M. Bannister and M.F. Ashby, "The Deformation and Fracture of Constrained Metal Sheets", Acta Metall. Mater., 39, 2575-2582, (1991).

12. A.G. Evans, M. Rühle, B.G. Dalgliesh and P.G. Charalambides, "The Fracture Energy of Bimaterial Interfaces", Met. Trans. A, 21A, 2419-2429, (1990).

ACKNOWLEDGEMENTS

The authors would like to acknowledge the assistance of Mr. R.A. Nardi and Mrs. R.L. Casey for Magnetron sputtering of the laminates, Mssrs. D.A. Gibson and R.A. Rosa for sample fabrication and heat treatment and Mssrs. D.A. Catharine and C. Canestraro for tensile testing. Part of this research was performed under Air Force Wright Laboratories Contract F33615-91-C-5613, WRDC, Wright-Patterson AFB, OH.

MICROSTRUCTURES AND PROPERTIES OF REFRACTORY METAL-SILICIDE EUTECTICS

B.P. BEWLAY, K-M. CHANG, J.A. SUTLIFF, and M.R. JACKSON

G.E.-C.R.D, Schenectady, N.Y. 12301

ABSTRACT

New materials with improved high temperature properties are required for higher efficiency gas turbines. One group of alloys which show promise in this area are the eutectic systems Cr-Cr_3Si, Nb-Nb_3Si, and V-V_3Si. Crystals of these refractory metal-silicide eutectics were directionally solidified using Czochralski crystal growth in conjunction with cold crucible levitation melting. The crystals were examined using scanning electron microscopy and transmission electron microscopy. In the present paper, these systems are examined with respect to the directionally solidified eutectic microstructures, phase equilibria, crystallography and mechanical properties.

INTRODUCTION

The present paper describes a study of three silicide strengthened directionally solidified eutectics, Cr-Cr_3Si, Nb-Nb_3Si, and V-V_3Si. Eutectic composites based on refractory metals are presently under investigation as high temperature structural materials [1-7] because of their high melting temperatures (>1600°C), high temperature strengths and low densities, as shown in Table I. In these three eutectic systems, the first component is the refractory metal (R), and the second is the silicide (R_3Si). Phase volume fraction calculations from the assessed phase diagrams are shown in Table I.

One approach to increase the toughness of intermetallics is to generate a two phase microstructure of the intermetallic with a ductile phase. The role of eutectics in the development of ductile phase-intermetallic composites has recently been emphasized by Mazdiyasni and Miracle [7]. Directionally solidified eutectics containing a strong ductile phase with a strong intermetallic possess three advantages. First, there is intrinsic thermodynamic stability between the matrix and the reinforcement. Second, the microstructure can be influenced by solidification conditions. Third, component fabrication is generally easier than for synthetic composites.

The Cr-Si binary phase diagram has been reviewed recently by Gokhale and Abbaschian [8, 9]; the eutectic is at 15 at.%Si (all compositions in the present paper are in at.%) and is between Cr (with 9.5 % Si in solid solution at the eutectic temperature) and Cr_3Si (22.5 %Si). Si concentrations in Cr at the eutectic temperature as low as 2.3 %Si and as high as 10.7 %Si have been reported. In addition, Cr_3Si compositions between 15 and 24 %Si have been published.

The binary phase diagram of Nb-Si indicates a eutectic point at 18.7 %Si for Nb-Nb_3Si eutectic at 1883°C [9]. The silicide phase is a line compound and undergoes eutectoid decomposition at 1720°C [1]. The solid solubility of Si in Nb at the eutectic temperature is ~4.8 %Si. The V-Si binary phase diagram [10, 11] shows a eutectic reaction between V and V_3Si at 13 %Si and 1870°C. The solid solubility of Si in V at the eutectic temperature is 7 %Si. V_3Si has a homogeneity range between 19 and 29 % Si.

Nb-Si alloys have good high temperature strength, but they possess low room temperature fracture toughness and very poor oxidation resistance [3]. On the other hand, Aitken [12] has suggested that the limiting temperature for oxidation resistance of Cr-Si alloys is 1200°C. Mazdiyasni and Miracle [7] also reported good oxidation resistance of Cr-Cr_3Si eutectic at 800°C and 1200°C. Anton and Shah [2] reported that stoichiometric Cr_3Si has cyclic oxidation resistance similar to $MoSi_2$ at 1100°C.

Table I Melting temperatures, compositions and densities of refractory metal-silicide binary eutectics based on the assessed phase diagrams [9]. Matrix phases and volume fractions of the second phases are also shown.

Refractory Metal	Silicide	Eutectic Point		Density (g/cm^3)	Expected Structure	
		Temperature (^oC)	Composition (at.%Si)		Matrix Phase	Vol. Fract. of R_3Si
V	V_3Si	1870	13.0	5.6	V_3Si	0.51
Nb	Nb_3Si*	1883	18.7	7.4	Nb_3Si	0.32
Cr	Cr_3Si	1705	15.0	6.5	Cr	0.45

*is not stable below 1720°C.

At present, there is only limited mechanical property data on Cr-Cr_3Si eutectic alloys. Anton and Shah [2] provided some data on the high temperature yield stress and creep strength of Cr_3Si. They reported that at 1200°C, Cr_3Si fractured in a transgranular manner and did not display a ductile-to-brittle transition below 1200°C. Newkirk and Sago [5] reported a yield strength for Cr-Si alloys of 359 MPa (52 ksi) at 1200°C.

The aim of the present study was to grow Cr-Cr_3Si, Nb-Nb_3Si and V-V_3Si crystals with aligned eutectic structures. Crystal growth conditions, solidification microstructures, including phase morphologies and volume fractions, and and some properties of these three eutectics are described. Property evaluations are under continuing investigation.

EXPERIMENTS

Crystals of the eutectics were grown using Czochralski cold crucible crystal growth, as described in detail elsewhere [3]. Alloys were prepared prior to crystal growth by induction melting 99.99% pure Cr, Nb, or V with semiconductor grade Si in a sixteen segment water-cooled copper crucible. The charges were triple melted in order to improve homogeneity of the alloy. The crystal growth chamber was evacuated to 10^{-4} Torr and then back-filled with argon. A rare gas purifier was used to reduce impurity concentrations in the gas to ultra high purity levels (<0.75 molar ppm). Induction melting was employed to levitate the melt from the hearth of the crucible during crystal growth. Crystals were grown using growth rates between $8x10^{-3}$ mms^{-1} and 0.25 mms^{-1}, depending on the alloy and the desired microstructure. A constant growth rate was employed for each experiment and the furnace power was adjusted to maintain a stable melt temperature distribution. Crystals up to ~10 mm diameter and ~100mm long were grown.

Crystals were prepared for SEM and TEM using standard techniques. A scanning electron microscope equipped with a back scatter electron detector and an X-ray analyzer were employed. A commercially available software program was used for quantitative corrections of matrix effects in the x-ray spectra. By employing the overall chemical composition of the crystal as the reference standard for EDAX, the phase compositions were determined to an accuracy of ± 0.2 at.%.

Notched bend bars with a cross section of 3.8x5.1 mm were used for the fracture toughness measurements. Specimens were cut using electro-discharge machining so that the longitudinal axis was parallel to the growth direction of the crystals. The specimens were tested by four point bending. The strain rate was 8.5 $x10^{-3}mms^{-1}$.

RESULTS AND DISCUSSION

Microstructures of Cr-Cr_3Si, Nb-Nb_3Si and V-V_3Si Eutectics

Figure 1 shows a scanning electron micrograph taken using the backscattered electron signal (BSE) of the longitudinal section of a Cr-Cr_3Si eutectic crystal grown at 8.3×10^{-2} mms^{-1}. The eutectic has a lamellar morphology; this is a negative image and shows the Cr_3Si darker than the Cr-rich phase. The inter-lamellar spacing was 2.2±0.3μm. Lamellar faults were also observed. The volume fraction of Cr_3Si was 0.62±0.1, as shown in Table II. Most lamellae had a length to width ratio greater than 100. There were no Cr-rich or Cr_3Si dendrites indicating that the 15 %Si composition is at or close to the eutectic composition.

A Bright field transmission electron micrograph of the transverse section of a Cr-Cr_3Si eutectic crystal is shown in Figure 2. Two crystallographic orientation relationships were observed, as determined by electron diffraction using (TEM), and are described by :

Type I : $[111]_{Cr} \parallel [001]_{Cr_3Si}$ and $(110)_{Cr} \parallel (210)_{Cr_3Si}$

Type II : $[001]_{Cr} \parallel [001]_{Cr_3Si}$ and $(110)_{Cr} \parallel (210)_{Cr_3Si}$

It was also determined that the lamellar interface plane was always $(110)_{Cr}$ and that the growth direction of the eutectic was always parallel to $[001]_{Cr_3Si}$. The Type II orientation relationship dominated foils prepared from longitudinal sections of the crystals. Additional details of microstructural analyses have been reported elsewhere [3].

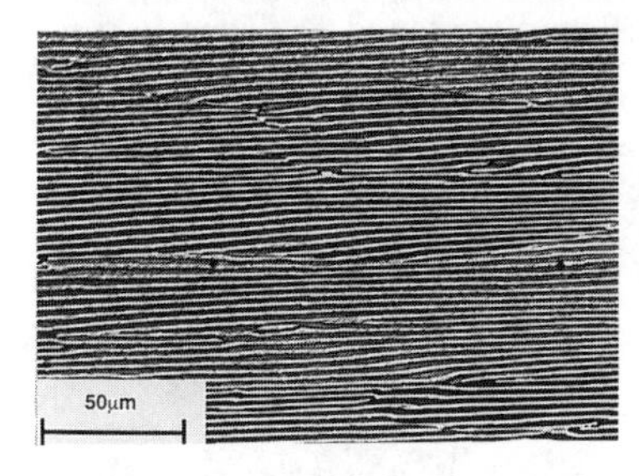

Figure 1 Scanning electron micrograph (BSE negative) of the longitudinal section of a Cr-Cr_3Si crystal (Cr-15 at.%Si) grown at 8×10^{-2} mms^{-1}.

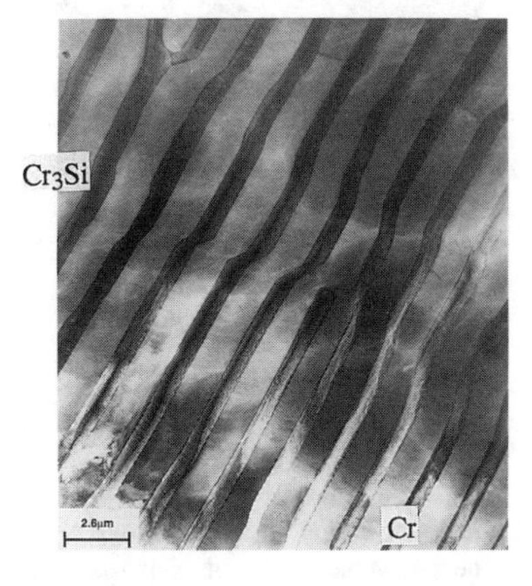

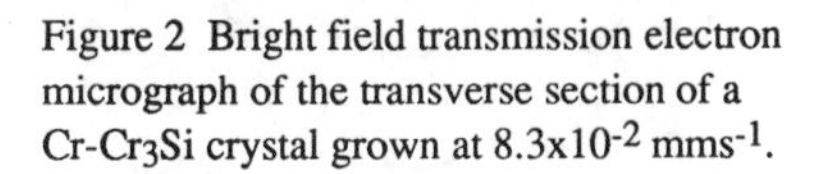

Figure 2 Bright field transmission electron micrograph of the transverse section of a Cr-Cr_3Si crystal grown at 8.3×10^{-2} mms^{-1}.

Table II Measured eutectic compositions, together with second phase volume fractions and morphologies for refractory metal-silicide crystals.

Refractory Metal	Silicide	Eutectic at.%Si	Matrix Phase	Faceted	Second phase Morphology	Faceted	Vol. Fract. of R_3Si
V	V_3Si	12.6	V	No	rod	No	0.48
Nb	Nb_3Si	18.2	Nb_3Si	Yes	rod/plate	No	0.36
Cr	Cr_3Si	15.0	Cr_3Si	No	Lamellar		0.62

Comparison of the inter-planar distances in Cr and Cr_3Si for the low index planes indicates that the (110) plane in Cr has almost exactly the same inter-planar spacing as the (210) plane in Cr_3Si. This is consistent with the preferred lamellae interface plane identified from the diffraction patterns. Regular lamellar structures, as observed in the Cr-Cr_3Si eutectic crystals, are generally only observed in non-faceting/non-faceting eutectics [13]. Faceting was not observed in the Cr-Cr_3Si microstructures.

A scanning electron micrograph (BSE) of the transverse section of a Nb-18.7 %Si crystal grown at $8.3x10^{-2}$ mms^{-1} is shown in Figure 3. The Nb_3Si phase is dark and the Nb is light. Three morphologically distinct areas were observed; Nb_3Si dendrites, eutectic cells, and cell boundaries. Highly faceted Nb_3Si dendrites were observed at the center of eutectic cells. These L-shaped dendrites had a width of 20±4 μm. A layer of Nb (~0.5 μm thick) was observed around these dendrites. The cell size was 75±10μm. In the eutectic areas, fibrous Nb rods and plates were uniformly distributed in the Nb_3Si matrix. An aligned fibrous structure was developed in the Nb-Nb_3Si eutectic regions.

A scanning electron micrograph (BSE) of the transverse section of a V-13 %Si crystal grown at $8.3x10^{-2}$ mms^{-1} is shown in Figure 4. BSE imaging showed the fibrous V_3Si as the light phase in a V-rich matrix. The fibres were generally aligned close to the growth direction. The aspect ratios of the fibers were generally greater than 20 and sometimes as high as several hundred. The fibre diameter was 1.6±0.2μm. The volume fraction of V was 0.52, as shown in Table II. A small number of non-faceted V_3Si dendrites were visible at lower magnifications. There was a suggestion that some of the V_3Si rods had a hexagonal cross-section, but this was not a dominant feature.

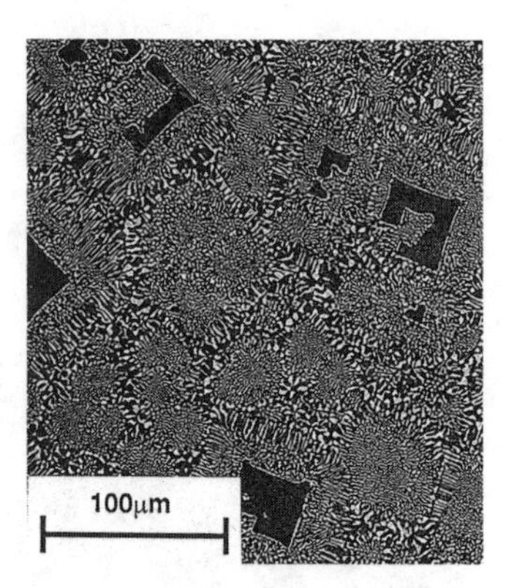

Figure 3 Scanning electron micrograph (BSE) of the transverse section of a Nb-18.7 at. %Si crystal grown at $8.3x10^{-2}$ mms^{-1}.

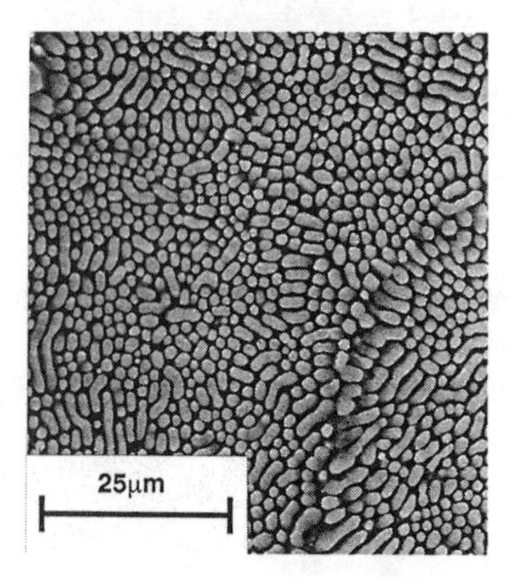

Figure 4 Scanning electron micrograph (BSE) of the transverse section of a V-V_3Si crystal grown at $8.3x10^{-2}$ mms^{-1}.

Phase Equilibria

The Cr-Si phase diagram predicts that the volume fraction of Cr_3Si is 0.45. Densities of the Cr-rich and Cr_3Si phases were calculated to be 6.91g/cc and 6.47 g/cc respectively [8]. The predicted volume fraction is significantly less than the observed volume fraction of 0.62. There are four possible reasons for this difference. First, the actual densities of the Cr-rich and Cr_3Si phases at the eutectic temperature may be different from the calculated densities. Second, the Si compositions of the individual phases may continue to adjust during post solidification cooling. Third, the actual compositions of the Cr-rich and Cr_3Si phases at the eutectic temperature may not be as stated by the phase diagram. Fourth, due to volatilization of Cr or Si the composition of the melt may have shifted from the eutectic.

The last two effects are probably more significant and are presently being explored [14] by performing heat treatments at temperatures from 1100 to 1600°C, followed by microprobe analysis. In the Cr-15 %Si sample which was heat treated at 1400°C for 100

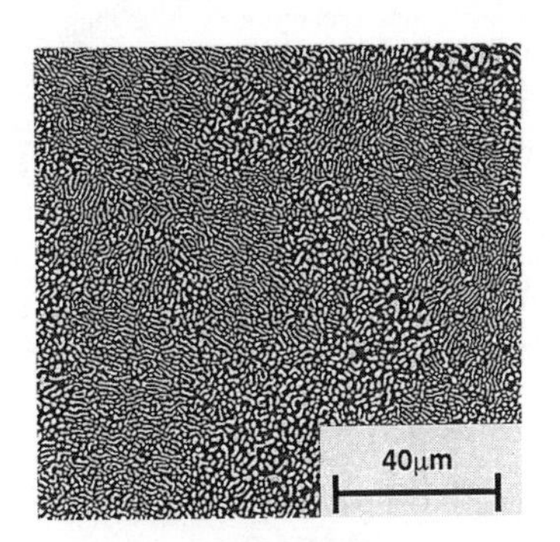

Figure 5 Scanning electron micrograph (BSE) of the transverse section of a Nb-18.2 at. %Si crystal grown at $8.3x10^{-3}$ mms^{-1}.

hours, the composition of the Cr-rich phase was 6.7 %Si and that of Cr_3Si was 21.8 %Si. The phase diagram compositions at 1400°C for Cr and Cr_3Si are 7.5 %Si and 22.6 %Si respectively. Calculation of the volume fractions of the phases for the measured compositions at 1400°C indicates that the volume fraction of Cr_3Si is 0.58, which is closer to the measured value of 0.62. The volume fraction of Cr_3Si at 1400°C, based on the phase diagram [8], is 0.48. Thus the microprobe data indicate that both Cr and Cr_3Si contain less Si than indicated by the phase diagram. This is the most likely explanation for Cr_3Si being the majority phase in the composite. However, additional microprobe data from samples heat treated at temperatures closer to the eutectic temperature are required.

Clearly, the Nb-18.7 %Si composition contained more Si than the eutectic. In order to identify the eutectic composition more accurately, crystals with compositions between 17.4 and 18.4 %Si were grown and examined. Microstructural analysis and EDAX indicated that the eutectic point was 18.2 %Si; 0.5 % less than reported [9]. The Nb-18.2 %Si composition was then directionally solidified. Figure 5 shows a scanning electron micrograph (BSE) of the transverse section of a crystal grown at $8.3x10^{-2}$ mm/s. The microstructure consists of Nb rods and plates in a Nb_3Si matrix. The rods had a diameter of 1.2±0.3 μm. The volume fraction of Nb in the Nb-Nb_3Si eutectic for the measured eutectic composition of 18.2 %Si was calculated as 0.34, and the observed volume fraction was 0.36.

Crystals of V-13 %Si contained V_3Si dendrites. The eutectic composition was determined by growing several V-Si crystals containing 11 to 13.3 %Si. The chemical compositions of the crystals were determined using emission spectroscopy and the bulk compositions were used as the reference standard for EDAX. The eutectic point of V-V_3Si was measured as 12.6 %Si. The predicted volume fraction of V_3Si was 0.51, close to the observed value. Densities were calculated from lattice parameter data provided [10, 11]. The volume fraction of rods is greater than the theoretical limit of 0.32 [15]. There are two possible reasons for the rod type eutectic. First, the growth rates investigated may not have been suitable for growth of lamellae. Second, there might be a large difference in the average interfacial free energies in the lamellar and rod structures [15].

Fracture Toughness

Fracture toughness data of V-V_3Si, Cr-Cr_3Si and Nb-Nb_3Si eutectics at 20°C and 400°C are compared in Table III. At 20°C, fracture toughness of all the eutectics was low. However, at 400°C the fracture toughness of V-V_3Si is significantly higher than Nb-Nb_3Si. Although the metallic phases of the Cr, Nb, and V eutectics are more ductile than the intermetallic phases, due to the relatively high DBTT's of the metallic phases the room temperature fracture toughness of the eutectics is still low. Newkirk and Sago [5] measured the fracture toughness of Cr-15.5 %Si as 7MPa√m; this is similar to that measured in the present study. V-Si alloys have lower DBTT's than the Nb-Si and Cr-Si alloys [16] and may

TABLE III Fracture Toughness of Cr-Cr_3Si, Nb-Nb_3Si, and V-V_3Si Eutectics

Eutectic	Temperature (°C)	K_Q (MPa√m)
Nb-Nb_3Si	20	6.0
V-V_3Si	20	6.8
Cr-Cr_3Si	20	7.3
Nb-Nb_3Si	400	5.8
V-V_3Si	400	16.5

offer an avenue to improving room temperature ductility.

CONCLUSIONS

The present paper describes the microstructures of directionally solidified crystals of Cr-Cr_3Si, Nb-Nb_3Si and V-V_3Si binary eutectics. Although the eutectic component phases have similar crystal structures the resultant microstructures are significantly different. Crystals of the Cr-Cr_3Si eutectic have a lamellar structure. At a growth rate of 0.08 mms^{-1}, the inter-lamellar spacing was 2.2±0.3μm. The matrix phase was Cr_3Si with a volume fraction of 0.62; this is much greater than predicted. The likely causes for this discrepancy are, that the Cr-rich phase contains less than 9.5 %Si, and Cr_3Si contains less that 22.5 %Si. Microprobe analysis at 1400°C indicated that the Si concentrations in the individual phases was less than indicated by the phase diagram. In the Nb-Nb_3Si eutectic, Nb_3Si is the matrix phase. Faceted dendrites were observed in the Nb-18.7 %Si crystal. EDAX quantitative analysis indicated that the Nb-Nb_3Si eutectic composition was 18.2 %Si. In the eutectic grown at $8.3x10^{-2}$ mms^{-1}, rods and plates of Nb were observed. The volume fraction of Nb was 0.36. The V-V_3Si eutectic consists of V_3Si fibres in a V matrix. The fibre diameter was 1.8±0.2 μm at a growth rate of $8.3x10^{-2}$ mms^{-1}.

ACKNOWLEDGEMENTS

The authors would like to thank D.J. Dalpe for the crystal growth and R. Rosa for the SEM. We are also very grateful to Prof. H.A. Lipsitt for his comments.

REFERENCES

[1] M.G. Mendiratta and D.M. Dimiduk, Mat. Res. Soc. Symp. Proc., 133, 441 (1989).

[2] D.L. Anton and D.M. Shah, in "Proceedings of the International Symposium on Intermetallic Compounds-Structure and Mechanical Properties (JIMIS-6)", ed. by O. Izumi (The Japan Institute of Metals, 1991), 379.

[3] K-M. Chang, B.P. Bewlay, J. A. Sutliff and M.R. Jackson, J. of Metals, 44 (6), 34 (1992).

[4] B. Cockeram, H.A. Lipsitt, R. Srinivasan and I. Weiss, Scripta Met. and Mater., 25, 2109 (1991).

[5] J.W. Newkirk and J.A. Sago, Mat. Res. Soc. Symp. Proc., 194, pp. 183-189 (1990).

[6] R.L. Fleischer, Mat. Res. Soc. Symp. Proc., 194, 251 (1990).

[7] S. Mazdiyasni and D.B. Miracle, Mat. Res. Soc. Symp. Proc., 194, 155 (1990).

[8] A.B. Gokhale and G.J. Abbaschian, Bull. of Alloy Phase Diagrams, 8, 474 (1987).

[9] T.B. Massalski, Binary Alloy Phase Diagrams, ASM Metals Park, Ohio (1991).

[10] J.F. Smith, Bull. of Alloy Phase Diagrams, 6, (3), 266 (1985).

[11] J.L. Jorda and J. Muller, J. Less Common Met., 84, 39 (1982).

[12] E.A. Aitken, in "Intermetallic Compounds". ed. by J.H. Westbrook, (Huntington, N.Y.: Krieger Publishing Co., (1977), 513.

[13] J.D. Hunt and K.A. Jackson, Trans. Met. Soc. AIME, 239, 864 (1967).

[14] H.A. Lipsitt, J.A. Sutliff, B.P. Bewlay and K.-M. Chang, unpublished work, 1992.

[15] J.D. Hunt and J.P. Chilton, J. Inst. Met., 91, 338 (1962).

[16] T.E. Tietz and J.W. Wilson, Behavior and Properties of Refractory Metals, (Stanford University Press, Stanford, CA, 1965), 32.

STRENGTH AND TOUGHNESS OF $Be_{12}Nb$, $Be_{17}Nb_2$ AND TWO-PHASE $Be_{12}Nb$-$Be_{17}Nb_2$

Stephen M. Bruemmer, Bruce W. Arey and Charles H. Henager, Jr.
Pacific Northwest Laboratory, Richland, WA

ABSTRACT

Bend strength, compression strength, and fracture toughness of niobium beryllide intermetallic compounds have been assessed at temperatures from ambient to 1200°C. Hot-isostatically-pressed (HIP) $Be_{12}Nb$ showed significantly improved low- and high-temperature mechanical properties over vacuum-hot-pressed (VHP) material. Strengths at 20°C were 250 MPa in bending and 2750 MPa in compression with a fracture toughness of ~4 MPa√m, much higher than previously measured for this compound. High-temperature ($\geq$ 1000°C) mechanical properties were also improved with bend strengths of 250 MPa at 1200°C as compared to only 70 to 100 MPa for the VHP material. However, severe pest embrittlement was detected in the HIP material at temperatures between 650 and 1000°C.

HIP $Be_{17}Nb_2$ exhibited poorer low-temperature strength, but much better high-temperature strength, than the HIP $Be_{12}Nb$. Bend strengths were one-half that measured for $Be_{12}Nb$ at 20°C, but were several times greater at 1100 to 1200°C. Fracture toughness for $Be_{17}Nb_2$ changed little with temperature below 1100°C, exhibiting values of 2.2 to 2.5 MPa√m. Compression strengths for $Be_{17}Nb_2$ were similar to those measured for VHP $Be_{12}Nb$, consistently lower than for HIP $Be_{12}Nb$. No improvement in mechanical properties were observed for the VHP two-phase, $Be_{12}Nb$ + $Be_{17}Nb_2$ heat. Behavior was similar to VHP $Be_{12}Nb$ at temperatures below 1100°C, with low strengths and fracture toughnesses over this temperature range. However, strengths increased at temperatures above 1100°C. Results indicate that a two-phase microstructure may improve mechanical properties, but that variables such as grain size and oxide dispersion have a dominant influence on behavior.

INTRODUCTION

Refractory metal beryllides, such as $Be_{12}X$, $Be_{13}X$ and $Be_{17}X_2$, have significant high-temperature strength-to-weight benefits over most other intermetallics, metallic alloys or ceramics [1-8]. Much of these benefits stem from their very low densities (2.7 to 5.1 g/cm^3), but beryllides also exhibit high elastic moduli and excellent oxidation resistance. As with many high-melting intermetallic compounds, the structural application of beryllides is limited by low-temperature brittleness due to their complex crystal structures. Classical slip deformation processes appear to be restricted until temperatures are reached where diffusion processes become rapid (about one-half the melting temperature). However, mechanisms of embrittlement are essentially unknown since direct measurements of deformation and fracture behavior in these compounds are extremely limited.

The most studied of the beryllide compounds has been the body-centered-tetragonal $Be_{12}X$ ($Mn_{12}Th$ structure type, I4/mmm space group, D^{17}_{4h} Schoenflies notation).

Mat. Res. Soc. Symp. Proc. Vol. 273.

Recent work [7,8] has evaluated the mechanical properties and deformation behavior [9,10] of vacuum-hot-pressed (VHP) $Be_{12}Nb$ in some detail. Toughness-limited strength was observed at temperatures <1100°C in bending and <800°C in compression. Fracture toughness was 2.4 MPa√m at ambient temperature and reached only 3.4 MPa√m at 1000°C. Higher strengths and improved toughness was observed at temperatures corresponding to the onset of macroscopic plasticity. Considerable ductility and extensive macrodeformation (many active slip systems) was documented during deformation at 1200°C. [9] However, the slip character was observed to change, and ductility eliminated, as test temperatures were reduced. [8,10]

The present work examines the strength and fracture toughness of hot-isostatically-pressed (HIP) $Be_{12}Nb$, HIP $Be_{17}Nb_2$ and a VHP two-phase material ($Be_{11}Nb$) containing both $Be_{12}Nb$ and $Be_{17}Nb_2$. Four-point bend, chevron notched bend and uniaxial compression tests were used to evaluate mechanical behavior at temperatures from 20 to 1200°C. Results are compared to similar experiments performed on a VHP $Be_{12}Nb$ heat.

EXPERIMENTAL PROCEDURE

The beryllide intermetallic heats were produced by Brush-Wellman, Inc., Elmore, OH, by VHP or by VHP and HIP powder-metallurgy methods. Compositions and grain sizes for the four heats examined are listed in Table 1. The "$Be_{11}Nb$" heat was produced by combining $Be_{12}Nb$ (66.7%) and $Be_{17}Nb_2$ (33.4%) powders in order to obtain a two-phase microstructure. Attempts to HIP this material were unsuccessful. A high density of BeO particles was present in all heats. A major difference between VHP and HIP heats is the location of these oxides. The VHP materials show particles primarily on grain boundaries, whereas oxides are aligned along prior powder particle interfaces in the HIP materials. Initial powder particle sizes were larger for $Be_{17}Nb_2$ (~40 μm) than for $Be_{12}Nb$ (~10 μm). The HIP heats exhibited a uniform grain size distribution, as compared to bimodal distributions in the VHP heats. Densities for the VHP and HIP heats were reported to be >99% of theoretical, but separate immersion density measurements indicated a lower density (~93%) for the VHP $Be_{12}Nb$ heat.

Bend specimens (4 x 4 x 40 mm) were prepared by electro-discharge machining and tested using a SiC four-point-bend fixture with an upper span of 20 mm and a lower span of 40 mm. The load was applied through bearing cylinders transverse to the specimen longitudinal axis. Displacement rates for most tests were 0.5 mm/min. Chevron-notch bend (CNB) specimens were prepared by notching bend bars with two 30° cuts to produce a 60° chevron-shaped ligament. The stress intensity (K_{IC}) was determined from the equation

$$K_{IC} = P \cdot Y_{min}/(B\sqrt{w}) \tag{1}$$

where P is the load, Y_{min} is a geometry factor (calculated to be 11.9) for the CNB specimen, B is the specimen width and w is the specimen thickness. [11] Compression tests were performed on small cylinder specimens (3-mm in dia x 3-mm in height) using SiC platens and a crosshead speed of 1.7×10^{-4} mm/sec. All bend and compression tests were performed in air (relative humidity of ~60%).

Table 1. Compositions of Be-Nb Heats, wt% or wppm

Element	VHP $Be_{12}Nb$	HIP $Be_{12}Nb$	HIP $Be_{17}Nb_2$	VHP $Be_{11}Nb$
Nb	46.9%	46%	55%	49.4%
O	0.29%	0.90%	0.69%	0.43%
C	0.03%	0.07%	0.11%	0.06%
N	85	140	110	205
Fe	520	315	265	285
Si	285	190	150	235
Al	180	195	95	180
Mg	65	290	20	~5
Ni	95	50	35	30
Grain Size, μm	15-40	5	8	10-40

RESULTS

HIP $Be_{12}Nb$

Four-point bend specimens failed in a brittle manner (exhibiting only elastic behavior) at test temperatures from 20 to 1000°C. Limited plasticity was observed in bending at 1100°C, which increased to several percent plastic strain before failure at 1200°C. Bend strengths of 210 MPa (30 ksi) or greater were measured during testing from 20 to 650°C and from 1000 to 1200°C. Surprisingly, the *highest* bend strengths (~250 MPa) were recorded during tests at the *lowest* temperatures (20 to 300°C). Strength dropped sharply from 650 to 900°C, and then increased to 240-250 MPa at higher temperatures as shown in Figure 1. The decrease at intermediate temperatures is due to environment-induced embrittlement referred to as the "pest"

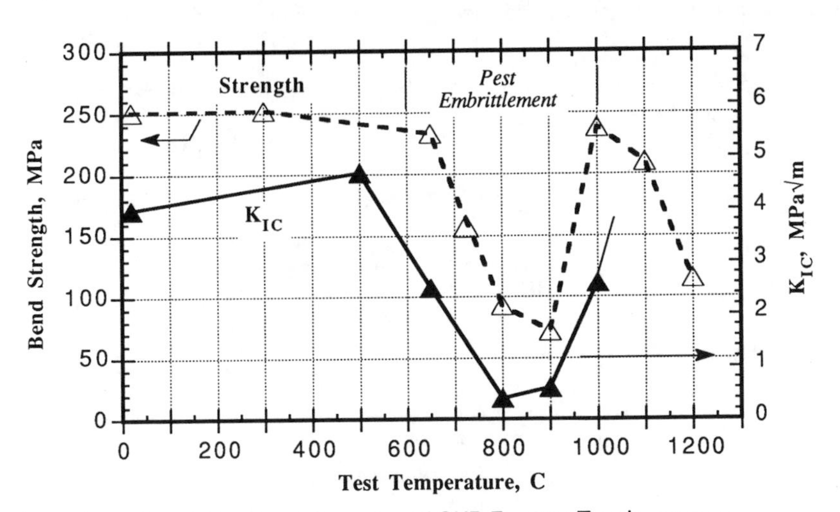

Figure 1. Four-Point Bend Strength and CNB Fracture Toughness as a Function of Temperature for VHP and HIP $Be_{12}Nb$ Tested in Air.

phenomena. [1-6,12] This embrittlement was also produced by pretreatment of a specimen at 800°C before lower temperature testing. For example, the ambient temperature bend strength dropped from 250 to 70 MPa as a result of the 800°C pretreatment, comparable to that measured at 800-900°C. Temperature effects on fracture toughness from CNB tests closely matched the bend strength results (Figure 1). Low-temperature K_{IC} ranged from 4 to 4.6 MPa√m corresponding to the highest measured bend strengths. Toughness values decreased to less than 0.5 Mpa√m in the pest temperature regime before increasing at temperatures above 900°C.

Fracture morphologies for the bend and CNB specimens changed slightly with test temperature. At low temperatures, a fine granular surface appearance is observed consisting of intergranular (IG) facets and small cleavage regions (across individual grains). The amount of IG facets and the observed porosity in the fracture surface increased at temperatures from 725 to 1000°C. The overall surface morphology of became very flat and featureless. Individual cleavage regions extended over larger distances than those at low temperatures.

The maximum uniaxial compressive yield strength (reported strengths are the proportional limit, not 0.2% offset strengths) of 2750 MPa was measured at 20°C. Strength remained quite high (2380 MPa) at 800°C and then decreased rapidly at temperatures ≥1000°C (Figure 2). Specimens failed in a brittle manner when compressed at temperatures below 1000°C, while tests at higher temperatures revealed measurable plastic strain.

<u>HIP $Be_{17}Nb_2$</u>

The HIP $Be_{17}Nb_2$ exhibited brittle cleavage fracture in bend and CNB tests at temperatures up to 1100°C. Bend strengths increased slightly with temperature from 110 MPa at 20°C to 200 MPa at 800°C, then more sharply to a maximum of 740 MPa at 1100°C. Measured strengths for HIP $Be_{17}Nb_2$ at high temperatures were much greater than for the other beryllide heats as shown in Figure 3. However, low-

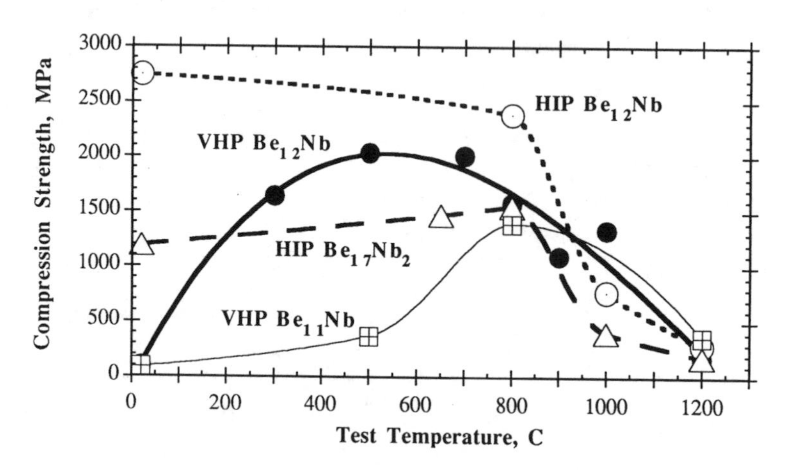

Figure 2. Compressive Strength as a Function of Temperature for Beryllide Heats.

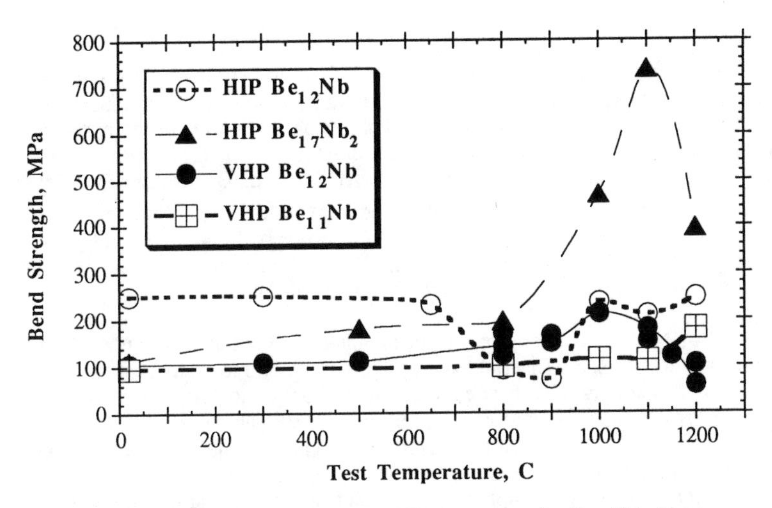

Figure 3. Bend Strength as a Function of Temperature for Beryllide Heats.

temperature strengths were inferior to the HIP $Be_{12}Nb$, particularly at 20°C. Toughness-limited strength was indicated by the CNB measurements of K_{IC}s of $\leq$2.5 MPa√m, even at test temperatures up to 1000°C. Fracture toughness values did not increase from 20 to 1000°C, even though the measured bend strength increased by ~4 times over this range. The fracture surfaces were very flat, almost featureless (except for the IG dispersion of large oxides which intersect the transgranular fracture path) up to test temperatures of 1100°C.

Compression tests for HIP $Be_{17}Nb_2$ also revealed toughness-limited strength at low temperatures. Strength increased with temperature reaching a maximum value of 1550 MPa at 800°C (Figure 2). At temperatures below 1000°C, measured strengths were significantly less than for the HIP $Be_{12}Nb$ and comparable to the VHP $Be_{12}Nb$.

VHP $Be_{11}Nb$

The two-phase, $Be_{11}Nb$ ($Be_{12}Nb$+$Be_{17}Nb_2$) material exhibited mechanical behavior similar to the VHP $Be_{12}Nb$. Bend strengths (Figure 3), toughnesses (<3 MPa√m), and compressive strengths (Figure 2) were equal to, or less than, VHP $Be_{12}Nb$ at all temperatures from 20 to 1100°C. Unlike the monolithic materials, bend strengths and toughnessess did not increase significantly at either 1000°C (for $Be_{12}Nb$) or 1100°C (for $Be_{17}Nb_2$). Bend strengths at these temperatures were ~3 times less than for $Be_{12}Nb$ and ~7 times less than for $Be_{17}Nb_2$. The two-phase material did exhibit an increase in bend strength at 1200°C, reaching a stress of 185 MPa, which is more than double the VHP $Be_{12}Nb$ strength. Compression strength at 1200°C was also ~2 times higher for $Be_{11}Nb$ than for $Be_{12}Nb$ (VHP or HIP) or $Be_{17}Nb_2$. Load-deflection curves in bending and in compression reveal only elastic behavior at all temperatures below 1200°C. A mixture of IG and TG features were present in the brittle fracture surfaces.

DISCUSSION

The improved mechanical properties of the HIP $Be_{12}Nb$ over the VHP material most likely results from its reduced grain size and intragranular oxide distribution. Grain size has been shown to have a significant influence on strength for many materials, including some limited work on beryllides. Stonehouse et al. [2] found much poorer high-temperature strength for several VHP beryllides (e.g., $Be_{12}Nb$ and $Be_{13}Zr$) with a grain size of ~50 μm versus heats with sizes from 10 to 25 μm. Bend strengths at 1200°C measured for the HIP heat in this study are comparable to strengths reported for fine-grained VHP heats by Stonehouse. Thus, at high temperatures, grain size appears to be a primary variable controlling bend strength.

The HIP $Be_{12}Nb$ heat also shows a significant improvement in low-temperature strength and toughness. This behavior appears to be due to factors more than grain size alone. Ambient temperature bend strengths from 80 to 150 MPa have been measured by various investigators [2,4,6,9] on VHP $Be_{12}Nb$ heats, which is significantly below the 250 MPa measured for the HIP material. Strength values for all heats are flaw dominated and are well below the intrinsic strength. The HIP $Be_{12}Nb$ reaches a higher strength because of its higher K_{IC} values, i.e. 4 versus 2.5 $MPa\sqrt{m}$ for the VHP heat and its reduced internal flaw size. The lower K_{IC} values for VHP materials is believed to result from its high density of IG oxides and IG porosity which promote brittle IG separation. Compression data corroborates the high low-temperature strength of the HIP $Be_{12}Nb$, where an ambient temperature strength of 2750 MPa is much greater than the other beryllide heats examined.

Bend strength versus temperature data (Figure 3) for HIP $Be_{17}Nb_2$ is quite consistent with the VHP heats examined by Stonehouse, et al. [2] in the late 1950s. High-temperature strengths are improved (e.g., 740 versus 480 MPa at 1100°C), but the HIP $Be_{17}Nb_2$ did not show an increase in low-temperature strength over prior work on VHP heats. The poor low-temperature strength for $Be_{17}Nb_2$ may be related to a low cleavage strength. Flat fracture morphologies show only minor changes as the material cleaves across various grains. This is different than the fracture morphology observed in HIP $Be_{12}Nb$ where the path shows distinct grain-to-grain reorientation. In both cases the fracture is brittle, but a more torturous path is present for the HIP $Be_{12}Nb$.

The two-phase, $Be_{12}Nb$ + $Be_{17}Nb_2$ material did not reveal improved properties over the monolithic beryllides. Bend strength and toughness was inferior to the HIP heats over the entire temperature range. This poor behavior results from a VHP microstructure consisting of large $Be_{17}Nb_2$ grains clumped within somewhat smaller $Be_{12}Nb$ grains. Intergranular oxides and porosity were similar to the previous VHP $Be_{12}Nb$ heat and prompted similar mechanical properties. A small positive influence on properties results from the $Be_{17}Nb_2$ second phase at 1200°C where bend and compression strengths are greater that the VHP $Be_{12}Nb$. An improved dispersion of $Be_{17}Nb_2$ in $Be_{12}Nb$ is required to produce a microstructure that exhibits the better low-temperature strength and toughness of $Be_{12}Nb$ along with the better high-temperature strength of $Be_{17}Nb_2$.

CONCLUSIONS

Hot-isostatically-pressed (HIP) $Be_{12}Nb$ showed significantly improved low-temperature mechanical properties over vacuum-hot-pressed (VHP) material. Room-temperature bend strength increased from 105 to 250 MPa, compression strength from 100 to 2750 MPa, and fracture toughness from about 2.5 to 4 MPa√m for the HIP versus VHP $Be_{12}Nb$. High-temperature (≥ 1000°C) mechanical properties were also improved with bend strengths of 250 MPa at 1200°C as compared to only 70 to 100 MPa for the VHP material. Pest embrittlement was detected in the HIP material at temperatures between 650 and 1000°C. Bend strengths were decreased by 3 times, and fracture toughness by 9 times, at 800 and 900°C compared to values at 20°C.

The HIP $Be_{17}Nb_2$ heat exhibited poorer low-temperature strength, but much better high-temperature strength, than the HIP $Be_{12}Nb$. Bend strengths were ~2 times greater for $Be_{12}Nb$ at 20°C and ~5 times greater for $Be_{17}Nb_2$ at 1100°C. The $Be_{17}Nb_2$ material retained bend strengths of 740 and 400 MPa at 1100 and 1200°C, respectively. Fracture toughness for $Be_{17}Nb_2$ was effectively independent of temperature below 1100°C, exhibiting values of ~2.5 MPa√m. Compression strengths for the HIP $Be_{17}Nb_2$ were similar to those measured for the VHP $Be_{12}Nb$ and was consistently lower than for the HIP $Be_{12}Nb$.

Mechanical properties for the VHP two-phase, $Be_{12}Nb + Be_{17}Nb_2$ heat were similar to VHP $Be_{12}Nb$ properties at temperatures below 1100°C. Low bend strengths (~110 MPa) and fracture toughnesses (~2.5 MPa√m) were measured over this temperature range. However, bend strengths increased at temperatures above 1100°C, reaching a value 2 to 3 times that of VHP $Be_{12}Nb$ at 1200°C. Compression strengths showed a similar trend with the two-phase material strength exceeding that for $Be_{12}Nb$ at high temperatures. Results indicate that a two-phase microstructure may improve mechanical properties, but that microstructural variables, such as grain size and oxide dispersion have a dominant influence on behavior.

REFERENCES

1. J.R. Lewis, J. Metals, 13 (1961) 357 and 829.
2. A.J. Stonehouse, R.M. Paine and W.W. Beaver, *Mechanical Properties of Some Transition Element Beryllides, Chapter 13,* Mechanical Properties of Intermetallic Compounds, ed. J.H. Westbrook, John Wiley & Sons, NY, 1960, p. 297.
3. A.J. Stonehouse, Mater. Design Eng., Feb., (1962).
4. E.A. Aitken and J.P. Smith, *Properties of Beryllium Intermetallic Compounds,* GEMP-105, General Electric Co., 1961.
5. E.G. Kendall, *Intermetallic Materials - Carbides, Borides, Beryllides, Nitrides and Silicides, Chapter 5,* Ceramic Advanced Technology, 1965, p. 143.
6. R.F. Kirby, PhD Dissertation, University of Arizona, 1969.
7. S.M. Bruemmer, B.W. Arey, R.E. Jacobson and C.H. Henager, Jr., *High-Temperature Ordered Intermetallic Alloys IV*, Materials Research Society, Vol. 213, 1991, p. 475.
8. C. H. Henager, Jr., R. E. Jacobson and S. M. Bruemmer, Mat. Sci. Eng., in press.

9. S.M. Bruemmer, L.A. Charlot, J.L. Brimhall, C.H. Henager, Jr., and J.P. Hirth, *Dislocation Structures in $Be_{12}Nb$ After High-Temperature Deformation*, submitted for publication in Phil. Mag.
10. S.M. Bruemmer, L.A. Charlot, J.L. Brimhall, C.H. Henager, Jr., and J.P. Hirth, in preparation.
11. P. A. Whithey and P. Bowen, Int. J. of Fracture, 46 (1990) 55-59.
12. E.A. Aitken and J.P. Smith, J. Nucl. Mat., 6-1 (1962) 119.

ACKNOWLEDGMENTS

Contributions of L.A. Charlot and J.L. Brimhall, as well as helpful discussions with J.P. Hirth, are acknowledged. Work is supported by the Defense Advanced Research Projects Agency through the Office of Naval Research and under U.S. Department of Energy contract DE-AC06-76RL0 1830 with Pacific Northwest Laboratory, which is operated by Battelle Memorial Institute.

FRACTURE AND FATIGUE BEHAVIOR IN Nb_3Al+Nb INTERMETALLIC COMPOSITES

L. MURUGESH, K. T. VENKATESWARA RAO, L. C. DeJONGHE and R. O. RITCHIE
Department of Materials Science and Mineral Engineering, University of California, Berkeley, CA 94720.

ABSTRACT

Model high-melting point Nb_3Al + Nb intermetallic composites have been fabricated *in situ* by vacuum hot pressing and reaction sintering elemental powders mixed in the ratio Nb + 7wt.% Al. In both cases, microstructures feature islands of ductile Nb solid solution (~20 vol.%) in a brittle Nb_3Al intermetallic matrix. Thermal treatment for 24 h at 1800°C results in a lamellar microstructure containing a uniform and fine distribution of filamentary Nb in a Nb_3Al matrix following the massive peritectic transformation. In this paper, the fatigue and fracture resistance of these two microstructures are examined and compared to pure Nb_3Al and Nb. Preliminary results suggest that the Nb phase can provide significant toughening to Nb_3Al via crack bridging, plastic stretching and interfacial debonding mechanisms. Measured plane-strain fracture toughness values for the as hot-pressed and fully-aged microstructures are ~6-8 MPa$\sqrt{m}$ compared to ~2 MPa$\sqrt{m}$ for pure Nb_3Al. However, under cyclic loading, the composites tend to show a strong dependence on applied stress-intensity level; fatigue thresholds range between 2-3 MPa$\sqrt{m}$.

INTRODUCTION

Niobium-based alloys have been the subject of renewed research interest since the mid-1980's as the materials community continues to seek new high-performance materials for advanced high-temperature applications in aerospace propulsion systems [1]. One such alloy, Nb_3Al, possesses the properties that are required for an intermetallic compound to be useful as a high temperature structural material [2,3]. This alloy is preferred over Nb_2Al and Al_3Nb due to its higher melting temperature. However, like many intermetallics that exhibit high melting points and elastic moduli, Nb_3Al also has a low-symmetry, complex crystal structure (A-15). Although this structure is responsible for its high-temperature strength, it concurrently limits the number of slip systems, increases the slip vector, restricts cross slip and transfer of slip across grain boundaries [4,5], all factors which contribute to its limited ductility, toughness and tendency for brittle fracture at ambient temperatures; accordingly, Nb_3Al currently has limited structural use.

One approach to improving the ductility and toughness of brittle intermetallics is to incorporate a ductile phase so as to impede crack advance by ductile-ligament bridging [6]. This may be accomplished either by artificially hybridizing the microstructure through powder-metallurgy techniques or by *in situ* precipitation reactions. In the case of the binary Nb-Al system, the peritectic reaction at 2060°C (Fig. 1), involving the precipitation of Nb_3Al from the liquid and high-temperature Nb solid solution (Nb_{ss}), can be utilized for fabricating two-phase microstructures of Nb_{ss} in a brittle Nb_3Al matrix:

$$\text{Liquid} + Nb_{ss} \rightarrow Nb_3Al \quad . \tag{1}$$

Previous studies have shown this to be a sluggish transformation, such that the Nb solid solution can be fully retained at room temperature with only moderate undercooling [7-9]. Subsequent heat-treatment precipitates the ordered Nb_3Al phase. The precipitation reaction is shown to be a massive transformation resulting in a highly uniform and fine distribution of a filamentary Nb within a non-contiguous matrix. Such a duplex Nb/Nb_3Al microstructure can potentially enhance crack-growth resistance by crack bridging with an additional contribution from the deformation of the ductile phase.

Model high-melting temperature Nb_3Al + Nb composites have been fabricated *in situ* by vacuum hot pressing and reaction sintering elemental powders mixed in the ratio Nb + 7wt.%Al [10,11]. Microstructures feature islands of ductile Nb solid solution (~20 vol.%) in a brittle Nb_3Al

matrix. Thermal treatment for 24 h at 1800°C results in a lamellar microstructure containing a uniform and fine distribution of filamentary Nb in a Nb_3Al matrix. In this paper, the fatigue and fracture resistance of the as hot-pressed and fully-aged microstructure are examined and compared to pure Nb_3Al and Nb.

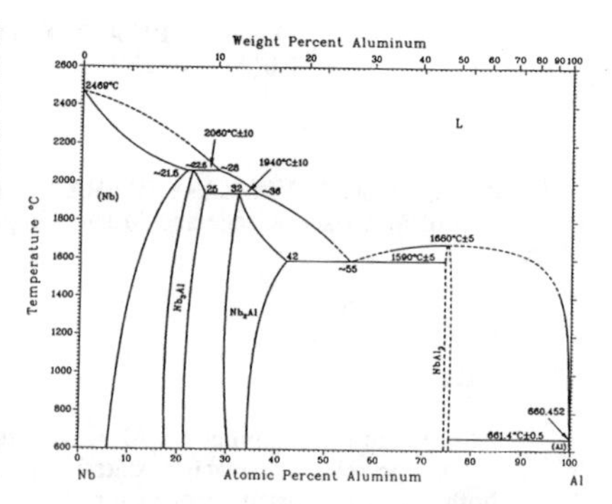

Fig. 1: The Nb-Al phase diagram

EXPERIMENTAL PROCEDURE

Materials

Microstructures: Figure 2(a) shows the microstructure obtained upon hot-pressing the synthesized Nb-7wt.%Al powders at 1650°C. The microstructure consists of a Nb_3Al matrix with elemental Nb distributed as the second phase (~20 vol.%). A small amount of elemental Al is also present due to incomplete synthesis of the powders. The Nb phase is distributed as islands with a size of ~2-10 μm. The fully-aged microstructure (Fig. 2(b)) consists of lamellae of Nb_3Al with a "stringy" second phase of the Nb solid solution. The Nb solid solution is rather fine, on the order of a few microns thick and uniformly distributed throughout the matrix.

Influence of Thermal Treatment: The formation of primary and secondary dendrites of Nb_3Al later in the aging sequence is believed to be due to the sluggish nature of the phase transformation which may be achieved at temperatures as low as 1000°C. Figure 3 shows transmission electron micrographs of the development and growth of the lamellar Nb_3Al phase. Nucleation appears to be heterogeneous with nucleation of the Nb_3Al colonies at grain boundaries. Trace analysis from selected area diffraction (SAD) patterns indicate that colony growth occurs in a direction of <110>Nb_3Al. As seen in Fig. 3(c), the growth has progressed to a point that the lamellae have fused into small, elongated grains with a 3-5° misorientation between them.

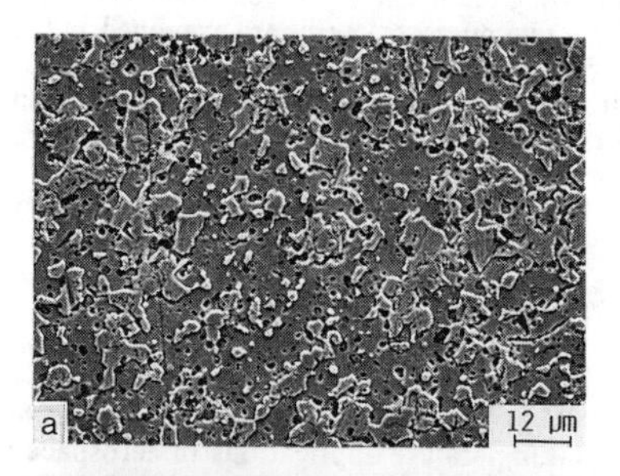

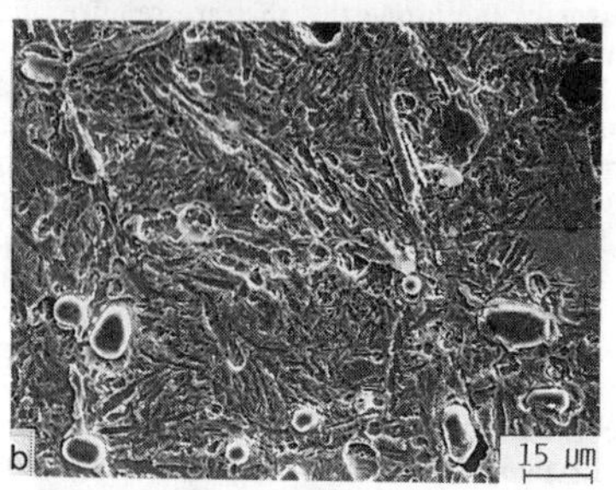

Fig. 2: Microstructural features of the (a) as hot-pressed Nb-7wt.% Al sample showing Nb_3Al matrix with islands of Nb phase, and (b) fully-aged (1800°C for 24 h) sample showing elongated Nb_3Al grains and "stringy" Nb filaments.

Fracture Toughness and Fatigue Crack-Growth Testing

Fatigue-crack growth behavior under cyclic tension loading was examined using 38mm-diameter, 28mm-wide and 2.5mm-thick disc-shaped DC(T) specimens, with a wedge-shaped starter notch to assist precracking. Specimens were cut using electrical-discharge machining. Tests were performed manually on electro-servo-hydraulic testing machines operating under displacement control, in room temperature air (22°C and ~45% relative humidity), at a load ratio ($R = K_{min}/K_{max}$) of 0.1 and a frequency of 25 Hz (sine wave). Growth rates over the range of ~10^{-10} to 10^{-6} m/cycle were obtained under load-shedding conditions, with crack lengths continuously monitored *in situ* (resolution ±2 μm) by measuring the

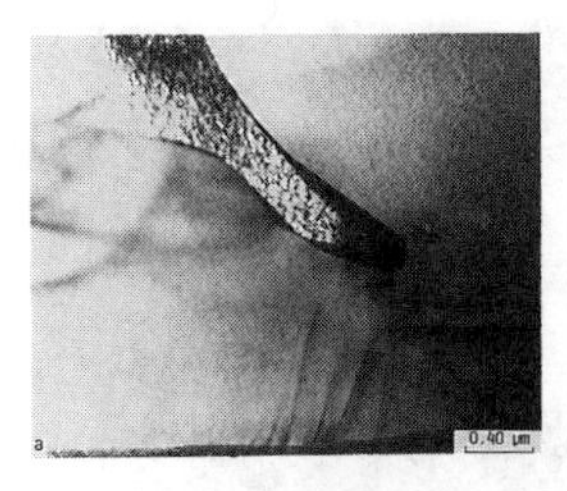

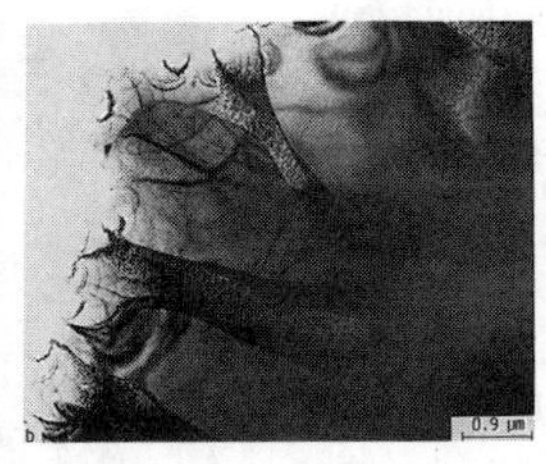

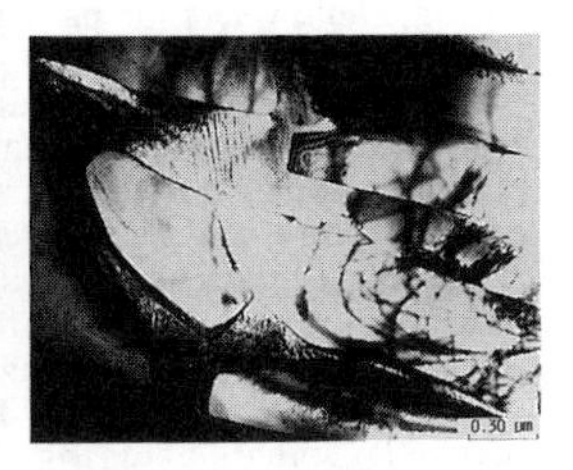

Fig. 3: (a) High- and (b) low-magnification transmission electron micrographs of the growth of Nb_3Al colonies from grain boundaries with growth direction of <110> Nb_3Al during early stages of the peritectic transformation, and (c) the fully-transformed condition. Note the trapped Nb_{ss} between Nb_3Al lamellae in (c). Micrographs taken at various regions in the hot-pressed sample subsequently aged 24 h at 1800°C.

electrical resistance of thin metallic foils bonded onto the specimen surface; techniques are similar to those developed for cyclic fatigue of ceramics. Results are presented in the form of crack-growth rates per cycle (da/dN) as a function of the applied stress-intensity range, ($\Delta K = K_{max}-K_{min}$).

Corresponding fracture-toughness behavior was evaluated by subsequently monotonically loading the fatigue-cracked samples to failure in room temperature air in order to determine resistance curves (R-curves), characterized by the crack-growth resistance (K_R) as a function of crack extension (Δa).

RESULTS AND DISCUSSION

Fracture Toughness

Measurements of the fracture toughness of the as hot-pressed samples indicate an increase in toughness from the inclusion of ~20 vol.% Nb; compared to a K_{Ic} value of ~2 MPa√m for monolithic Nb_3Al, the toughness of the as hot-pressed composite was ~6.5 MPa√m. Crack initiation was coincident with instability, with no evidence of increasing crack-growth resistance with crack extension. Using indentation techniques, the fracture toughness of the fully-aged sample was 7.5 MPa√m.

Metallographic observations of the monotonically loaded fracture surfaces indicate that ~50% of the Nb particles intersect the crack path and provide impedence to crack advance. A profile of the crack path under monotonic loading (Fig. 4(a)) shows that the crack is bridged by the unbroken ductile Nb phase in the wake of the crack tip, thus producing a crack-bridging zone with some additional plastic deformation of the ductile phase (Fig. 4(b)). Accordingly, fracture surfaces under monotonic loading (Fig. 5) show a few plastically stretched Nb particles where the ductile phase has formed dimples due to microvoid coalescence. Additionally, in some cases, the crack circumvents the particle

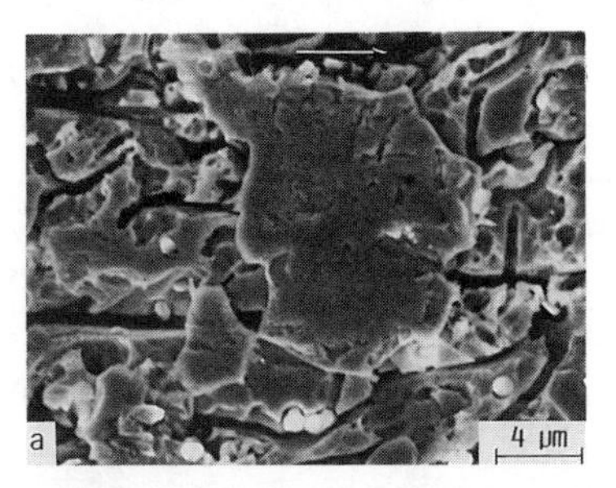

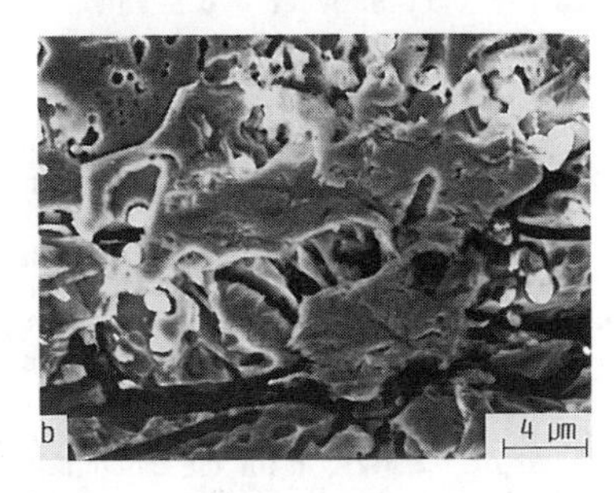

Fig. 4: High-magnification SEM micrographs of the crack profile showing (a) bridging of the crack and (b) plastic-stretching of the Nb particle under monotonic loading.

leaving a hole where the particle should have been.

The increase in toughness of the brittle intermetallic by the reinforcement of ductile phases can be modelled by considering the increase in the strain-energy release rate caused by the intact ligaments. Provided the crack path intercepts the reinforcing phase, the dominant contribution to toughening arises from tractions produced by the unbroken ductile ligaments bridging the two crack surfaces, thus partially shielding the crack tip from remote loads. In addition, toughening contributions from crack deflection, crack trapping, crack renucleation and decohesion along the particle-matrix interface further accentuate this effect. The extent of toughening depends on the length of the bridging zone in the wake of the crack tip; at steady state, where the zone is at a maximum length governed by ductile ligament rupture at a critical crack-opening displacement u*, the steady-state fracture energy increase, ΔG_c, has been estimated in terms of the area fraction f of ductile ligaments intersecting the crack path, their uniaxial yield strength, σ_y, and a representative cross-sectional radius, r, as [13]:

$$\Delta G_c = f \sigma_y r \chi , \quad (2)$$

where χ is a dimensionless function representing the work of rupture, given by [13]:

$$\chi = \int_0^{u^*} \frac{\sigma(u)du}{\sigma_y \cdot r} , \quad (3)$$

which can vary between ~0.5 and ~8, depending upon the degree of interface debonding and constitutive properties of the reinforcement phase [14-16]. Note that this formulation only applies when the size of the bridging zone is small compared to crack length and specimen dimensions.

For the as hot-pressed microstructure, estimating the area fraction of ductile Nb phase that is intercepted by the crack path as f~0.1 and using σ_y for Nb ~200 MPa, particle size r~3 μm and χ~2, the increase in toughness from crack bridging is ~4.5 MPa√m. For a toughness of ~2 MPa√m for the monolithic Nb_3Al, the increase in toughness by the incorporation of the ductile phase gives a composite toughness of ~6.5 MPa√m, which agrees well with experimental results obtained.

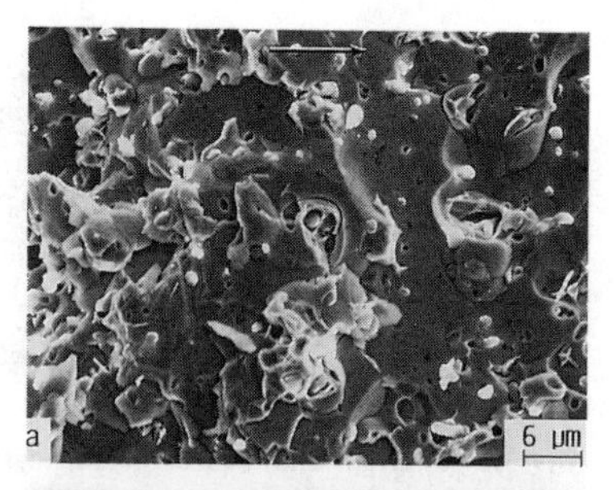

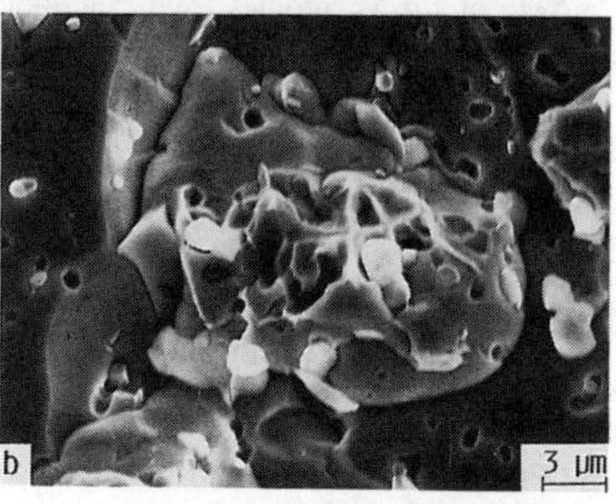

Fig. 5: SEM micrographs of fracture surfaces in the as hot-pressed Nb_3Al + Nb composite under monotonic loading, showing (a) cracking mostly confined to matrix with the Nb particles intersecting the crack path, and (b) plastically stretched Nb particle showing dimples and voids. Arrow indicates direction of crack advance.

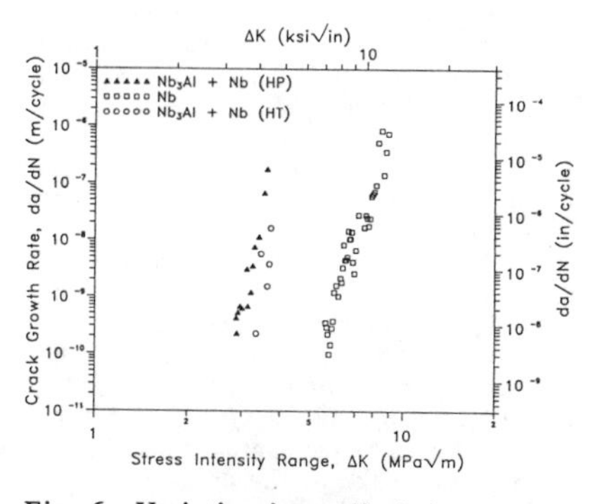

Fig. 6: Variation in cyclic fatigue-crack propagation rate, da/dN, as a function of the applied stress-intensity range, ΔK, in the as hot-pressed composite, fully-aged microstructure and a pure Nb sample, in controlled room temperature air at a load ratio R of 0.1 and test frequency of 25 Hz.

Fatigue Crack-Growth Behavior

The variation in fatigue-crack growth rates with applied ΔK for the as hot-pressed, fully-aged and a pure Nb sample is shown in Fig. 6. Growth rates are strongly dependent upon ΔK; expressed in terms of a simple Paris power-law da/dN versus ΔK relationship, the exponents on ΔK are ~25, 26 and 13 for the as hot-pressed, fully-aged and pure Nb sample respectively, compared to values between 2 and 4 typically observed for metals [e.g., ref. 17].

Crack-particle interactions under cyclic loading are different than those observed under monotonic loading. In general, fatigue cracks tend to avoid the ductile Nb phase; the mechanism of failure of the Nb particles that are intercepted by the crack is predominantly by a shear mechanism (Fig. 7). Profiles (Fig. 8) show that the effect of crack bridging is diminished in fatigue; there is, however, some evidence of crack deflection and bifurcation within the ductile Nb phase.

The absence of any significant crack-particle interactions in the as hot-pressed Nb_3Al composite under cyclic loading can result from several factors. Nb is embrittled easily by interstitial impurities such as carbon, hydrogen, nitrogen and oxygen and is thus especially prone to fatigue cracking [18]; Fig. 9 shows fatigue data for Nb with hydrogen taken from [18]. Moreover, since fatigue-crack advance can occur at lower stress-intensity levels (~3 MPa$\sqrt{m}$), the toughening contribution from the deformation of unbroken bridges under cyclic loading would naturally be much smaller than that prevalent at the large crack-opening displacements associated with monotonic crack-growth mechanisms at the higher stress intensities (5-6 MPa$\sqrt{m}$).

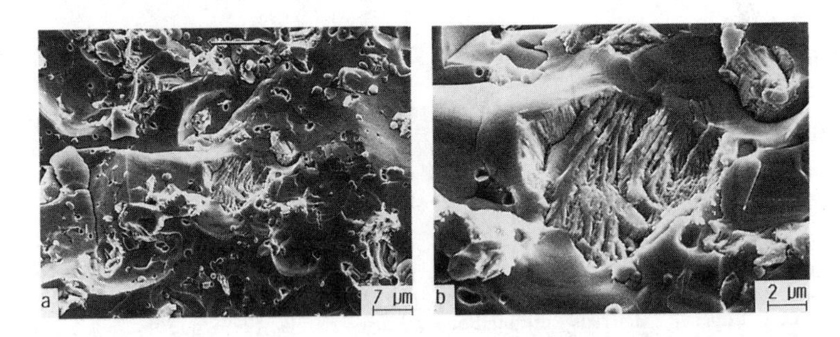

Fig. 7: SEM micrographs of fracture surfaces in the as hot-pressed Nb_3Al + Nb composite under cyclic loading showing (a) fracture surfaces with reduced crack-particle interaction, and (b) brittle-shear cracking in the few Nb particles that are intercepted by the crack. Arrow indicates direction of crack advance.

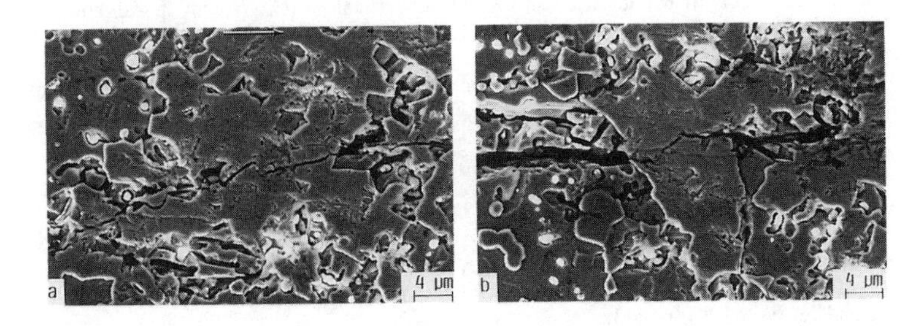

Fig. 8: High-magnification SEM micrographs of the crack profile showing brittle cracking of the Nb particles under cyclic loading. Some crack deflection and bifurcation is seen within the Nb phase.

CONCLUSIONS

The current results demonstrate that composite microstructures developed for superior toughness may not necessarily result in optimum resistance to fatigue. An increase in fracture toughness was obtained for the as hot-pressed as well as fully-aged microstructures by crack bridging from the incorporation of a ductile Nb phase. Under cyclic loading the effect of crack bridging is diminished and fatigue cracks grow at lower stress-intensity levels. Apart from substituting ductile phases with better fatigue properties, the design of more fatigue resistant, composite microstructures may very well involve weakened interfaces at high aspect-ratio ductile particles, where the resulting crack deflection and delamination at the particle/matrix boundary would delay cracking in the ductile phase, thereby promoting bridging at lower stress-intensity levels.

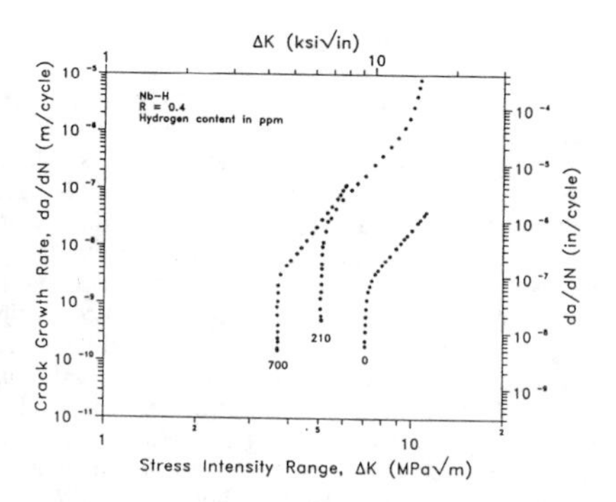

Fig. 9: The effect of hydrogen on the cyclic fatigue-crack propagation rate, da/dN, as a function of the applied stress-intensity range, ΔK, for a Nb sample. Tests were conducted in room temperature air at a load ratio R of 0.4 [from ref. 18].

Acknowledgments

This work was supported by the U.S. Air Force Office of Scientific Research under Grant No. AFOSR-90-0167.

References

1. J. J. Stephens, J. Metals 42 (8), 22 (1990).
2. D. L. Anton, D. M. Shah, D. N. Duhl and A. F. Giamei, J. Metals 41, 12 (1989).
3. D. L. Anton and D. M. Shah, in High Temperature Structural Intermetallic Compounds, edited by C. T. Liu et al. (Materials Research Society, Pittsburgh, PA, 1989), p. 361.
4. C. T. Liu and J. O. Stiegler, Science 226, 636 (1984).
5. E. M. Schulson, Int. J. Powder Met. 23 (1), 25 (1987).
6. I. Baker and P. R. Munroe, J. Metals 40 (2), 28 (1988).
7. C. E. Lundin and A. S. Yamamoto, Trans. Met. Soc. AIME 236, 863 (1966).
8. L. Kohot, R. Horyn and N. Iliev, J. Less Common Met. 44, 215 (1976).
9. L. Jorda, R. Flukinger and J. Miller, J. Less Common Met. 75, 227 (1980).
10. K. T. Venkateswara Rao, L. Murugesh, L. C. DeJonghe and R. O. Ritchie, "Micromechanisms of Monotonic and Cyclic Subcritical Crack Growth in Advanced High Melting Point Low-Ductility Intermetallics", Report No. UCB/R/91/A1072 to AFOSR, May 1991 (unpublished).
11. L. Murugesh, K. T. Venkateswara Rao, L. C. DeJonghe and R. O. Ritchie, in Developments in Ceramic and Metal-Matrix Composites, edited by K. Upadhya (TMS, Warrendale, PA, 1991), p. 65.
12. M. Hong and J. W. Morris, J. Appl. Phys. Lett. 37 (11), 1044 (1980).
13. M. F. Ashby, F. J. Blunt and M. Bannister, Acta Metall. 37, 1847 (1989).
14. H. E. Dève, A. G. Evans, G. R. Odette, R. Mehrabian, M. L. Emiliani and R. J. Hecht, Acta Metall. Mater. 38, 1491 (1990).
15. G. R. Odette, H. E. Dève, C. K. Elliott, A. Hasigawa and G. E. Lucas, in Interfaces in Ceramic Metal Composites, edited by R. J. Arsenault *et al.*, Fishman (TMS-AIME, Warrendale, PA, 1990), p. 443.
16. H. C. Cao, B. J. Dalgleish, H. E. Dève, C. Elliott, A. G. Evans, R. Mehrabian and G. R. Odette, Acta Metall. 37, 2969 (1989).
17. R. O. Ritchie, Int. Met. Rev. 20, 205 (1979).
18. S. Fariabi, A. L. W. Collins and K. Salama, Metall. Trans. A 14A, 701 (1983).

Author Index

Subject Index